Evidence-based Anticancer Complementary and Alternative Medicine

For further volumes:
http://www.springer.com/series/8883

William C. S. Cho
Editor

Evidence-based Anticancer Materia Medica

Springer

Editor
Dr. William C. S. Cho
Department of Clinical Oncology
Queen Elizabeth Hospital
13/F, Block R
30 Gascoigne Road
Kowloon
Hong Kong SAR
chocs@ha.org.hk

ISBN 978-94-007-0525-8 e-ISBN 978-94-007-0526-5
DOI 10.1007/978-94-007-0526-5
Springer Dordrecht Heidelberg London New York

Library of Congress Control Number: 2011924782

Cover design: deblik, Berlin

Printed on acid-free paper

Springer is part of Springer Science+Business Media (www.springer.com)

Preface

Interest in anticancer materia medica is growing in the world, but some clinicians and scientists have queried about the scientific validity of anticancer materia medica due to the lack of scientific evidence. As presenting in my previous book entitled *Supportive Cancer Care with Chinese Medicine*, a number of laboratory evidences and clinical trials have demonstrated the effectiveness and efficacies of Chinese medicine for supportive cancer care. Accumulating experiments show that some materia medica may have anticancer action by inducing apoptosis and differentiation, enhancing the immune system, inhibiting angiogenesis and proliferation, as well as reversing multidrug resistance. Clinical trials in cancer patients also demonstrate that some materia medica can prolong survival, increase tumor response, improve quality of life, and reduce chemotherapy or radiotherapy toxicity. These findings reveal the potential of materia medica as a promising anticancer therapeutics. Thorough understanding of materia medica and its safety is apparently a prerequisite for all professionals involved and interested in this field. These stimulating interests drove me to compile this unique volume.

William C. S. Cho
February 2011

Contents

Contributors

Jędrzej Antosiewicz Department of Bioenergetics and Physiology of Exercise, Medical University of Gdańsk, Gdańsk, Poland

Seung Joon Baek Department of Pathobiology, College of Veterinary Medicine, University of Tennessee, 2407 River Dr, Knoxville, TN 37996, USA

G. Bar-Sela Division of Oncology, Rambam-Health Care Campus and Faculty of Medicine, Technion-Israel Institute of Technology, POB 9602, 31096 Haifa, Israel

Barrie R. Cassileth Integrative Medicine Service, Memorial Sloan-Kettering Cancer Center, 1429 First Avenue, New York, NY 10021, USA

Tzyy-Wen Chiou Department of Life Science and Graduate Institute of Biotechnology, National Dong Hwa University, No 1, Sec 2, Da Hsueh Road, Shoufeng, Hualien, Taiwan

Karthik Garapati Department of Urology, School of Medicine, The University of Texas Health Science Center, 7703 Floyd Curl Drive, San Antonio, TX, USA

Rita Ghosh Department of Urology, School of Medicine, The University of Texas Health Science Center, 7703 Floyd Curl Drive, San Antonio, TX, USA

Guoqing Gong Department of Pharmacology, School of Pharmacy, China Pharmaceutical University, Nanjing, Jiangsu, People's Republic of China

Heather Graham Department of Urology, School of Medicine, The University of Texas Health Science Center, 7703 Floyd Curl Drive, San Antonio, TX, USA

Jyothirmai Gubili Integrative Medicine Service, Memorial Sloan-Kettering Cancer Center, 1429 First Avenue, New York, NY 10021, USA

Horng-Jyh Harn Department of Pathology, China Medical University and Hospital, 2 Yuh-Der Road, Taichung 40447, Taiwan

Anna Herman-Antosiewicz Department of Molecular Biology, University of Gdańsk, Kładki 24, 80-822 Gdańsk, Poland

Yuko Horii Institute for Protein Research, Osaka University, Suita, Osaka, Japan ANBAS Corporation, Toyosaki, Kita-Ku, Osaka, Japan

Damao Huang Xiangya Hospital of Central South University, Changsha, Hunan, People's Republic of China

Lee Jia Developmental Therapeutics Program, DCTD, National Cancer Institute/ NIH, Bethesda, MD, USA

Hongchi Jiang Department of General Surgery, The First Affiliated Hospital of Harbin Medical University, Key Laboratory of Hepatosplenic Surgery, Ministry of Education, No 23 Youzheng Street, Harbin, 150001 Heilongjiang, People's Republic of China

Jing Jing School of Pharmacy, The University of Queensland, Brisbane, QLD, Australia

Anna Kawiak Department of Biotechnology, Intercollegiate Faculty of Biotechnology, University of Gdańsk and Medical University of Gdańsk, Gdańsk, Poland

Laboratory of Human Physiology, Faculty of Health Sciences with Subfaculty of Nursing, Medical University of Gdańsk, Gdańsk, Poland

Masahiro Kizaki Department of Hematology, Saitama Medical Center, Saitama Medical University, 1981 Kamoda, Kawagoe, Saitama 350-8550, Japan

Addanki P. Kumar Department of Urology, School of Medicine, The University of Texas Health Science Center, 7703 Floyd Curl Drive, San Antonio, TX, USA

Joshua D. Lambert The Department of Food Science, 332 Food Science Building, The Pennsylvania State University, University Park, PA 16802, USA

Po-Cheng Lin Department of Pathology, Center for Neuropsychiatry, China Medical University and Hospital, 2 Yuh-Der Road, Taichung 40447, Taiwan

Lianxin Liu Department of General Surgery, The First Affiliated Hospital of Harbin Medical University, Key Laboratory of Hepatosplenic Surgery, Ministry of Education, No 23 Youzheng Street, Harbin, 150001 Heilongjiang, People's Republic of China

Danijela Maksimović-Ivanić Department of Immunology, Institute for Biological Research "Sinisa Stankovic", Belgrade University, Belgrade, Serbia

Wenhua Mei The Third Affiliated Hospital of Jinan University, 79 Kangning Road, 519000 Zhuhai, Guangdong, People's Republic of China

Sanja Mijatović Department of Immunology, Institute for Biological Research "Sinisa Stankovic", Belgrade University, Belgrade, Serbia

Katsuya Nagai Institute for Protein Research, Osaka University, Suita, Osaka, Japan; ANBAS Corporation, Toyosaki, Kita-Ku, Osaka, Japan

Harendra S. Parekh School of Pharmacy, Pharmacy Australia Centre of Excellence, The University of Queensland, Brisbane, QLD, Australia

Keduo Qian Division of Medicinal Chemistry and Natural Products, Eshelman School of Pharmacy, University of North Carolina at Chapel Hill, NC, USA

Craig Robson Department of Urology, School of Medicine, The University of Texas Health Science Center, 7703 Floyd Curl Drive, San Antonio, TX, USA

Morihiko Sagawa Department of Hematology, Saitama Medical Center, Saitama Medical University, Kawagoe, Saitama, Japan

M. Schaffer Division of Oncology, Rambam-Health Care Campus and Faculty of Medicine, Technion-Israel Institute of Technology, Haifa, Israel

Jun Shao Department of Pharmaceutical Sciences, College of Pharmacy and Allied Health Professions, St. John's University, 8000 Utopia Parkway, Queens, NY 11439, USA

Lee-Yan Sheen Graduate Institute of Food Science and Technology, National Taiwan University, No 1, Sec 4, Roosevelt Road, 106 Taipei, Taiwan

Jiao Shen ANBAS Corporation, Nakajima Building, 4-12-17, Toyosaki, Kita-Ku, Osaka 531-0072 Japan

Stanislava Stošić-Grujičić Department of Immunology, Institute for Biological Research "Sinisa Stankovic", Belgrade University, Bulevar Despota Stefana 142, 11060 Belgrade, Serbia

Zheng-Yuan Su Graduate Institute of Food Science and Technology, National Taiwan University, Taipei, Taiwan

Faqing Tang The Third Affiliated Hospital of Jinan University, 79 Kangning Road, 519000 Zhuhai, Guangdong, People's Republic of China; Xiangya Hospital, Central South University, 41008, Changsha, Hunan, People's Republic of China

Mamoru Tanida Institute for Protein Research, Osaka University, Suita, Osaka, Japan; ANBAS Corporation, Toyosaki, Kita-Ku, Osaka, Japan

Daofa Tian The First Affiliated Hospital, Traditional Chinese Medicine University of Hunan, Changsha, Hunan, People's Republic of China

Louis Trombetta Department of Pharmaceutical Sciences, College of Pharmacy and Allied Health Professions, St John's University, Queens, NY, USA

Chong-Zhi Wang NIH/NCCAM Center for Herbal Research on Colorectal Cancer, Tang Center for Herbal Medicine Research, The University of Chicago, Chicago, IL, USA; Department of Anesthesia and Critical Care, The University of Chicago, Chicago, IL, USA

Paulus S. Wang Department of Physiology, School of Medicine, National Yang-Ming University, Shih-Pai, Taipei 11221, Taiwan; Department of Medicine Research and Education, Taipei Veterans General Hospital, Taipei, Taiwan

Shyi-Wu Wang Department of Physiology and Pharmacology, School of Medicine, Chang Gung University, Taoyuan, Taiwan

Nichelle C. Whitlock Laboratory of Environmental Carcinogenesis, Department of Pathobiology, College of Veterinary Medicine, University of Tennessee, Knoxville, TN, USA

Brian Y. Y. Wong Department of Biology, Pacific Union College, 1 Angwin Avenue, Angwin, CA 94508, USA

Hannah H. L. Wong Department of Pathology, School of Medicine, Loma Linda University, Loma Linda, CA, USA

Jia-fei Yao Department of Gynecology, The First Affiliated Hospital, China Medical University, Shenyang, Liaoning, People's Republic of China

Jiun-Yih Yeh Department of Medicine Research and Education, Taipei Veterans General Hospital, Taipei, Taiwan

K. Simon Yeung Integrative Medicine Service, Memorial Sloan-Kettering Cancer Center, 1429 First Avenue, New York, NY 10021, USA

Dalong Yin Department of General Surgery, The First Affiliated Hospital of Harbin Medical University, Key Laboratory of Hepatosplenic Surgery, Ministry of Education, No 23 Youzheng Street, Harbin, 150001 Heilongjiang, People's Republic of China

Ching-Han Yu Department of Physiology, School of Medicine, Chung Shan Medical University, Taichung, Taiwan

Chun-Su Yuan Tang Center for Herbal Medicine Research, University of Chicago, 5841 South Maryland Avenue, MC 4028, Chicago, IL 60637, USA

Tongsen Zheng Department of General Surgery, The First Affiliated Hospital of Harbin Medical University, Key Laboratory of Hepatosplenic Surgery, Ministry of Education, No 23 Youzheng Street, Harbin, 150001 Heilongjiang, People's Republic of China

Introduction

Cancer is one of the leading killers in the world and the incidence is rising, but most cancer patients and cancer survivors suffer much from the disease and its conventional treatments' side effects. Complementary and alternative medicine (CAM) has been increasingly utilized by cancer patients in developed countries. Among the various forms of CAM, some traditional medicines have well constructed theoretical frameworks and established treatment approaches for diseases including cancer. Recent researches have revealed growing evidences suggesting that CAM is effective in the supportive care of cancer patients during and after major conventional cancer treatments.

However, some clinicians and scientists have queried about the anticancer abilities of CAM due to the lack of scientific evidence. There is great demand in the knowledge gap to explore the scientific and evidence-based knowledge of CAM in the anticancer field. With this aim, a book series is needed to structurally deliver the knowledge to readers.

This book series seeks to provide scientific evidence supporting the study of anticancer CAM modalities, particularly traditional medicines. It emphasizes health outcome, while documenting biological mechanisms of action. Gathering international opinion leaders' views, this book series is devoted to the advancement of science in the field of basic research, clinical study, methodology, and scientific theory in diverse areas of evidence-based anticancer CAM. Each volume discusses different research and clinical evidences that support the usage of various CAM treatment strategies including materia medica, acupuncture, Tai Chi, massage, mind-body medicine, and diet therapy as an anticancer treatment. The volumes contain concrete evidence-based anticancer CAM information and I envision this book series will assist healthcare providers and patients for the battle against cancer.

Chapter 1
An Overview of Anticancer Herbal Medicines

**Addanki P. Kumar, Heather Graham, Craig Robson,
Karthik Garapati and Rita Ghosh**

Abstract The use of medicinal plants to help sustain good health and vitality and to reduce inflammation has an ancient and respected history. Approximately 38% of American adults are using some form of complementary and alternative medicine, resulting in $34 billion dollars spent annually. Herbal medicine compositions offer a potential advantage in that they usually comprise multiple components that interact and act simultaneously through multiple molecular targets and signaling pathways. These complex and often synergistic botanicals may also decrease toxicity and increase bioavailability and offer potential as strategies for cancer management. The quality and content of the active supplement depends on collection, processing and composition of the raw material and extraction procedures. In addition, despite evidence for usefulness of complex mixtures in cell culture and pre-clinical animal models, there is no formal regulation of natural supplements by the United States Food and Drug Administration, however, which creates conflict in the medical community, resulting in reluctance to recommend the use of herbals and alternative medicines for patient care. Clinicians should be aware of these alternatives given their future potential in oncology and a potential role in treatment when standard medical cures and treatment have been attempted and there are no clinical trials available to patients with advanced disease. Further investigation and clinical trials should focus on natural herbs and medicines in the chemo-prevention of chronic diseases and certain cancers. The advantages and limitations of potential use of various natural products into mainstream medical practice are discussed.

A. P. Kumar (✉)
Department of Urology, School of Medicine, The University of Texas Health Science Center, 7703 Floyd Curl Drive, San Antonio, TX, USA
e-mail: kumara3@uthscsa.edu

W. C. S. Cho (ed.), *Evidence-based Anticancer Materia Medica,* Evidence-based Anticancer Complementary and Alternative Medicine,
DOI 10.1007/978-94-007-0526-5_1,

1.1 Cancer Etiology and Management Approaches

Chronic diseases such as cardiovascular, many cancers, diabetes and obesity are among the most prevalent, costly to treat yet very preventable through life-style changes. They affect more than 90 million Americans, account for approximately 70% of all deaths and cost more than $1.4 trillion in medical care (http://www.cdc.gov/nccdphp). Of these, cancer is a growing public health problem both in the United States (US) as well as worldwide, and the Center for Diseases Control reports $74 billion was spent on cancer care in 2004 (http://www.cancer.org). Since the inception of the "War on Cancer" by President Nixon in 1950, there has been significant progress in cancer management. Despite this, there is only a marginal benefit of increased survival of approximately 1.5 months at a cost of $90,000 for some of the anticancer drugs like Avastin that is being used for treatment of metastatic cancer (Merlo et al. 2006). In addition Medicare spent $7.3 billion on inpatient cancer alone excluding chemotherapy treatment in 2004 (Merlo et al. 2006). Further it has been estimated that global cancer death rate could increase by 45% from 7.9 million in 2007 to 11.5 million by 2030. One of the reasons for this increased cancer death rate is that many of the available chemo-therapeutic agents are not uniformly successful. Such chemo-therapeutic agents are not only expensive but their long term use is associated with adverse toxic side effects including cancer recurrence and resistance to treatment (Merlo et al. 2006; http://www.cancer.org). Therefore, alternative cost-effective and non-toxic strategies/approaches that can reverse such therapeutic resistance, cancer recurrence and improve the quality of life for cancer patients are in great demand. Despite the great deal of heterogeneity in cancer between patients, a common thread is involved in the transformation of normal cell into a tumor cell which in general involves deregulation of multiple signaling pathways (Croce 2008; Reid 2008; Seyfried and Shelton 2010). Neoplastic transformation includes progression from a pre-cancerous lesion (typically more amenable to treatment) to malignant tumors (typically resistant to treatment strategies), caused by combination of both internal and external factors. For example, (i) exposure to external environmental factors such as radiation (UV, gamma and X-ray) damages cellular DNA and can lead to skin cancer, (ii) infectious microorganisms such as *Helicobacter pylori* can cause gastric cancer; viruses such as papilloma virus causes cervical and hepatitis B causes liver cancer, (iii) carcinogenic agents generated either endogenously through metabolic activities and smoking can contribute to the carcinogenic process (Yokota 2000; DeBerardinis 2008). In addition physiological oxidant-antioxidant homeostasis is important in basic events of cell regulation. When the balance between oxidants and antioxidants is disrupted and more oxidants are produced, a cell or organism is said to be in oxidative stress (Bankson et al. 1993; McCord 1993). For example, an overproduction of oxygen free radicals and the deficiency of or overwhelming antioxidant defense or repair mechanisms can lead to oxidative stress. Under these conditions, reactive oxygen intermediates can damage DNA, which in turn can disrupt normal transcription and replication and induce mutations (Ames et al. 1995; Beckman and Ames 1998; Cavalieri et al.

2000; Evans et al. 2004). Such oxidative damage to macromolecules can induce inflammatory signaling which in turn leads to cancer development (Hietanen et al. 2004; Kawanishi and Hiraku 2006; Sawa and Ohshima 2006).

1.2 Diet and Cancer

Epidemiological and laboratory studies strongly suggest reduction of cancer risk related to consumption of specific types of fruits and vegetables suggestive of cancer preventive properties associated with plant-based food products (phytoceuticals) (Surh 2003; Amin et al. 2009; Ma and Chapman 2009). Candidate dietary components that influence human carcinogenesis include fat, calories, fruits, vegetables, antioxidants, and various micronutrients. For example, epidemiological evidence suggests that prostate cancer risk is high in Western population consuming high fat and low fiber diet compared to Oriental populations where fruits and vegetables are consumed in large portions (Kelloff et al. 1999; Yip et al. 1999; Chan et al. 2005b; Carlsen et al. 2010; Huang and Cai 2010). Interestingly, migration of this low-risk population to Western countries increases their risk of cancer. These studies not only suggest the potential influence of dietary lifestyle in cancer incidence but also implicate a potential role for developing such dietary factors as cancer preventive agents. Due to the long latency involved in the development of cancer, life-style modulation through nutritional intervention in early stages of disease process to reduce the carcinogen-induced risk is a viable cancer prevention approach. Mechanistically, such preventive or protective potential of health promoting phytochemicals is due to combination of antioxidant, anti-inflammatory, immune enhancing and anti-hormone effects (Wargovich 1999; Yu and Kong 2007; Sarkar and Li 2009; Kumar et al. 2010a; Trottier et al. 2010). Further, these dietary chemicals can also modulate drug metabolizing enzymes, cell cycle regulatory proteins, as well as the processes of cell differentiation, apoptosis, proliferation, and angiogenesis (Kelloff et al. 1999; Wargovich 1999; Yip et al. 1999; Chan et al. 2005b; Surh 2003; Yu and Kong 2007; Amin et al. 2009; Ma and Chapman 2009; Sarkar and Li 2009; Carlsen et al. 2010; Gupta et al. 2010; Huang and Cai 2010; Kumar et al. 2010a; Trottier et al. 2010). Published data using phytoceuticals also suggests the potential benefit of plant-derived complex mixtures or combination of agents in reducing cancer incidence and mortality (Kelloff et al. 1999; Wargovich 1999; Yip et al. 1999; Surh 2003; Chan et al. 2005a; Yu and Kong 2007; Amin et al. 2009; Ma and Chapman 2009; Sarkar and Li 2009; Carlsen et al. 2010; Gupta et al. 2010; Huang and Cai 2010; Trottier et al. 2010; Kumar et al. 2010a). A number of excellent review articles describing the mechanism of action of various phytoceuticals have been published and is therefore not a subject of this review (Kelloff et al. 1999; Wargovich 1999; Yip et al. 1999; Surh 2003; Chan et al. 2005b; Yu and Kong 2007; Amin et al. 2009; Liu et al. 2009; Ma and Chapman 2009; Sarkar and Li 2009; Carlsen et al. 2010; Gupta et al. 2010; Huang and Cai 2010; Kumar et al. 2010a; Trottier et al. 2010).

1.3 Importance of Plant-derived Natural Products in Cancer Drug Discovery

There are texts surviving from the ancient cultures of China, Mesopotamia, Egypt, and India that describe the use of medicinal plants to help sustain good health, vitality, treat gastrointestinal diseases, analgesia, and inflammation (Wang et al. 2010d). In fact opium, cannabis, myrrh, frankincense, fennel, cassia, thyme, henna, juniper, aloe, linseed, castor, caraway seeds, marjoram, spearmint and peppermint leaves, basil, hibiscus, calendula, anise seeds, parsley, cumin, licorice root, chamomile, dill, and cloves of garlic have been referenced in ancient Egyptian herbal remedies, implicating a prominent role for herbs in Egyptian medicine. Similar information is available in traditional Chinese medicine as well as Indian Ayurveda (Patwardhan and Mashelkar 2009). A vast majority of approved anticancer drugs are (i) natural products, (ii) derived semi-synthetically from natural products, or (iii) synthetic products based on natural products (Cragg et al. 1997; Fabricant and Farnsworth 2001; Ichikawa et al. 2007; Newman and Cragg 2007; Liu et al. 2009; Patwarhan and Mashelkar 2009). It has been reported that up to 1994, 62% of the available anticancer drugs are either of natural origin or are modeled based on natural products (Cragg et al. 1997; Fabricant and Farnsworth 2001; Ichikawa et al. 2007; Newman and Cragg 2007). Further, it has been shown that 69% of the approved anticancer drugs between 1980 and 2002 are either natural products or drugs developed based on knowledge gained from natural products (Newman and Cragg 2007). It has also been reported that about 25% of US prescriptions dispensed contain active ingredients from plants. Consistent with this, according to a government survey conducted in 2007, approximately 38% of adults use some form of complementary and alternative medicine including herbs for treating a variety of diseases including cancer and spend approximately $34 billion (Ernst and Cassileth 1998; Sparber et al. 2000; Bernstein and Grasso 2001; Armstrong and Gilbert 2008; Cassileth et al. 2008; Gansler et al. 2008; Gratus et al. 2009; Li et al. 2009; Shord et al. 2009; Supoken et al. 2009; Wu et al. 2009; Bishop et al. 2010; Roberts 2010; Tian et al. 2010; Yildirim 2010; http://www.cancer.org). For example, 43–80% of patients with prostate cancer are on some form of alternative therapy. These patients include those with strong family history of prostate cancer, those on active surveillance, and those who have failed active prostate cancer treatment or are on androgen deprivation and seek to delay disease progression by natural means (Trottier et al. 2010). Despite the wealth of knowledge regarding nature's bounty, only 5–15% of the approximately 250,000 species of higher plant have been systematically investigated for the presence of bioactive compounds (Balandrin et al. 1993). Therefore, there is a tremendous potential to exploit nature for development of novel anticancer drugs (Harvey 1999; Gordaliza 2007; Mazzio and Soliman 2010; Monneret 2010). Unlike Western medicine that generally uses purified compounds and targets a single physiological endpoint, traditional herbal medicine compositions usually comprise multiple herbs and components that interact and act simultaneously through multiple molecular targets and signaling pathways (Harvey 1999; McCarty 2004; Gordaliza

2007; Mazzio and Soliman 2010; Monneret 2010). Therefore using a combination of herbs or complex botanicals may serve different functions while others may decrease toxicity while yet others may increase bioavailability. Although several drugs have been developed based on natural product chemistry, to the best of our knowledge natural products *per se* have not been integrated into medical practice. However several clinical trials are currently ongoing using natural products which will be discussed later in this article.

Of the numerous approved anticancer drugs, several are derived from natural products including doxorubicin, daunomycin, mithramycin, paclitaxel, vinblastine, and vincristine (Table 1.1) (Lee et al. 1975; Cortese et al. 1977; Horwitz et al. 1982;

Table 1.1 Approved anticancer agents from plant-derived natural products

Compound	Natural product source	Mechanism	Use in clinic
Paclitaxel	Pacific yew tree	Accelerate tubulin polymerization	Breast cancer, hormone-refractory prostate cancer, ovarian cancer, Kaposi's sarcoma, non-small cell lung cancer[a,b,c,d]
Vinblastine	Madagascar periwinkle	Inhibit tubulin polymerization	Bladder cancer, lymphoma, Kaposi's sarc oma[a,e,b,c,d]
Vincristine	Madagascar periwinkle	Inhibit tubulin polymerization	Leukemia, multiple myeloma, lymphoma[a,b,c,d]
Etoposide	American mayapple	Topoisomerase II inhibitor	Small cell lung cancer, testicular cancer[a,f,d]
Teniposide	Podophyllotoxin	Causes dose-dependent single- and double-strand breaks in DNA	Acute lymphoblastic leukemia[a,g,h,d]
Doxorubicin	*Streptomyces peucetius*	Intercalating agent; topoisomerase II inhibitor	Acute lymphoblastic leukemia, breast cancer, gastric cancer, Hodgkin's lymphoma, neuroblastoma, ovarian cancer, small cell lung cancer, thyroid cancer, bladder cancer, bone sarcomas[a,d]
Daunomycin	*Streptomyces peucetius*	Intercalating agent; topoisomerase II inhibitor	Acute myeloid leukemia, acute lymphoblastic leukemia[a,d]

[a] http://www.cancer.gov/cancertopics/drugdictionary
[b] Horwitz et al. 1982
[c] Hearn et al. 2006
[d] Newman and Cragg 2007
[e] Lee et al. 1975
[f] Martincic and Hande 2005
[g] Canel et al. 2000
[h] Cortese et al. 1977

Canel et al. 2000; Siemann et al. 2004; Young and Chaplin 2004; Martincic and Hande 2005; Srivastava et al. 2005; Hearn et al. 2006; Lampson 2007; http://www.cancer.gov/cancertopics/druginfo). Vinblastine and vincristine are vinca alkaloids isolated from *Catharanthus roseus* (Madagascar periwinkle) and currently used in combination with other chemo-therapeutic drugs for treatment of leukemia, lymphoma, testicular, breast, and lung cancers. Both have been shown to inhibit tubulin polymerization and had originally been used for treatment of diabetes. Etoposide and teniposide are semi-synthetic derivatives of the natural product epipodophyllotoxin isolated from the roots of the *Podophyllum peltatum* (American mandrake or mayapple), used in the treatment of lymphoma, bronchial, and testicular cancers. These compounds inhibit topoisomerase II that plays an important role in DNA replication. Paclitaxel and its analogue docetaxel work through promotion of tubulin polymerization and microtubule stabilization and are used to treat breast, ovarian, and non-small cell lung cancers. It is naturally derived from *Taxus brevifolia* (bark of Pacific Yew) that has traditionally been used by Native Americans for treating non-cancerous conditions. In addition, leaves of this plant have been used in Ayurveda for a variety of ailments including cancer. Topotecan is a semi-synthetic derivative of Camptothecin isolated from *Camptotheca acuminata* (Chinese ornamental tree). Camptothecin inhibits topoisomerase I and is used for treatment of ovarian and small cell lung cancers. Irinotecan, a derivative of topotecan is used for treatment of colorectal cancer (Lee et al. 1975; Cortese et al. 1977; Horwitz et al. 1982; Canel et al. 2000; Siemann et al. 2004; Young and Chaplin 2004; Martincic and Hande 2005; Srivastava et al. 2005; Hearn et al. 2006; Lampson 2007; http://www.cancer.gov/cancertopics/druginfo).

In addition, using knowledge gained from natural product chemistry several synthetic agents including flavopiridol, combretastatin, roscovitine, etc have been developed as anticancer agents (Table 1.2) (Young and Chaplin 2004; Srivastava et al. 2005). Most of these agents have been shown to exhibit growth inhibitory activities in cell culture and pre-clinical models through multiple mechanisms including modulation of proliferation, apoptosis, invasion, metastasis, receptor signaling and transcriptional regulation (Table 1.2) (Tanaka et al. 2000; Tian et al. 2001; Granado-Serrano et al. 2007; Rettig et al. 2008; Stan et al. 2008; Adhami et al. 2009; De Stefano et al. 2009; Fulda and Kroemer 2009; Kraft et al. 2009; Lee et al. 2009a, b, 2010; Adams et al. 2010; Brown et al. 2010; Li et al. 2010a, b, c; Liu et al. 2010; Reagan-Shaw et al. 2010; Senthilkumar et al. 2010; Sturgeon and Ronnenberg 2010; Yuan et al. 2010). Extensive information is available regarding the mechanism of action of some of these agents (Table 1.2). Although several such active ingredients from natural products have been identified, we will discuss the potential role of two such compounds, namely curcumin and eugenol, which are key components of spices used in culinary practices in cancer management. In addition, we will discuss the promising role of resveratrol, a polyphenolic phytoalexin found in the skin of various fruits and nuts.

Table 1.2 Examples of plant-derived synthetic agents in pre-clinical use

Natural product	Source	Mechanism
Allyl sulfides	Garlic	Growth inhibition, increased reactive oxygen species (ROS), cell cycle arrest, apoptotic induction *via* mitochondrial signaling[a]
Apple peel extract	Apples	Cell cycle arrest, inhibited proliferation, increased levels of mapsin[b]
Betulinic acid	White birch *Betula pubescens*	Anti-retroviral, inhibition of topoisomerase, induction of apoptosis, production of ROS, activation of caspases[c]
Capsaicin	Chili peppers	Cell cycle arrest & growth inhibition *via* TRPV1 & E2F pathways[d]
Chlorogenic acid	Caffeic acid & quinic acid	Modulation of ROS[e]
Combrestatin	African bush willow *Combretum caffrum*	Microtubule depolymerization, inhibition of angiogenesis (thereby inducing tumor necrosis)[f]
DIM (diindolyl-methane)	Cruciferous vegetables	Growth inhibition, cell cycle arrest, apoptotic induction, prevention of tumor development *in vivo*[g]
Ganoderma lucidum	Mushrooms	Decreased proliferation, inhibited tumor growth *in vivo*, induced autophagy *via* p38 MAPK pathway, apoptotic induction[h]
Korean angelica gigas Nakai	Root of Korean angelica	Decreased proliferation & angiogenesis, increased apoptosis *in vivo*[i]
Flavopiridol	Rohitukine	Tyrosine kinase inhibitor, induces cell cycle arrest; apoptotic induction[j]
Limonoids	Oranges & lemons	Apoptotic induction[k]
Myricetin	Bayberry *Myrica cerifera*	Inhibition of matrix metalloproteinases (MMPs)[l]
Phloridzin	Apple, cherry & fruit trees	Increased expression of tyrosinase genes *via* cyclic adenosine mono-phosphate[m]
Pomegranate extract	Pomegranate	Inhibited proliferation, apoptotic induction, inhibition of NF-κB activity[n]
Quercetin	Fruits & vegetables	Pro-apoptosis, modulation of insulin-like growth factors, reduced proliferation & angiogenesis[o]
Retinoic acid	Fruits & vegetables (carrots, sweet potatoes)	Growth inhibition, decreased telomerase activity, apoptotic induction[p]
Roscovitine	Olumucine (radish)	Inhibition of cyclin-dependent kinases, inhibits survivin[q]
Sanguinarine	Certain plants (bloodroot, Mexican prickly poppy, opium poppy)	Reduced proliferation, angiogenesis, & tumor development *in vivo*, apoptotic induction & caspase activation[r]
Silibinin	Milk thistle	Down-regulation of tumor necrosis factor induced MMP-9 *via* MEK/ERK, inhibition of HIF-1[s]

Table 1.2 (continued)

Natural product	Source	Mechanism
Sulforaphane	Broccoli	Modulation of transcription induced by PTEN deletion *in vivo*, inhibition of viability & proliferation *in vitro*[t]
Theaflavin	Black tea	Inhibition of MMP-2, apoptotic induction *via* Fas/caspase-8 & Akt/pBad[u]
Withaferin A	Winter cherry *Withania somnifera*	Apoptotic induction, targeting of Hsp90, cell cycle arrest, down-regulation of pAkt[v]

[a] Malki et al. 2009; Liao et al. 2009; Stan and Singh 2009; Zhou et al. 2009a; Herman-Antosiewicz et al. 2010; Nagaraj et al. 2010; Shrotriya et al. 2010; Wang et al. 2010a, e

[b] Ding et al. 2004; Reagan-Shaw et al. 2010

[c] Drag et al. 2009; Drag-Zalesinska et al. 2009; Fulda et al. 2008, 2009; Kommera et al. 2010a, b, c; Kommera et al. 2009, 2010b; Liu et al. 2009; Nakagawa-Goto et al. 2009; Ahmad et al. 2010; Csuk et al. 2010; Li et al. 2010c; Mullauer et al. 2010

[d] Mori et al. 2006; Sánchez et al. 2007; Kim et al. 2009a; Malagarie-Cazenave et al. 2009; Wang et al. 2009a; Baez et al. 2010; Brown et al. 2010; Choi et al. 2010; Ghosh and Basu 2010; Li et al. 2010c; Maity et al. 2010; Oyagbemi et al. 2010; Thoennissen et al. 2010; Yang et al. 2010

[e] Belkaid et al. 2006; Xie et al. 2009; Park et al. 2010b; Reddivari et al. 2010

[f] Young and Chaplin 2004; Srivastava et al. 2005

[g] Ahmad et al. 2009; Banerjee et al. 2009; Bhatnagar et al. 2009; Khwaja et al. 2009; Rahman et al. 2009; Vivar et al. 2009; Acharya et al. 2010; Cho et al. 2010; Heath et al. 2010; Jin et al. 2010; Kandala and Srivastava 2010; Kim et al. 2010a; Fares et al. 2010; Rahimi et al. 2010

[h] Wang et al. 2009b; Weng et al. 2009; Zhou et al. 2009b; Oka et al. 2010; Thyagarajan et al. 2010; Weng and Yen 2010

[i] Jiang et al. 2006; Lee et al. 2009a

[j] Bright et al. 2010; Dickson et al. 2010; Ferkrazad et al. 2010; Nitta et al. 2010

[k] Akihisa et al. 2009; Harish et al. 2009; Patil et al. 2009; El-Readi et al. 2010; Priyadarsini et al. 2010

[l] Jung et al. 2008; Shih et al. 2009; Siegelin et al. 2009; Zhang et al 2010

[m] Shoji et al. 1997; Andlauer et al. 2004; Jung et al. 2009a

[n] Pantuck et al. 2006; Rettig et al. 2008; Koyama et al. 2010

[o] Lee et al. 2008; Du et al 2010; Ekstrom et al. 2010; Gupta et al. 2010; Howells et al. 2010; Noori-Daloii et al. 2010; Senthilkumar et al. 2010; Singh et al. 2010; Tan et al. 2010; Thangasamy et al. 2010; Tokalov et al. 2010; Vargas and Burd 2010; Yuan et al. 2010; Zhou et al. 2010

[p] He et al. 2009; Sadikoglou et al. 2009; Touma et al. 2009; Bryan et al. 2010; Bushue and Wan 2010; Ciolino et al. 2010; DiPaola et al. 2010; Holzel et al. 2010; Jung et al. 2010; Karabulut et al. 2010; Koay et al. 2010; Montesinos et al. 2010; Perri et al. 2010; Pirouzpanah et al. 2010

[q] Maurer et al. 2009; De Leon et al. 2010; Wesierska-Gadek et al. 2009

[r] Choi et al. 2008, 2009a, b; Matkar et al. 2008; Serafim et al. 2008; De Stefano et al. 2009; Jang et al. 2009; Larsson et al. 2010; Park et al. 2010a

[s] Handorean et al. 2009; Jiang et al. 2009; Jung et al. 2009b; Kim et al. 2009b, c, 2010b; Brandon-Warner et al. 2010; Cheung et al. 2010; Deep and Agarwal 2010; Flaig et al. 2010; Kaur et al. 2010; Mateen et al. 2010; Rajamanickam et al. 2010; Sangeetha et al. 2010; Velmurugan et al. 2010; Wang et al. 2010b, c; Wu et al. 2010

[t] Chambers et al. 2009; Gibbs et al. 2009; Hunakova et al. 2009; Kim and Singh 2009; Singh et al. 2009; Tarozzi et al. 2009; Xiao et al. 2009; Bryant et al. 2010; Do et al. 2010; Hahm and Singh 2010; Kaminski et al. 2010a, b; Li et al. 2010c; Meeran et al. 2010; Nair et al. 2010; Nishikawa et al. 2010; Shan et al. 2010; Sharma et al. 2010; Shirasugi et al. 2010; Traka et al. 2010

[u] Lu et al. 2008; Yamaoka et al. 2009; Adhikary et al. 2010; Kumar et al. 2010b; Lahiry et al. 2010; Sil et al. 2010

[v] Oh et al. 2008; Mandal et al. 2008; Stan et al. 2008; Kalthur et al. 2009; Lee et al. 2009b; Manoharan et al. 2009; Panjamurthy et al. 2009; Koduru et al. 2010; Lee et al. 2010; Yu et al. 2010

1.4 Curcumin

Curcumin (diferuloylmethane), the major active yellow pigment of the rhizome of *Curcuma longa* (turmeric plant) is widely used as a food-flavoring agent by Asians. Traditionally, turmeric has been described in Ayurveda as a potent anti-inflammatory agent (Hatcher et al. 2008; Aggarwal and Harikumar 2009). Curcumin has been studied extensively for its chemo-preventative and chemo-therapeutic properties. There is a large volume of published literature regarding its anti-tumorigenic activity in breast, colorectal, liver, pancreatic, and prostate cancers both *in vitro* and *in vivo* (Aggarwal et al. 2007; Nonn et al. 2007; Barve et al. 2008; Iqbal et al. 2009; Kim et al. 2008; Swamy et al. 2008; Yu et al. 2008). While the exact mechanism is unknown, many biological processes including cell proliferation, angiogenesis, apoptosis, cell cycle, and invasion have been implicated in the action of curcumin. However, prospective clinical utility is seriously limited owing to its rapid metabolism and therefore availability in the target tissue. This shortcoming does not bring forth the desired chemo-preventive effect. Several analogues and derivatives have been synthesized by various groups including our own (Kumar et al. 2000, 2003). Many of these derivatives are more soluble than curcumin while potentially retaining curcumin's favorable biological activities such as not being subject to the type of reductive metabolism documented with curcumin *in vivo*. However, to the best of our knowledge none of these derivatives or analogues has passed beyond preclinical testing. Despite these limitations, data from Phase I clinical trials shows that doses as high as 12 g per day are well-tolerated in humans (Sharma et al. 2004; Garcea et al. 2005; Dhillon et al. 2008; Vareed et al. 2008). Considering its pharmacologic safety in humans at doses as high as 12 g per day over long periods of time and its cost-effectiveness, curcumin could be a suitable chemo-preventive agent (Sharma et al. 2004; Garcea et al. 2005; Dhillon et al. 2008; Vareed et al. 2008).

1.5 Eugenol

Eugenol (2-methoxy-4-2-(2-propenyl) phenol) is found in the essential oils of *Syzygium aromaticum* (clove), *Pimenta racemosa* (bay leaf), and *Cinnamomum verum* (cinnamon leaf) that are used as flavoring agents in culinary practices all over the world. Medicinal applications of eugenol include its use as an antiseptic, anti-bacterial, analgesic agent, and as a cavity filling agent in dentistry (Thomson et al. 1984; Phillips 1990; Sukumaran et al. 1994). Further, eugenol has been shown to be effective as an anticancer agent. Three times weekly cutaneous applications of eugenol for 63 weeks to a group of female ICR/HA Swiss mice, inhibited the formation of skin carcinomas. Eugenol was partially effective in inhibiting tumors when benzo[a]pyrene was used as the carcinogen. Eugenol inhibited DMBA-croton oil-induced papilloma by about 84%. Eugenol was found to inhibit metastasis of melanoma tumors to the lung of mice in a xenograft study (Ghosh et al. 2005). Eugenol has been negative as a carcinogen in the Ames assay (mutation), and the urine of rats treated with eugenol was not mutagenic. Two-year carcinogenesis studies done in rodents (mice and rats) in which they were fed a diet containing eugenol (3,000–12,500 ppm). In these studies, final body weights remained comparable to control animals. No evidence of carcinogenicity was seen in male and female rats fed on a eugenol diet for 103 weeks (Thomson et al. 1984; Phillips 1990; Sukumaran et al. 1994). These data show that eugenol is (i) absorbed, (ii) not mutagenic, and (iii) not carcinogenic in rodents. Further, eugenol has a safe dose for human consumption according to the evaluation of the joint Food and Agriculture Organization of the United Nations/World Health Organization Expert Committee on Food Additives that showed the daily intake level of up to 2.5 mg/kg body weight cause no toxicological effect in humans. Although the existing literature on the mechanism through which eugenol acts as an anticancer agent is limited, it appears that in melanoma cells it inhibits cell cycle progression and induces apoptosis by targeting the important regulator of cell cycle regulation namely E2F1. Collaborative studies from our laboratory showed that eugenol inhibits melanoma tumor growth through modulation of transcriptional activation of E2F1 (Ghosh et al. 2005). At the time of writing this review, we were not aware of any clinical testing using eugenol.

1.6 Resveratrol

Resveratrol (3,5,4-tridhydroxystilbene, RES) is a polyphenolic phytoalexin found in the skin of various fruits such as red grapes, raspberries, mulberries, peanuts, blueberries, cranberries, and plums (Jang et al. 1997; Holmes-McNary and Baldwin 2000; Sanders et al. 2000; Bråkenhielm et al. 2001; She et al. 2001, 2002; Wolter et al. 2001; Shih et al. 2002; Ulrich et al. 2005; Fulda and Deabatin 2006; Athar et al. 2007; Harper et al. 2007). It is also produced naturally as a phytoalexin by several plants in response to bacterial and fungal infection. RES is nontoxic, sold as nutritional supplement and has been shown to demonstrate anti-inflammatory, anti-tumorigenic, immunomodulatory, cardioprotective, antioxidative, and chemopreventive activities (Jang et al. 1997; Sanders et al. 2000; Holmes-McNary and Baldwin 2000; Bråkenhielm et al. 2001; She et al. 2001, 2002; Wolter et al. 2001; Shih et al. 2002; Ulrich et al. 2005; Fulda and Deabatin 2006; Athar et al. 2007; Harper et al. 2007). RES has also been shown to modulate expression of genes involved in numerous signal transduction pathways including cell proliferation, apoptosis, angiogenesis, inflammation, and cell cycle regulation (Jang et al. 1997; Holmes-McNary and Baldwin 2000; Sanders et al. 2000; Bråkenhielm et al. 2001; She et al. 2001, 2002; Wolter et al. 2001; Shih et al. 2002; Ulrich et al. 2005; Fulda and Deabatin 2006; Athar et al. 2007; Harper et al. 2007).

Pre-clinical studies demonstrate that administration of RES inhibits growth of tumors in various rodent cancer models including skin, breast, colon, and prostate (Jang et al. 1997; Bhat et al. 2001; Schneider et al. 2001; Harper et al. 2007). Topical application of RES reduced the number of skin tumors per mouse by up to 98%

(Jang et al. 1997). Administration of 100 mg/kg RES in rats delayed NMU-induced mammary tumors significantly in rats (Bhat et al. 2001); 15 mg/kg/day RES (0.01%) in drinking water has been reported to reduce adenoma load in APC min mouse model by 70% (Schneider et al. 2001); 200 g/kg/day RES administered through diet decreased number of aberrant colonic crypt foci in azoxymethane-induced rat colon carcinogenesis model by 40% and multiplicity by 50% (Tessitore et al. 2000); intraperitoneal administration of 1 mg/kg RES decreased tumor multiplicity by 52% in NMBA induced formation of esophageal cancer in rats (Li et al. 2002), and dietary administration of 1 mg/kg RES decreased tumor incidence by 45% and multiplicity by 55% in DMBA-induced mammary carcinogenesis rat model (Banerjee et al. 2002). The data from pre-clinical studies cited above indicate a wide variation in the doses of RES ranging from 100 ng to 1,500 mg/kg body weight in animals for its cancer preventing activity in pre-clinical models (Baur and Sinclair 2006). Such high doses of RES may be required as RES is known to undergo rapid metabolism by glucouronidation and sulfation (Walle et al. 2004; Baur and Sinclair 2006). Most of the RES has been shown to be converted to sulfates and to a lesser extent to glucuronides in humans within 30 minutes of ingestion. The half-life of total RES in serum has been shown to be less than 10 hours (Walle et al. 2004).

Studies using radiolabeled RES in mice and rats suggest it is well absorbed with recovery of radioactivity reported from stomach, liver, kidney, intestine, bile, and urine in mice (Gesher and Steward 2003). Further, RES has been shown to undergo rapid glucouronidation and sulfonation both in the liver and intestinal epithelial cells (Gesher and Steward 2003; Wenzel et al. 2005). Taken together these *in vivo* pharmacokinetic studies suggest that availability of RES is poor due to its rapid metabolism involving glucouronidation and sulfonation. It is not clear whether RES is more active biologically in the conjugated or unconjugated form. Additionally, a recent clinical study in humans reported a plasma concentration of free RES of approximately 21 nM (less than 5%) following an oral dose of 25 mg (Goldberg et al. 2003; Boocock et al. 2007). Despite this poor bioavailability, RES demonstrated *in vivo* tumor preventing activities in various models including skin, breast, prostate, and colon (Tessitore et al. 2000; Bhat et al. 2001; Schneider et al. 2001; Wolter et al. 2001; Banerjee et al. 2002; Li et al. 2002). Accordingly, studies are being conducted to develop novel derivatives of RES (Bishayee 2009). Such attempts have resulted in the identification of hexahydroxystilbene as an effective compound to inhibit tumor growth including metastatic spread in pre-clinical models (Szekeres et al. 2010). In addition, several studies reported combinatorial effects of RES in combination with curcumin, dacarbacine, Genistein, roscovitine and vitamin E in multiple tumor types (Komina and Wesierska-Gadek 2008; Leone et al. 2008; Snyder et al. 2008; Wesierska-Gadek et al. 2008; Harper et al. 2009; Majumdar et al. 2009; Narayanan et al. 2009). Taken together, these data suggest a promising role for developing RES either alone or in combination with other natural products. Further clinical trials testing the beneficial effects of RES on colon cancer prevention and follicular lymphoma are ongoing (Tables 1.3 and 1.4) in addition to completed Phase I dose-escalation and pharmacokinetics of RES in healthy volunteers (http://Clinicaltrials.gov).

Table 1.3 Examples of completed clinical trials using natural products

Clinical trial no.	Product & study purpose	End point(s)
NCT00200824	Effect of soy compounds on prostate and breast cancer patients	Surrogate markers, bone density markers, quality of life
NCT00010829	Effect of phytoestrogens on hormone-dependent cancers, cardiovascular disease, osteoporosis	Surrogate markers, estrogen metabolites
NCT00625391	Effect of green tea on bone health, prevention of bone loss, & problems in post-menopausal women with low bone mass	Bone biomarkers associated with bone remodeling; oxidative stress
NCT00214032	Effect of pycnogenol (French maritime pine bark extract) on arm lymph-edema in women following treatment for breast cancer	Bioelectric impedence
NCT00064857	Effect of pycnogenol (French maritime pine bark extract) on arm lymph-edema in women following treatment for breast cancer	Bioelectric impedence
NCT00026117	Effect of shark cartilage extract on colorectal & breast cancer	Quality of life
NCT00005838	Effect of shark cartilage extract (AE-941) on non-small cell lung cancers	Combine with radiation therapy
NCT00094562	Effect of fish oil supplements (AAFA) in patients with disease-related weight loss including cancer, chronic obstructive pulmonary disease, rheumatoid arthritis, chronic heart failure	Maintain body weight and improve quality of life
NCT00034450	Effect of L-carnitine on fatigue in cancer patients	Change in fatigue
NCT00200837	Plants from Central & South America oral pain & inflammation relief	Observational study
NCT00079794	Effect of iscar on lung cancer	Quality of life
NCT00037154	Effect of saw palmetto extract on benign prostatic hyperplasia	Quality of life
NCT00064272	Use of ginger to prevent nausea & vomiting in cancer patients receiving chemotherapy	Decreased nausea & vomiting
NCT00033878	Use of freeze-dried noni fruit extract to treat cancer patients	Define toxicity, determine maximum tolerated dose & efficacy, identify antitumor properties of extract
NCT0098969	Use of resveratrol in preventing cancer	Determine safety, concentration & metabolites

Table 1.4 Examples of ongoing clinical trials using plant-derived natural products

Clinical trial no.	Product & study purpose	End points
NCT00765310	Effect of dietary lipoic acid on heart disease	Markers of inflammation & obesity
NCT00579540	Effect of flax seed, fish oil, soybean oil on polycystic ovarian syndrome patients	Blood sugar and insulin levels
NCT00006392	Effect of vitamin E & selenium on prostate cancer patients	Quality of life
NCT00049608	Effect of gemcitabine with mistletoe on breast, pancreatic, and non-small cell lung cancer patients with advanced solid tumors	Pharmacokinetics & tumor response
NCT00612560	Effect of flaxseed & aromatase inhibitor therapy on breast cancer	Markers of Ki67, apoptosis, ER, PgR, HER2
NCT00109980	Effect of medicinal plants on inflammation	*In vitro* studies
NCT01018615	Effect of silymarin & green tea extract on patients with chronic hepatitis C infection	Safety, metabolism, and antioxidant activity
NCT00579904	Effect of walnuts, walnut oil, almonds, fish oil on patients with polycystic ovarian syndrome (improve insulin resistance)	Prolong patient survival, evaluate blood sugar and insulin levels
NCT00246767	Effect of selected vegetables & herb mix on non-small cell lung cancer	Prolong patient survival, quality of life, minimal toxicity
NCT00003851	Effect of gemcitabine with pancreatic enzyme therapy plus specialized diet (Gonzalez regimen) on pancreatic cancer	Survival, quality of life
NCT00519779	Effect of fish oil (omega-3 intervention) on medical students to improve immune function & mood	Proinflammatory cytokine levels
NCT00118846	Effect of isoflavone soy on post-menopausal women	Breast tissue density, bone mineral density & anti-arteriosclerotic
NCT00578396	Effect of resveratrol in colon cancer prevention	Expression of beta-catenin in intestinal mucosa side effects
NCT00455416	Effect of resveratrol in patients with follicular lymphoma	Markers of proliferation, apoptosis, oxidative stress, host immune cell infiltrate
NCT00719563	Effect of ginseng on fatigue in cancer patients	Fatigue, stress, toxicity
NCT01136928	Effect of ginseng in healthy volunteers	Effect of ginseng treatment on pharmacokinetic profile of efavirenz
NCT01105130	Effect of ginseng on prostate cancer survivors	Quality of life, erectile function
NCT00631852	Effect of ginseng in breast cancer patients	Tumor & inflammatory biomarkers

1.7 Complex Plant-derived Products and Pre-clinical Studies

Our personal view of the limitation of current cancer treatment approach is the focus on a mono-targeted approach. Cancer occurs due to changes in multiple genes and signaling pathways, therefore targeting multiple signaling pathways (targets) would be a preferred approach. In this context, "herbal medicine or complex mixtures/botanicals" offer tremendous potential. In the last decade, about 51 new drugs have been approved by the US Food and Drug Administration for treatment of various types of cancers; some of which include approvals for use of the same drug for a different cancer type (Fig. 1.1 and Table 1.5). These drugs include recently approved cabazitaxel and provenge for treatment of hormone-refractory prostate cancer (approved in 2010), Avastin for renal cancer (2009), cervix for cervical (2009), etc. This is disproportionate to the rate at which cancer incidence has increased in that time. In addition, a majority of these drugs have only marginally increased the overall survival of patients. Further this benefit comes at high expense. Several natural products (complex mixtures) have shown promise in pre-clinical models including green tea polyphenols, pomegranate, soy, tomato paste, and an herbal combination product PC-SPES (Kumar et al. 2010a). In addition, the efficacy of tomato paste or broccoli alone, and in combination (tomato and broccoli) was compared with lycopene in rodent models of prostate cancer (Kumar et al. 2010a). These studies were recently reviewed by us elsewhere (Kumar et al. 2010a). We discuss the utility of novel promising candidate complex mixtures of plant origin in the following section.

Nexrutine is a commercially available herbal extract from the Chinese plant *Phellodendron amurense* (cork tree in Greek), which is widely used for the treat-

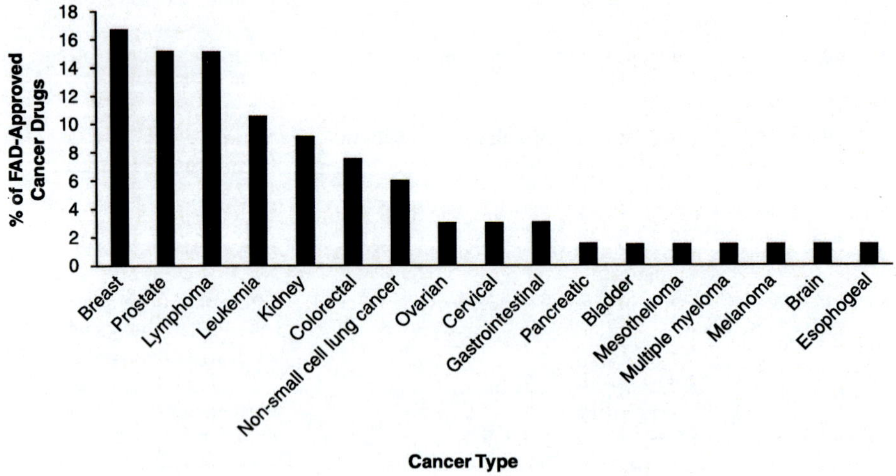

Fig. 1.1 United States Food and Drug Administration-approved drugs from 1996–2010

Table 1.5 United States Food and Drug Administration-approved drugs from 1996–2010

Cancer type	Number of approved drugs	% total approved
Breast	11	16.7
Prostate	10	15.2
Lymphoma	10	15.2
Leukemia	7	10.6
Kidney	6	9.1
Colorectal	5	7.6
Non-small cell lung cancer	4	6.1
Ovarian	2	3.0
Cervical	2	3.0
Gastrointestinal	2	3.0
Pancreatic	1	1.5
Bladder	1	1.5
Mesothelioma	1	1.5
Multiple myeloma	1	1.5
Melanoma	1	1.5
Brain	1	1.5
Esophageal	1	1.5
Total	66	

ment of inflammation, gastroenteritis, abdominal pain, and diarrhea in folk-lore medicine (Uchiyama et al. 1989; Mori et al. 1995). This tree is native to Asia and has been reported to contain isoquinoline alkaloids, phenolic compounds, and flavone glycosides. Nexrutine is sold as an over the counter anti-inflammatory herbal supplement in tablet and capsule form in the US. In addition, these extracts have been used as anti-rheumatic, aphrodisiac, bitter stomachic, cholagogue, diuretic, expectorant, febrifuge, hypoglycemic, ophthalmic, skin, vasodilator, and tonic. It is taken internally for the treatment of acute diarrhea, dysentery, jaundice, vaginal infections including Trichomonas, acute urinary tract infections, enteritis, boils, abscesses, night sweats, and skin diseases. Although Nexrutine is not well studied as an anticancer agent, studies from our laboratory show that administration of Nexrutine through diet prevents the development and progression of prostate tumors in a pre-clinical animal model (Kumar et al. 2007; Ghosh et al. 2010). Investigation of molecular mechanism showed that Nexrutine inhibits proliferation of prostate cancer cells through inhibition of multiple targets including Akt, transcription factors CREB and NF-κB (Garcia et al. 2006; Muralimanoharan et al. 2009). In addition using activity-guided fractionation, we have identified butanol fraction (F3) that recapitulates the anti-proliferative activity of Nexrutine. Ultra performance liquid chromatography analysis identified berberine or berberine-like compounds in the F3 fraction and pure berberine inhibited proliferation of prostate cancer cells. Our data thus far suggest that agents targeting multiple signal transduction pathways have tremendous translational potential due to multifactorial nature of cancer progression. In addition, since the identified signaling pathways (Akt/CREB/NF-κB) play an important role in other cancer types, Nexrutine may have utility in those cancers.

Another complex mixture that has tremendous potential for clinical development especially for colorectal and esophageal cancers is black raspberry extract (Duncan et al. 2009; Zikri et al. 2009). Lyophilized black raspberries are a source of multiple nutrient and non-nutrient compounds including vitamins A, C, E, folic acid, calcium, selenium, ellagic acid, ferulic acid, quercetin, and anthocyanins. Most of these agents have demonstrated chemo-preventive activity both *in vitro* and *in vivo* through a variety of mechanisms including inhibition of carcinogen activation, stimulation of carcinogen detoxification, down-regulation of genes associated with development of cancer, stimulation of apoptosis and inhibition of angiogenesis. Recent studies demonstrate that sustained delivery and uptake of freeze-dried black raspberry occurs at the target site (oral cavity) in healthy volunteers (Ugalde et al. 2009; Desai et al. 2010).

Astragalus membranaceus (milk vetch root or Huangqi in traditional Chinese medicine) is another natural product that has been used for immune stimulation and is recognized as a possible adjuvant therapy in the treatment of lung cancer (Cho and Leung 2007a, b). Results from meta-analysis that reviewed 34 randomized studies using multiple herbal formulae containing astragalus as the major component suggested improved survival, increased tumor response, reduced chemotherapy toxicity, and enhanced effectiveness of platinum based chemotherapy for advanced non-small cell lung cancer. However, a limitation of this study was that most of the clinical trials evaluated were of low power and conducted in China. Nevertheless, the results are encouraging and warrant more systematic and multicenter trials (McCulloch et al. 2006). Astragalus extracts may contribute to cancer regression by potentiating host immune function through the stimulation of macrophage and natural killer cell activity thus activating host antitumor immune mechanisms and may potentiate cell cytotoxicity of interleukin-2 (IL-2) thereby reducing the therapeutic dose of recombinant IL-2 (Chu et al. 1988). It is not certain whether the extract increases the immune mediated antitumor activity of IL-2 or if the astragalus extracts stimulates increased IL-2 production as a study of 31 patients with end-stage renal disease demonstrated that with treatment the astragalus extract resulted in a significantly higher level of IL-2 compared to placebo group (Qun et al. 1999). There is also data to suggest a hepatoprotective effect as astragalus extracts were shown to delay chemical-induced hepato-carcinogenesis in chemically-induced liver injury model using diethyl nitrosamine and 2-acetylaminofluorene in rats (Cui et al. 2003).

In addition, compounds containing ginseng have been reported to be effective in the management of obesity. Given the relationship between obesity and cancer, such compounds may potentially be effective in cancer management. There are at least four ongoing clinical trials currently in various phases of patient recruitment to investigate the effect of ginseng (Table 1.4) (http://cancerclinicaltrials.gov). We will await the results of these trials to determine the suitability of using ginseng as an anticancer agent in humans.

1.8 Natural Products and Standard of Care

As discussed previously, it is well known that one of the problems with cancer treatment is the development of resistance to chemo-therapeutic drugs or radiation. A common reason for such resistance is due to defective apoptotic pathway which could result in clonal expansion of transformed cells resistant to apoptosis induction in response to treatment (Kashkar 2010). For example, docetaxel-based chemotherapy is used as standard of care for patients with androgen independent prostate cancer to induce tumor cell death in androgen independent cells before the manifestation of clinically significant disease. Two recent randomized trials demonstrated a small but consistent survival benefit for androgen independent prostate cancer patients receiving docetaxel therapy (Fizazi et al. 2007; Berthold et al. 2008; Shepard and Dreicer 2010; Vishnu and Tan 2010). Although docetaxel treatment demonstrated improved survival, it did not eliminate late-stage disease. Further, tumor cells develop resistance to cytotoxic drugs including docetaxel. It has been shown that transcription factor NF-κB can contribute to chemo-resistance in prostate cancer cells in addition to predicting high risk of relapse in patients with localized prostate cancer (Karin 2009; Oeckinghaus and Ghosh 2009). Therefore, down-regulation of NF-κB activation using natural products could potentiate the effects of docetaxel or other chemo-therapeutic agents synergistically or additively. Such strategies have the advantage of (i) simultaneously targeting multiple signaling pathways important in carcinogenesis; (ii) using lower concentrations of individual compounds; (iii) reducing toxicity associated problems; (iv) reducing the chances of developing drug resistance that arise from using high concentrations of individual compounds. Such studies will have immediate impact for combining natural products with existing standard-of-care practices including radiotherapy and chemotherapy. Although such combinatorial approaches have been successfully used in the treatment of tuberculosis, acquired immunodeficiency syndrome, and leukemia (Torrance et al. 2000), they have not been examined in depth in cancer management to achieve maximum efficacy. We investigated if intervention with Nexrutine would potentiate cytotoxic effects of radiotherapy. Such a combination showed potential to (i) increase tumor sensitivity to both radiation and anticancer drugs (potentially allowing lower doses of both); (ii) reduce the cytotoxic effects of radiation on normal cells; and (iii) improve therapeutic index of the chemo-radiotherapy. In a study conducted in our laboratory, 20 week-old TRAMP mice (age at which they display invasive carcinoma) were fed pelleted diet containing 600 mg/kg *Phellodendron amurense* bark extract 6 weeks prior to radiation therapy and compared to a group that received radiation alone. Efficacy was evaluated by histological analysis of prostate tissue at the termination of the experiment. Results demonstrate that a majority of animals that received *Phellodendron amurense* bark extract for 6 weeks prior to radiation therapy had no overt cancer but exhibited features consistent with high grade prostatic intraepithelial neoplasia. In contrast, animals that received radiation alone exhibited features consistent with well to poorly differentiated adenocarcinoma.

A Phase II study of the combination of curcumin and gemcitabine in patients with advanced pancreatic cancer showed promising results. One patient who participated on this trial showed partial response; 4 stable disease and tumors in 6 patients progressed (Epelbaum et al. 2008). Therefore, despite poor bioavailability, curcumin appears to possess biological activity in humans that warrant more elaborate multi-center investigations to develop this natural product as a potential cancer preventive/therapeutic agent.

Further, a combination of four herbs traditional Chinese medicine formulation called PHY906 was tested in Phase I studies in patients with advanced solid tumors who failed standard therapy. Patients received PHY906 with increasing does of capecitabine until maximum tolerated dose was achieved. Due to the lack of adverse toxic effects, a Phase II study is ongoing in patients with gemcitabine-refractory pancreatic cancer (http://www.clinicaltrials.gov). Given the limited treatment options for prostate cancer patients who have undergone primary therapy (either radiotherapy or prostatectomy), strategies to delay clinically significant prostate cancer progression or extend the interval from treatment failure to hormonal ablation would be noteworthy. Natural products including curcumin or Nexrutine may have potential utility in such patients. Along these lines, a Phase II clinical trial to test the potential of pomegranate in men with rising prostate-specific antigen after radiotherapy or prostatectomy showed significant prolongation of serum prostate-specific antigen doubling time (Albrecht et al. 2004). *Pinica granatum* (pomegranate fruit) has been used for centuries in ancient human culture for its medicinal purposes. This has been shown to possess anti-proliferative, apoptotic, angiogenic, and NF-κB inhibitory activities in various tumor types including prostate (Pantuck et al. 2006; Rettig et al. 2008; Koyama et al. 2010). Pre-clinical studies from our laboratory demonstrate that a combination of antioxidant intervention can inhibit development of prostatic intraepithelial neoplasia formation by about 70% in Noble rats. All these data suggest the potential for developing natural products either alone or in combination with other agents including standard of care chemo-therapeutic agents for successful cancer management. Such approaches could inhibit tumor development synergistically or additively due to modulation of multiple targets by these natural products as opposed to current mono-targeted approach. Such encouraging preliminary results warrant more thorough studies using pre-clinical animal models and clinical trials to demonstrate the usefulness of such combinatorial strategies.

1.9 Future Prospects

Some of natural products have been tested in clinical trials to assess their efficacy on parameters ranging from quality of life, surrogate biomarkers associated with bone remodeling and oxidative stress, toxicity and efficacy studies in cancer patients (Table 1.3). Although not conclusive, these preliminary studies hold much hope for the utility of such natural products for cancer management. However, more thorough and randomized trials need to be conducted. As shown in Table 1.4, several clinical trials are ongoing using natural products. For example, studies are examining the effect of dietary lipoic acid on markers of inflammation and obesity. Although this study does not involve cancer patients, if successful results from this study will have an impact on cancer patients since both inflammation and obesity are risk factors for certain types of cancers. Similarly studies to test the efficacy of silimarin and green tea extract on patients with chronic hepatitis C infection and vegetables on non-small cell lung cancer are ongoing.

The use of natural products both single compounds and complex mixtures in clinical use also have their set of challenges (Miller et al. 2008; Tomlinson et al. 2008; Yeung et al. 2008). For example, the composition of the raw material (bark, leaves, fruits, or vegetables) depends on species and on subtype as well as on the environmental (geographic location, climate), farming, production, and storage conditions. In addition, the quality and content of the final product also depends on factors such as sun exposure, soil quality, agricultural practices, harvesting time, and length of time between harvest and preservation methods. Therefore, in order to provide consistent treatment benefit procurement of raw material should be regulated so that these medicinal plants are harvested using uniform agricultural and handling practices. Nevertheless, we believe that because of excellent structural diversity, natural products offer invaluable resources for anticancer drug discovery. We made an attempt to review information regarding some successful drugs derived from natural products in their original or synthetic form as well as several new compounds in this chapter. However, herbal remedies are yet to be integrated into mainstream medical practice mainly due to lack of experimental and clinical studies on their safety, efficacy, and pharmacological mechanisms. Careful *in vitro* and *in vivo* studies are essential to evaluate their efficacy and safety before clinical trials can be contemplated. Nevertheless, oncologists should view use of natural products (when such data are available) as promising agents either *per se* or as adjuncts to standard of care for successful cancer management.

Acknowledgments Financial support from ACS RSG-04-169 and NIH (CA 98744, 135451, 137518, and 142025) and VA-Merit Award (I101 BX000766-01) is acknowledged. Contribution of Gretchen Garcia, Ashima Gupta, Shylesh Bhaskaran, and Manonmani Ganapathy during the course of some of these studies is also acknowledged. We thank Jingjing Gong and Praveen Jampala for help with preparation of bibliography. We sincerely thank Dr. William Cho for his excellent editorial assistance including bibliography.

References

A Phase I/II multi-center, open label dose escalation, safely and efficacy study of PHY 906 plus Capecitabine in patients with advanced pancreatic carcinoma. http://www.clinicaltrials.gov. Accessed 10 July 2010.

Acharya, A., Das, I., Singh, S., et al. (2010). Chemopreventive properties of indole-3-carbinol, diindolylmethane and other constituents of cardamom against carcinogenesis. *Recent Patents on Food, Nutrition & Agriculture, 2,* 166–177.

Adams, L. S., Zhang, Y., Seeram, N. P., et al. (2010). Pomegranate ellagitannin-derived compounds exhibit antiproliferative and antiaromatase activity in breast cancer cells in vitro. *Cancer Prevention Research, 3,* 108–113.

Adhami, V. M., Khan, N., & Mukhtar, H. (2009). Cancer chemoprevention by pomegranate: Laboratory and clinical evidence. *Nutrition and Cancer, 61,* 811–815.

Adhikary, A., Mohanty, S., Lahiry, L., et al. (2010). Theaflavins retard human breast cancer cell migration by inhibiting NF-kappaB via p53-ROS cross-talk. *FEBS letters, 584,* 7–14.

Aggarwal, B. B., & Harikumar, K. B. (2009). Potential therapeutic effects of curcumin, the anti-inflammatory agent, against neurodegenerative, cardiovascular, pulmonary, metabolic, autoimmune and neoplastic diseases. *The International Journal of Biochemistry & Cell Biology, 41,* 40–59.

Aggarwal, B. B., Banerjee, S., Bharadwaj, U., et al. (2007). Curcumin induces the degradation of cyclin E expression through ubiquitin-dependent pathway and up-regulates cyclin-dependent kinase inhibitors p21 and p27 in multiple human tumor cell lines. *Biochemical Pharmacology, 73,* 1024–1032.

Ahmad, A., Kong, D., Wang, Z., et al. (2009). Down-regulation of uPA and uPAR by 3,3′-diindolylmethane contributes to the inhibition of cell growth and migration of breast cancer cells. *Journal of Cellular Biochemistry, 108,* 916–925.

Ahmad, F.B., Ghaffari Moghaddam, M., Basri, M., et al. (2010). Anticancer activity of 3-O-acylated betulinic acid derivatives obtained by enzymatic synthesis. *Bioscience, Biotechnology, and Biochemistry, 74,* 1025–1029.

Akihisa, T., Noto, T., Takahashi, A., et al. (2009). Melanogenesis inhibitory, anti-inflammatory, and chemopreventive effects of limonoids from the seeds of Azadirachta indicia A. Juss. (neem). *Journal of Oleo Science, 58,* 581–594.

Albrecht, M., Jiang, W., Kumi-Diaka, J., et al. (2004). Pomegranate extracts potently suppress proliferation, xenograft growth and invasion of human prostate cancer cells. *Journal of Medicinal Food, 7,* 274–283.

American Cancer Society. (2010). Cancer Facts & Figures 2010. http://www.cancer.org/Research/CancerFactsFigures/CancerFactsFigures/cancer-facts-and-figures-2010. Accessed 19 July 2010.

American Cancer Society. (6 Nov 2009). Complementary and Alternative Medicine. http://www.cancer.org/Treatment/TreatmentsandSideEffects/ComplementaryandAlternativeMedicine/index. Accessed 11 June 2008.

Ames, B. N., Gold, L. S., & Willett, W. C. (1995). The causes and prevention of cancer. *Proceedings of the National Academy of Sciences of the United States of America, 92,* 5258–5265.

Amin, A. R., Kucuk, O., & Khuri, F. R. (2009). Perspectives for cancer prevention with natural compounds. *Journal of Clinical Oncology: Official Journal of the American Society of Clinical Oncology, 27,* 2712–2725.

Andlauer, W., Kolb, J., & Furst, P. (2004). Phloridzin improves absorption of genistin in isolated rat small intestine. *Clinical Nutrition (Edinburgh, Scotland), 23,* 989–995.

Armstrong, T. S., & Gilbert, M. R. (2008). Use of complementary and alternative medical therapy by patients with primary brain tumors. *Current Neurology and Neuroscience Reports, 8,* 264–268.

Athar, M., Back, J. H., Tang, X., et al. (2007). Resveratrol: A review of preclinical studies for human cancer prevention. *Toxicology and Applied Pharmacology, 224,* 274–283.

Baez, S., Tsuchiya, Y., Calvo, A., et al. (2010). Genetic variants involved in gallstone formation and capsaicin metabolism, and the risk of gallbladder cancer in Chilean women. *World Journal of Gastroenterology: WJG, 16,* 372–378.

Balandrin, M. F., Kinghorn, A. D., & Farnsworth, N. (1993). Human medicinal agents from plants kinghorn. *ACS symposium series 534* (pp. 2–12). Washington, DC: American Chemical Society.

Banerjee, S., Bueso-Ramos, C., & Aggarwal, B. B. (2002). Suppression of 7,12-dimethylbenz(a)anthracene-induced mammary carcinogenesis in rats by resveratrol: Role for nuclear factor kappa B, cyclooxygenase2 and matrix metalloprotease 9. *Cancer Research, 62,* 4945–4954.

Banerjee, S., Wang, Z., Kong, D., et al. (2009). 3,3′-Diindolylmethane enhances chemosensitivity of multiple chemotherapeutic agents in pancreatic cancer. *Cancer Research, 69,* 5592–5600.

Bankson, D. D., Kestin, M., & Rifai, N. (1993). Role of free radicals in cancer and atherosclerosis. *Clinics in Laboratory Medicine, 13,* 463–480.

Barve, A., Khor, T. O., Hao, X., et al. (2008). Murine prostate cancer inhibition by dietary phytochemicals—curcumin and phenyethylisothiocyanate. *Pharmaceutical Research, 25,* 2181–2189.

Baur, J. A., & Sinclair, D. A. (2006). Therapeutic potential of resveratrol: The in vivo evidence. *Nature Reviews, 5,* 493–506.

Beckman, K. B., & Ames, B. N. (1998). The free radical theory of aging matures. *Physiological Reviews, 78,* 547–581.

Belkaid, A., Currie, J. C., Desgagnes, J., et al. (2006). The chemopreventive properties of chlorogenic acid reveal a potential new role for the microsomal glucose-6-phosphate translocase in brain tumor progression. *Cancer Cell International, 6,* 7.

Bernstein, B. J., & Grasso, T. (2001) Prevalence of complementary and alternative medicine use in cancer patients. *Oncology, 15,* 1267–1272.

Berthold, D. R., Pond, G. R., & Roessner, M., et al. (2008). Treatment of hormone-refractory prostate cancer with docetaxel or mitoxantrone: Relationships between prostate-specific antigen, pain, and quality of life response and survival in the tax-327 study. *Clinical Cancer Research: An Official Journal of the American Association for Cancer Research, 14,* 2763–2767.

Bhat, K. P., Lantvit, D., Christov, K., et al. (2001). Estrogenic and antiestrogenic properties of resveratrol in mammary tumor models. *Cancer Research, 61,* 7456–7463.

Bhatnagar, N., Li, X., Chen, Y., et al. (2009). 3,3′-diindolylmethane enhances the efficacy of butyrate in colon cancer prevention through down-regulation of survivin. *Cancer Prevention Research (Philadelphia, Pa.), 2,* 581–589.

Bishayee, A. (2009). Cancer prevention and treatment with resveratrol: From rodent studies to clinical trials. *Cancer Prevention Research (Philadelphia, Pa.), 2,* 409–418.

Bishop, F. L., Prescott, P., Chan, Y. K., et al. (2010). Prevalence of complementary medicine use in pediatric cancer: A systematic review. *Pediatrics, 125,* 768–776.

Boocock, D. J., Faust, G. E., Patel, K. R., et al. (2007). Phase I dose escalation pharmacokinetic study in healthy volunteers of resveratrol, a potential cancer chemopreventive agent. *Cancer Epidemiology Biomarkers and Prevention, 16,* 1246–1252.

Bråkenhielm, E., Cao, R., & Cao, Y. (2001). Suppression of angiogenesis, tumor growth, and wound healing by resveratrol, a natural compound in red wine and grapes. *The FASEB Journal: Official Publication of the Federation of American Societies for Experimental Biology, 15,* 1798–1800.

Brandon-Warner, E., Sugg, J. A., Schrum, L. W., et al. (2010). Silibinin inhibits ethanol metabolism and ethanol-dependent cell proliferation in an in vitro model of hepatocellular carcinoma. *Cancer Letters, 291,* 120–129.

Bright, S. A., Campiani, G., Deininger, M. W., et al. (2010). Sequential treatment with flavopiridol synergistically enhances pyrrolo-1,5-benzoxazepine-induced apoptosis in human chronic myeloid leukaemia cells including those resistant to imatinib treatment. *Biochemical Pharmacology, 80,* 31–38.

Brown, K. C., Witte, T. R., Hardman, W. E., et al. (2010). Capsaicin displays anti-proliferative activity against human small cell lung cancer in cell culture and nude mice models via the E2F pathway. *PLoS ONE, 5,* e10243.

Bryan, M., Pulte, E. D., Toomey, K. C., et al. (2010). A pilot phase II trial of all-trans retinoic acid (Vesanoid) and paclitaxel (Taxol) in patients with recurrent or metastatic breast cancer. *Investigational New Drugs* (in press).

Bryant, C. S., Kumar, S., Chamala, S., et al. (2010). Sulforaphane induces cell cycle arrest by protecting RB-E2F-1 complex in epithelial ovarian cancer cells. *Molecular Cancer, 9,* 47.

Bushue, N., & Wan, Y. J. (2010). Retinoid pathway and cancer therapeutics. *Advanced Drug Delivery Reviews, 62,* 1285–1298.

Canel, C., Moraes, R. M., Dayan, F. E., et al. (2000). Podophyllotoxin. *Phytochemistry, 54,* 115–120.

Carlsen, M. H., Halvorsen, B. L., Holte, K., et al. (2010). The total antioxidant content of more than 3100 foods, beverages, spices, herbs and supplements used worldwide. *Nutrition Journal, 9,* 1–11.

Cassileth, B., Yeung, K. S., & Gubili, J. (2008). Herbs and other botanicals in cancer patient care. *Current Treatment Options in Oncology, 9,* 109–116.
Cavalieri, E., Frenkel, K., Liehr, J. G., et al. (2000). Estrogens as endogenous genotoxic agents—DNA adducts and mutations. *Journal of the National Cancer Institute. Monographs, 27,* 75–93.
Centers for Disease Control and Prevention. Chronic disease prevention and health promotion. http://www.cdc.gov/nccdphp. Accessed 10 July 2010.
Chambers, K. F., Bacon, J. R., Kemsley, E. K., et al. (2009). Gene expression profile of primary prostate epithelial and stromal cells in response to sulforaphane or iberin exposure. *The Prostate, 69,* 1411–1421.
Chan, J. M., Elkin, E. P., Silva, S. J., et al. (2005a). Total and specific complementary and alternative medicine use in a large cohort of men with prostate cancer. *Adult Urology, 66,* 1223–1228.
Chan, J. M., Gann, P. H., & Giovannucci, E. L. (2005b). Role of diet in prostate cancer development and progression. *Journal of Clinical Oncology: Official Journal of the American Society of Clinical Oncology, 23,* 8152–8160.
Cheung, C. W., Gibbons, N., Johnson, D. W., et al. (2010). Silibinin—A promising new treatment for cancer. *Anti-cancer Agents in Medicinal Chemistry, 10,* 186–195.
Cho, W. C., & Leung, K. N. (2007a). In vitro and in vivo anti-tumor effects of Astragalus membranaceus. *Cancer Letters, 252,* 43–54.
Cho, W. C., & Leung, K. N. (2007b). In vitro and in vivo immunomodulating and immunorestorative effects of Astragalus membranaceus. *Journal of Ethnopharmacology, 113,* 132–141.
Cho, S. D., Lee, S. O., Chintharlapalli, S., et al. (2010). Activation of nerve growth factor-induced B alpha by methylene-substituted diindolylmethanes in bladder cancer cells induces apoptosis and inhibits tumor growth. *Molecular Pharmacology, 77,* 396–404.
Choi, W. Y., Kim, G. Y., Lee, W. H., et al. (2008). Sanguinarine, a benzophenanthridine alkaloid, induces apoptosis in MDA-MB-231 human breast carcinoma cells through a reactive oxygen species-mediated mitochondrial pathway. *Chemotherapy, 54,* 279–287.
Choi, W. Y., Jin, C. Y., Han, M. H., et al. (2009a). Sanguinarine sensitizes human gastric adenocarcinoma AGS cells to TRAIL-mediated apoptosis via down-regulation of AKT and activation of caspase-3. *Anticancer Research, 29,* 4457–4465.
Choi, Y. H., Choi, W. Y., Hong, S. H., et al. (2009b). Anti-invasive activity of sanguinarine through modulation of tight junctions and matrix metalloproteinase activities in MDA-MB-231 human breast carcinoma cells. *Chemico-Biological Interactions, 179,* 185–191.
Choi, C. H., Jung, Y. K., & Oh, S. H. (2010). Autophagy induction by capsaicin in malignant human breast cells is modulated by p38 and extracellular signal-regulated mitogen-activated protein kinases and retards cell death by suppressing endoplasmic reticulum stress-mediated apoptosis. *Molecular Pharmacology, 78,* 114–125.
Chu, D. T., Lepe-Zuniga, J., Wong, W. L., et al. (1988). Fractionated extract of astragalus membranaceus, a Chinese medicinal herb, potentiates lak cell cytotoxicity generated by a low dose of recombinant interleukin-2. *Journal of Clinical & Laboratory Immunology, 26,* 183–187.
Ciolino, H. P., Dai, Z., & Nair, V. (2010). Retinol inhibits aromatase activity and expression in vitro. *The Journal of Nutritional Biochemistry.* doi:10.1016/j.jnutbio.2010.04.004.
Cortese, F., Bhattacharyya, B., & Wolff, J. (1977) Podophyllotoxin as a probe for the colchicine binding site of tubulin. *The Journal of Biological Chemistry, 252,* 1134–1140.
Cragg, G. M., Newman, D. J., & Snader, K. M. (1997). Natural products in drug discovery and development. *Journal of Natural Products, 60,* 52–60.
Croce, C. M. (2008). Oncogenes and cancer. *The New England Journal of Medicine, 358,* 502–511.
Csuk, R., Barthel, A., Schwarz, S., et al. (2010). Synthesis and biological evaluation of antitumor-active gamma-butyrolactone substituted betulin derivatives. *Bioorganic & Medicinal Chemistry, 18,* 2549–2558.
Cui, R., He, J., Wang, B., et al. (2003). Suppressive effect of astragalus membranaceus bunge on chemical hepatocarcinogenesis in rats. *Cancer Chemotherapy and Pharmacology, 51,* 75–80.
De Leon, G., Cavino, M., D'Angelo, M., et al. (2010). PNUTS knockdown potentiates the apoptotic effect of Roscovitine in breast and colon cancer cells. *International Journal of Oncology, 36,* 1269–1275.

De Stefano, I., Raspaglio, G., Zannoni, G. F., et al. (2009). Antiproliferative and antiangiogenic effects of the benzophenanthridine alkaloid sanguinarine in melanoma. *Biochemical Pharmacology, 78,* 1374–1381.

DeBerardinis, R. J. (2008). Is cancer a disease of abnormal cellular metabolism? New angles on an old idea. *Genetics in Medicine: Official Journal of the American College of Medical Genetics, 10,* 767–777.

Deep, G., & Agarwal, R. (2010). Antimetastatic efficacy of silibinin: Molecular mechanisms and therapeutic potential against cancer. *Cancer Metastasis Reviews, 29,* 447–463.

Desai, K. G., Olsen, K. F., Mallery, S. R., et al. (2010). Formulation and in vitro-in vivo evaluation of black raspberry extract-loaded PLGA/PLA injectable millicylindrical implants for sustained delivery of chemopreventive anthocyanins. *Pharmaceutical Research, 27,* 628–643.

Dhillon, N., Aggarwal, B. B., Newman, R. A., et al. (2008). Phase II trial of curcumin in patients with advanced pancreatic cancer. *Clinical Cancer Research: An Official Journal of the American Association for Cancer Research, 14,* 4491–4499.

Dickson, M. A., Rathkopf, D. E., Carvajal, R. D., et al. (2010). A phase I pharmacokinetic study of pulse-dose vorinostat with flavopiridol in solid tumors. *Investigational New Drugs* (in press).

Ding, M., Lu, Y., Bowman, L., Huang, C., et al. (2004). Inhibition of AP-1 and neoplastic transformation by fresh apple peel extract. *Journal of Biological Chemistry, 279,* 10670–10676.

DiPaola, R. S., Chen, Y. H., Stein, M., et al. (2010). A randomized phase II trial of mitoxantrone, estramustine and vinorelbine or bcl-2 modulation with 13-cis retinoic acid, interferon and paclitaxel in patients with metastatic castrate-resistant prostate cancer: ECOG 3899. *Journal of Translational Medicine, 8,* 20.

Do, D. P., Pai, S. B., Rizvi, S. A., et al. (2010). Development of sulforaphane-encapsulated microspheres for cancer epigenetic therapy. *International Journal of Pharmaceutics, 386,* 114–121.

Drag, M., Surowiak, P., Drag-Zalesinska, M., et al. (2009). Comparision of the cytotoxic effects of birch bark extract, betulin and betulinic acid towards human gastric carcinoma and pancreatic carcinoma drug-sensitive and drug-resistant cell lines. *Molecules (Basel, Switzerland), 14,* 1639–1651.

Drag-Zalesinska, M., Kulbacka, J., Saczko, J., et al. (2009) Esters of betulin and betulinic acid with amino acids have improved water solubility and are selectively cytotoxic toward cancer cells. *Bioorganic & Medicinal Chemistry Letters, 19,* 4814–4817.

Du, G., Lin, H., Yang, Y., et al. (2010). Dietary quercetin combining intratumoral doxorubicin injection synergistically induces rejection of established breast cancer in mice. *International Immunopharmacology, 10,* 819–826.

Duncan, F. J., Martin, J. R., Wulff, B. C., et al. (2009). Topical treatment with black raspberry extract reduces cutaneous UVB-induced carcinogenesis and inflammation. *Cancer Prevention Research, 2,* 665–672.

Ekstrom, A. M., Serafini, M., Nyren, O., et al. (2010). Dietary quercetin intake and risk of gastric cancer: Results from a population-based study in Sweden. *Annals of Oncology: Official Journal of the European Society for Medical Oncology/ESMO, 22,* 438–443.

El-Readi, M. Z., Hamdan, D., Farrag, N., et al. (2010). Inhibition of P-glycoprotein activity by limonin and other secondary metabolites from Citrus species in human colon and leukaemia cell lines. *European Journal of Pharmacology, 626,* 139–145.

Eng, J., Ramsum, D., Verhoef, M., et al. (2003) A population-based survey of complementary and alternative medicine use in men recently diagnosed with prostate cancer. *Integrative Cancer Therapies, 2,* 212–216.

Epelbaum, R., Vizel, B., & Bar-Sela, G. (2008). Phase II study of curcumin and Gemcitabine in patients with advanced pancreatic cancer. *Journal of Clinical Oncology: Official Journal of the American Society of Clinical Oncology, 26*(Suppl.), 15619.

Ernst, E., & Cassileth, B. R. (1998). The prevalence of complementary/alternative medicine in cancer: A systematic review. *Cancer, 83,* 777–782.

Evans, M. D., Dizdaroglu, M., & Cooke, M. S. (2004). Oxidative DNA damage and disease: Induction, repair and significance. *Mutation Research, 567,* 1–61.

Fabricant, D. S., & Farnsworth, N. R. (2001). The value of plants used in traditional medicine for drug discovery. *Environmental Health Perspectives, 109,* 69–75.

Fares, F., Azzam, N., Appel, B., et al. (2010) The potential efficacy of 3, 3′-diindolylmethane in prevention of prostate cancer development. *European Journal of Cancer Prevention: The Official Journal of the European Cancer Prevention Organisation (ECP), 19,* 199–203.

Fizazi, K., Le Maitre, A., Hudes, G., et al. (2007). Addition of estramustine to chemotherapy and survival of patients with castration-refractory prostate cancer: A meta-analysis of individual patient data. *The Lancet Oncology, 8,* 994–1000.

Flaig, T. W., Glode, M., Gustafson, D., et al. (2010). A study of high-dose oral silybin-phytosome followed by prostatectomy in patients with localized prostate cancer. *The Prostate, 70,* 848–855.

Fulda, S. (2008). Betulinic acid for cancer treatment and prevention. *International Journal of Molecular Sciences, 9,* 1096–1107.

Fulda, S., & Debatin, K. M. (2006). Resveratrol modulation of signal transduction in apoptosis and cell survival: A mini-review. *Cancer Detection and Revention, 30,* 217–223.

Fulda S, & Kroemer G. (2009). Targeting mitochondrial apoptosis by betulinic acid in human cancers. *Drug Discovery Today, 14,* 885–890.

Gansler, T., Kaw, C., Crammer, C., et al. (2008). A population-based study of prevalence of complementary methods use by cancer survivors. *American Cancer Society, 113,* 1048–1057.

Garcea, G., Berry, D. P., Jones, D. J., et al. (2005). Consumption of the putative chemopreventive agent curcumin by cancer patients: Assessment of curcumin levels in the colorectum and their pharmacodynamic consequences. *Cancer Epidemiology, Biomarkers & Prevention: A Publication of the American Association for Cancer Research, Cosponsored by the American Society of Preventive Oncology, 14,* 120–125.

Garcia, G. E., Nicole, A., Bhaskaran, S., et al. (2006). Akt-and CREB-mediated prostate cancer cell proliferation inhibition by NexrutineR, a Phellodendron amurense extract. *Neoplasia (New York, N.Y.), 6,* 523–533.

Gesher, A. J., & Steward, W. P. (2003). Relationship between mechanisms, bioavailability, and preclinical chemopreventive efficacy of resveratrol: A conundrum. *Cancer Epidemiology, Biomarkers & Prevention: A Publication of the American Association for Cancer Research, Cosponsored by the American Society of Preventive Oncology, 12,* 953–957.

Ghosh, A. K., & Basu, S. (2010). Fas-associated factor 1 is a negative regulator in capsaicin induced cancer cell apoptosis. *Cancer Letters, 287,* 142–149.

Ghosh, R., Nadiminty, N., Fitzpatrick, J. E., et al. (2005). Eugenol causes melanoma growth suppression through inhibition of E2F1 transcriptional activity. *The Journal of Biological Chemistry, 280,* 5812–5819.

Ghosh, R., Graham, H., Rivas, P., et al. (2010). Phellodendron amurense bark extract prevents progression of prostate tumors in transgenic adenocarcinoma of mouse prostate: Potential for prostate cancer management. *Anticancer Research, 30,* 857–865.

Gibbs, A., Schwartzman, J., Deng, V., et al. (2009). Sulforaphane destabilizes the androgen receptor in prostate cancer cells by inactivating histone deacetylase 6. *Proceedings of the National Academy of Sciences of the United States of America 2009, 106,* 16663–16668.

Goldberg, D. M., Yan, J., & Soleas, G. J. (2003). Absorption of three wine-related polyphenols in three different matrices by healthy subjects. *Clinical Biochemistry, 36,* 79–87.

Gordaliza, M. (2007). Natural products as leads to anticancer drugs. *Clinical & Translational Oncology: Official Publication of the Federation of Spanish Oncology Societies and of the National Cancer Institute of Mexico, 9,* 767–776.

Granado-Serrano, A. B., Martín, M. A., Izquierdo-Pulido, M., et al. (2007). Molecular mechanisms of (-)-epicatechin and chlorogenic acid on the regulation of the apoptotic and survival/proliferation pathways in a human hepatoma cell line. *Journal of Agricultural and Food Chemistry, 55,* 2020–2027.

Gratus, C., Wilson, S., Greenfield, S. M., et al. (2009). The use of herbal medicines by people with cancer: A qualitative study. *BMC Complementary and Alternative Medicine, 9,* 14.

Gupta, S. C., Kim, J. H., Prasad, S., et al. (2010). Regulation of survival, proliferation, invasion, angiogenesis, and metastasis of tumor cells through modulation of inflammatory pathways by nutraceuticals. *Cancer Metastasis Reviews, 29,* 405–434.

Hahm, E. R., & Singh, S. V. (2010). Sulforaphane inhibits constitutive and interleukin-6-induced activation of signal transducer and activator of transcription 3 in prostate cancer cells. *Cancer Prevention Research (Philadelphia, Pa.), 3,* 484–494.

Handorean, A. M., Yang, K., Robbins, E. W., et al. (2009). Silibinin suppresses CD44 expression in prostate cancer cells. *American Journal of Translational Research, 1,* 80–86.

Harish, K. G., Chandra, M. K., Jagannadha, R. A., et al. (2009). Nimbolide a limonoid from Azadirachta indica inhibits proliferation and induces apoptosis of human choriocarcinoma (BeWo) cells. *Investigational New Drugs, 27,* 246–252.

Harper, C. E., Patel, B. B., Wang, J., et al. (2007). Resveratrol suppresses prostate cancer progression in transgenic mice. *Carcinogenesis, 28,* 1946–1953.

Harper, C. E., Cook, L. M., Patel, B. B., et al. (2009). Genistein and resveratrol, alone and in combination, suppress prostate cancer in sv-40 tag rats. *The Prostate, 69,* 1668–1682.

Harvey, A. L. (1999). Medicines from nature: Are natural products still relevant to drug discovery? *Trends in Pharmacological Sciences, 20,* 196–198.

Hatcher, H., Planalp, R., Cho, J., et al. (2008). Curcumin: From ancient medicine to current clinical trials. *Cellular and Molecular Life Sciences: CMLS, 65,* 1631–1652.

He, M., Vanaja, D. K., Karnes, R. J., et al. (2009). Epigenetic regulation of Myc on retinoic acid receptor beta and PDLIM4 in RWPE1 cells. *The Prostate, 69,* 1643–1650.

Hearn, B. R., Shaw, S. J., & Myles, D. C. (2006). Microtubule targeting agents. *Comprehensive Medicinal Chemistry II, 7,* 81–110.

Heath, E. I., Heilbrun, L. K., Li, J., et al. (2010). A phase I dose-escalation study of oral BR-DIM (BioResponse 3,3′-Diindolylmethane) in castrate-resistant, non-metastatic prostate cancer. *American Journal of Translational Research, 2,* 402–411.

Herman-Antosiewicz, A., Kim, Y. A., Kim, S. H., et al. (2010). Diallyl trisulfide-induced G2/M phase cell cycle arrest in DU145 cells is associated with delayed nuclear translocation of cyclin-dependent kinase 1. *Pharmaceutical Research, 27,* 1072–1079.

Hietanen, E., Bartsch, H., Béréziat, J. C., et al. (2004). Diet and oxidative stress in breast, colon and prostate cancer patients: A case-control study. *European Journal of Clinical Nutrition, 48,* 575–586.

Holmes-McNary, M., & Baldwin, A., Jr. (2000) Chemopreventive properties of trans-resveratrol are associated with inhibition of activation of the IkappaB kinase. *Cancer Research, 60,* 3477–3483.

Holzel, M., Huang, S., Koster, J., et al. (2010). NF1 is a tumor suppressor in neuroblastoma that determines retinoic acid response and disease outcome. *Cell, 142,* 218–229.

Horwitz, S. B., Parness, J., Schiff, P. B., et al. (1982). Taxol: A new probe for studying the structure and function of microtubules. *Cold Spring Harbor Symposia on Quantitative Biology, 46*(Pt 1), 219–226.

Howells, L. M., Britton, R. G., Mazzoletti, M., et al. (2010). Preclinical colorectal cancer chemopreventive efficacy and p53-modulating activity of 3',4',5'-trimethoxyflavonol, a quercetin analogue. *Cancer Prevention Research (Philadelphia, Pa.), 3,* 929–939.

Huang, W., & Cai, Y. (2010). Natural phenolic compounds from medicinal herbs and dietary plants: Potential use for cancer prevention. *Nutrition and Cancer, 62,* 1–20.

Hunakova, L., Sedlakova, O., Cholujova, D., et al. (2009). Modulation of markers associated with aggressive phenotype in MDA-MB-231 breast carcinoma cells by sulforaphane. *Neoplasma, 56,* 548–556.

Ichikawa, H., Nakamura, Y., Kashiwada, Y., et al. (2007). Anticancer drugs designed by Mother Nature: Ancient drugs but modern targets. *Current Pharmaceutical Design, 13,* 3400–3416.

Iqbal, M., Okazaki, Y., & Okada, S. (2009). Curcumin attenuates oxidative damage in animals treated with a renal carcinogen, ferric nitrilotriacetate (Fe-NTA): Implications for cancer prevention. *Molecular and Cellular Biochemistry, 324,* 157–164.

Jang, M., Cai, L., Udeani, G. O., et al. (1997). Cancer chemopreventive activity of resveratrol, a natural product derived from grapes. *Science, 275,* 218–220.

Jang, B. C., Park, J. G., Song, D. K., et al. (2009). Sanguinarine induces apoptosis in A549 human lung cancer cells primarily via cellular glutathione depletion. *Toxicology In vitro: An International Journal Published in Association with BIBRA, 23,* 281–287.

Jiang, C., Lee, H. J., Li, G. X., et al. (2006). Potent antiandrogen and androgen receptor activities of an angelica gigas-containing herbal formulation: Identification of decursin as a novel and active compound with implications for prevention and treatment of prostate cancer. *Cancer Research, 66,* 453–463.

Jiang, Y. Y., Wang, H. J., Wang, J., et al. (2009). The protective effect of silibinin against mitomycin C-induced intrinsic apoptosis in human melanoma A375-S2 cells. *Journal of Pharmacological Sciences, 111,* 137–146.

Jin, Y., Zou, X., & Feng, X. (2010). 3,3′-Diindolylmethane negatively regulates Cdc25A and induces a G2/M arrest by modulation of microRNA 21 in human breast cancer cells. *Anti-cancer Drugs, 21,* 814–822.

Jung, S. K., Lee, K. W., Byun, S., et al. (2008). Myricetin suppresses UVB-induced skin cancer by targeting Fyn. *Cancer Research, 68,* 6021–6029.

Jung, E., Lee, J., Huh, S., et al. (2009a). Phloridzin-induced melanogenesis is mediated by the cAMP signaling pathway. *Food and Chemical Toxicology: An International Journal Published for the British Industrial Biological Research Association, 47,* 2436–2440.

Jung, H. J., Park, J. W., Lee, J. S., et al. (2009b). Silibinin inhibits expression of HIF-1alpha through suppression of protein translation in prostate cancer cells. *Biochemical and Biophysical Research Communications, 390,* 71–76.

Jung, E. U., Yoon, J. H., Lee, Y. J., et al. (2010). Hypoxia and retinoic acid-inducible NDRG1 expression is responsible for doxorubicin and retinoic acid resistance in hepatocellular carcinoma cells. *Cancer Letters, 298,* 9–15.

Kalthur, G., Mutalik, S., & Pathirissery, U. D. (2009). Effect of Withaferin A on the development and decay of thermotolerance in B16F1 melanoma: A preliminary study. *Integrative Cancer Therapies, 8,* 93–97.

Kaminski, B. M., Loitsch, S. M., Ochs, M. J., et al. (2010a). Isothiocyanate sulforaphane inhibits protooncogenic ornithine decarboxylase activity in colorectal cancer cells via induction of the TGF-beta/Smad signaling pathway. *Molecular Nutrition & Food Research, 54,* 1486–1496.

Kaminski, B. M., Weigert, A., Brüne, B., et al. (2010b). Sulforaphane potentiates oxaliplatin-induced cell growth inhibition in colorectal cancer cells via induction of different modes of cell death. *Cancer Chemotherapy and Pharmacology.* doi:10.1007/s00280-010-1413-y.

Kandala, P. K., & Srivastava, S. K. (2010). Activation of checkpoint kinase 2 by 3,3′-diindolylmethane is required for causing G2/M cell cycle arrest in human ovarian cancer cells. *Molecular Pharmacology, 78,* 297–309.

Karabulut, B., Karaca, B., Atmaca, H., et al. (2010). Regulation of apoptosis-related molecules by synergistic combination of all-trans retinoic acid and zoledronic acid in hormone-refractory prostate cancer cell lines. *Molecular Biology Reports, 38,* 249–259.

Karin, M. (2009). NF-κB as a critical link between inflammation and cancer. *Cold Spring Harbor Perspectives in Biology, 1,* a000141.

Kashkar, H. (2010). XIAP: A chemoresistance factor or a hollow promise. *Clinical Cancer Research: An Official Journal of the American Association for Cancer Research, 16,* 4496–4502.

Kaur, M., Velmurugan, B., Tyagi, A., et al. (2010). Silibinin suppresses growth of human colorectal carcinoma SW480 cells in culture and xenograft through down-regulation of beta-catenin-dependent signaling. *Neoplasia (New York, N.Y.), 12,* 415–424.

Kawanishi, S., & Hiraku, Y. (2006). Oxidative and nitrative DNA damage as biomarker for carcinogenesis with special reference to inflammation. *Antioxidants & Redox Signaling, 8,* 1047–1058.

Kelloff, G. J., Lieberman, R., Steele, V. E., et al. (1999). Chemoprevention of prostate cancer: Concepts and strategies. *European Journal of Urology, 35,* 342–350.

Khwaja, F. S., Wynne, S., Posey, I., et al. (2009). 3,3′-diindolylmethane induction of p75NTR-dependent cell death via the p38 mitogen-activated protein kinase pathway in prostate cancer cells. *Cancer Prevention Research (Philadelphia, Pa.), 2,* 566–571.

Kim, S. H., & Singh, S. V. (2009). D, L-Sulforaphane causes transcriptional repression of androgen receptor in human prostate cancer cells. *Molecular Cancer Therapeutics, 8,* 1946–1954.

Kim, H. I., Huang, H., Cheepala, S., et al. (2008). Curcumin inhibition of integrin (alpha6beta4)-dependent breast cancer motility and invasion. *Cancer Prevention Research (Philadelphia, Pa.), 1,* 385–391.

Kim, M. Y., Trudel, L. J., & Wogan, G. N. (2009a). Apoptosis induced by capsaicin and resveratrol in colon carcinoma cells requires nitric oxide production and caspase activation. *Anticancer Research, 29,* 3733–3740.

Kim, S., Choi, M. G., Lee, H. S., et al. (2009b). Silibinin suppresses TNF-alpha-induced MMP-9 expression in gastric cancer cells through inhibition of the MAPK pathway. *Molecules (Basel, Switzerland), 14,* 4300–4311.

Kim, S., Kim, S. H., Hur, S. M., et al. (2009c). Silibinin prevents TPA-induced MMP-9 expression by down-regulation of COX-2 in human breast cancer cells. *Journal of Ethnopharmacology, 126,* 252–257.

Kim, E. J., Park, H., Kim, J., et al. (2010a). 3,3′-diindolylmethane suppresses 12-O-tetradecanoylphorbol-13-acetate-induced inflammation and tumor promotion in mouse skin via the downregulation of inflammatory mediators. *Molecular Carcinogenesis, 49,* 672–683.

Kim, S., Lee, H. S., Lee, S. K., et al. (2010b). 12-O-Tetradecanoyl phorbol-13-acetate (TPA)-induced growth arrest is increased by silibinin by the down-regulation of cyclin B1 and cdc2 and the up-regulation of p21 expression in MDA-MB231 human breast cancer cells. *Phytomedicine: International Journal of Phytotherapy and Phytopharmacology, 17,* 1127–1132.

Koay, D. C., Zerillo, C., Narayan, M., et al. (2010). Anti-tumor effects of retinoids combined with trastuzumab or tamoxifen in breast cancer cells: Induction of apoptosis by retinoid/trastuzumab combinations. *Breast Cancer Research: BCR, 12,* R62.

Koduru, S., Kumar, R., Srinivasan, S., et al. (2010). Notch-1 inhibition by Withaferin-A: A therapeutic target against colon carcinogenesis. *Molecular Cancer Therapeutics, 9,* 202–210.

Komina, O., & Wesierska-Gadek, J. (2008). Action of resveratrol alone or in combination with roscovitine, a cdk inhibitor, on cell cycle progression in human hl-60 leukemia cells. *Biochemical Pharmacology, 76,* 1554–1562.

Kommera, H., Kaluderović, G. N., Kalbitz, J., et al. (2009). Lupane Triterpenoids-Betulin and Betulinic acid derivatives induce apoptosis in tumor cells. *Investigational New Drugs*. 10.1007/s10637-009-9358-x.

Kommera, H., Kaluderović, G. N., Bette, M., et al. (2010a). In vitro anticancer studies of alpha- and beta-D-glucopyranose betulin anomers. *Chemico-Biological Interactions, 185,* 128–136.

Kommera, H., Kaluderović, G. N., Dittrich, S., et al. (2010b). Carbamate derivatives of betulinic acid and betulin with selective cytotoxic activity. *Bioorganic & Medicinal Chemistry Letters, 20,* 3409–3412.

Kommera, H., Kaluderović, G. N., Kalbitz, J., et al. (2010c). Synthesis and anticancer activity of novel betulinic acid and betulin derivatives. *Archiv der Pharmazie, 343,* 449–457.

Koyama, S., Cobb, L. J., Mehta, H. H., et al. (2010). Pomegranate extract induces apoptosis in human prostate cancer cells by modulation of the IGF-IGFBP axis. *Growth Hormone & Igf Research: Official Journal of the Growth Hormone Research Society and the International IGF Research Society, 20,* 55–62.

Kraft, T. E., Parisotto, D., Schempp, C., et al. (2009). Fighting cancer with red wine? Molecular mechanisms of resveratrol. *Critical Reviews in Food Science and Nutrition, 49,* 782–799.

Kumar, S., Dubey, K. K., Tripathi, S., et al. (2000). Design and synthesis of curcumin-bioconjugates to improve systemic delivery. *Nucleic Acids Symposium Series, 44,* 75–76.

Kumar, A. P., Garcia, G. E., Ghosh, R., et al. (2003). 4-Hydroxy3-methoxybenzoic acid methyl ester: A curcumin derivative targets Akt/NFκB cell survival signaling pathway: Potential for prostate cancer management. *Neoplasia (New York, N.Y.), 5,* 255–266.

Kumar, A. P., Bhaskaran, S., Ganapathy, M., et al. (2007). Akt/cAMP-responsive element binding protein/cyclin D1 network: A novel target for prostate cancer inhibition in transgenic adenocarcinoma of mouse prostate model mediated by NexrutineR, a Phellodendron amurense bark extract. *Clinical Cancer Research: An Official Journal of the American Association for Cancer Research, 9,* 2784–2794.

Kumar, A. P., Graham, H., Robson, C., et al. (2010a). Natural products: Potential for developing phellodendron amurense bark extract for prostate cancer management. *Mini Reviews in Medicinal Chemistry, 10,* 388–397.

Kumar, G., Pillare, S. P., & Maru, G. B. (2010b). Black tea polyphenols-mediated in vivo cellular responses during carcinogenesis. *Mini Reviews in Medicinal Chemistry, 10,* 492–505.

Lahiry, L., Saha, B., Chakraborty, J., et al. (2010). Theaflavins target Fas/caspase-8 and Akt/pBad pathways to induce apoptosis in p53-mutated human breast cancer cells. *Carcinogenesis, 31,* 259–268.

Lampson, M. A. (2007). Using small molecules to unravel biological mechanisms. *Chemistry & Biology, 1,* 71–94.

Larsson, D. E., Wickstrom, M., Hassan, S., et al. (2010). The cytotoxic agents NSC-95397, brefeldin A, bortezomib and sanguinarine induce apoptosis in neuroendocrine tumors in vitro. *Anticancer Research, 30,* 149–156.

Lee, J. C., Harrison, D., & Timasheff, S. N. (1975). Interaction of vinblastine with calf brain microtubule protein. *The Journal of Biological Chemistry, 250,* 9276–9282.

Lee, D. H., Szczepanski, M., & Lee, Y. J. (2008). Role of Bax in quercetin-induced apoptosis in human prostate cancer cells. *Biochemical Pharmacology, 75,* 2345–2355.

Lee, H. J., Lee, H. J., Lee, E. O., et al. (2009a). In vivo anti-cancer activity of Korean Angelica gigas and its major pyranocoumarin decursin. *The American Journal of Chinese Medicine, 37,* 127–142.

Lee, T. J., Um, H. J., Min, do. S., et al. (2009b). Withaferin A sensitizes TRAIL-induced apoptosis through reactive oxygen species-mediated up-regulation of death receptor 5 and down-regulation of c-FLIP. *Free Radical Biology & Medicine 2009, 46,* 1639–1649.

Lee, J., Hahm, E. R., & Singh, S. V. (2010). Withaferin A inhibits activation of signal transducer and activator of transcription 3 in human breast cancer cells. *Carcinogenesis, 31,* 1991–1998.

Leone, S., Fiore, M., Lauro, M. G., et al. (2008). Resveratrol and X rays affect gap junction intercellular communications in human glioblastoma cells. *Molecular Carcinogenesis, 47,* 587–598.

Li, Z. G., Hong, T., Shimada, Y., et al. (2002). Suppression of N-nitrosomethylbenzylamine (NMBA)- induced esophageal tumorgenesis in F344 rats by resveratrol. *Carcinogenesis (Lond), 23,* 1531–1536.

Li, H. Y., Cui, L., & Cui, M. (2009). Hot topics in Chinese herbal drugs research documented in PubMed/MEDLINE by authors inside China and outside of China in the past 10 years: Based on co-word cluster analysis. *Journal of Alternative and Complementary Medicine (New York, N.Y.), 15,* 779–785.

Li, Q., Wang, X. H., Yang, Z. H., et al. (2010a). Induction of cell cycle arrest in bladder cancer RT4 cells by capsaicin. *Zhonghua Yi Xue Za Zhi, 90,* 1230–1233.

Li, Y., He, K., Huang, Y., et al. (2010b). Betulin induces mitochondrial cytochrome c release associated apoptosis in human cancer cells. *Molecular Carcinogenesis, 49,* 630–640.

Li, Y., Zhang, T., Korkaya, H., et al. (2010c). Sulforaphane, a dietary component of broccoli/broccoli sprouts, inhibits breast cancer stem cells. *Clinical Cancer Research: An Official Journal of the American Association for Cancer Research, 16,* 2580–2590.

Liao, Q. J., Su, J., He, J., et al. (2009). Effect of diallyl disulfide on cell cycle arrest of human colon cancer SW480 cells. *Ai Zheng, 28,* 138–141.

Liu, E. H., Qi, L. W., Wu, Q., et al. (2009). Anticancer agents derived from natural products. *Mini Reviews in Medicinal Chemistry, 9,* 1547–1555.

Liu, P. L., Tsai, J. R., Charles, A. L., et al. (2010). Resveratrol inhibits human lung adenocarcinoma cell metastasis by suppressing heme oxygenase 1-mediated nuclear factor-kappaB pathway and subsequently downregulating expression of matrix metalloproteinases. *Molecular Nutrition & Food Research, 54*(Suppl. 2), S196–S204.

Lu, J., Gosslau, A., Liu, A. Y., et al. (2008). PCR differential display-based identification of regulator of G protein signaling 10 as the target gene in human colon cancer cells induced by black tea polyphenol theaflavin monogallate. *European Journal of Pharmacology, 601,* 66–72.

Ma, R. W., & Chapman, K. (2009). A systematic review of the effect of diet on prostate cancer prevention and treatment. *Journal of Human Nutrition and Dietetics: The Official Journal of the British Dietetic Association, 22,* 187–199.

Maity, R., Sharma, J., & Jana, N. R. (2010). Capsaicin induces apoptosis through ubiquitin-proteasome system dysfunction. *Journal of Cellular Biochemistry, 109,* 933–942.

Majumdar, A. P., Banerjee, S., Nautiyal, J., et al. (2009). Curcumin synergizes with resveratrol to inhibit colon cancer. *Nutrition and Cancer, 61,* 544–553.

Malagarie-Cazenave, S., Olea-Herrero, N., Vara, D., et al. (2009). Capsaicin, a component of red peppers, induces expression of androgen receptor via PI3 K and MAPK pathways in prostate LNCaP cells. *FEBS Letters, 583,* 141–147.

Malki, A., El-Saadani, M., & Sultan, A. S. (2009). Garlic constituent diallyl trisulfide induced apoptosis in MCF7 human breast cancer cells. *Cancer Biology & Therapy, 8,* 2175–2185.

Mandal, C., Dutta, A., Mallick, A., et al. (2008). Withaferin A induces apoptosis by activating p38 mitogen-activated protein kinase signaling cascade in leukemic cells of lymphoid and myeloid origin through mitochondrial death cascade. *Apoptosis: An International Journal on Programmed Cell Death, 13,* 1450–1464.

Manoharan, S., Panjamurthy, K., Menon, V. P., et al. (2009). Protective effect of Withaferin-A on tumour formation in 7,12-dimethylbenz[a]anthracene induced oral carcinogenesis in hamsters. *Indian Journal of Experimental Biology, 47,* 16–23.

Martincic, D., & Hande, K. R. (2005). Topoisomerase ii inhibitors. *Cancer Chemotherapy and Biological Response Modifiers, 22,* 101–121.

Mateen, S., Tyagi, A., Agarwal, C., et al. (2010). Silibinin inhibits human nonsmall cell lung cancer cell growth through cell-cycle arrest by modulating expression and function of key cell-cycle regulators. *Molecular Carcinogenesis, 49,* 247–258.

Matkar, S. S., Wrischnik, L. A., & Hellmann-Blumberg, U. (2008). Sanguinarine causes DNA damage and p53-independent cell death in human colon cancer cell lines. *Chemico-Biological Interactions, 172,* 63–71.

Maurer, M., Komina, O., & Wesierska-Gadek, J. (2009). Roscovitine differentially affects asynchronously growing and synchronized human MCF-7 breast cancer cells. *Annals of the New York Academy of Sciences, 1171,* 250–256.

Mazzio, E. A., & Soliman, K. F. (2010). In vitro screening of tumoricidal properties of international medicinal herbs: Part II. *Phytotherapy Research, 24,* 1813–1824.

McCarty, M. F. (2004). Targeting multiple signaling pathways as a strategy for managing prostate cancer: Multifocal signal modulation therapy. *Integrative Cancer Therapies, 3,* 349–380.

McCord, J. M. (1993). Human disease, free radicals, and the oxidant/antioxidant balance. *Clinical Biochemistry, 26,* 351–357.

McCulloch, M., See, C., Shu, X. J., et al. (2006). Astragalus-based chinese herbs and platinum-based chemotherapy for advanced non-small-cell lung cancer: Meta-analysis of randomized trials. *Journal of Clinical Oncology: Official Journal of the American Society of Clinical Oncology, 24,* 419–430.

Meeran, S. M., Patel, S. N., & Tollefsbol, T. O. (2010). Sulforaphane causes epigenetic repression of hTERT expression in human breast cancer cell lines. *PLoS One, 5,* e11457.

Merlo, L, M., Pepper, J. W., Reid, B. J., et al. (2006). Cancer as an evolutionary and ecological process. *Nat Rev Cancer, 6,* 924–935.

Miller, S., Stagl, J., Wallerstedt, D. B., et al. (2008). Botanicals used in complementary and alternative medicine treatment of cancer: Clinical science and future perspectives. *Expert Opinion on Investigational Drugs, 17,* 1353–1364.

Monneret, C. (2010). Current impact of natural products in the discovery of anticancer drugs. *Annales Pharmaceutiques Francaises, 68,* 218–232.

Montesinos, P., Gonzalez, J. D., Gonzalez, J., et al. (2010). Therapy-related myeloid neoplasms in patients with acute promyelocytic leukemia treated with all-trans-retinoic acid and anthracycline-based chemotherapy. *Journal of Clinical Oncology: Official Journal of the American Society of Clinical Oncology, 28,* 3872–3879.

Mori, H., Fuchigami, M., Inoue, N., et al. (1995). Principle of the bark phellodendron amurense to suppress the cellular immune response: Effect of phellodendrine on cellular and humoral immune response. *Planta Medica, 61,* 45–49.

Mori, A., Lehmann, S., O'Kelly J., et al. (2006). Capsaicin, a component of red peppers, inhibits the growth of androgen-independent, p53 mutant prostate cancer cells. *Cancer Research, 66,* 3222–3229.

Mullauer, F. B., Kessler, J. H., & Medema, J. P. (2010). Betulinic acid, a natural compound with potent anticancer effects. *Anti-cancer Drugs, 21,* 215–227.

Muralimanoharan, S. B., Kunnumakkara, A. B., Shylesh, B., et al. (2009). Butanol fraction containing berberine or related compound from NexrutineR inhibits NFkB signaling and induces apoptosis in prostate cancer cells. *The Prostate, 69,* 494–504.

Nagaraj, N. S., Anilakumar K. R., & Singh, O. V. (2010). Diallyl disulfide causes caspase-dependent apoptosis in human cancer cells through a Bax-triggered mitochondrial pathway. *The Journal of Nutritional Biochemistry, 21,* 405–412.

Nair, S., Barve, A., Khor, T. O., et al. (2010). Regulation of Nrf2- and AP-1-mediated gene expression by epigallocatechin-3-gallate and sulforaphane in prostate of Nrf2-knockout or C57BL/6J mice and PC-3 AP-1 human prostate cancer cells. *Acta Pharmacologica Sinica, 31,* 1223–1240.

Nakagawa-Goto, K., Yamada, K., Taniguchi, M., et al. (2009). Cancer preventive agents 9. Betulinic acid derivatives as potent cancer chemopreventive agents. *Bioorganic & Medicinal Chemistry Letters, 19,* 3378–3381.

Narayanan, N. K., Nargi, D., Randolph, C., et al. (2009). Liposome encapsulation of curcumin and resveratrol in combination reduces prostate cancer incidence in pten knockout mice. *International Journal of Cancer. Journal International du Cancer, 125,* 1–8.

National Cancer Institute (2010). Cancer drug information. http://www.cancer.gov/cancertopics/druginfo. Accessed 23 August 2010.

National Cancer Institute. NCI drug dictionary. http://www.cancer.gov/drugdictionary. Accessed 23 August 2010.

Newman, D. J., & Cragg, G. M. (2007). Natural products as sources of new drugs over the last 25 years. *Journal of Natural Products, 70,* 461–477.

Nishikawa, T., Tsuno, N. H., Okaji, Y., et al. (2010). Inhibition of autophagy potentiates sulforaphane-induced apoptosis in human colon cancer cells. *Annals of Surgical Oncology: The Official Journal of the Society of Surgical Oncology, 17,* 592–602.

Nitta, N., Sonoda, A., Seko, A., et al. (2010). A combination of cisplatin-eluting gelatin microspheres and flavopiridol enhances anti-tumour effects in a rabbit VX2 liver tumour model. *The British Journal of Radiology, 83,* 428–432.

Nonn, L., Duong D, & Peehl, D. M. (2007). Chemopreventative anti-inflammatory activities of curcumin and other phytochemicals mediated by MAP kinase phosphatase-5 in prostate cells. *Carcinogenesis, 28,* 1188–1196.

Noori-Daloii, M. R., Momeny, M., Yousefi, M., et al. (2010). Multifaceted preventive effects of single agent quercetin on a human prostate adenocarcinoma cell line (PC-3): Implications for nutritional transcriptomics and multi-target therapy. *Medical Oncology (Northwood, London, England).* doi:10.1007/s12032-010-9603-3.

Oeckinghaus, A., & Ghosh, S. (2009). The NF-κB family of transcription factors and its regulation. *Cold Spring Harbor Perspectives in Biology, 1,* a000034.

Oh, J. H., Lee, T. J., Kim, S. H., et al. (2008). Induction of apoptosis by withaferin A in human leukemia U937 cells through down-regulation of Akt phosphorylation. *Apoptosis: An International Journal on Programmed Cell Death, 13,* 1494–1504.

Oka, S., Tanaka, S., Yoshida, S., et al. (2010). A water-soluble extract from culture medium of Ganoderma lucidum mycelia suppresses the development of colorectal adenomas. *Hiroshima Journal of Medical Sciences, 59,* 1–6.

Oyagbemi, A. A., Saba, A. B., & Azeez, O. I. (2010). Capsaicin: A novel chemopreventive molecule and its underlying molecular mechanisms of action. *Indian Journal of Cancer, 47,* 53–58.

Panjamurthy, K., Manoharan, S., Nirmal, M. R., et al. (2009). Protective role of Withaferin-A on immunoexpression of p53 and bcl-2 in 7,12-dimethylbenz(a)anthracene-induced experimental oral carcinogenesis. *Investigational New Drugs, 27,* 447–452.

Pantuck, A. J., Leppert, J. T., Zomorodian, N., et al. (2006). Phase II study of pomegranate juice for men with rising prostate specific antigen following surgery or radiation for prostate cancer. *Clinical Cancer Research: An Official Journal of the American Association for Cancer Research 2006, 12,* 4018–4026.

Park, H., Bergeron, E., Senta, H., et al. (2010a). Sanguinarine induces apoptosis of human osteosarcoma cells through the extrinsic and intrinsic pathways. *Biochemical and Biophysical Research Communications, 399,* 446–451.

Park, H. J., Davis, S. R., Liang, H. Y., et al. (2010b). Chlorogenic acid differentially alters hepatic and small intestinal thiol redox status without protecting against azoxymethane-induced colon carcinogenesis in mice. *Nutrition and Cancer, 62,* 362–370.

Patil, J. R., Chidambara Murthy, K. N., Jayaprakasha, G. K., et al. (2009). Bioactive compounds from Mexican lime (Citrus aurantifolia) juice induce apoptosis in human pancreatic cells. *Journal of Agricultural and Food Chemistry, 57,* 10933–10942.

Patwardhan, B., & Mashelkar, R. A. (2009). Traditional medicine-inspired approaches to drug discovery: Can ayurveda show the way forward? *Drug Discovery Today, 14,* 804–811.

Perri, M., Pingitore, A., Cione, E., et al. (2010). Proliferative and anti-proliferative effects of retinoic acid at doses similar to endogenous levels in Leydig MLTC-1/R2C/TM-3 cells. *Biochimica et Biophysica Acta, 1800,* 993–1001.

Phillips, D. H. (1990). Further evidence that eugenol does not bind to DNA in vivo. *Mutation Research, 245,* 23–26.

Pirouzpanah, S., Taleban, F. A., Atri, M., et al. (2010). The effect of modifiable potentials on hypermethylation status of retinoic acid receptor-beta2 and estrogen receptor-alpha genes in primary breast cancer. *Cancer Causes Control, 21,* 2101–2111.

Priyadarsini, R. V., Murugan, R. S., Sripriya, P., et al. (2010). The neem limonoids azadirachtin and nimbolide induce cell cycle arrest and mitochondria-mediated apoptosis in human cervical cancer (HeLa) cells. *Free Radical Research, 44,* 624–634.

Qun, L., Luo, Q., Zhang, Z. Y., et al. (1999). Effects of astragalus on IL-2/IL-2R system in patients with maintained hemodialysis. *Clinical Nephrology, 52,* 333–334.

Rahimi, M., Huang, K. L., & Tang, C. K. (2010). 3,3′-Diindolylmethane (DIM) inhibits the growth and invasion of drug-resistant human cancer cells expressing EGFR mutants. *Cancer Letters, 295,* 59–68.

Rahman, K. M., Banerjee, S., Ali, S., et al. (2009). 3,3′-Diindolylmethane enhances taxotere-induced apoptosis in hormone-refractory prostate cancer cells through survivin down-regulation. *Cancer Research, 69,* 4468–4475.

Rajamanickam, S., Velmurugan, B., Kaur, M., et al. (2010). Chemoprevention of intestinal tumorigenesis in APCmin/+ mice by silibinin. *Cancer Research, 70,* 2368–2378.

Reagan-Shaw, S., Eggert, D., Mukhtar, H., et al. (2010). Antiproliferative effects of apple peel extract against cancer cells. *Nutrition and Cancer, 62,* 517–524.

Reddivari, L., Vanamala, J., Safe, S. H., et al. (2010). The bioactive compounds alpha-chaconine and gallic acid in potato extracts decrease survival and induce apoptosis in LNCaP and PC3 prostate cancer cells. *Nutrition and Cancer, 62,* 601–610.

Reid, B. J. (2008). Cancer risk assessment and cancer prevention: Promises and challenges. *Cancer Prevention Research (Philadelphia, Pa.), 1,* 229–232.

Rettig, M. B., Heber, D., An, J., et al. (2008). Pomegranate extract inhibits androgen-independent prostate cancer growth through a nuclear factor-kappaB-dependent mechanism. *Molecular Cancer Therapeutics, 7,* 2662–2671.

Roberts, H. (2010). Safety of herbal medicinal products in women with breast cancer. *Maturitas, 66,* 363–369.

Sadikoglou, E., Magoulas, G., Theodoropoulou, C., et al. (2009). Effect of conjugates of all-trans-retinoic acid and shorter polyene chain analogues with amino acids on prostate cancer cell growth. *European Journal of Medicinal Chemistry, 44,* 3175–3187.

Sánchez, A. M., Malagarie-Cazenave, S., Olea, N., et al. (2007). Apoptosis induced by capsaicin in prostate PC-3 cells involves ceramide accumulation, neutral sphingomyelinase, and JNK activation. *Apoptosis: An International Journal on Programmed Cell Death, 12,* 2013–2024.

Sanders, T. H., McMichael, R. W. Jr., & Hendrix, K. W. (2000). Occurrence of resveratrol in edible peanuts. *Journal of Agricultural and Food Chemistry, 48,* 1243–1246.

Sangeetha, N., Aranganathan, S., Panneerselvam, J., et al. (2010). Oral supplementation of silibinin prevents colon carcinogenesis in a long term preclinical model. *European Journal of Pharmacology, 643,* 93–100.

Sarkar, F. H., & Li, Y. (2009). Harnessing the fruits of nature for the development of multi-targeted cancer therapeutics. *Cancer Treatment Reviews, 35,* 597–607.

Sawa, T., & Ohshima, H. (2006). Nitrative DNA damage in inflammation and its possible role in carcinogenesis. *Nitric Oxide: Biology and Chemistry/Official Journal of the Nitric Oxide Society, 14,* 91–100.

Schneider, Y., Duranton, B., Gosse, F., et al. (2001). Reveratrol inhibits intestinal tumorgenesis and modulates host-defense-related gene expression in an animal model of human familial adenomatous polypois. *Nutrition and Cancer, 39,* 102–107.

Senthilkumar, K., Elumalai, P., Arunkumar, R., et al. (2010). Quercetin regulates insulin like growth factor signaling and induces intrinsic and extrinsic pathway mediated apoptosis in androgen independent prostate cancer cells (PC-3). *Molecular and Cellular Biochemistry, 344,* 173–184.

Serafim, T. L., Matos, J. A., Sardao, V. A., et al. (2008). Sanguinarine cytotoxicity on mouse melanoma K1735-M2 cells—Nuclear vs. mitochondrial effects. *Biochemical Pharmacology, 76,* 1459–1475.

Seyfried, T. N., & Shelton, L. M. (2010). Cancer as a metabolic disease. *Nutrition and Metabolism, 7,* 7.

Shan, Y., Wang, X., Wang, W., et al. (2010). p38 MAPK plays a distinct role in sulforaphane-induced up-regulation of ARE-dependent enzymes and down-regulation of COX-2 in human bladder cancer cells. *Oncology Reports, 23,* 1133–1138.

Sharma, R. A., Euden, S. A., Platton, S. L., et al. (2004). Phase I clinical trial of oral curcumin: Biomarkers of systemic activity and compliance. *Clinical Cancer Research: An Official Journal of the American Association for Cancer Research, 10,* 6847–6854.

Sharma, R., Sharma, A., Chaudhary, P., et al. (2010). Role of lipid peroxidation in cellular responses to D, L-sulforaphane, a promising cancer chemopreventive agent. *Biochemistry, 49,* 3191–3202.

She, Q. B., Bode, A. M., Ma, W. Y., et al. (2001). Resveratrol-induced activation of p53 and apoptosis is mediated by extracellular-signal-regulated protein kinases and p38 kinase. *Cancer Research, 61,* 1604–1610.

She, Q. B., Huang, C., Zhang, Y., et al. (2002). Involvement of c-jun NH(2)-terminal kinases in resveratrol-induced activation of p53 and apoptosis. *Molecular Carcinogenesis, 33,* 244–250.

Shepard, D. R., & Dreicer, R. (2010). Zibotentan for the treatment of castrate-resistant prostate cancer. *Expert Opinion on Investigational Drugs, 19,* 899–908.

Shih, A., Davis, F. B., Lin, H. Y., et al. (2002). Resveratrol induces apoptosis in thyroid cancer cell lines via a MAPK- and p53-dependent mechanism. *The Journal of Clinical Endocrinology and Metabolism, 87,* 1223–1232.

Shih, Y. W., Wu, P. F., Lee, Y. C., et al. (2009). Myricetin suppresses invasion and migration of human lung adenocarcinoma A549 cells: Possible mediation by blocking the ERK signaling pathway. *Journal of Agricultural and Food Chemistry, 57,* 3490–3499.

Shirasugi, I., Kamada, M., Matsui, T., et al. (2010). Sulforaphane inhibited melanin synthesis by regulating tyrosinase gene expression in B16 mouse melanoma cells. *Bioscience, Biotechnology, and Biochemistry, 74,* 579–582.

Shoji, T., Kobori, M., Shinmoto, H., et al. (1997). Progressive effects of phloridzin on melanogenesis in B16 mouse melanoma cells. *Bioscience, Biotechnology, and Biochemistry, 61,* 1963–1967.

Shord, S. S., Shah, K., & Lukose, A. (2009). Drug-botanical interactions: A review of the laboratory, animal, and human data for 8 common botanicals. *Integrative Cancer Therapies, 8,* 208–227.

Shrotriya, S., Kundu, J. K., Na, H. K., et al. (2010). Diallyl trisulfide inhibits phorbol ester-induced tumor promotion, activation of AP-1, and expression of COX-2 in mouse skin by blocking JNK and Akt signaling. *Cancer Research, 70,* 1932–1940.

Siegelin, M. D., Gaiser, T., Habel, A., et al. (2009). Myricetin sensitizes malignant glioma cells to TRAIL-mediated apoptosis by down-regulation of the short isoform of FLIP and bcl-2. *Cancer Letters, 283,* 230–238.

Siemann, D. W., Chaplin, D. J., & Horsman, M. R. (2004). Vascular-targeting therapies for treatment of malignant disease. *Cancer, 100,* 2491–2499.

Sil, H., Sen, T., Moulik, S., et al. (2010). Black tea polyphenol (theaflavin) downregulates MMP-2 in human melanoma cell line A375 by involving multiple regulatory molecules. *Journal of Environmental Pathology, Toxicology and Oncology: Official Organ of the International Society for Environmental Toxicology and Cancer, 29,* 55–68.

Singh, S. V., Warin, R., Xiao, D., et al. (2009). Sulforaphane inhibits prostate carcinogenesis and pulmonary metastasis in TRAMP mice in association with increased cytotoxicity of natural killer cells. *Cancer Research, 69,* 2117–2125.

Singh, B., Mense, S. M., Bhat, N. K., et al. (2010). Dietary quercetin exacerbates the development of estrogen-induced breast tumors in female ACI rats. *Toxicology and Applied Pharmacology, 247,* 83–90.

Snyder, R. M., Yu, W., Jia, L., et al. (2008). Vitamin E analog alpha-tea, methylseleninic acid, and trans-resveratrol in combination synergistically inhibit human breast cancer cell growth. *Nutrition and Cancer, 60,* 401–411.

Sparber, A., Bauer, L., Curt, G., et al. (2000). Use of complementary medicine by adult patients participating in clinical cancer trials. *Oncology Nursing Forum, 27,* 623–630.

Srivastava, V., Negi, A. S., Kumar, J. K., et al. (2005). Plant-based anticancer molecules: A chemical and biological profile of some important leads. *Bioorganic & Medicinal Chemistry, 13,* 5892–5908.

Stan, S. D., & Singh, S. V. (2009). Transcriptional repression and inhibition of nuclear translocation of androgen receptor by diallyl trisulfide in human prostate cancer cells. *Clinical Cancer Research: An Official Journal of the American Association for Cancer Research, 15,* 4895–4903.

Stan, S. D., Zeng, Y., & Singh, S. V. (2008). Ayurvedic medicine constituent withaferin a causes G2 and M phase cell cycle arrest in human breast cancer cells. *Nutrition and Cancer, 60*(Suppl. 1), 51–60.

Sturgeon, S. R., & Ronnenberg, A. G. (2010). Pomegranate and breast cancer: Possible mechanisms of prevention. *Nutrition Reviews, 68,* 122–128.

Sukumaran, K., Unnikrishnan, M. C., & Kuttan, R. (1994). Inhibition of tumour promotion in mice by eugenol. Indian Journal of Physiology and Pharmacology, *38,* 306–308.

Supoken, A., Chaisrisawatsuk, T., & Chumworathayi, B. (2009). Proportion of gynecologic cancer patients using complementary and alternative medicine. *Asian Pacific Journal of Cancer Prevention: APJCP, 10,* 779–782.

Surh, Y. J. (2003). Cancer chemoprevention with dietary phytochemicals. *Nature Reviews. Cancer, 3,* 768–780.

Swamy, M. V., Citineni, B., Patlolla, J. M., et al. (2008). Prevention and treatment of pancreatic cancer by curcumin in combination with omega-3 fatty acids. *Nutrition and Cancer, 60,* 81–89.

Szekeres, T., Fritzer-Szekeres, M., Saiko, P., et al. (2010). Resveratrol and resveratrol analogues—Structure-activity relationship. *Pharmaceutical Research, 27,* 1042–1048.

Tan, J., Zhu, L., & Wang, B. (2010). From GC-rich DNA binding to the repression of survivin gene for quercetin nickel (II) complex: Implications for cancer therapy. *Biometals: An International Journal on the Role of Metal Ions in Biology, Biochemistry, and Medicine, 23,* 1075–1084.

Tanaka, T., Kohno, H., Tsukio, Y., et al. (2000). Citrus limonoids obacunone and limonin inhibit azoxymethane-induced colon carcinogenesis in rats. *BioFactors (Oxford, England), 13,* 213–218.

Tarozzi, A., Morroni, F., Merlicco, A., et al. (2009). Sulforaphane as an inducer of glutathione prevents oxidative stress-induced cell death in a dopaminergic-like neuroblastoma cell line. *Journal of Neurochemistry, 111,* 1161–1171.

Tessitore, L., Davit, A., Sarotto, I., et al. (2000). Resveratrol depresses the growth of colorectal aberrant crypt foci by affecting bax and p21 (CIP) expression. *Carcinogenesis (Lond), 21,* 1619–1622.

Thangasamy, T., Sittadjody, S., Mitchell, G. C., et al. (2010). Quercetin abrogates chemoresistance in melanoma cells by modulating deltaNp73. *BMC Cancer, 10,* 282.

Thoennissen, N. H., O'Kelly, J., Lu, D., et al. (2010). Capsaicin causes cell-cycle arrest and apoptosis in ER-positive and -negative breast cancer cells by modulating the EGFR/HER-2 pathway. *Oncogene, 29,* 285–296.

Thomson, D., Norbeck, K., Olson, L., et al. (1984). Peroxidase-catalyzed oxidation of eugenol: Formation of a cytotoxic metabolite(s). *The Journal of Biological Chemistry, 264,* 1016–1021.

Thyagarajan, A., Jedinak, A., Nguyen, H., et al. (2010). Triterpenes from Ganoderma Lucidum induce autophagy in colon cancer through the inhibition of p38 mitogen-activated kinase (p38 MAPK). *Nutrition and Cancer, 62,* 630–640.

Tian, G., Guo, L., & Gao, W. (2010). Use of compound Chinese medicine in the treatment of lung cancer. *Current Drug Discovery Technologies, 7,* 32–36.

Tian, Q., Miller, E. G., Ahmad, H., et al. (2001). Differential inhibition of human cancer cell proliferation by citrus limonoids. *Nutrition and Cancer, 40,* 180–184.

Tokalov, S. V., Abramyuk, A. M., & Abolmaali, N. D. (2010). Protection of p53 wild type cells from taxol by genistein in the combined treatment of lung cancer. *Nutrition and Cancer, 62,* 795–801.

Tomlinson, B., Hu, M., & Lee, V. W. (2008). In vivo assessment of herb-drug interactions: Possible utility of a pharmacogenetic approach? *Molecular Nutrition & Food Research, 52,* 799–809.

Torrance, C. J., Jackson, P. E., Montgomery, E., et al. (2000). Combinatorial chemoprevention of intestinal neoplasia. *Nature Medicine, 6,* 1024–1028.

Touma, S. E., Perner, S., Rubin, M. A., et al. (2009). Retinoid metabolism and ALDH1A2 (RALDH2) expression are altered in the transgenic adenocarcinoma mouse prostate model. *Biochemical Pharmacology, 78,* 1127–1138.

Traka, M. H., Spinks, C. A., Doleman, J. F., et al. (2010). The dietary isothiocyanate sulforaphane modulates gene expression and alternative gene splicing in a PTEN null preclinical murine model of prostate cancer. *Molecular Cancer, 9,* 189.

Trottier, G., Bostrom, P. J., Lawrentschuk, N., et al. (2010) Neutraceuticals and prostate cancer prevention: A current review. *Nature Reviews. Urology, 7,* 21–30.

Uchiyama, T., Kamikawa, H., Ogiat, Z., et al. (1989). Anti-ulcer effect of extract from phellodendri cortex. *Jakugaku Zasshi, 109,* 672–676.

Ugalde, C. M., Liu, Z., Ren, C., et al. (2009). Distribution of anthocyanins delivered from a bioadhesive black raspberry gel following topical intraoral application in normal healthy volunteers. *Pharmaceutical Research, 26,* 977–986.

Ulrich, S., Wolter, F., & Stein, J. M. (2005). Molecular mechanisms of the chemopreventive effects of resveratrol and its analogs in carcinogenesis. *Molecular Nutrition & Food Research, 49,* 452–461.

Vareed, S. K., Kakarala, M., Ruffin, M. T., et al. (2008). Pharmacokinetics of curcumin conjugate metabolites in healthy human subjects. *Cancer Epidemiology, Biomarkers & Prevention: A Publication of the American Association for Cancer Research, Cosponsored by the American Society of Preventive Oncology, 17,* 1411–1417.

Vargas, A. J., & Burd, R. (2010). Hormesis and synergy: Pathways and mechanisms of quercetin in cancer prevention and management. *Nutrition Reviews, 68,* 418–428.

Velmurugan, B., Gangar, S. C., Kaur, M., et al. (2010). Silibinin exerts sustained growth suppressive effect against human colon carcinoma SW480 xenograft by targeting multiple signaling molecules. *Pharmaceutical Research, 27,* 2085–2097.

Vishnu, P., & Tan, W. W. (2010). Update on options for treatment of metastatic castration-resistant prostate cancer. *OncoTargets and Therapy, 3,* 39–51.

Vivar, O. I., Lin, C. L., Firestone, G. L., et al. (2009). 3,3′-Diindolylmethane induces a G(1) arrest in human prostate cancer cells irrespective of androgen receptor and p53 status. *Biochemical Pharmacology, 78,* 469–476.

Walle, T., Hsieh, F., DeLegge, M. H., et al. (2004). High absorption but very low bioavailability of oral resveratrol in humans. *Drug Metabolism and Disposition, 32,* 1377–1382.

Wang, H. M., Chueh. P. J., Chang, S. P., et al. (2009a). Effect of capsaicin on tNOX (ENOX2) protein expression in stomach cancer cells. *BioFactors (Oxford, England), 34,* 209–217.

Wang, J., Zhang, L., Yu, Y., et al. (2009b). Enhancement of antitumor activities in sulfated and carboxymethylated polysaccharides of Ganoderma lucidum. *Journal of Agricultural and Food Chemistry, 57,* 10565–10572.

Wang, H. C., Yang, J. H., Hsieh, S. C., et al. (2010a). Allyl sulfides inhibit cell growth of skin cancer cells through induction of DNA damage mediated G2/M arrest and apoptosis. *Journal of Agricultural and Food Chemistry, 58,* 7096–7103.

Wang, H. J., Jiang, Y. Y., Wei, X. F., et al. (2010b). Silibinin induces protective superoxide generation in human breast cancer MCF-7 cells. *Free Radical Research, 44,* 90–100.

Wang, H. J., Wei, X. F., Jiang, Y. Y., et al. (2010c). Silibinin induces the generation of nitric oxide in human breast cancer MCF-7 cells. *Free Radical Research, 44,* 577–584.

Wang, S., Penchala, S., Prabhu, S., et al. (2010d). Molecular basis of traditional Chinese medicine in cancer chemoprevention. *Current Drug Discovery Technologies 7,* 67–75.

Wang, Y. B., Qin, J., Zheng, X. Y. et al. (2010e). Diallyl trisulfide induces Bcl-2 and caspase-3-dependent apoptosis via downregulation of Akt phosphorylation in human T24 bladder cancer cells. *Phytomedicine: International Journal of Phytotherapy and Phytopharmacology, 17,* 363–368.

Wargovich, M. J. (1999). Nutrition and cancer: The herbal revolution. *Current Opinion in Gastroenterology, 15,* 177–180.

Weng, C. J., & Yen, G. C. (2010). The in vitro and in vivo experimental evidences disclose the chemopreventive effects of Ganoderma lucidum on cancer invasion and metastasis. *Clinical & Experimental Metastasis, 27,* 361–369.

Weng, C. J., Chau, C. F., Yen, G. C., et al. (2009). Inhibitory effects of Ganoderma lucidum on tumorigenesis and metastasis of human hepatoma cells in cells and animal models. *Journal of Agricultural and Food Chemistry, 57,* 5049–5057.

Wenzel, E., Soldo, T., Erbersdobler, H., et al. (2005). Bioactivity and metabolism of trans-resveratrol orally administered to Wistar rats. *Molecular Nutrition & Food Research, l49,* 482–494.

Wesierska-Gadek, J., Kramer, M. P., & Maurer, M. (2008). Resveratrol modulates roscovitine-mediated cell cycle arrest of human mcf-7 breast cancer cells. *Food and Chemical Toxicology: An International Journal Published for the British Industrial Biological Research Association, 46,* 1327–1333.

Wesierska-Gadek, J., Borza, A., Komina, O., et al. (2009). Impact of roscovitine, a selective CDK inhibitor, on cancer cells: Bi-functionality increases its therapeutic potential. *Acta Biochimica Polonica, 56,* 495–501.

Wolter, F., Akoglu, B., Clausnitzer, A., et al. (2001). Downregulation of the cyclin D1/Cdk4 complex occurs during resveratrol-induced cell cycle arrest in colon cancer cell lines. *The Journal of Nutrition, 131,* 2197–2203.

Wu, P., Dugoua, J. J., Eyawo, O., et al. (2009). Traditional Chinese medicines in the treatment of hepatocellular cancers: A systematic review and meta-analysis. *Journal of Experimental & Clinical Cancer Research, 128,* 112.

Wu, K., Zeng, J., Li, L., et al. (2010). Silibinin reverses epithelial-to-mesenchymal transition in metastatic prostate cancer cells by targeting transcription factors. *Oncology Reports, 23,* 1545–1552.

Xiao, D., Powolny, A. A., Antosiewicz, J., et al. (2009). Cellular responses to cancer chemopreventive agent D,L-sulforaphane in human prostate cancer cells are initiated by mitochondrial reactive oxygen species. *Pharmaceutical Research, 26,* 1729–1738.

Xie, Y. Y., Yuan, D., Yang, J. Y., et al. (2009). Cytotoxic activity of flavonoids from the flowers of Chrysanthemum morifolium on human colon cancer Colon205 cells. *Journal of Asian Natural Products Research, 11,* 771–778.

Yamaoka, Y., Ohguchi, K., Itoh, T., et al. (2009). Effects of theaflavins on melanin biosynthesis in mouse b16 melanoma cells. *Bioscience, Biotechnology, and Biochemistry, 73,* 1429–1431.

Yang, Z. H., Wang, X. H., Wang, H. P., et al. (2010). Capsaicin mediates cell death in bladder cancer T24 cells through reactive oxygen species production and mitochondrial depolarization. *Urology, 75,* 735–741.

Yeung, K. S., Gubili, J., & Cassileth, B. (2008). Evidence-based botanical research: Applications and challenges. *Hematology/Oncology Clinics of North America, 22,* 661–670.

Yildirim, Y. (2010). Patterns of the use of complementary and alternative medicine in women with metastatic cancer. *Cancer Nursing, 33,* 194–200.

Yip, I., Heber, D., & Aronson, W. (1999). Nutrition and prostate cancer. *The Urologic clinics of North America, 26,* 403–411.

Yokota, J. (2000). Tumor progression and metastasis. *Carcinogenesis, 21,* 497–503.

Young, S. L., & Chaplin, D. J. (2004). Combretastatin a4 phosphate: Background and current clinical status. *Expert Opinion on Investigational Drugs, 13,* 1171–1182.

Yu, S., & Kong, A. N. (2007). Targeting carcinogen metabolism by dietary cancer preventive compounds. *Current Cancer Drug Targets, 7,* 416–424.

Yu, S., Shen, G., Khor, T. O., et al. (2008) Curcumin inhibits Akt/mammalian target of rapamycin signaling through protein phosphatase-dependent mechanism. *Molecular Cancer Therapeutics, 7,* 2609–2620.

Yu, Y., Hamza, A., Zhang, T., et al. (2010). Withaferin A targets heat shock protein 90 in pancreatic cancer cells. *Biochemical Pharmacology, 79,* 542–551.

Yuan, H., Young, C. Y., Tian, Y., et al. (2010). Suppression of the androgen receptor function by quercetin through protein-protein interactions of Sp1, c-Jun, and the androgen receptor in human prostate cancer cells. *Molecular and Cellular Biochemistry, 339,* 253–262.

Zhang, X., Ling, Y., Yu, H., et al. (2010). Studies on mechanism of myricetin-induced apoptosis in human hepatocellular carcinoma HepG-2 cells. *Zhongguo Zhong Yao Za Zhi, 35,* 1046–1050.

Zhou, C., Mao, X. P., Guo Q, et al. (2009a). Diallyl trisulphide-induced apoptosis in human melanoma cells involves downregulation of Bcl-2 and Bcl-xL expression and activation of caspases. *Clinical and Experimental Dermatology, 34,* e537–e543.

Zhou, G. Q., Zhao, H. Y., & Lu, C. (2009b). Effect of Ganoderma lucidum polysaccharides on intestinal mucosal immune system in H22 liver cancer bearing mice. *Zhongguo Zhong Xi Yi Jie He Za Zhi, 29,* 335–339.

Zhou, W., Kallifatidis, G., Baumann, B., et al. (2010). Dietary polyphenol quercetin targets pancreatic cancer stem cells. *International Journal of Oncology, 37,* 551–561.

Zikri, N. N., Rieldi, K. M., Wang, L. S., et al. (2009). Black raspberry components inhibit proliferation, induce apoptosis and modulate gene expression in rat esophageal epithelial cells. *Nutrition and Cancer, 61,* 816–826.

Chapter 2
An Evidence-based Perspective of Arsenic Trioxide (As_2O_3) for Cancer Patients

Hongchi Jiang, Lianxin Liu, Tongsen Zheng and Dalong Yin

Abstract The ancient drug, arsenic, has remarkable efficacy in the treatment of relapsed acute promyelocytic leukemia (APL) and it is highly likely that a regimen for treatment of APL that does not require any traditional chemotherapy drugs will be developed in the future. Arsenic trioxide (white arsenic or As_2O_3) was approved by the United States Food and Drug Administration for being used in the treatment of relapsed/refractory APL in 2000. This success has led to exploration of its use in other malignancies. As_2O_3 interacts with multiple molecular targets and signaling pathways. The resultant effect depends on factors, including cell type, and the dose and duration of As_2O_3 exposure. Understanding the molecular and biological basis of these effects will promote the rational and optimal application of As_2O_3 in diseases other than APL. A series of clinical trials with As_2O_3 has confirmed its benefit in the therapy of APL, although its role in the treatment of other malignancies remains to be determined. Careful attention to the clinical management of patients on As_2O_3 therapy can significantly lessen the risk of major side effects. The administration of As_2O_3 can be done safely if careful attention to electrolyte abnormalities and electrocardiographic monitoring is undertaken. Here we provide an overview of the mechanism of action of arsenic and summarize its development in the treatment of APL and other malignant disorders.

2.1 Introduction

Arsenic trioxide (white arsenic or As_2O_3) has been used in medicine for more than 2,400 years for a variety of ailments including ulcers, the plague, and malaria (Waxman and Anderson 2001). In 1878, potassium arsenite was reported to have an anti-

---

H. Jiang (✉)
Department of General Surgery, The First Affiliated Hospital of Harbin Medical University, Key Laboratory of Hepatosplenic Surgery, Ministry of Education, No 23 Youzheng Street, Harbin, 150001 Heilongjiang, People's Republic of China
e-mail: jianghc@vip.163.com

W. C. S. Cho (ed.), *Evidence-based Anticancer Materia Medica,* Evidence-based Anticancer Complementary and Alternative Medicine,
DOI 10.1007/978-94-007-0526-5_2,

leukemic effect and was used for this purpose in the late nineteenth and early twentieth centuries until it was replaced by busulfan in the 1950s (Kwong and Todd 1997). In the modern era, interest in arsenic as a chemotherapy was rekindled after it was identified as an active ingredient in traditional medicines in China (Sears et al. 1988; Cutler and Bradford 1878). Researchers evaluated arsenic compounds for the treatment of various cancers and published the results of a trial in which intravenous administration of As_2O_3 produced a complete response (CR) in 21 (66%) of 32 patients with acute promyelocytic leukemia (APL) (Wang and Chen 2008; Sun et al. 1992). In subsequent studies, Zhang et al. (1996) reported that As_2O_3 induced a CR in 22 (73%) of 30 newly diagnosed and 22 (52%) of 42 relapsed APL patients, and Shen et al. (1997) observed a CR in nine (90%) of 10 relapsed APL patients. As_2O_3 induces a high rate of complete remissions (87%) in this disease, As_2O_3 was approved by the United States Food and Drug Administration for use in the treatment of relapsed/refractory APL in 2000 (Waxman and Anderson 2001; Grimwade et al. 2009). Treatment of APL patients with As_2O_3 is associated with the disappearance of the promyelocytic leukemia (PML)-RARα fusion transcript, the gene product of the chromosomal translocation t(15;17) characteristic of APL (Grimwade et al. 2009; Miller et al. 2002).

In contrast to Western medicine, traditional Chinese medicine (TCM) does not focus on a single target but multiple targets involved in a particular disease condition by applying diverse modalities. One major reason for reluctance of Western academia towards TCM is due to the lack of clinical studies of TCM recipes. This situation is changing recently, and a number of clinical studies were conducted on As_2O_3 providing convincing evidence for the first time to gain credibility and reputation outside China. Clinical trials with As_2O_3 remedies focus on three major fields in cancer research: (1) improvement of poor treatment response rates towards standard chemotherapy and radiotherapy, (2) reduction of severe adverse effects of standard cancer therapy, and (3) unwanted interactions of standard therapy with As_2O_3. Appropriate quality assurance and control of As_2O_3 products as well as sustainable production methods are pre-conditions for the implementation of As_2O_3 in cancer therapy at an international level. As_2O_3 has also shown efficacy in other hematological malignancies and solid tumors (Emadi and Gore 2010). Therapeutic doses of As_2O_3 are well tolerated, with no evidence of long-term toxicity. Adverse events include APL differentiation syndrome, electrocardio-graphic abnormalities, and mild elevations in liver enzymes. This chapter highlights trials investigating the role of As_2O_3 in induction and consolidation for different cancers in view of the evidence based medicine. The chemistry, mechanisms of action, and clinical side effects of As_2O_3 are also discussed.

2.2 History of Arsenicals Medicinal Use and Their Metabolism

Arsenic is a semimetal commonly found in soil, water, and air. Common inorganic and organic forms of arsenic are listed in Table 2.1. Often, arsenic complexes with sulfur as red arsenic (As_2S_2), also called realgar or sandaraca, or as yellow arsenic

Table 2.1 Arsenic in nature

Arsenic forms	Chemical formula	Other names
Red arsenic	As_2S_2	Realgar, sandaraca
Yellow arsenic	As_2S_3	Arsenikon, aurumpigmentum, orpiment
White arsenic	As_2O_3	–
Phenylarsineoxide	C_6H_5AsO	–

(As_2S_3), also called orpiment or auripigment. White arsenic (As_2O_3) is produced by heating realgar. Arsenic is rarely found in a pure state; rather, it exists in both trivalent and pentavalent oxidation states as a chemically unstable sulfide or oxide, or as a salt of sodium, potassium, or calcium. Trivalent arsenicals, including sodium arsenite and the more soluble As_2O_3, inhibit many enzymes by reacting with biological ligands that possess available sulfur groups. Being pentavalent, arsenic is recognized as an uncoupler of mitochondrial oxidative phosphorylation. Although most inorganic arsenic that is ingested is eliminated fairly rapidly in the urine, a small amount may be modified by methylation to monomethylarsonic acid and to dimethylarsinic acid, a process referred to as biotransformation. Whereas these enzymatic reactions are considered to be detoxifying, some organic arsenic metabolites may actually contribute to the cytotoxicity of arsenic (Mantadakis et al. 2008; Antman 2001).

Arsenic has also long been demonstrated to have anticancer activity in some cases in the old days in China. In TCM, As_2O_3 was recorded in the Compendium of Materia Medica by one of the greatest physicians and pharmacologists Li Shi-Zhen (1518–1593). Medical use of arsenic and its derivatives dates back more than 2,400 years to ancient Greece, China, and Rome. Arsenic was considered as both a therapeutic agent and a poison. Hippocrates administered orpiment and realgar as an ulcer remedy; Dioscorides used orpiment as a depilatory. Arsenic has also been used to treat the plague, malaria, and cancer, and to promote sweating. Physicians prescribed arsenic for both external and internal use throughout the eighteenth century. Arsenides and arsenic salts were key ingredients in antiseptics, antispasmodics, antiperiodics, caustics, cholagogues, hematinics, sedatives, and tonics (Miller et al. 2002). Approximately 60 different arsenic preparations have been developed and distributed during the lengthy history of this agent. More than 20 of these preparations were still in use at the end of the nineteenth century, including Aiken's Tonic Pills, Andrew's Tonic, and Arsenauro. When physicians first boiled arsenous acid with an alkali in the late 1700s and produced a water-soluble compound, the administration of medicinal arsenic changed radically from generally external to primarily internal (Haller 1975; Jackson and Grainge 1975; Chan and Huff 1997; Bernstam and Nriagu 2000). In 1786, Thomas Fowler, a physician to the General Infirmary of the County of Stafford, England, recommended use of potassium arsenite for the treatment of intermittent fever. Fowler's Solution gained great renown and was used to treat many conditions, including paralytic afflictions, rheumatism, hypochondriasis, epilepsy, hysteria, melancholia, dropsy, rachitis, heart palpitations, convulsions, syphilis, ulcers, cancer, and dyspepsia. In 1911, Fowler's Solution was used as a treatment for pernicious anemia, asthma, psoriasis, pemphigus, and eczema. In 1910, additional experimentation with the properties of arsenic led Paul Ehrlich, the German physician and founder of chemotherapy, to the discovery of an organic arsenical, salvarsan. Arsphenamine was the standard therapy for syphilis for nearly 40 years before it was replaced by penicillin (Haller 1975; Jackson and Grainge 1975; Chan and Huff 1997; Bernstam and Nriagu 2000). Arsphenamine was also believed to be an effective treatment for trypanosomiasis. As above mentioned, arsenic has been proven recently to be highly effective in the treatment of APL; its use induces and maintains complete remissions, with a less toxic profile than traditional chemotherapy (Waxman and Anderson 2001). The role of arsenic in the treatment of other malignancies and hematologic disorders is much less clear but continues to be explored in combination with other modalities of therapy.

Arsenic was nicknamed "The Mule" not only for its dependability in many therapeutic regimens but also for the stubborn persistency with which it was used and the inconstant nature of its toxic capacity (Haller 1975; Jackson and Grainge 1975; Chan and Huff 1997; Bernstam and Nriagu 2000). Although arsenic was found to be beneficial in many disease states and side effects or later repercussions of therapy were inconsistent from patient to patient, concerns among medical professionals about toxicities associated with arsenic use, especially long-term use, surfaced in later years. The International Agency for Research on Cancer (IARC) first evaluated the carcinogenicity of arsenic and arsenic compounds in 1973. It found a "causal relationship between skin cancer and heavy exposure to inorganic arsenic in drugs, drinking water with a high arsenic content, or in the occupational environment". However, experimental studies with arsenic in animals were considered inadequate, and the causative role of arsenic remained largely unclear. In 1979, the agency classified arsenic and certain arsenic compounds as Group 1 carcinogens, those that were "carcinogenic to humans", on the basis of epidemiological studies of the relationship between exposure to arsenic compounds and skin cancer *via* occupation, ingestion, or medical use. In 1980, re-evaluation of the data determined that exposure to arsenic and arsenic-containing compounds were a cause of lung cancer in humans. Finally, in 1987, support for arsenic as a cause of human cancers was judged as adequate on the basis of "limited evidence of carcinogenicity in experimental animals" (Haller 1975; Jackson and Grainge 1975; Chan and Huff 1997; Bernstam and Nriagu 2000). Dermatological manifestations of arsenic use were commonly observed in patients with prolonged arsenic exposure. Long-term ingestion of Fowler's Solution produces signs of chronic arsenic intoxication. In one study, palmar and plantar keratoses occurred in all of the patients evaluated, although the minimum time to onset was 15 years (median, 24 years) after beginning arsenic treatment. Basal cell carcinoma developed in a majority of these patients, with the same time to onset. Other manifestations included carcinoma *in situ*, squamous cell carcinoma, breast adenocarcinoma, and colon carcinoma, all appearing within a minimum of 28–63 years after treatment. The precise amounts of arsenic these patients received over time is not known, because small amounts of arsenic are present in the environment, and there are any number of possible routes of exposure. Nonetheless, despite the anecdotal nature of the evidence, the keratoses and carcinomas that developed were attributed to past and prolonged arsenic intake (Haller 1975; Jackson and Grainge 1975; Chan and Huff 1997; Bernstam and Nriagu 2000).

2.3 Mechanisms of Action of As_2O_3

Exactly how As_2O_3 mediates its clinical efficacy is not fully understood. Two main mechanisms of action of As_2O_3 have been identified from both *in vivo* and *in vitro* studies: promotion of APL cell differentiation observed at low levels of As_2O_3 and induction of apoptosis observed at high levels of As_2O_3 (Fig. 2.1). It has become clear that As_2O_3 interferes with a variety of cellular processes by targeting numerous different intracellular molecules and thereby disrupting key signal transduction mechanisms and producing programmed cell death. These findings underscore the importance of understanding how differences in cell types or cellular environments might affect the action of As_2O_3 (Davison et al. 2002).

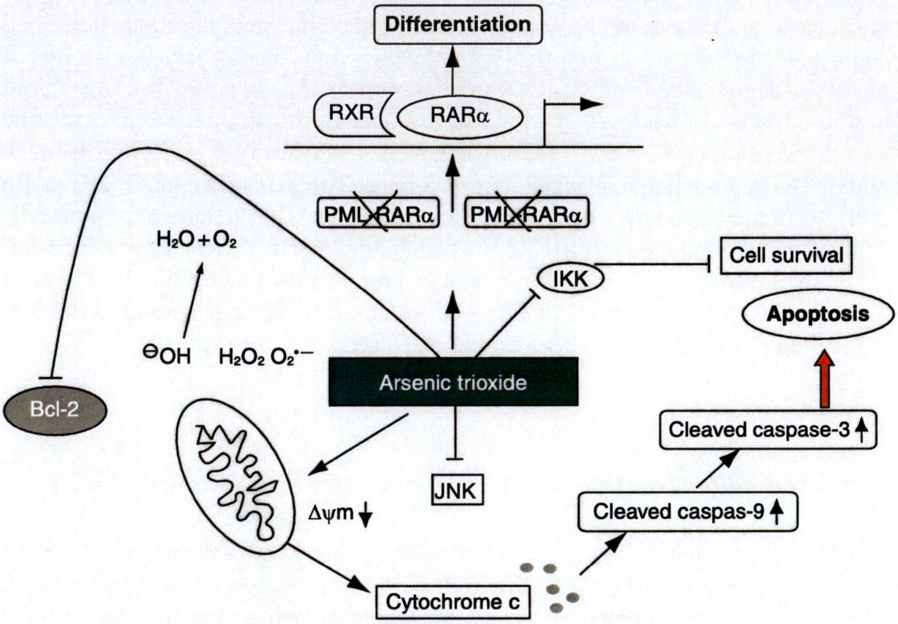

Fig. 2.1 Cellular targets of Arsenic trioxide (As_2O_3). As_2O_3 targets multiple pathways in malignant cells resulting in apoptosis or in the promotion of the differentiation program

2.3.1 Differentiation and Oncogenic Proteins

As_2O_3 has potent cytotoxic and antitumor activities *in vitro* and *in vivo*, there has been extensive research focused on the identification of the mechanisms by which it generates its effects on target cells. Degradation of the fusion protein, PML-RARα, is most likely the mechanism by which As_2O_3 induces cell differentiation in APL cells. Degradation of PML-RARα allows malignant promyelocytes to overcome their maturation block (Novick and Warrell 2000). In clinical trials, cell samples taken from APL patients treated with As_2O_3 suggest that partial differentiation of the maturation-arrested leukemia cells contributes to the therapeutic effect (Soignet et al. 1998). As_2O_3 degrades the fusion protein and induces differentiation in APL cells whether or not they are resistant to retinoic acid. However, there is conflicting evidence whether As_2O_3 can synergize with retinoic acid *in vitro* or *in vivo*. The partial differentiation effects of As_2O_3 appear to be unique to APL because of the direct effect of As_2O_3 on PML-RARα degradation (Chen et al. 1997; Jing et al. 2001). Empirically, treatment of APL cells with As_2O_3 leads to their terminal differentiation *in vitro* and *in vivo*. APL cells are uniquely sensitive due to the expression of the PML-RARα fusion protein; however, the mechanism by which As_2O_3 treatment induces terminal differentiation remains somewhat speculative. In normal myeloid cells, PML protein is localized to macromolecular structures in the nucleus (nuclear bodies), where PML antagonizes many processes required for the initiation and progression of malignancy. In leukemic cells, the PML-RARα fusion protein blocks the expression of genes required for normal differentiation. The fusion protein disrupts the nuclear bodies, and the PML protein is dispersed into smaller organelles. PML contains a cysteine-rich region that is hypothesized to interact with As III, resulting in the degradation of PML-RARα fusion protein. Furthermore, As_2O_3-induced histone acetylation has been reported to promote differentiation *via* alteration in gene transcription (Dyck et al. 1994; Chen et al. 1996; Wang et al. 1998; List et al. 2003). The degradation of PML-RARα is a major mechanism of As_2O_3 activity in APL, but it may also target other oncogenic proteins. As_2O_3 treatment of human T-cell leukemia lymphoma virus (HTLV-1) containing acute T-cell leukemia cells resulted in degradation of Tax (a transactivator protein activated by HTLV-1) leading to apoptosis. Decreased levels of Bcr-Abl protein in K562 cells and other Bcr-Abl expressing cell lines has also been documented after As_2O_3 treatment.

2.3.2 Apoptosis

As_2O_3-induced apoptosis, in contrast, occurs *via* a variety of mechanisms, which appear to be independent of the presence of PML-RARα. The apoptotic effects of As_2O_3 occur, in part, through direct effects on mitochondria. An important initial cellular event that occurs during treatment of target cells with As_2O_3 involves elevation of reactive oxygen species (ROS). Such generation of ROS appears to be regulated, at least in part, by activation of nicotinamide adenine dinucleotide phosphate oxidase and nitric oxide synthase isozymes. Also, arsenic-containing compounds are potent modulators of the thioredoxin system that includes thioredoxin, thioredoxin reductase, and nicotinamide adenine dinucleotide phosphate. The thioredoxin system controls, to a large extent, intracellular redox reactions, regulates apoptosis, and protects cells from stress damage, and the ability of arsenic-containing compounds to target and block thioredoxin reductase may be important in the induction of its pro-apoptotic effects.

2.3.2.1 Reactive Oxygen Species

Over-production of ROS is linked to the induction of apoptosis by As_2O_3. Accumulation of hydrogen peroxide leads to decreases in the mitochondrial membrane potential, resulting in cytochrome *c* release and activation of the caspase cascade. This appears to be a common mechanism of induction of cell death in diverse cellular backgrounds. There is extensive evidence implicating arsenic-dependent, ROS-mediated activation of caspases in various types of malignant cells. These include cells of APL origin, human T cell lymphotrophic virus I-infected T cell lines and primary adult T cell leukemia cells (Nishikawa 2008; Dai et al. 1999; Sen 1998), multiple myeloma cells, and different types of solid tumor cells. However, caspase-independent death pathways have been also reported to be activated by arsenic in myeloma cells and may mediate pro-apoptotic signals (Nishikawa 2008; Dai et al. 1999; Sen 1998). Other recent work has implicated the c-Jun N-terminal kinase (JNK) as an essential component of As_2O_3-dependent apoptosis (Nishikawa 2008; Dai et al. 1999; Sen 1998). It was demonstrated that activation of JNK occurs in an As_2O_3-inducible manner in cells of APL origin and that As_2O_3 resistance correlates with defective activation of the JNK pathway. Notably, in these studies, it was also shown that pharmacological inhibition of JNK significantly decreases As_2O_3-dependent growth inhibition and apoptosis, but it does not protect cells from the effects of chemotherapy (doxorubicin) (Nishikawa 2008; Dai et al. 1999; Sen 1998).

2.3.2.2 Bcl-2

A further mitochondria-related effect is the ability of As_2O_3 to down-regulate *Bcl-2* expression. This has been associated with As_2O_3-mediated growth inhibition and apoptosis in a variety of cell types, although it has not been a universal finding (Lu et al. 1999; Zhang et al. 1998).

2.3.2.3 NF-κB

NF-κB is a transcriptional factor promoting cell survival with an important role in many cancer cells. Activation of NF-κB depends on the integrity of the IκB kinase (IKK); upon phosphorylation by IKK, the inhibitory protein IκB releases NF-κB for translocation to the nucleus. As_2O_3 has been shown to inhibit IKK by binding to cysteine-179 in the activation loop of the enzyme catalytic subunit. Although cysteine-179 is not located in the vicinity of another cysteine within the IKK primary structure, it has been suggested that another cysteine may come within a critical distance of cysteine-179 upon the folding of the polypeptide chain or the dimerization of the catalytic subunits, thus providing a high-affinity target for arsenite (Kapahi et al. 2000).

2.3.3 *Anti-proliferative/Cell Cycle Effects*

Interference with cell cycle progression is a major factor in the growth inhibition caused by As_2O_3 as it impairs proliferation and may lead to apoptosis or sensitize cells to differentiation. The cell cycle effects of As_2O_3 vary considerably depending on the concentration and cell type. It has been associated with both a prolongation of the cell cycle as well as cell cycle arrest of malignant hematopoietic cells in the G1 and G2/M phases (Park et al. 2001).

2.3.4 *Angiogenesis Inhibition*

Angiogenesis is a vital process for the growth and survival of solid tumors and a consistent pathological feature in hematological malignancies. The bone marrow of patients with conditions including acute and chronic leukemias, myelodysplastic syndrome, non-Hodgkin lymphoma (NHL), and multiple myeloma demonstrate increased microvascular density and vascular endothelial growth factor (VEGF) levels. In a mouse fibrosarcoma model, a single dose of As_2O_3 induced apoptosis of new blood vessel endothelial cells and ischemic necrosis of the tumor while sparing normal tissues. Human umbilical vein endothelial cells have also been used as a model to study the anti-angiogenic effects of As_2O_3. In this system, As_2O_3 induced the activation of endothelial cells, up-regulation of endothelial cell adhesion molecules, inhibition of capillary tubule growth and vessel branching, apoptosis of endothelial cells, and inhibition of VEGF production. As_2O_3 also inhibited VEGF production in the human erythroleukemia leukemic cell line. It is thought that a reciprocal positive feedback loop exists between leukemic cells producing VEGF and the stimulated, rapidly proliferating endothelial cells producing leukemic cell growth factors (including granulocyte macrophage colony stimulating factor, interleukin IL-6, IL-7, and IL-10), and that As_2O_3 may disrupt this loop (Roboz et al. 2000).

2.4 From Bedside to Bench, Then to Bedside Again

The clinical results achieved with As_2O_3 in APL have prompted investigations into the mechanisms of action by which arsenic produces clinical benefit. This "bedside to bench" approach has shown that As_2O_3 has multiple targets. Chen et al. (1996) determined that, at low concentrations, As_2O_3 promotes differentiation of APL cells and, at higher concentrations, triggers apoptosis and down-regulates Bcl-2 expression. Subsequently, an intense research effort focused on whether these effects were unique to APL cells or to a more general response by different types of neoplastic cells (Xu et al. 2009; Jutooru et al. 2010; Lee et al. 2010; Subbarayan et al. 2010; Tingting et al. 2010; Wu et al. 2010). Considerable pre-clinical evidence supports the potential of As_2O_3 against a number of different malignancies. Studies in cultured cells show that it inhibits growth and promotes apoptosis in myeloid leukemia, multiple myeloma, lymphoma, and lymphocytic leukemia, and solid tumor cell lines. As_2O_3 exerts its cytotoxic effects on neoplastic cells by inhibiting proliferation, inducing apoptosis, and promoting cellular differentiation. Anecdotal and preliminary clinical data from China suggest a broad therapeutic potential for As_2O_3 in the treatment of cancer. These findings support further investigation of the clinical utility of As_2O_3 for the treatment of hematologic and solid tumor malignancies other than APL (Wang et al. 1998; Perkins et al. 2000; Yoo et al. 2009; Zhang et al. 2009; Zhao et al. 2009).

2.4.1 *Clinical Trials in Hematologic Malignancies*

There are several diseases in which As_2O_3 have been studied in clinical trials. Hematologic malignancies studied include APL, multiple myeloma, acute myeloid leukemia (AML), acute lymphoblastic leukemia, adult T-cell leukemia/lymphoma (ATL), myelodysplastic syndrome, NHL, and chronic myeloid leukemia (Dai et al. 1999; Zhu et al. 1999; Zhou et al. 2010; Munshi 2001). In a recent study, a total of 19 children (< or = 15 years of age) with newly diagnosed APL were treated with single-agent As_2O_3 for remission induction and post-remission therapy. The results suggested that seventeen of the children (89.5%) achieved complete hematologic remission, and two early deaths occurred from intracranial hemorrhage. As_2O_3-induced leukocytosis was observed in 13 (68.4%) patients. Other As_2O_3-related toxicities were minimal and transient. Post-remission As_2O_3 therapy has continued for 3 years; the most commonly side effect was As_2O_3-induced neutropenia. With a median follow-up of 53 months (range, 23–76 months), the calculated 5-year overall survival and event-free survival were 83.9% and 72.7%, respectively, which are comparable with results achieved by the use of all-*trans* retinoic acid (ATRA) plus chemotherapy, which is the standard therapy for APL. No chronic arsenic toxicity or second malignancies were found during the follow-up period, and arsenic retention was not significant in patients off treatment more than 24 months. As_2O_3 resistance was observed in only one patient with a complex karyotype. The results indicate the high efficacy and safety of single-agent As_2O_3 regimens in the treatment of children with *de novo* APL.

Here we also summarize the clinical trials of As_2O_3 in other hematologic malignancies (Table 2.2). ATL is resistant to chemotherapy and carries a dismal prognosis particularly for the acute and lymphoma subtypes. Chronic ATL has a relatively better outcome, but poor long-term survival is noted when patients are managed with a watchful-waiting policy or with chemotherapy. Clinically, arsenic/interferon therapy

Table 2.2 Clinical studies of Arsenic trioxide (As_2O_3) in hematologic malignancies other than acute promyelocytic leukemia

Hematologic malignancies	Phase	N	Regimen	Adverse effects	Results	References
AML	I/II	64	As_2O_3 0.25 mg iv + low-dose cytarabine twice daily	Neutropenic fever was observed in >80% of patients and 41% of patients had bacteremia. Non-hematologic toxicity generally was mild and reversible, included fatigue, nausea, diarrhea, rash, peripheral edema, and elevated transaminases	Improve responses in elderly AML patients compared with either agent alone	Roboz et al. (2008)
Relapsed or refractory lymphoid malignancies	II	17	As_2O_3 0.25 mg/kg iv + AA 1,000 mg iv for 5 days/1st week followed by twice weekly infusions during week 2–6 (repeated every 8 week)	Hematologic toxicities were the most commonly reported	Generally well tolerated but had limited activity in patients	Chang et al. (2009)
Relapsed or refractory MM	I/II	22	As_2O_3 0.125 or 0.250 mg/kg + bortezomib 0.7, 1.0, 1.3 mg/m^2 + AA 1 g iv on day 1, 4, 8, 11 of a 21-day cycle for max 8 cycles	One occurrence of grade 4 thrombocytopenia was observed. One patient had asymptomatic arrhythmia and withdrew from the study	Well tolerated by most patients and produced preliminary signs of efficacy with 27% objective response rate	Berenson et al. (2007)
Relapsed or refractory MM	II	31	Melphalan 0.1 mg/kg po + As_2O_3 0.25 mg/kg iv + AA 1 g iv on day 1–4 of week 1, As_2O_3 + AA twice weekly during week 2–5 and no treatment during week 6 of cycle 1	Specific grade 3/4 haematological (3%) or cardiac adverse events occurred infrequently. Frequent grade 3/4 non-haematological adverse events included fever/chills (15%), pain (8%) and fatigue (6%)	A new therapeutic option for patients with relapsed/refractory MM	Berenson et al. (2006)
MM	II	20	As_2O_3 + dexamethasone + AA	The regimen was well tolerated with most adverse events being mild or moderate	Showed the clinical efficacy and tolerability of this combination	Abou-Jawde et al. (2006)

Table 2.2 (continued)

Hematologic malignancies	Phase	N	Regimen	Adverse effects	Results	References
MM	II	24	As_2O_3 0.25 mg/kg/day for 5 days/1st 2 week of each 4-week cycle	Sixteen patients had grade 3 or 4 neutropenia and one required antibiotics	Active and reasonably well tolerated as a single-agent salvage therapy, even in patients with late-stage, relapsed/refractory MM	Hussein et al. (2004)
Myeloid leukemia	II	11	As_2O_3 0.25 mg/kg/day	Myelosuppression was the major adverse effect, most likely due to disease progression rather than drug-related	All subjects had progressive disease and there was no direct treatment-related mortality	Parmar et al. (2004)
MM	II	14	As_2O_3 iv	Although well tolerated, in these patients with extensive prior therapy, 11 developed cytopenia, five associated with infectious complications and three developed deep vein thromboses	Support further investigation of this novel drug for the treatment of patients with relapsed/refractory MM	Munshi et al. (2002)
Relapsed/refractory myeloma	I/II	6	As_2O_3 0.25 mg/kg/day+AA 1,000 mg/day for 25 days over a 35-day period	One episode of grade 3 hematological toxicity (leukopenia) and no grade 3 non-hematological toxicities (in particular cardiac) were observed	Two patients with thalidomide- refractory disease had partial responses; four patients had stable disease. This regimen has acceptable toxicity and promising evidence of activity in refractory/relapsed myeloma	Bahlis et al. (2002)

AA: ascorbic acid, AML: acute myeloid leukemia, As_2O_3: Arsenic trioxide, MM: multiple myeloma exhibits some efficacy in refractory aggressive ATL patients. Promising results were obtained from Phase II study of the efficacy and safety of the combination of As_2O_3, interferon alpha, and zidovudine in newly diagnosed chronic ATL. Among 10 newly diagnosed chronic ATL patients, an impressive 100% response rate was observed, including seven complete remissions, two complete remissions but with more than 5% circulating atypical lymphocytes, and one partial response. Responses were rapid and no relapse was noted. Side effects were moderate and mostly hematologic. The results suggested that treatment of chronic ATL with arsenic, interferon-alpha, zidovudine is feasible and exhibits an impressive response rate with moderate toxicity (Kchour et al. 2009). The addition of As_2O_3 to low-dose cytarabine was also reported to improve responses in elderly patients who had AML compared with either agent alone, and a randomized trial of the combination *versus* single-agent low-dose cytarabine is ongoing.

The utility of As_2O_3 in the treatment of a number of lymphoid malignancies is also being evaluated as well. At Memorial Sloan-Kettering Cancer Center, patients with relapsed or refractory intermediate- or high-grade NHL are being treated with As_2O_3, 0.25 mg/kg/day, 5 days a week for 5 weeks. In a trial at Mount Sinai Hospital (New York, NY, USA), patients with relapsed or refractory low-grade NHL or chronic lymphoid leukemia (CLL) are eligible for treatment with As_2O_3. Investigators at Northwestern University (Evanstown, IL, USA) have undertaken a trial of As_2O_3 0.25 mg/kg/day for 60 days in patients with relapsed or refractory Hodgkin's disease. MD Anderson Cancer Center has investigated As_2O_3 in patients with relapsed or refractory CLL. Clinical evaluation of As_2O_3 for the treatment of multiple myeloma is under way too. Combination therapy with As_2O_3 and ascorbic acid is also being studied in patients with relapsed or refractory multiple myeloma. The addition of ascorbic acid is linked to observations that, at least in part, the glutathione redox system mediates the inhibition of growth and induction of apoptosis that follow exposure to As_2O_3. Earlier work showed that ascorbic acid decreases glutathione levels and renders malignant cells, but not normal cells, more sensitive to As_2O_3-induced apoptosis. In a mouse model, ascorbic acid enhanced the anti-lymphoma effect observed in response to arsenic treatment, without additional toxicity. Abundant pre-clinical evidence shows that As_2O_3 inhibits growth and promotes apoptosis in many different cancer cell types. This broad mechanism of action supports a potential for clinical activity in numerous neoplastic diseases other than APL and provides a basis for further clinical evaluation of As_2O_3 in hematologic malignancies and solid tumors. The dose and dosing regimen required for clinical response in such cancers may be different from those for APL (Dai et al. 1999; Zhu et al. 1999; Zhou et al. 2010; Munshi 2001).

2.4.2 *Clinical Trials in Solid Tumors*

Based on promising pre-clinical data, clinical trials to examine the potential of As_2O_3 for the treatment of solid tumors are under way or in the final planning stages (Table 2.3). But almost all the clinical trials indicated single-agent As_2O_3

Table 2.3 Clinical studies of Arsenic trioxide (As_2O_3) in solid tumors

Tumor type	Phase	N	Regimen	Adverse effects	Results	References
Metastatic melanoma	II	10	As_2O_3 0.25 mg/kg/day for 5 days followed by 0.35 mg/kg/day twice a week	Grade 3 toxicity included neutropenia, fatigue, abdominal pain and arthralgia. Grade 4 toxicity did not occur	Generally well tolerated, but no tumor regression was observed	Kim et al. (2005)
Advanced metastatic melanoma	II	21	As_2O_3 0.32 mg/kg/day for 4 days in week 1 followed by 0.25 mg/kg/day twice a week for 6 weeks followed by 1 week rest	Possible treatment- related grade 3 of 4 toxicities included idiopathic thrombocytopenic purpura (1) and elevated lactate dehydrogenase (1)	Well tolerated and had limited activity in patients with metastatic melanoma	Tarhini et al. (2008)
Metastatic melanoma	II	11	As_2O_3 0.25 mg/kg/day+AA 1,000 mg for 5 days and then twice weekly at 0.35 mg/kg during an 8-week cycle	Common grade 1 and 2 side effects included nausea and vomiting (10), fatigue (6), edema (6), rash (6) and elevated AST or ALT (6). Grade 3 and 4 side effects included nausea and vomiting (3), elevated AST or ALT (2), seizure (1) and renal failure (1)	No responses were seen in the 1st 10 evaluable patients leading to early closure of the study	Bael et al. (2008)
Pancreatic adenoCA refractory to gemcitabine	II	13	As_2O_3 0.3 mg/kg for 5 days every 28 days	Anemia (50%) and leukopenia (25%), fatigue and thrombosis (17%), prolongation of the QTc interval (1) occurred in patients	Has no activity in pancreatic cancer patients who develop progressive disease after gemcitabine	Kindler et al. (2008)

Table 2.3 (continued)

Tumor type	Phase	N	Regimen	Adverse effects	Results	References
Hepatocellular carcinoma	II	29	As_2O_3 0.16–0.24 mg/kg/day for 5–6 days/week for 3–4 week followed by 1-week rest	Fatigue, rash, abdominal pain and nausea are reported	Not active against advanced hepatocellular carcinoma	Lin et al. (2007)
Metastatic renal cell carcinoma	II	16	As_2O_3 0.3 mg/kg/day iv for 5 days every 4 week	The most common toxicity observed was grade 2 elevations in liver function tests (36%), anemia (21%), renal insufficiency (14%), rash (7%) and diarrhea (7%)	Best response was stable disease in three patients, but did not achieve a complete or partial response in metastatic renal cell carcinoma	Vuky et al. (2002)
Advanced head and neck cancer	I	11	As_2O_3 10, 20, 30 mg/week a day prior to hyperthermia	No amplification of toxicities due to radiation or hyperthermia was evident	Patients who received 30 mg of As_2O_3 weekly showed non-serious acute toxicities and patients without prior treatment showed better response	Huilgol (2006)

AA: ascorbic acid, AdenoCA: adenocarcinoma, As_2O_3: Arsenic trioxide was generally well tolerated; however, no tumor regression was observed in most of the solid tumor patient situation. Future clinical trials should evaluate As_2O_3 in combination with other anticancer drugs that may improve its clinical activity in solid tumor. As_2O_3 cytotoxicity and apoptosis induction has been demonstrated with numerous solid tumor cell lines, including human melanoma, hepatocellular carcinoma (HCC), gastric carcinoma, etc. A recent second-line, Phase II, single-arm study of As_2O_3 was conducted recently in patients with inoperable American Joint Committee on Cancer stage IV melanoma. Twenty-one patients (median age, 63.8 years) were accrued in the study. All had stage IV melanoma including M1a (two patients), M1b (six patients), and M1c (13 patients) disease. Among 17 evaluable patients, one patient (6%; 95% confidence interval (CI), 0–29%) achieved a partial response lasting 7 months, and 10 patients (59%) had disease stabilization after at least one cycle, but all eventually developed disease progression. The median time to disease progression was 17 weeks (95% CI, 11–38 weeks) and the median survival was 13 months (95% CI, 12–26 months). Their results suggested that As_2O_3 as tested in the current trial was found to be well tolerated and had limited activity in patients with metastatic melanoma. The application of this agent in combination with either chemotherapy or agents that target recognized critical signaling and anti-apoptotic pathways of melanoma has not yet been performed (Tarhini et al. 2008). There is no effective therapy for patients with metastatic pancreatic cancer who fail initial therapy with gemcitabine. Previous studies had indicated that As_2O_3 has potent anti-proliferative and pro-apoptotic effects in pancreatic cancer cell lines. A multicenter Phase II trial in patients with advanced pancreatic cancer who experienced disease progression on or after a gemcitabine-containing regimen was conducted in USA. Thirteen patients were enrolled between December 2002 and November 2003. Twenty-four cycles were administered (median 2; range 1–2). In this study, As_2O_3 0.3 mg/kg was administered intravenously over 1 hour daily for 5 consecutive days every 28 days. Restaging computed tomography scans were obtained every two cycles. Their results indicated that 13 patients were enrolled between December 2002 and November 2003. Twenty-four cycles were administered (median 2; range 1–2). There were no grade 3/4 hematologic toxicities; grade 1/2 anemia and leukopenia occurred in 50% and 25% of patients, respectively. Grade 3 toxicities included fatigue and thrombosis in 17% of patients. Only one patient developed a prolongation of the QTc interval. There were no objective responses. Median progression-free survival was 1.6 months (95% CI, 1.2–1.9). Median survival was 3.8 months (95% CI, 1.6–6.8) (Kindler et al. 2008). Despite promising *in vitro* data, As_2O_3 has no activity in pancreatic cancer patients who develop progressive disease after gemcitabine.

Early reports also indicated that single-agent As_2O_3 was not active against advanced HCC (Lin et al. 2007). As all the clinical trials indicated, despite promising *in vitro* data, As_2O_3 has no activity in solid tumor patients who develop progressive disease. Multicenter Phase II trials are feasible in this patient population, and novel agents are clearly needed.

2.5 Mechanisms of Resistance to As_2O_3

As an ancient TCM, As_2O_3 has been successfully used as a therapeutic agent for leukemia. Drug resistance and toxicity are major concerns with the treatment. Although a significant anticancer effect of As_2O_3 had been reported in pre-clinical and clinical trials, failure of chemotherapy was reported frequently in patients and some cell lines. Among the reasons anticancer drugs fail, the most common is acquired drug resistance. Resistance of cancer cells to As_2O_3 continues to be a major clinical obstacle to the successful treatment of cancer.At present, the anticancer drug resistance is considered as a multifactorial phenomenon involving several major mechanisms, such as decreased uptake of water-soluble drugs, increased repair of DNA damage, reduced apoptosis, altered metabolism of drugs, and increased energy-dependent efflux of chemo-therapeutic drugs that diminish the ability of cytotoxic agents to kill cancer cell, changes in glutathione transferase expression and topoisomerase II.Causes of cancer-specific drug resistance are currently believed to be linked to the random drug-induced mutational events (genetic hypothesis), to the drug-induced non-mutational alterations of gene function (epigenetic hypothesis), and recently, to the drug-induced karyotypic changes.Unfortunately, the key determinants of this phenomenon remain largely unknown (Lehnert 1998; Andrew et al. 2004; Glasspool et al. 2006; Iwasa et al. 2006; Duesberg et al. 2007; Fojo 2007; Roberti et al. 2006).

Increasing efflux of anticancer drugs is most commonly encountered in laboratories and clinical medication mediated by membrane transporters. The ABC (ATP-binding cassette) transporter family is well known in this area, especially such members as ABCB1 (P-gp, P-glycoprotein), MDR1 (multidrug-resistance gene 1), ABCC1 (multidrug-resistant protein 1, MRP1), ABCC2 (MRP2) and ABCG2 (breast cancer resistance protein, BCRP). The MRPs have been functionally characterized as the property of ATP-dependent export pumps for conjugates with glutathione (GSH), glucuronate, or sulfate. With respect to arsenic compounds, GSH S-transferase (GST) has been reported to conjugate GSH to arsenicals and several GST isoforms and genetic polymorphisms might influence arsenic metabolism and susceptibility. It should be pointed out that the arsenic-GSH conjugates are the substrates of some ABC transporter proteins, which efflux them outside mammalian cells. In As_2O_3-resistant human leukemia cell line K562/AS-3, the main mechanism of tolerance seems to be drug efflux by MRP1. Chronic exposure to low-concentration As_2O_3 or methylated arsenicals can increase the expression of P-gp, MRP1 and MRP2 at both the protein and mRNA level. However, there is no established arsenic-resistant human solid tumor cell line (Keppler et al. 1998; Kala et al. 2000; Ejendal and Hrycyna 2002; Deeley et al. 2006; Schläwicke Engström et al. 2007). Therefore, which of the ABC members contribute to arsenic resistance by actively pumping arsenic compounds outside solid tumors derived from the human digestive tract is still uncertain.

It is known that As_2O_3 induces DNA damage and p53 accumulation and that As_2O_3-induced apoptosis is required for p53 (Jiang et al. 2001; Filippova and Duerksen-Hughes 2003). Reducing the expression or inhibiting the function of p53 may abrogate the apoptosis of tumor cells induced by As_2O_3. Several signaling pathways are involved in the regulation of p53; of these, ubiquitylation is the most common way for the degradation of p53 through the binding of MDM2 to p53 (Yang et al. 2004). Previously, studies further verify that gankyrin promotes both mono- and poly-ubiquitylation of p53 by MDM2 in a p300-independent manner (Higashitsuji et al. 2005). Direct interaction between gankyrin and MDM2 is necessary to enhance p53 ubiquitylation and gankyrin facilitates the binding of MDM2 to p53 *in vivo* and *in vitro*. This suggests that gankyrin could be a key point in the regulation of p53 and may contribute to the establishment of resistance to As_2O_3 in tumor cells. In a previous study, Chen et al. (2009) established two arsenic-resistant solid tumor cell lines, HepG2/AS and SGC7901/AS to investigate the reasons of resistance. Their results indicate that HepG2/AS seems to efflux arsenic compounds mainly by ABCB1, ABCC1, and ABCC2, whereas SGC7901/AS seems to efflux arsenic compounds mainly by ABCB1. Inactivation of ABCB1 by higher concentration of verapamil can restore the sensitivity of As_2O_3 in HepG2/AS and SGC7901/AS. Increasing the levels of gankyrin and MDM2 may enhance the degradation of p53 and the phosphorylation of Rb, resulting in abrogation of apoptosis (Fig 2.2). A better

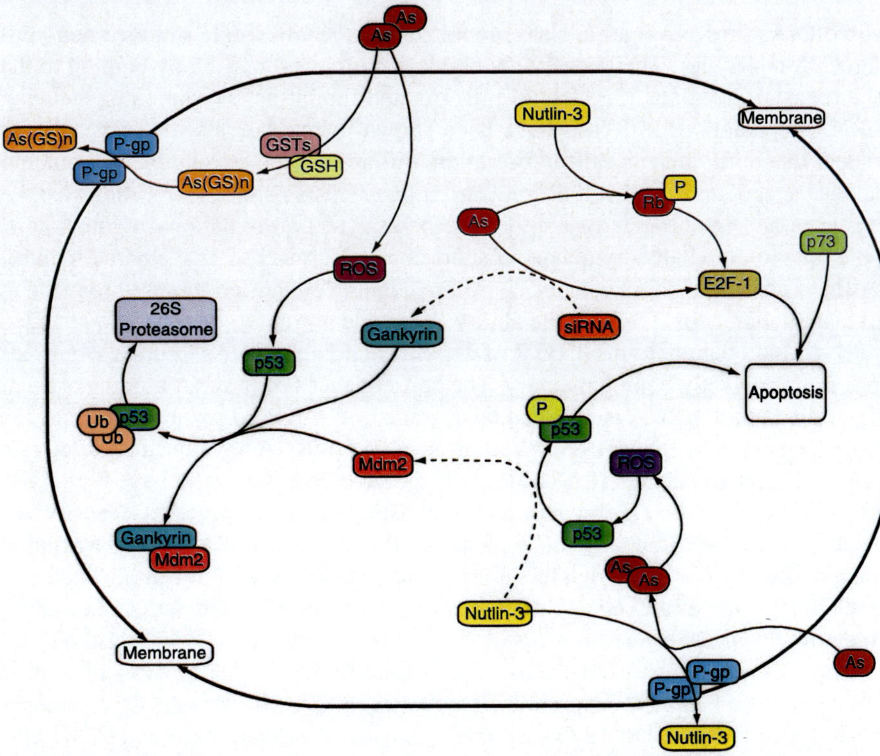

Fig. 2.2 The probable mechanisms of Arsenic trioxide (As_2O_3) in solid tumor cells (e.g., HepG2/As) and the role of Nutlin-3 or siRNA gankyrin in overcoming this resistance

understanding of arsenic resistance in solid tumor cells may provide novel targets for treating arsenic-resistant tumors and promote screening of suitable patients. To evaluate whether inhibition of transporters or activation of p53 enhance the sensitivity to As_2O_3 in liver cancer cells, Chen et al. (2009) used siRNA against gankyrin and molecular inhibitor nutlin-3 to reactivate p53. It was found the down-regulation of gankyrin or nutlin-3 could not inhibit cell proliferation in arsenic resistant cells. The combination of nutlin-3 with As_2O_3 showed significant lethal effect in sensitive and resistant cell lines. Nutlin-3 increased intracellular arsenicals through the inhibition of drug efflux of P-gp and induced significant apoptosis in sensitive and resistant cell lines when combined with As_2O_3. The combination of nutlin-3 and As_2O_3 induced the aggregation of p-p53 (ser392) in nucleus and activated E2F-p73 pathway in resistant cells. This therapeutic alliance showed effective anticancer effect in all the liver cancer cells and could be a more potential way in reversal of drug resistance (unpublished data).

2.6 MicroRNA and As_2O_3 in Cancer Research

MicroRNAs (miRNAs) are a class of endogenous, small, single stranded non-coding RNA molecules. They negatively regulate protein expression by binding to the 3' untranslated regions (3'UTR) of mRNAs and inhibiting translation or inducing mRNA degradation. MiRNAs have been shown to regulate a wide range of biological functions such as cell proliferation, differentiation, development, signaling pathways, apoptosis, and cell death. Currently, extensive studies have indicated the existence and importance of another mechanism of non-mutational regulation of gene function mediated by means of short non-coding RNA. As the name implies, miRNAs are small RNAs usually 19–23 bp in length or shorter, that are produced in all mammalian cells. Lacking the ability to encode a protein, these single-stranded miRNAs bind mainly to the 3'UTR of protein encoding mRNAs through sequences that are imperfectly complementary. The consequences of miRNA binding are that either the bound mRNA is silenced or degraded, resulting in reduced levels of the protein encoded by the mRNA. Aberrant levels of miRNA have been reported in a variety of human cancers (Cho 2010a). They have been shown to have both diagnostic and prognostic significance and to constitute a novel target for cancer treatment (Cho 2010b). Recently the evidence of the roles for miRNAs in determining drug sensitivity/resistance has been emerging (Bartel 2004; Lim et al. 2005; Lu et al. 2005; Calin and Croce 2006; Sevignani et al. 2006; Bushati and Cohen 2007; Barbarotto et al. 2008).

In a previous study, Meng et al. (2010) used miRNA microarrays and studied the alteration of miRNA expression profile in HCC cells after As_2O_3 treatment. Their results showed that As_2O_3 did alter specific miRNA expression in HCC cells, among which we found *miR-29a* might play an important role. Furthermore, they predicted and verified target genes of *miR-29a* which might be involved in the mechanisms of As_2O_3 therapy. In this study, treatment of HepG2 cells with As_2O_3

increased the expression of *miR-29a* and also down-regulated the expression of its target gene *PPM1D*, and the anti-*miR-29a* inhibitor suppressed the effect of As_2O_3 on *PPM1D*, which indicated that one of the important pathways of As_2O_3 on cancer cells is As_2O_3 → *miR-29a* → *PPM1D* → *Wip-1* → *p53* → cancer cell growth inhibition and apoptosis. It is exciting to speculate that *miR-29a* may represent the first non-coding miRNA to be related to As_2O_3 treatment in cancers. High level of *miR-29a* may render the cancer cells more susceptibility to As_2O_3 treatment (Meng et al. 2010). *MiR-21*, one of the most prominent miRNAs in the genesis and progression of many human cancers, has been rarely characterized in myelogenous leukemia. As_2O_3 was successfully used in the treatment of APL. However, cytotoxicity or insensitivity is a major concern in the successful treatment of leukemia. Using a specific precursor pre-*miR-21* or anti-*miR-21* oligonucleotide (AMO-*miR-21*), a group studied the sensitivity of HL60 and K562 cells to As_2O_3. It was found that there was somewhat synergistic effect of AMO-*miR-21* and As_2O_3 in growth inhibition and apoptosis promotion. Meanwhile, enforced pre-*miR-21* expression increased resistance to As_2O_3, nevertheless not affecting cell growth alone. So, this suggested *miR-21* by targeting *PDCD4* may play a functional role in modulating As_2O_3-induced cell death, and strategy using AMO-*miR-21* and its combination with As_2O_3 may be useful as a myelogenous leukemia therapy. Further study suggested that AMO-*miR-21* sensitized leukemic K562 cells to As_2O_3 by inducing apoptosis partially due to its up-regulation of PDCD4 protein level. The combination of As_2O_3 and As_2O_3-*miR-21* presents therapeutic potential for CML (Gu et al. 2011; Li et al. 2010).

2.7 As_2O_3 and Cancer Stem Cells

The existence of a small population of "cancer-initiating cells" is responsible for tumor maintenance has been firmly demonstrated in leukemia. This concept is currently being tested in solid tumors. Leukemia-initiating cells, particularly those that are in a quiescent state, are thought to be resistant to chemotherapy and targeted therapies, resulting in disease relapse (Ito et al. 2008). AML is a stem cell disease. The inefficient targeting of the leukemic stem cells is considered responsible for relapse after the induction of complete hematologic remission in AML. APL is a subtype of AML characterized by the t(15;17) translocation and expression of the PML/RARα fusion protein. Treatment of APL with ATRA induces complete hematologic remission, but not molecular remission (CMR), because the fusion transcript remains detectable, followed by relapse within a few months. Arsenic induces high rates of complete hematologic remission and CMR followed by a long relapse-free survival. After comparing the effects of ATRA and arsenic on PML/RARα-positive stem cell compartments, it was found that in contrast to ATRA, arsenic abolishes the aberrant stem cell capacity of PML/RARα-positive stem cells. Arsenic had no apparent influence on the proliferation of PML/RARα-positive stem cells, whereas ATRA greatly increased the proliferation of these cells. Furthermore, ATRA induces proliferation of APL-derived stem cells, whereas arsenic inhibits their growth. These data suggest a relationship between the capacity of a compound to target the leukemia-initiating cell and its ability to induce long relapse-free survival (Zheng et al. 2007). Cancer stem-like cells are potential targets for treatment of glioblastoma multiforme due to their role in tumorigenesis and recurrence. In a recent study, the inhibitory effect of As_2O_3 on cancer stem-like cells (CSLCs) of glioblastoma multiforme was investigated in human glioma cell lines (U87MG, U251MG, and U373MG) *in vivo* and *in vitro*. Immuno-fluorescence staining and flow cytometry revealed that the percentage of Nestin-positive cells in the aforementioned cell lines was diminished by 12%, 14%, and 7%, respectively, after treatment with 2 μM As_2O_3. Furthermore, they used soft-agar in U87MG and tumor xenografts in nude mice to demonstrate the ability of As_2O_3 to inhibit the formation of tumor in the three cell lines. These results indicate the negative regulation of CSLCs by As_2O_3. In addition, a Western blot analysis revealed decreased levels of Notch1 and Hes1 proteins due to As_2O_3 treatment. It was concluded that As_2O_3 has a remarkable inhibitory effect on CSLCs in glioma cell lines *in vivo* and *in vitro*; in addition, the mechanism of CSLC inhibition involves the deregulation of Notch activation (Zhen et al. 2010).

2.8 Synergy of Arsenic with Other Agents

Because of the many pathways involved in mediating the effects of arsenic, the potential exists for synergism with other agents to provide enhanced therapeutic benefits. As mentioned earlier, As_2O_3 shows effects distinct from those of ATRA in APL cells and has clinical efficacy in patients with ATRA-resistant APL. The combination of ATRA and As_2O_3 may be synergistic or antagonistic *in vitro*, whereas *in vivo* the combination or sequential use of the agents has been reported to accelerate tumor regression by enhancing both differentiation and apoptosis in some but not all of the models (Zhou et al. 2007). Furthermore, combination therapy may allow for administration of lower doses of As_2O_3, minimizing toxicity and potential drug antagonism.

Observations that perturbations in cellular methyl metabolism modulate the cytotoxicity of arsenic have led to the suggestion that methotrexate may act synergistically with As_2O_3. The combination of As_2O_3 with IFN-α has activity in adult T cell leukemia cell lines *in vivo* and *in vitro* (Mahieux and Hermine 2005). Potential synergy of As_2O_3 and vitamin C has been shown *in vitro* and *in vivo* by several groups (Berenson et al. 2007; Subbarayan et al. 2007), and additional studies of the mechanism both *in vitro* and in the clinic appear warranted. Ascorbic acid is not the only agent affecting intracellular redox to synergize with arsenic *in vitro*. The profound increases in arsenic sensitivity *in vitro* associated with glutathione depletion by buthionine sulfoximine suggest a need for experiments in animal models using this or related agents as well (Han et al. 2008).

Enhancement of radiation response is possible to be achieved by combination treatment with therapeutic drugs. Previously, it has been shown that As_2O_3 in combination with ionizing radiation enhanced radiation response in cervical cancer cells. In a study, molecular mechanism of synergistic enhancement of radiation response in combination with As_2O_3 was further investigated. The combination treatment of HeLa cells induced translocation of Bax to the mitochondria and a marked phosphorylation of Bcl-2. p38 MAPK and JNK were also found to be activated in response to the combination treatment. Pre-treatment of PD169316, a p38 MAPK specific inhibitor, completely attenuated the combination treatment-induced mitochondrial relocalization of Bax, and altered Bcl-2 phosphorylation. Moreover, pre-treatment of SP600125, JNK specific inhibitor, clearly attenuated Bcl-2 phosphorylation, but did not affect Bax translocation to the mitochondria. In addition, N-acetyl-L-cysteine, a thiol-containing antioxidant, completely blocked p38 MAPK and JNK activations, Bax relocalization, and Bcl-2 phosphorylation. These results indicate that activation of p38 MAPK is specifically required for translocation of Bax to the mitochondria, and both JNK and p38 MAPK are involved in phosphorylation of Bcl-2 in response to combination treatment with gamma-radiation and As_2O_3, and that ROS is a critical regulator of p38 MAPK and JNK activations. The molecular mechanism elucidated in this study may provide insight into the design of future combination cancer therapies to cells intrinsically less sensitive to radiation treatment (Chun et al. 2002; Kang and Lee 2008).

2.9 Safety and Tolerability

Arsenic is well known as a toxic agent. Inorganic arsenic has been classified by the US Department of Health and Human Services, the International Agency for Research on Cancer, and the US Environmental Protection Agency as a known carcinogen. Chronic arsenic exposure can result in a variety of skin manifestations including hyperpigmentation, keratosis, squamous cell carcinoma, and bowenoid lesions. Other potential signs of arsenic poisoning include peripheral neuropathy, cardiomyopathy, and renal failure. Despite its reputation as a poison, as a therapeutic entity As_2O_3 has been generally well tolerated. When administered intravenously at a dosage of 0.15 mg/kg/day, leukocytosis, gastrointestinal disorders (e.g. nausea, vomiting, and diarrhea), fatigue, fever, headache, cough, and dyspnea are commonly observed. Common potentially serious toxicities include APL differentiation syndrome (APLDS) and electrocardiogram (ECG) abnormalities. Previously known as "retinoic-acid syndrome," APLDS may present during remission induction with ATRA or As_2O_3 therapies as a complex of signs and symptoms, including fever, dyspnea, hypotension, weight gain, acute renal failure, and lung infiltrates, and is usually treated with high-dose corticosteroids. In the pivotal trial of As_2O_3 in APL, APLDS occurred in 10 (25%) patients; in three of these patients, APLDS was considered to be serious. Therapy with As_2O_3 was briefly interrupted (for 1–5 days) in eight patients. Notably, all patients affected by APLDS achieved a CR. A similar incidence and severity of APLDS was reported with combined ATRA and As_2O_3; in the MD Anderson Phase II trial, 13 patients (16%) developed APLDS, and all cases were successfully managed by withholding ATRA and administering corticosteroids.

In a study comparing As_2O_3, ATRA, and the combination, the incidence of APLDS-associated hyperleukocytosis was not increased with As_2O_3 plus ATRA; however, one death in the combination arm was attributed to APLDS. ECG abnormalities, including prolonged QT interval and complete atrioventricular block, have been reported with As_2O_3 treatment. QT prolongation (defined as ≥450 ms for males and ≥470 ms for females) was seen in 63% of patients in the pivotal trial and led to temporary discontinuation of As_2O_3 therapy in one patient (3%). In the Phase II study by Ravandi et al. (2009), As_2O_3 was discontinued in five patients (6%) due to adverse cardiac events including a trial arrhythmias and myocardial infarction. ECG and electrolyte monitoring is recommended prior to and during arsenic therapy. Serum potassium levels should be kept above 4 meq/l and magnesium concentrations above 1.8 mg/dL (Soignet et al. 2001; Shen et al. 2004; Hu et al. 2009; Ravandi et al. 2009).

Liver function test abnormalities can commonly occur and typically cause hepatitis, with increases in aminotransferases (rarely greater than five times normal) beginning about 5–10 days after drug administration. Peripheral neuropathy in a "glove and stocking" distribution can occur in up to 10% of patients. Carcinogenesis is a major concern associated with long-term exposure to arsenic. In the study published in PNAS in 2008 on long-term efficacy and safety of ATRA/As_2O_3-based therapy in newly diagnosed APL patients who started being accrued to the trial in 2001, no secondary carcinoma (including skin cancer) was observed. One male transiently tested positive for carcinoembryogenic antigen, and a mild unsustained increase in CA125 in a female patient was recorded. Moreover, arsenic concentrations in the urine of patients who had ceased As_2O_3 treatment for 24 months were below the safety limits recommended by government agencies in several countries or regions. Interestingly, As_2O_3 therapy has been associated with frequent varicella zoster reactivation, which developed in 26% of patients within a year of treatment with no preferential dermatomal involvement or systemic spread. Another study reported an increased incidence of herpes simplex infection after As_2O_3 therapy. This increased incidence of viral infections is thought to be related to apoptosis of T helper lymphocytes by As_2O_3. Finally, there is concern that chronic exposure to arsenic could increase the risk of secondary malignancies as a result of DNA damage. An increased number of structural chromosomal abnormalities have been seen in rats chronically exposed to arsenic. A report of solid tumors (including nasopharyngeal carcinoma and colonic adenocarcinoma) developing after treatment with As_2O_3 has elevated this concern, although retrospective analysis suggested that these cancers may have been present prior to or shortly after the start of As_2O_3 therapy (Soignet et al. 2001; Shen et al. 2004; Hu et al. 2009; Ravandi et al. 2009).

2.10 Conclusions and Future Perspectives

The therapeutic uses and potential of arsenic-containing compounds have been evolving over centuries, starting with the empiric use of arsenic in ancient times up to the current United States Food and Drug Administration approval of As_2O_3 for the treatment of APL in humans. Despite the well-known toxicities and side effects of arsenic compounds, the prospects for arsenic use in the treatment of human diseases remain high. The evolution of our understanding of how arsenic mediates biological responses over the last decade has led to new studies aimed at establishing conditions for the selective enhancement of its antitumor properties *in vitro* and *in vivo*. It is possible that the next phase in the medical use of arsenic compounds will involve selective applications to malignancies with distinct molecular profiles that define arsenic sensitivity and/or combinations with other agents that target cellular pathways which negatively control arsenic responses. The usefulness of such approaches remains to be established over the next several years. Independently of the outcome of such studies and on the basis of historical considerations and the ongoing evolution in the field, one can argue with a degree of certainty that arsenicals will continue to be the focus of intense research investigations in the near and distant future.

References

Abou-Jawde, R. M., Reed, J., Kelly, M., et al. (2006). Efficacy and safety results with the combination therapy of arsenic trioxide, dexamethasone, and ascorbic acid in multiple myeloma patients: A phase 2 trial. *Medical Oncology, 23,* 263–272.

Andrew, B., Daniel, H., & Tim, M. (2004). Systemic treatment and liver transplantation for hepatocellular carcinoma: Two ends of the therapeutic spectrum. *The Lancet Oncology, 5,* 409–501.

Antman, K. H. (2001). Introduction: The history of arsenic trioxide in cancer therapy. *The Oncologist, 6,* 1–2.

Bael, T. E., Peterson, B. L., & Gollob, J. A. (2008). Phase II trial of arsenic trioxide and ascorbic acid with temozolomide in patients with metastatic melanoma with or without central nervous system metastases. *Melanoma Research, 18,* 147–151.

Bahlis, N. J., McCafferty-Grad, J., Jordan-McMurry, I., et al. (2002). Feasibility and correlates of arsenic trioxide combined with ascorbic acid-mediated depletion of intracellular glutathione for the treatment of relapsed/refractory multiple myeloma. *Clinical Cancer Research, 8,* 3658–3668.

Barbarotto, E., Schmittgen, T. D., & Calin, G. A. (2008). MicroRNAs and cancer: Profile, profile, profile. *International Journal of Cancer. Journal international du cancer, 122,* 969–977.

Bartel, D. P. (2004). MicroRNAs: Genomics, biogenesis, mechanism, and function. *Cell, 116,* 281–297.

Berenson, J. R., Boccia, R., Siegel, D., et al. (2006). Efficacy and safety of melphalan, arsenic trioxide and ascorbic acid combination therapy in patients with relapsed or refractory multiple myeloma: a prospective, multicentre, phase II, single-arm study. *Br J Haemato, 135,*174–183.

Berenson, J. R., Matous, J., Swift, R. A., et al. (2007). A phase I/II study of arsenic trioxide/bortezomib/ascorbic acid combination therapy for the treatment of relapsed or refractory multiple myeloma. *Clinical Cancer Research, 13,* 1762–1768.

Bernstam, L., & Nriagu, J. (2000). Molecular aspects of arsenic stress. *Journal of Toxicology and Environmental Health. Part B, Critical Reviews, 3,* 293–322.

Bushati, N., & Cohen, S. M. (2007). microRNAs functions. *Annual Review of Cell and Developmental Biology, 23,* 175–205.

Calin, G. A., & Croce, C. M. (2006). MicroRNA signatures in human cancers. *Nature Reviews. Cancer, 6,* 857–866.

Chan, P. C., & Huff, J. (1997). Arsenic carcinogenesis in animals and in humans: Mechanistic, experimental, and epidemiological evidence. *Journal of Environmental Science and Health. Part C, Environmental Carcinogenesis & Ecotoxicology Reviews, 15,* 83–122.

Chang, J. E., Voorhees, P. M., Kolesar, J. M., et al. (2009). Phase II study of arsenic trioxide and ascorbic acid for relapsed or refractory lymphoid malignancies: A Wisconsin oncology network study. *Hematological Oncology, 27,* 11–16.

Chen, G. Q., Zhu, J., Shi, X. G., et al. (1996). In vitro studies on cellular and molecular mechanisms of arsenic trioxide (As_2O_3) in the treatment of acute promyelocytic leukemia: As_2O_3 induces NB4 cell apoptosis with downregulation of Bcl-2 expression and modulation of PML-RAR alpha/PML proteins. *Blood, 88,* 1052–1061

Chen, G. Q., Shi, X. G., Tang, W., et al. (1997). Use of arsenic trioxide (As_2O_3) in the treatment of acute promyelocytic leukemia (APL): As_2O_3 exerts dose-dependent dual effects on APL cells. *Blood, 89,* 3345–3353.

Chen, X., Zhang, M., & Liu, L. X. (2009). The overexpression of multidrug resistance-associated proteins and gankyrin contribute to arsenic trioxide resistance in liver and gastric cancer cells. *Oncology Reports, 22,* 73–80.

Cho, W. C. (2010a). MicroRNAs in cancer—From research to therapy. *Biochimica et biophysica acta, 1805,* 209–217.

Cho, W. C. (2010b). MicroRNAs: Potential biomarkers for cancer diagnosis, prognosis and targets for therapy. *The International Journal of Biochemistry & Cell Biology, 42,* 1273–1281

Chun, Y. J., Park, I. C., Park, M. J., et al. (2002). Enhancement of radiation response in human cervical cancer cells in vitro and in vivo by arsenic trioxide (As_2O_3). *FEBS Letters, 519,* 195–200.

Cutler, E. G., & Bradford, E. H. (1878). Action of iron, cod-liver oil, and arsenic on the globular richness of the blood. *The American Journal of the Medical Sciences, 75,* 74–84

Dai, J., Weinberg, R. S., Waxman, S., et al. (1999). Malignant cells can be sensitized to undergo growth inhibition and apoptosis by arsenic trioxide through modulation of the glutathione redox system. *Blood, 93,* 268–277

Davison, K., Mann, K. K., & Miller, W. H., Jr. (2002). Arsenic trioxide: Mechanisms of action. *Seminars in Hematology, 39,* 3–7.

Deeley, R. G., Westlake, C., & Cole, S. P. (2006). Transmembrane transport of endo- and xenobiotics by mammalian ATP-binding cassette multidrug resistance proteins. *Physiological Review, 86,* 849–899.

Duesberg, P., Li, R., Sachs, R., Fabarius, A., et al. (2007). Cancer drug resistance: The central role of the karyotype. *Drug Resistance Updates, 10,* 51–58.

Dyck, J. A., Maul, G. G., Miller, W. H., Jr., et al. (1994). A novel macromolecular structure is a target of the promyelocyte-retinoic acid receptor oncoprotein. *Cell, 76,* 333–343.

Ejendal, K. F., & Hrycyna, C. A. (2002). Multidrug resistance and cancer: The role of the human ABC transporter ABCG2. *Current Protein & Peptide Science, 3,* 503–511.

Emadi, A., & Gore, S. D. (2010). Arsenic trioxide—An old drug rediscovered. *Blood Reviews, 24*(4–5), 191–199.

Filippova, M., & Duerksen-Hughes, P. J. (2003). Inorganic and dimethylated arsenic species induce cellular p53. *Chemical Research in Toxicology, 16,* 423–431.

Fojo, T. (2007). Multiple paths to a drug resistance phenotype: Mutations, translocations, deletions and amplification of coding genes or promoterregions, epigenetic changes and microRNAs. *Drug Resistance Updates, 10,* 59–67.

Glasspool, R. M., Teodoridis, J. M., & Brown, R. (2006). Epigenetics as a mechanism driving polygenic clinical drug resistance. *British Journal of Cancer, 94,* 1087–1092.

Grimwade, D., Mistry, A. R., Solomon, E., et al. (2009). Acute promyelocytic leukemia: A paradigm for differentiation therapy. *Cancer Treatment and Research, 145,* 219–235.

Gu, J., Zhu, X., Li, Y., et al. (2011). MiRNA-21 regulates arsenic-induced anti-leukemia activity in myelogenous cell lines. *Medical Oncology, 28,* 211–218.

Haller, J. S. (1975). Therapeutic mule: The use of arsenic in the nineteenth century material medica. *Pharmacy in History, 17,* 87–100.

Han, Y. H., Kim, S. Z., Kim, S. H., et al. (2008). Induction of apoptosis in arsenic trioxide-treated lung cancer A549 cells by buthionine sulfoximine. *Molecules and Cells, 26,* 158–164.

Higashitsuji, H., Higashitsuji, H., Itoh, K., et al. (2005). The oncoprotein gankyrin binds to MDM2/HDM2, enhancing ubiquitylation and degradation of p53. *Cancer Cell, 8,* 75–87.

Hu, J., Liu, Y. F., Wu, C. F., et al. (2009). Long-term efficacy and safety of all-trans retinoic acid/arsenic trioxide-based therapy in newly diagnosed acute promyelocytic leukemia. *Proceedings of the National Academy of Sciences of the United States of America, 106,* 3342–3347.

Huilgol, N. G. (2006). A phase I study to study arsenic trioxide with radiation and hyperthermia in advanced head and neck cancer. *Int J Hyperthermia, 22,* 391–397.

Hussein, M. A., Saleh, M., Ravandi, F., et al. (2004). Phase 2 study of arsenic trioxide in patients with relapsed or refractory multiple myeloma. *British Journal of Haematology, 125,* 470–476.

Ito, K., Bernardi, R., Morotti, A., et al. (2008). PML targeting eradicates quiescent leukaemia-initiating cells. *Nature, 453,* 1072–1078.

Iwasa, Y., Nowak, M. A., & Michor, F. (2006). Evolution of resistance during clonal expansion. *Genetics, 172,* 2557–2566.

Jackson, R., & Grainge, J. W. (1975). Arsenic and cancer. *Canadian Medical Association Journal, 113,* 396–401.

Jiang, X. H., Wong, B. C., Yuen, S. T., et al. (2001). Arsenic trioxide induces apoptosis in human gastric cancer cells through up-regulation of p53 and activation of caspase-3. *International Journal of Cancer. Journal international du cancer, 91,* 173–179.

Jing, Y., Wang, L., Xia, L., et al. (2001). Combined effect of all-trans retinoic acid and arsenic trioxide in acute promyelocytic leukemia cells in vitro and in vivo. *Blood, 97,* 264–269.

Jutooru, I., Chadalapaka, G., Sreevalsan, S., et al. (2010). Arsenic trioxide downregulates specificity protein (Sp) transcription factors and inhibits bladder cancer cell and tumor growth. *Experimental Cell Research, 316,* 2174–2188.

Kala, S. V., Neely, M. W., Kala, G., et al. (2000). The MRP2/cMOAT transporter and arsenic-glutathione complex formation are required for biliary excretion of arsenic. *The Journal of Biological Chemistry, 275,* 33404–33408.

Kang, Y. H., & Lee, S. J. (2008). Role of p38 MAPK and JNK in enhanced cervical cancer cell killing by the combination of arsenic trioxide and ionizing radiation. *Oncology Reports, 20,* 637–643.

Kapahi, P., Takahashi, T., Natoli, G., et al. (2000). Inhibition of NF-kappa B activation by arsenite through reaction with a critical cysteine in the activation loop of Ikappa B kinase. *The Journal of Biological Chemistry, 275,* 36062–36066.

Kchour, G., Tarhini, M., Kooshyar, M. M., et al. (2009). Phase 2 study of the efficacy and safety of the combination of arsenic trioxide, interferon alpha, and zidovudine in newly diagnosed chronic adult T-cell leukemia/lymphoma (ATL). *Blood, 113,* 6528–6532.

Keppler, D., Leier, I., Jedlitschky, G., et al. (1998). ATP-dependent transport of glutathione S-conjugates by the multidrug resistance protein MRP1 and its apical isoform MRP2. *Chemico-Biological Interactions, 111–112,* 153–161.

Kim, K. B., Bedikian, A. Y., Camacho, L. H., et al. (2005). A phase II trial of arsenic trioxide in patients with metastatic melanoma. *Cancer, 104,* 1687–1692.

Kindler, H. L., Aklilu, M., Nattam, S., et al. (2008). Arsenic trioxide in patients with adenocarcinoma of the pancreas refractory to gemcitabine: A phase II trial of the University of Chicago phase II consortium. *American Journal of Clinical Oncology, 31,* 553–556.

Kwong, Y. L., & Todd, D. (1997). Delicious poison: Arsenic trioxide for the treatment of leukemia. *Blood, 89,* 3487–3488.

Lee, P. C., Kakadiya, R., Su, T. L., et al. (2010). Combination of bifunctional alkylating agent and arsenic trioxide synergistically suppresses the growth of drug-resistant tumor cells. *Neoplasia, 12,* 376–387.

Lehnert, M. (1998). Chemotherapy resistance in breast cancer. *Anticancer Research, 18,* 2225–2226.

Li, Y., Zhu, X., Gu, J., et al. (2010). Anti-miR-21 oligonucleotide sensitizes leukemic K562 cells to arsenic trioxide by inducing apoptosis. *Cancer Science, 101,* 948–954.

Lim, L. P., Lau, N. C., Garrett-Engele, P., et al. (2005). Microarray analysis shows that some microRNAs downregulate large numbers of target mRNAs. *Nature, 433,* 769–773.

Lin, C. C., Hsu, C., Hsu, C. H., et al. (2007). Arsenic trioxide in patients with hepatocellular carcinoma: A phase II trial. *Investigational New Drugs, 25,* 77–84.

List, A., Beran, M., DiPersio, J., et al. (2003). Opportunities for Trisenox® (Arsenic trioxide) in the treatment of myelodysplastic syndromes. *Leukemia, 17,* 1499–1507.

Lu, M., Levin, J., Sulpice, E., et al. (1999). Effect of arsenic trioxide on viability, proliferation, and apoptosis in human megakaryocytic leukemia cell lines. *Experimental Hematology, 27,* 845–852.

Lu, J., Getz, G., Miska, E. A., et al. (2005). MicroRNA expression profiles classify human cancers. *Nature, 435,* 834–838.

Mahieux, R., & Hermine, O. (2005). In vivo and in vitro treatment of HTLV-1 and HTLV-2 infected cells with arsenic trioxide and interferon-alpha. *Leukemia & Lymphoma, 46,* 347–355.

Mantadakis, E., Samonis, G., & Kalmanti, M. (2008). A comprehensive review of acute promyelocytic leukemia in children. *Acta Haematologica, 119,* 73–82.

Meng, X. Z., Zheng, T. S., Chen, X., et al. (2010). MicroRNA expression alteration after arsenic trioxide treatment in HepG-2 cells. *Journal of Gastroenterology and Hepatology*, in press.

Miller, W. H., Jr., Schipper, H. M., Lee, J. S., et al. (2002). Mechanisms of action of arsenic trioxide. *Cancer Research, 62,* 3893–3903.

Munshi, N. C. (2001). Arsenic trioxide: An emerging therapy for multiple myeloma. *The Oncologist, 6,* 17–21.

Munshi, N. C., Tricot, G., Desikan, R., et al. (2002). Clinical activity of arsenic trioxide for the treatment of multiple myeloma. *Leukemia, 16,* 1835–1837.

Nishikawa, M. (2008). Reactive oxygen species in tumor metastasis. *Cancer Letters, 266,* 53–59.

Novick, S. C., & Warrell, R. P., Jr. (2000). Arsenicals in hematologic cancers. *Seminars in Oncology, 27,* 495–501.

Park, J. W., Choi, Y. J., Jang, M. A., et al. (2001). Arsenic trioxide induces G2/M growth arrest and apoptosis after caspase-3 activation and bcl-2 phosphorylation in promonocytic U937 cells. *Biochemical and Biophysical Research Communications, 286,* 726–734.

Parmar, S., Rundhaugen, L. M., Boehlke, L., et al. (2004). Phase II trial of arsenic trioxide in relapsed and refractory acute myeloid leukemia, secondary leukemia and/or newly diagnosed patients at least 65 years old. *Leukemia Research, 28,* 909–919.

Perkins, C., Kim, C. N., Fang, G., et al. (2000). Arsenic induces apoptosis of multidrug-resistant human myeloid leukemia cells that express Bcr-Abl or overexpress MDR, MRP, Bcl-2, or Bcl-xL. *Blood, 95,* 1014–1022.

Ravandi, F., Estey, E., Jones, D., et al. (2009). Effective treatment of acute promyelocytic leukemia with all-trans-retinoic acid, arsenic trioxide, and gemtuzumab ozogamicin. *Journal of Clinical Oncology, 27,* 504–510.

Roberti, A., La Sala, D., & Cinti, C. (2006). Multiple genetic and epigenetic interacting mechanisms contribute to clonally selection of drug-resistant tumors: Current views and new therapeutic prospective. *Journal of Cell Physiology, 207,* 571–581.

Roboz, G. J., Dias, S., Lam, G., et al. (2000). Arsenic trioxide induces dose- and time-dependent apoptosis of endothelium and may exert an antileukemic effect via inhibition of angiogenesis. *Blood, 96,* 1525–1530.

Roboz, G. J., Ritchie, E. K., Curcio, T., et al. (2008). Arsenic trioxide and low-dose cytarabine in older patients with untreated acute myeloid leukemia, excluding acute promyelocytic leukemia. *Cancer, 113,* 2504–2511.

Schläwicke Engström, K., Broberg, K., Concha, G., et al. (2007). Genetic polymorphisms influencing arsenic metabolism: Evidence from Argentina. *Environmental Health Perspectives, 115,* 599–605.

Sears, D. A. (1988). History of the treatment of chronic myelocytic leukemia. *The American Journal of the Medical Sciences, 296,* 85–86.

Sen, C. K. (1998). Redox signaling and the emerging therapeutic potential of thiol antioxidants. *Biochemical Pharmacology, 55,* 1747–1748.

Sevignani, C., Calin, G. A., Siracusa, L. D., et al. (2006). Mammalian micro-RNAs: A small world for fine-tuning gene expression. *Mammalian Genome, 17,* 189–202.

Shen, Z. X., Chen, G. Q., Ni, J. H., et al. (1997). Use of arsenic trioxide (As_2O_3) in the treatment of acute promyelocytic leukemia (APL): II. Clinical efficacy and pharmacokinetics in relapsed patients. *Blood, 89,* 3354–3360.

Shen, Z. X., Shi, Z. Z., Fang, J., et al. (2004). All-trans retinoic acid/As_2O_3 combination yields a high quality remission and survival in newly diagnosed acute promyelocytic leukemia. *Proceedings of the National Academy of Sciences of the United States of America, 101,* 5328–5335.

Soignet, S. L., Maslak, P., Wang, Z.-G., et al. (1998). Complete remission after treatment of acute promyelocytic leukemia with arsenic trioxide. The *New England Journal of Medicine, 339,* 1341–1348.

Soignet, S. L., Frankel, S. R., Douer, D., et al. (2001). United States multicenter study of arsenic trioxide in relapsed acute promyelocytic leukemia. *Journal of Clinical Oncology, 19,* 3852–3860.

Subbarayan, P. R., Lima, M., & Ardalan, B. (2007). Arsenic trioxide/ascorbic acid therapy in patients with refractory metastatic colorectal carcinoma: A clinical experience. *Acta Oncologica, 46,* 557–561.

Subbarayan, P. R., Lee, K., & Ardalan, B. (2010). Arsenic trioxide suppresses thymidylate synthase in 5-FU-resistant colorectal cancer cell line HT29 in vitro re-sensitizing cells to 5-FU. *Anticancer Research, 30,* 1157–1162.

Sun, H. D., Ma, L., Hu, Z., et al. (1992). Arsenic trioxide treated 32 cases of acute promyelocytic leukemia. *Chinese Journal of Integrated Traditional and Western Medicine, 12,* 170–172.

Tarhini, A. A., Kirkwood, J. M., Tawbi, H., et al. (2008). Safety and efficacy of arsenic trioxide for patients with advanced metastatic melanoma. *Cancer, 112,* 1131–1138.

Tingting, R., Wei, G., Changliang, P., et al. (2010). Arsenic trioxide inhibits osteosarcoma cell invasiveness via MAPK signaling pathway. *Cancer Biology & Therapy, 10,* 251–257.

Vuky, J., Yu, R., Schwartz, L., et al. (2002). Phase II trial of arsenic trioxide in patients with metastatic renal cell carcinoma. *Investigational New Drugs, 20,* 327–330.

Wang, Z. Y., & Chen, Z. (2008). Acute promyelocytic leukemia: From highly fatal to highly curable. *Blood, 111,* 2505–2515.

Wang, Z. G., Ruggero, D., Ronchetti, S., et al. (1998). PML is essential for multiple apoptotic pathways. *Nature Genetics, 20,* 266–272.

Waxman, S., & Anderson, K. C. (2001). History of the development of arsenic derivatives in cancer therapy. *The Oncologist, 6,* 3–10.

Wu, D. D., Xiao, Y. F., Geng, Y., et al. (2010). Antitumor effect and mechanisms of arsenic trioxide on subcutaneously implanted human gastric cancer in nude mice. *Cancer Genetics and Cytogenetics, 198,* 90–96.

Xu, Y., Zhang, Y., Liu, X., et al. (2009). The effects of ultrasound and arsenic trioxide on neurogliocytoma cells and secondary activation of macrophages. *Tumori, 95,* 780–788.

Yang, Y., Li, C. C., & Weissman, A. M. (2004). Regulating the p53 system through ubiquitination. *Oncogene, 23,* 2096–2106.

Yoo, D. R., Chong, S. A., & Nam, M. J. (2009). Proteome profiling of arsenic trioxide-treated human hepatic cancer cells. *Cancer Genomics & Proteomics, 6,* 269–274.

Zhang, P., Wang, S. Y., & Hu, X. H. (1996). Arsenic trioxide treated 72 cases of acute promyelocytic leukemia. *Chinese Journal of Hematology, 17,* 58–62.

Zhang, W., Ohnishi, K., Shigeno, K., et al. (1998). The induction of apoptosis and cell cycle arrest by arsenic trioxide in lymphoid neoplasms. *Leukemia, 12,* 1383–1391.

Zhang, N., Wu, Z. M., McGowan, E., et al. (2009). Arsenic trioxide and cisplatin synergism increase cytotoxicity in human ovarian cancer cells: Therapeutic potential for ovarian cancer. *Cancer Science, 100,* 2459–2464.

Zhao, X. S., Song, P. L., Sun, B., et al. (2009). Arsenic trioxide inhibits metastatic potential of mouse hepatoma H22 cells in vitro and in vivo. *Hepatobiliary & Pancreatic Diseases International: HBPD INT, 8,* 510–517.

Zhen, Y., Zhao, S., Li, Q., et al. (2010). Arsenic trioxide-mediated Notch pathway inhibition depletes the cancer stem-like cell population in gliomas. *Cancer Letters, 292,* 64–72.

Zheng, X., Seshire, A., Rüster, B., et al. (2007). Arsenic but not all-trans retinoic acid overcomes the aberrant stem cell capacity of PML/RARalpha-positive leukemic stem cells. *Haematologica, 92,* 323–331.

Zhou, G. B., Zhang, J., Wang, Z. Y., et al. (2007). Treatment of acute promyelocytic leukaemia with all-trans retinoic acid and arsenic trioxide: A paradigm of synergistic molecular targeting therapy. *Philosophical Transactions of the Royal Society of London. Series B, Biological Sciences, 362,* 959–971.

Zhou, J., Zhang, Y., Li, J., et al. (2010). Single-agent arsenic trioxide in the treatment of children with newly diagnosed acute promyelocytic leukemia. *Blood, 115,* 1697–1702.

Zhu, X. H., Shen, Y. L., Jing, Y. K., et al. (1999). Apoptosis and growth inhibition in malignant lymphocytes after treatment with arsenic trioxide at clinically achievable concentrations. *Journal of the National Cancer Institute, 91,* 772–778.

Chapter 3
An Evidence-based Review of *Astragalus membranaceus* (Astragalus) for Cancer Patients

K. Simon Yeung, Jyothirmai Gubili and Barrie R. Cassileth

Abstract *Astragalus membranaceus* (astragalus), originally described in the Shennong's Classic of Materia Medica two thousand years ago, is used as a Qi-tonifying herb in traditional Chinese medicine. It is an important ingredient in many herbal formulas used to treat a variety of symptoms and ailments including fatigue and rectal prolapse. The root of astragalus is rich in saponins and polysaccharides. Modern research suggests antioxidant, immunomodulatory, and cytostatic properties. Animal and anecdotal human data show that astragalus reduces immunosuppression, a side effect of chemotherapy and it may also enhance the effects of such treatments. Whereas oral and parenteral preparations have been developed in Asia, products containing astragalus are consumed as dietary supplements in the West. Several formulas containing astragalus have been studied in cancer patients. Data indicate that they are safe to use in conjunction with chemotherapy and reduce treatment associated adverse effects. Based on existing evidence, there is also substantial interest in developing astragalus-based preparations for certain cancers. Although all products studied to date contain astragalus as the main ingredient, the variation across formulas makes it difficult to draw definitive conclusions. Future studies should address this issue. Astragalus is generally considered safe for traditional use, but the potential for herb-drug interactions exists because botanicals contain biologically active compounds. This chapter presents information about the use of astragalus in traditional medicine and summarizes existing scientific evidence of its benefits and limitations as an adjuvant cancer treatment.

3.1 Introduction

Many cancer patients use dietary supplements to enhance the likelihood of cure and to control treatment-related symptoms. They are also used to improve various aspects of quality of life and to prevent cancer recurrence (Correa-Velez et al. 2005;

J. Gubili (✉)
Integrative Medicine Service, Memorial Sloan-Kettering Cancer Center, 1429 First Avenue, New York, NY 10021, USA
e-mail: gubilij@mskcc.org

W. C. S. Cho (ed.), *Evidence-based Anticancer Materia Medica,* Evidence-based Anticancer Complementary and Alternative Medicine,
DOI 10.1007/978-94-007-0526-5_3,

Verhoef et al. 2005; Evans et al. 2007). A majority of herbal supplements, including *Astragalus membranaceus* (astragalus) and formulas containing astragalus, are consumed based on their traditional use. Astragalus may benefit cancer patients in various ways. It demonstrates chemo-preventive effects by enhancing immune defenses against cancer and serves as an adjuvant to cancer treatments. The most common application, however, is to relieve the side effects produced by cancer treatments.

Although astragalus has been used for many centuries, there is a paucity of scientific evidence and only limited data on its safety and efficacy. Earlier studies often were limited to *in vitro*, animal studies and case reports. The culture of publishing only positive results (Vickers et al. 1998) and unsophisticated peer review process often in place cast doubt on the credibility of many of these publications. However, traditional herbal medicine has been subjected to modern evidence-based research to examine safety, to validate indications and for drug development, and as a result, several high quality and well-designed studies and meta-analyses were published during the past decade.

3.2 Applications in Traditional Medicine

Astragalus was originally described in Shennong's Classic of Materia Medica more than two thousand years ago. It is considered to have "sweet" and slightly "warm" properties that enter the "Spleen" and "Lung" channels. The indications for as-

Table 3.1 Astragalus in traditional Chinese medicine and modern medicine

Function according to traditional Chinese medicine (Chen and Chen 2004)	Implications in biomedicine
Tonify the "Spleen" and raise Yang	Treat blood loss and anemia (Chang et al. 2009) Treat adverse effects from chemotherapy and radiotherapy (Cho and Chen 2009)
Tonify Wei (defensive) Qi, consolidate the exterior	Strengthen the immune system (Cho and Leung 2007a, b)
Promote the discharge of pus and generate flesh	Promote wound healing after surgery (Gao et al. 2001; Han et al. 2009; Huh et al. 2009)
Regulate water circulation	Reduce edema caused by poor circulation (Zhang et al. 2006a, b, c)
Relieve numbness and pain	Reduce pain and neuropathic symptoms (Chan et al. 2009; Lu et al. 2010)
Reverse Xiao Ke (wasting and thirst syndrome)	Treat diabetes related symptoms (Zhang et al. 2007; Liu et al. 2010)

tragalus use are numerous. They include deficiency of Qi characterized by lack of strength, anorexia and loose stools; sinking of "Spleen" Qi manifested by chronic diarrhea, rectal prolapse, abnormal uterine bleeding, spontaneous sweating due to weakened superficial resistance, edema due to deficiency of Qi, abscesses that are difficult to burst or heal, anemia, wasting-thirst caused by internal heat, albuminuria in chronic nephritis and diabetes (Pharmacopoeia 2005; Cho 2009) (Table 3.1). In traditional use, astragalus is employed as the chief ingredient in herbal formulas that may contain up to twenty different herbs. It is believed that the herbs mutually enhance the actions or counter side effects. Most modern clinical studies often follow this principle: Herbal formulas containing astragalus, rather than astragalus alone, are employed in clinical trials.

3.3 Botany and Phytochemistry

The Chinese pharmacopeia lists two entries on astragalus—*Radix astragali* (Huangqi), the dried root, and *Radix astragali Praeparata cum Melle* (Zhihuangqi), the honey processed root from *Astragalus membranaceus* of the Leguminosae family. The herb is usually cultivated and collected in spring and autumn. Besides macro- and microscopic identification, astragalus can be standardized based on the amount of astragaloside IV, a saponin constituent (Pharmacopoeia 2005). Astragalus root can be obtained from many different sources. For example, researchers of one study collected 43 astragalus root samples from farms located in the north of China and Mongolia (Tanaka et al. 2008). Genetic fingerprinting can be used to identify the origin (Yip and Kwan 2006). While *Astragalus membranaceus* and *A. membranaceus* var. *mongolicus* are two commonly used species, many other species with similar phenotype are mixed in commercial products. The amount of the main constituents between species can vary greatly which in turn affects the quality and medicinal results (Ma et al. 2002).

Traditional methods of preparing astragalus include boiling slices of the whole root in hot water or soaking the herb in wine. Extracts used in previous studies had been processed by hot water, which yields mainly polysaccharides, or by alcohols, yields saponins. Using advanced chromatography techniques, isoflavonoids, including formononetin, calycosin, astragalosides, and α-(1-4)-D-glucans and other constituents were detected in astragalus root samples (Yu et al. 2005; Huang et al. 2009; Li 2009; Auyeung and Ko 2010).

Each constituent is bioactive and can act synergistically with other components. In evidence-based botanical research, it is important to identify and characterize the constituents for the selection of chemical or bioactivity markers and to help understand the biological effects.

Tragacanth, a polysaccharide gum derived from *A. gummifer*, often is employed in the pharmaceutical industry as an emulsifier and thickening agent, rather than used for its medicinal properties.

3.4 Dosage and Toxicity

3.4.1 Dosage

In herbal medicine, the traditional dosage usually is applied as dose escalation trials are rarely conducted. Astragalus has very low toxicity; therefore, very high doses can be tolerated. The dose of raw astragalus root is 9–30 g (Pharmacopoeia 2005), but a daily dose as high as 90 g was used safely in children with acute leukemia (Dong et al. 2005). Since astragalus products are not standardized based on their bioactivity, no equivalent doses have been established. Astragalus often is used as an immunostimulant, and studies with other immunomodulators show that the dose-curve may peak in a non-linear fashion. In some cases, higher dose may be less effective than lower (Hishida et al. 1988; Deng et al. 2009). Future studies are needed to determine optimal dosage prior to use in clinical trials (Vickers 2006).

3.4.2 Toxicity

Generally, astragalus is considered safe. It has been used for many centuries with no reports of major adverse effects. Animal studies conducted to date have not revealed significant toxicity. Up to 40 g/kg given to rats intraperitoneally produced no adverse reactions. This is the equivalent of 70 times the human dose (Yu et al. 2007).

There are concerns that saponins present in astragalus may have hemolytic activities. However, an *in vitro* study showed that saponins are safe in that regard (Yang et al. 2005). Several other species of astragalus, such as *A. lentiginosus* and *A. lusitanicus*, are known to be toxic to livestock, where cases of poisoning have been reported. These species are called locoweed or milkvetch, and are not used in traditional medicine.

3.5 *In Vitro* and Animal Studies

In vitro and animal studies to date indicate that astragalus and its active constituents have immunomodulatory, antioxidant, and anti-inflammatory effects. Astragalus helps reduce the side effects of chemotherapy probably through its hemopoietic activity and its protective effects on vital organs. It may also help improve quality of life by reducing stress and fatigue.

3.5.1 Immunomodulation

The immune system is long known to be associated with cancer development and progression. For example, lymphocytes and interleukins (ILs) play a major role in immunosurveillance against cancer (Dunn et al. 2004). Many studies have explored the mechanisms responsible for the modulation of immune defenses that help prevent and treat cancer. Astragalus has been shown to function as an immunomodulator. In a study of cultured peripheral blood mononuclear cells from lung cancer patients, it lowered the Th2 cytokine, which is associated with tumor growth (Wei et al. 2003).

A water extract of astragalus root reduced the suppression of cell proliferation induced by methotrexate in mouse spleen cells. It also modulated the expression of *IL-1α* and *IL-12p40* mRNA, which suggests that astragalus may help protect against the immunosuppression caused by chemotherapy (Lee et al. 2003).

In models of murine renal cell carcinoma and murine bladder tumor urological neoplasm, cancer cell survival was shown to be due to suppression of macrophage function, which was reversed by astragalus extract (Rittenhouse et al. 1991). Astragalus extract also increased lymphocyte cytotoxicity by promoting production of IL-2 and γ-IFN (Kurashige et al. 1999).

In another study, astragalus was found to induce lymphokine-activated killer cell activity in patients with cancer and AIDS. It also potentiated the effects of recombinant IL-2 treatment (Chu et al. 1994).

Various fractions of astragalus also were found to enhance immune reaction in mononuclear cells from cancer patients (Chu et al. 1988), to restore immune function in tumor-bearing mice and in mice treated with cyclophosphamide, and restore the lymphocyte blastogenic response (Cho and Leung 2007a, b).

Further, polysaccharides derived from astragalus enhanced the immunomodulatory effects of probiotics in animals by suppressing *E. coli* and by regulating the intestinal flora (Li et al. 2009b). They were shown also to activate mouse B cells membrane Ig in a TLR4-independent manner (Shao et al. 2004), and to stimulate spleen lymphocyte production in rats with stomach cancer (Li et al. 2009a). These mechanisms may help explain the benefits of astragalus in cancer treatment.

Astragalus also has been examined as a candidate for cancer vaccine. Astragalus injection can enhance the anti-metastatic action of mice dendritic cells pre-sensitized by a tumor antigen (Dong and Dong 2005). The ethanol extract of astragalus showed potent activity as an immunologic adjuvant when administered with vaccines of various types (Ragupathi et al. 2008). Further study to identify the components that are responsible for the adjuvant activity found that astragalosides II and IV were the active constituents but the toxicity of these two differed dramatically. Astragaloside IV caused very few side effects in animals and can potentially be used as an immunological adjuvant. Other flavonoids also showed significant adjuvant activity (Hong et al. 2010).

3.5.2 *Inflammation*

Inflammation can be a sign of an activated immune response, but chronic inflammation of the microenvironment has been associated with tumor development (Mantovani et al. 2010). Some studies demonstrated that astragalus exerts its anti-inflammatory effects by suppressing p38 and Erk1/2 through MKP-1 (mitogen-activated protein kinase phosphatase-1) mediation (Ryu et al. 2008). Astragalus also down-regulates the iNOS, P-selectin, and ICAM-1 (inter-cellular adhesion molecule 1) protein expression (Ko et al. 2005).

The regulated expression of adhesion molecules on the surface of endothelial cells is a key process in the pathogenesis of inflammation. Astragaloside IV significantly reduces this adhesion and also inhibits TNF-α in the NF-κB pathway (Zhang et al. 2003).

In a mouse asthma model, astragalus injection reduced inflammatory infiltration and mucus secretion by inhibiting Th2 cytokines (Shen et al. 2008), and reduced chemical induced colonic lesions. Human studies are needed to determine the effectiveness of astragalus in inflammatory conditions.

3.5.3 Antioxidant Effects

Antioxidants are thought to prevent oxidative damage and to exhibit chemo-protective effects although large studies found no benefit of antioxidant supplementation for cancer prevention (Meyer et al. 2008; Lippman et al. 2009). Further research is needed.

Antioxidants may also help protect healthy cells from the side effects of cancer therapies, but evidence is mixed (Lawenda et al. 2008). In a model of oxidative stress-induced endothelial dysfunction, astragalide IV was found to reverse the inhibition of nitric oxide (NO) synthase pathway and to enhance superoxide dismutase activity (Qiu et al. 2010). Of all of its constituents, calycosin showed the most potent antioxidant activity (Yu et al. 2005).

Glutathione is an endogenous antioxidant that helps to reduce free radicals. Astragalus preserves glutathione level (Ko et al. 2005). Danggui Buxue Decoction, a classical formula with astragalus as a major ingredient, was shown to protect cells against oxidant injury by increasing cellular gluthathione level (Chiu et al. 2007). It also affords protection against myocardial ischemia-reperfusion injury in rats in a dose-dependent manner by stimulating myocardial mitochondrial and red blood cell glutathione status (Mak et al. 2006).

3.5.4 Cytostatic and Cytotoxic Effects

In vitro studies indicate that astragalus can halt tumor growth. The active constituent of astragalus, formononetin, was shown to inhibit colon cancer cell growth by facilitating apoptosis through caspase activation and the suppression of Bcl-2 and Bcl-XL proteins (Auyeung and Ko 2010). Similar effects were seen on mesothelial cells as well (Na et al. 2009).

The saponins present in astragalus also show antitumor properties and stimulate apoptosis by up-regulating the *NAG-1* (NSAID-activated gene) (Auyeung et al. 2009), inhibit p21 expression and cyclin-dependent kinase activity that lead to accumulation in S phase and G2/M arrest. They may also promote apoptosis through caspase 3 activation and poly (ADP-ribose) polymerase cleavage (Tin et al. 2007). Astragalus extract induced apoptosis by up-regulating the Apaf-1 (apoptotic protease-activating factor 1), caspase-3, and acetylcholinesterase (Cheng et al. 2004).

Various fractions from astragalus increased cytostatic activity against cancer cells by inducing lymphokine-activated killer-like activity (Cho and Leung 2007a, b). The addition of high doses of astragalus enhanced the activity of 5-fluorouracil against chemically-induced gastric tumors in mice (Zhang et al. 2006a, b, c).

3.5.5 Hematopoiesis

A common adverse effect experienced by cancer patients is anemia due to myelosuppression caused by chemotherapy and radiation therapy. Formulas containing astragalus and *Angelica sinensis* (angelica) are used to improve energy and to enrich the blood in traditional Chinese medicine (TCM) (He et al. 1986). Modern studies indicate that astragalus can increase red blood cell production by stimulating erythroid differentiation (Cheng et al. 2004; Yang et al. 2010).

Danggui Buxue Decoction, when applied to cultured Hep3B human hepatocellular carcinoma cells, induced mRNA expression of erythropoietin in a dose-dependent manner, suggesting a hematopoietic function (Gao et al. 2008). In a mouse model, this combination increased the production of red blood cells and platelets by promoting the growth of megakaryocytes and stromal cells in the bone marrow (Yang et al. 2009).

In another study of rats with cyclophosphamide-induced anemia, a similar formula enhanced blood cell count by increasing erythropoietin mRNA expression (Chang et al. 2009), which suggests that astragalus may play a role in reversing cyclophosphamide-induced anemia.

3.6 Protective Effects on Organs and Tissues

There is evidence that astragalus has protective effects on vital organs such as the liver, kidneys, and heart. Following are a few examples.

3.6.1 Liver

Astragalus has been used to treat liver diseases in TCM, and experimental evidence indicates that its antitumor potential can delay chemically-induced hepatocarcinogenesis in rats (Cui et al. 2003). When used along with *Schisandrae chinensis*, astragalus protected rat liver from chronic injury *via* antioxidant effects (Yan et al. 2009).

Another study in mice also showed the hepatoprotective effects of a combination of *Paeonia lactiflora* and astragalus against injury caused by bacillus Calmette-Guerin and lipopolysaccharide by inhibiting pro-inflammation mediators, TNF-α and IL-1 (Sun et al. 2008).

When used in carbon tetrachloride-treated rats, a raw astragaloside fraction showed antioxidant effects, reduced TNF-α and TGF-β1 activity, and slowed the progression of liver fibrosis (Gui et al. 2006).

The saponins extracted from the root of astragalus also were found to protect the liver from chemical-induced injury in mice (Zhang et al. 1992).

3.6.2 Kidneys

Alpha glucans isolated from astragalus were found to reduce proteinuria in rats with glomerulonephritis, suggesting renoprotective properties (Li 2009). In another study conducted in rats, a combination of astragalus and angelica enhanced the anti-fibrotic effect of enalapril, an ACE inhibitor, in the kidneys. This formula may have a role in treating ureteral obstruction (Wojcikowski et al. 2010). A case study in humans reported similar findings (Ahmed et al. 2007). Astragalus also protects rats from cyclosporine induced kidney damage (El-Kenawy 2010).

A systematic review of animal studies concluded that astragalus can effectively reduce serum sugar level, urine albumin, and improve glomerular filtration (Zhang et al. 2009).

3.6.3 Heart

Astragalus has been used in TCM formulas for cardiovascular diseases. *In vitro*, astragaloside IV improved post-ischemic heart function and ameliorated perfusion arrhythmias. The cardioprotection afforded by astragaloside IV was accompanied by a significant increase in coronary flow both *in vivo* and *in vitro* probably due to astragaloside IV's antioxidative and NO-inducing properties (Zhang et al. 2006a, b, c).

Astragaloside IV also helps dilate aortic vessels through the endothelium-dependent NO and cGMP pathways in a dose-dependent manner. Further, it may play a role in calcium channel blocking and in the inhibition of angiotensin II (Zhang et al. 2006a, b, c).

Anthracyclines, a class of chemotherapy drugs, are known for their cardiotoxicity, which limits their use. But a recent study found that astragalus can protect heart cells from anthracycline-induced cardiotoxicity by exerting antioxidant effects (Luo et al. 2009).

3.6.4 Nervous System

Nerve damage is a common adverse effect associated with chemotherapy. Symptoms can range from peripheral neuropathy to memory loss. Some studies suggest that astragalus has protective effects on the nervous system. For example, astragaloside IV was shown to protect against nerve degeneration (Chan et al. 2009), and astragalus extract helps regenerate nerve fibers (Lu et al. 2010).

In a rat focal cerebral ischemia/reperfusion model, astragaloside IV was found to protect the brain by reducing blood-brain barrier permeability through the regulation of tight junction proteins, occludin and ZO-1 (zonae occludens-1) (Qu et al. 2009).

Buyang Huanwu Decoction, a TCM formula that contains astragalus as its main ingredient, is used for treating stroke-induced disability. In a study with rats, this formula improved neurologic function and reduced the infarction volume in ischemic brains by stimulating the progenitor cells at the hippocampus and sub-ventricular zone. It also increased the number of VEGF-positive and Flk1-positive cells in this region. These results suggest that this formula can assist the recovery of neurologic deficits (Cai et al. 2007).

3.6.5 Bone

In an *in vitro* study, formononetin affected the process of bone remodeling in normal and osteoarthritic osteoblasts (Huh et al. 2010). Whether this means that astragalus plays a role in affecting bone metastasis remains to be examined in future studies.

3.7 Astragalus for Stress and Glycemic Control

In animal studies, astragalus reduced stress-induced anxiety and improved learning and memory (Park et al. 2009). Animals with artificially induced fatigue also displayed improved immune function after being fed astragalus (Kuo et al. 2009). Human studies are needed.

In traditional medicine, astragalus is used for Xiao Ke, a wasting and thirsting syndrome commonly experienced by patients with diabetes. Many cancer patients also have poor glycemic control and experience similar symptoms. Modern research shows that astragalus can reduce the formation of proinflammatory chemicals produced by normal metabolism that can lead to diabetes (Motomura et al. 2009).

In a diabetic mice model, polysaccharides from astragalus reduced hyperglycemia and insulin resistance by regulating insulin signaling in skeletal muscle, and are recommended for use in the treatment of type 2 diabetes (Liu et al. 2010). They also reduced blood glucose, plasma lipid and microalbuminuria by improving renal function through the reduction of NF-κB. This suggests that astragalus may help prevent and treat diabetic nephropathy (Zhang et al. 2007).

3.8 Clinical Studies

Several clinical studies have been conducted to determine the beneficial effects of astragalus.

In a study of 43 patients with systemic lupus erythematosus and compromised kidney function, patients receiving standard cyclophosphamide treatment at 0.8 g per month were randomized to trial and control groups. The trial group received 20 ml of astragalus injection (equivalent to 40 g of raw herb) daily for 12 days each month. After 3 months, the trial group had significant reduction in active clinical symptoms ($P<0.05$) and reduced infection rates compared to the control group (4.4% *vs* 25%). Patients also had decreased urine protein and increased red blood cell count (Su et al. 2007). Although the dosage and frequency of cyclophosphamide differ in chemotherapy regimens, this study suggests that astragalus can reduce the toxicity of cyclophosphamide.

A study conducted in patients with leucopenia showed that astragalus can increase white blood cell count in a dose-dependent manner. A pure astragalus preparation when given twice daily at a concentration equivalent to 15 g astragalus was 65% more effective ($P<0.01$) than a lower concentration of 5 g astragalus in increasing white blood cell count after 8 weeks of treatment (Weng 1995).

In another study of patients undergoing hemodialysis, 31 patients were divided into a treatment group with daily 30 ml astragalus injection and a control group with no additional treatment. After 2 months, the treatment group had significant increase in serum IL-2 level (from 3.86–5.38 ng/ml) compared to the control group (3.72–3.85 ng/ml) ($P<0.001$). This demonstrates that astragalus can help enhance immune function patients undergoing hemodialysis (Qun et al. 1999).

Astragalus can also enhance the dendritic cell induction of mononuclear cells. In a study of 44 children with acute leukemia undergoing chemotherapy, 20 children in the treatment group were given large daily doses (up to 90 g) of astragalus while the others received only chemotherapy. After 1 month, the treatment group had significantly higher proportion of dendritic cells 4.4×10^6 per 2.5×10^6 of mononuclear cells *vs* 2.6×10^6 per 2.5×10^6 ($P<0.01$) of mononuclear cells in the control group (Dong et al. 2005).

These results support the traditional theory that astragalus has potent immunostimulant effects and therefore may be used as a biological response modifier (Sun et al. 1983).

Astragalus injection also has benefits in patients with advanced non-small cell lung cancer (NSCLC) when given with chemotherapy. In a study of 60 NSCLC patients (stage IIIb or stage V) randomized to astragalus injection (10 ml equal to the potency of 20 g of raw herb) daily with chemotherapy, or chemotherapy alone, patients in the former group had statistically significant response rate (40.0% *vs* 36.7%), higher survival rate in 1 year (46.7% *vs* 13.3%) ($P<0.05$), and improvement in quality of life compared to the control group 80.4% *vs* 43.3% ($P<0.01$) (Zou and Liu 2003).

The effectiveness of astragalus extract (Injectio Radici Astragali) was also tested by a point injection to reduce adverse effects from chemotherapy. In a study, 78 cancer patients all with stage III or higher lung, breast, liver, GI and other cancers and a leukocyte count lower than 4.0 G/L were randomized to 2 groups. The treatment group was given astragalus injection once a day into the acupoint Zusanli (ST-36) along with chemotherapy. The control group took a TCM preparation Gan Xue Bao orally 3 times a day. After 3 weeks, the treatment had a significant total effective rate defined by restoration of the total and differential leukocyte counts compared to the control group (82.2% *vs* 51.5%, $P=0.01$), Astragalus extract when injected into acupoint is more effective than oral supplements in improving immune function measured by natural killer cell activity (Chen et al. 2005). However, it is unclear whether the effect was due to the astragalus extract or acupuncture or both.

Astragalus is often consumed orally and is rarely used alone in traditional medicine. Herbalists generally customize formulas that contain multiple ingredients based on each patient's pattern presentation. The selection of ingredients is guided by descriptions in classical literature. Similarly, products used in clinical studies generally follow these traditional formulas. Many proprietary products have been developed, some of which claim to have unique activity; they are often patented based on special usage or manufacturing processes to protect commercial interest. Several such formulas have been studied in cancer patients.

A formula called Huangqi Zengmian Powder, consisting of astragalus, *Panax ginseng, Lycium barbarum, Ligustrum lucidum*, and *Cistanche salsa* was tested in esophageal cancer patients for interstitial response following surgery. Thirty-seven patients with stage I to stage IV esophageal cancer were treated with 10 g of this formula given 3 times a day starting 1 week before surgery and continued for 4 more weeks; 14 patients in the control group did not receive any herbal treatment. Results showed that the treatment group had an increase in interstitial mastocytes ($\chi^2=11.14$, $P<0.01$) and improvement in microvessel damage ($\chi^2=7.10$, $P<0.01$). The histological and immunological improvements suggest that this formula can accelerate wound healing and recovery after surgery in patients with esophageal cancer (Gao et al. 2001).

Formulas containing astragalus, have been studied as a treatment for fatigue and as hematopoietic agents.

Myelophil, an extract containing astragalus and *Salvia miltiorrhiza* roots reduced fatigue in a randomized double blind controlled trial. In this study, 36 patients with persistent fatigue for more than 6 months were randomized to low-dose group that consumed 1.5 g extract twice daily; high-dose group that took 3 g extract twice daily; or a placebo control group. Following 4 weeks of treatment, the low-dose group had significantly lower fatigue severity score compared to the controls ($P<0.05$) (Cho et al. 2009). Similarly, Huangqi Jianzhong Decoction, a formula containing astragalus, *Paeonia lactiflora, Zingiber officinale, Zizyphus spinosae, Glycyrrhizae uralensis, Cinnamomi cassiae*, and *Saccharum granorum*, reduced fatigue in athletes by increasing oxygen uptake and systemic utility. In this study, 12 male athletes were randomly divided into experimental and control groups. Athletes in the experimental group took Huangqi Jianzhong Decoction while those in the control group took a placebo made of fried starch. After 8 weeks, the anaerobic threshold, a marker for stamina building and fatigue recovery, was significantly increased in the experimental group ($P=0.02$) (Chen et al. 2002).

In another randomized, double-blind placebo-controlled study, 103 women with acute menopausal symptoms were enrolled. The patients were given 3 g daily of Danggui Buxue Decoction (a mixture of 5 parts of astragalus and 1 part of angelica) orally, or a placebo. After 6 months, this formula was found to be effective only for mild hot flushes. The number of mild hot flushes per month improved from 18.9 ± 23.5 at baseline to 8.6 ± 17.1 in the treatment group ($P<0.01$) and 26.0 ± 43.5 to 12.4 ± 17.6 in the placebo group ($P>0.05$). No overall changes were found in vasomotor symptoms (Haines et al. 2008).

Studies also show that some astragalus products can be used safely with chemotherapy. In a pharmacokinetic study of Jin Fu Kang, an astragalus-based oral herbal formula, researchers found that it did not cause significant interactions with docetaxel when used in NSCLC patients (Cassileth et al. 2009).

In a review of four clinical trials conducted to assess the effectiveness of astragalus compounds on the quality of life, side effects of chemotherapy, and on adverse effects in colorectal cancer patients, astragalus use was found to help reduce nausea and vomiting along with a decrease in the rate of leucopenia and an increase in CD3, CD4, and CD8 subsets of T-lymphocytes compared to those treated with chemotherapy alone. Use of Chinese herbal medicine along with chemotherapy appears promising for patients with colorectal cancer (Taixiang et al. 2005).

A recent meta-analysis of fourteen randomized controlled trials published from 1980–2008 on Aidi, a parenteral formula containing astragalus, found that it has therapeutic effects when used with radiation therapy or with navelbine and platinum compounds. Studies also suggest that Aidi can improve quality of life, immune function, and hematopoiesis. However, the formula did not improve survival (Ma et al. 2009).

Due to its Qi tonifying property, astragalus is used also to treat lung ailments. Platinum-based drugs, like cisplatin and carboplatin, often are used in conjunction with taxanes and vinca alkaloids in chemotherapy regimens for NSCLC. In addition, gemcitabine, premetrexate, and bevacizumab are also used. Several studies have examined the benefits of using both oral and parenteral astragalus-containing formulas along with these chemotherapy drugs. A meta-analysis of 34 randomized studies totaling 2,815 patients found such a combination to improve survival and improve tumor response to treatment (McCulloch et al. 2006). Further, a systemic review of 15 trials of oral astragalus formulas in patients with NSCLC reported quality of life improvement (Chen et al. 2010).

In a meta-analysis of TCM and chemotherapy for hepatocellular carcinoma (HCC), of 26 studies analyzed, 12 used herbal formulas that contained astragalus. Researchers concluded that these products improve survival and tumor response when used along with chemotherapy (Shu et al. 2005).

In another meta-analysis, thirty studies published over the last decade involving patients with HCC treated by transcatheter arterial chemo-embolization and 5-flurouracil based chemotherapy regimens were examined. Patients who used Chinese herbal therapies with standard treatments were found to have improved survival; improved quality of life; and fewer adverse effects. No benefits were seen in liver function tests or in short-term survival (>6-month). Of the 30 studies analyzed, 18 used herbal formulas that contained astragalus as major ingredient (Cho and Chen 2009).

Although all products studied to date contain astragalus as the main ingredient, the variation across formulas makes it difficult to draw definitive conclusions (Firenzuoli et al. 2006). Future studies should address this issue.

3.9 Herb-drug Interactions and Other Concerns

Although astragalus is generally considered safe for traditional use, the potential for herb-drug interactions exists because botanicals contain biologically active compounds (Meijerman et al. 2006; Yeung and Gubili 2007).

For example, saponins from astragalus can induce the hepatic microsomal cytochrome P-450 enzymes, which may account for its hepatoprotective effects but can also alter the metabolism of other herbs or drugs that are substrates of these enzymes (Zhang et al. 1992).

In an early study of a xenogeneic graft-*versus*-host reaction model, an astragalus fraction reversed the immunosuppressive effect of cyclophosphamide on mononuclear cells derived from cancer patients. This led to the use of astragalus as an immunomodulating agent to reduce the adverse effects of chemotherapy (Chu et al. 1988, 1989). However, it may also reverse the effects of immunosuppressants in patients following organ or bone marrow transplants.

A few studies showed that compounds present in astragalus may have estrogenic activities (Zhang et al. 2005; Huh et al. 2010), although it is unclear whether astragalus affects hormonal therapy in cancer patients. Many Western oncologists advise patients with hormone-sensitive cancers to avoid phytoestrogens, while others argue that phytoestrogens may actually provide a protective effect (Virk-Baker et al. 2010).

3.10 Future Research

There has been an increase in basic research and the number of clinical trials over the last two decades to determine the effectiveness of botanicals in the treatment and prevention of cancer. Several studies to date have demonstrated that astragalus plays a positive role in reducing the adverse effects when used with standard chemotherapy. However, the majority of studies used compound astragalus formulas, and the observed benefits may be due to synergistic effects contributed by other components. Whether astragalus alone would provide similar benefits needs to be determined.

Botanicals research involves unique issues ranging from product standardization to study design. From a drug development perspective, future clinical studies should focus on testing purified constituents. Using a single, well defined compound can provide consistent and measureable results for comparison of efficacy.

Astragalus products have not been used in dose escalation studies. It is important to determine an optimal dose in order to design meaningful large clinical trials. Clinical outcome may vary depending on the dose regimen.

Few studies have looked into how genetic variation in humans can affect the utility of astragalus. Given the current understanding of pharmacogenomics and polymorphism, such studies are particularly important in cancer treatment as genetics play a major role in cancer diagnosis and treatment.

Preliminary results also hold promise in developing astragalus as a cancer vaccine adjuvant, but the field of cancer vaccine treatment is still new. We believe that astragalus, with its unique constituents and immunomodulatory properties, is a potential candidate for this role.

3.11 Conclusions

Used in traditional medicine for millennia, astragalus continues to be an important constituent of traditional medicine formulas in Asia, and to a limited extent as a dietary supplement in the West. Its pharmacologic effects have not been fully elucidated but current research supports its use as an immunomodulator. Both classic and modern patented astragalus formulas are used in cancer care although human studies are limited. Use of astragalus products may help reduce adverse effects from chemotherapy. However, before they can be incorporated as standard therapy, products should be well characterized. Large clinical studies using standardized products with known optimal doses should be conducted. Using astragalus extract with a cancer vaccine is an exciting new frontier but this also requires more definitive clinical research. Until then, astragalus products will likely continue to be used by cancer patients in conjunction with standard therapies. Patients and practitioners should be aware of the benefits and risks of adverse effects and of potential astragalus-prescription drug interactions.

References

Ahmed, M. S., Hou, S. H., Battaglia, M. C., et al. (2007). Treatment of idiopathic membranous nephropathy with the herb Astragalus membranaceus. *American Journal of Kidney Diseases, 50,* 1028–1032.

Auyeung, K. K., & Ko, J. K. (2010). Novel herbal flavonoids promote apoptosis but differentially induce cell cycle arrest in human colon cancer cell. *Investigational New Drugs, 28,* 1–13.

Auyeung, K. K., Cho, C. H., & Ko, J. K. (2009). A novel anticancer effect of Astragalus saponins: Transcriptional activation of NSAID-activated gene. *International Journal of Cancer, 125,* 1082–1091.

Cai, G., Liu, B., Liu, W., et al. (2007). Buyang Huanwu Decoction can improve recovery of neurological function, reduce infarction volume, stimulate neural proliferation and modulate VEGF and Flk1 expressions in transient focal cerebral ischaemic rat brains. *Journal of Ethnopharmacol, 5*(113), 292–299.

Cassileth, B. R., Rizvi, N., Deng, G., et al. (2009). Safety and pharmacokinetic trial of docetaxel plus an Astragalus-based herbal formula for non-small cell lung cancer patients. *Cancer Chemotherapy and Pharmacology, 65,* 67–71.

Chan, W. S., Durairajan, S. S., Lu, J. H., et al. (2009). Neuroprotective effects of Astragaloside IV in 6-hydroxydopamine-treated primary nigral cell culture. *Neurochemistry International, 55,* 414–422.

Chang, M. S., Kim do, R., Ko, E. B., et al. (2009). Treatment with Astragali radix and Angelicae radix enhances erythropoietin gene expression in the cyclophosphamide-induced anemic rat. *Journal of Medicinal Food, 12,* 637–642.

Chen, J. K., & Chen, T. T. (2004). *Chinese medical herbology and pharmacology*. City of industry, CA: Art of Medicine Press.

Chen, K. T., Su, C. H., Hsin L. H., et al. (2002). Reducing fatigue of athletes following oral administration of huangqi jianzhong tang. *Acta Pharmacologica Sinica, 23,* 757–761.

Chen, L., Xie, C., & Wu, L. (2005). Point injection of injectio radici astragali for treatment of post-chemotherapy adverse reactions. *Journal of Traditional Chinese Medicine, 25,* 21–22.

Chen, S., Flower, A., Ritchie, A., et al. (2010). Oral Chinese herbal medicine (CHM) as an adjuvant treatment during chemotherapy for non-small cell lung cancer: A systematic review. *Lung Cancer, 68,* 137–145.

Cheng, X. D., Hou, C. H., Zhang, X. J., et al. (2004). Effects of Huangqi (Hex) on inducing cell differentiation and cell death in K562 and HEL cells. *Acta Biochimica et Biophysica Sinica, 36,* 211–217.

Chiu, P. Y., Leung, H. Y., Siu, A. H., et al. (2007). Dang-Gui Buxue Tang protects against oxidant injury by enhancing cellular glutathione in H9c2 cells: role of glutathione synthesis and regeneration. *Planta Medica, 73,* 134–141.

Cho, W. C. (2009). Astragalus, Astragalus membranaceus (Fisch) Bge. In A. S. Awaad, V. K. Singh, & J. N. Govil (Eds.), *Recent progress in medicinal plant* (Vol. 28, pp. 31–45). Ethnomedicine: Source and Mechanism II.

Cho, W. C., & Leung, K. N. (2007a). In vitro and in vivo immunomodulating and immunorestorative effects of Astragalus membranaceus. *Journal of Ethnopharmacology, 113,* 132–141.

Cho, W. C., & Leung, K. N. (2007b). In vitro and in vivo anti-tumor effects of Astragalus membranaceus. *Cancer Letters, 252,* 43–54.

Cho, W. C., & Chen, H. Y. (2009). Transcatheter arterial chemoembolization combined with or without Chinese herbal therapy for hepatocellular carcinoma: Meta-analysis. *Expert Opinion on Investigational Drugs, 18,* 617–635.

Cho, J. H., Cho, C. K., Shin, J. W., et al. (2009). Myelophil, an extract mix of Astragali Radix and Salviae Radix, ameliorates chronic fatigue: a randomised, double-blind, controlled pilot study. *Complementary Therapies in Medicine, 17,* 141–146.

Chu, D. T., Wong, W. L., & Mavligit, G. M. (1988). Immunotherapy with Chinese medicinal herbs. I. Immune restoration of local xenogeneic graft-versus-host reaction in cancer patients by fractionated Astragalus membranaceus in vitro. *Journal of Clinical & Laboratory Immunology, 25,* 119–123.

Chu, D. T., Sun, Y., & Lin, J. R. (1989). Immune restoration of local xenogeneic graft-versus-host reaction in cancer patients in vitro and reversal of cyclophosphamide-induced immune suppression in the rat in vivo by fractionated Astragalus membranaceus. *Zhong Xi Yi Jie He Za Zhi, 9,* 351–354, 326.

Chu, D. T., Lin, J. R., & Wong, W. (1994). The in vitro potentiation of LAK cell cytotoxicity in cancer and aids patients induced by F3—A fractionated extract of Astragalus membranaceus. *Zhonghua Zhong Liu Za Zhi (Chinese Journal of Oncology), 16,* 167–171.

Correa-Velez, I., Clavarino, A., & Eastwood, H. (2005). Surviving, relieving, repairing, and boosting up: reasons for using complementary/alternative medicine among patients with advanced cancer: A thematic analysis. *Journal of Palliative Medicine, 8,* 953–961.

Cui, R., He, J., Wang, B., et al. (2003). Suppressive effect of Astragalus membranaceus Bunge on chemical hepatocarcinogenesis in rats. *Cancer Chemotherapy and Pharmacology, 51,* 75–80.

Deng, G., Lin, H., Seidman, A., et al. (2009). A phase I/II trial of a polysaccharide extract from Grifola frondosa (Maitake mushroom) in breast cancer patients: Immunological effects. *Journal of Cancer Research and Clinical Oncology, 135,* 1215–1221.

Dong, J. C., & Dong, X. H. (2005). Comparative study on effect of astragalus injection and interleukin-2 in enhancing anti-tumor metastasis action of dendrite cells. *Zhongguo Zhong Xi Yi Jie He Za Zhi, 25,* 236–239.

Dong, J., Gu, H. L., Ma, C. T., et al. (2005). Effects of large dose of Astragalus membranaceus on the dendritic cell induction of peripheral mononuclear cell and antigen presenting ability of dendritic cells in children with acute leukemia. *Zhongguo Zhong Xi Yi Jie He Za Zhi, 25,* 872–875.

Dunn, G. P., Old, L. J., & Schreiber, R. D. (2004). The three Es of cancer immunoediting. *Annual Review of Immunology, 22,* 329–360.

El-Kenawy, A. E.-M. (2010). Investigating the protective effects of Astragalus membranaceus on nephrotoxicity in cyclosporine A-treated rats. *Kidney, 19,* 119–125.

Evans, M., Shaw, A., Thompson, E. A., et al. (2007). Decisions to use complementary and alternative medicine (CAM) by male cancer patients: Information-seeking roles and types of evidence used. *BMC Complementary and Alternative Medicine, 7,* 25.

Firenzuoli, F., Gori, L., Di Simone, L., et al. (2006). Important bias in the astragalus meta-analysis. *Journal of Clinical Oncology: Official Journal of the American Society of Clinical Oncology, 24,* 3215–3216.

Gao, X. P., Zhao, W. X., & Zhang, Z. L. (2001). Effect of huangqi zengmian powder on interstitial response in patients with esophageal cancer at peri-operational period. *Zhongguo Zhong Xi Yi Jie He Za Zhi, 21,* 171–173.

Gao, Q. T., Cheung, J. K., Choi, R. C., et al. (2008). A Chinese herbal decoction prepared from Radix Astragali and Radix Angelicae Sinensis induces the expression of erythropoietin in cultured Hep3B cells. *Planta Medica, 74,* 392–395.

Gui, S. Y., Wei, W., Wang, H., et al. (2006). Effects and mechanisms of crude astragalosides fraction on liver fibrosis in rats. *Journal of Ethnopharmacology, 103,* 154–159.

Haines, C. J., Lam, P. M., Chung, T. K, et al. (2008). A randomized, double-blind, placebo-controlled study of the effect of a Chinese herbal medicine preparation(Dang Gui Buxue Tang) on menopausal symptoms in Hong Kong Chinese women. *Climacteric, 11,* 244–251.

Han, D. O., Lee, H. J., & Hahm, D. H. (2009). Wound-healing activity of Astragali Radix in rats. *Methods and Findings in Experimental and Clinical Pharmacology, 31,* 95–100.

He, Z. P., Wang, D. Z., Shi, L. Y., et al. (1986). Treating amenorrhea in vital energy-deficient patients with angelica sinensis-astragalus membranaceus menstruation-regulating decoction. *Journal of Traditional Chinese Medicine, 6,* 187–190.

Hishida, I., Nanba, H., & Kuroda, H. (1988). Antitumor activity exhibited by orally administered extract from fruit body of Grifola frondosa (maitake). *Chemical & Pharmaceutical Bulletin, 36,* 1819–1827.

Hong, F., Xiao W., Ragupathi, G., et al. (2010). *The known immunologically active components of astragalus account for only a small proportion of the immunological adjuvant activity when combined with conjugate vaccines*. New York.

Huang, X., Liu, Y., Song, F., et al. (2009). Studies on principal components and antioxidant activity of different Radix Astragali samples using high-performance liquid chromatography/electrospray ionization multiple-stage tandem mass spectrometry. *Talanta, 78,* 1090–1101.

Huh, J. E., Kwon, N. H., Baek, Y. H., et al. (2009). Formononetin promotes early fracture healing through stimulating angiogenesis by up-regulating VEGFR-2/Flk-1 in a rat fracture model. *International Immunopharmacology, 9,* 1357–1365.

Huh, J. E., Seo, D. M., Baek, Y. H., et al. (2010). Biphasic positive effect of formononetin on metabolic activity of human normal and osteoarthritic subchondral osteoblasts. *International Immunopharmacology, 10,* 500–507.

Ko, J. K., Lam, F. Y., & Cheung, A. P. (2005). Amelioration of experimental colitis by Astragalus membranaceus through anti-oxidation and inhibition of adhesion molecule synthesis. *World Journal of Gastroenterology: WJG, 11,* 5787–5794.

Kuo, Y. H., Tsai, W. J., Loke, S. H., et al. (2009). Astragalus membranaceus flavonoids (AMF) ameliorate chronic fatigue syndrome induced by food intake restriction plus forced swimming. *Journal of Ethnopharmacology, 122,* 28–34.

Kurashige, S., Akuzawa, Y., & Endo, F. (1999). Effects of astragali radix extract on carcinogenesis, cytokine production, and cytotoxicity in mice treated with a carcinogen, N-butyl-N′-butanolnitrosoamine. *Cancer Investigation, 17,* 30–35.

Lawenda, B. D., Kelly, K. M., Ladas, E. J, et al. (2008). Should supplemental antioxidant administration be avoided during chemotherapy and radiation therapy? *Journal of the National Cancer Institute, 100,* 773–783.

Lee, Y. S., Han, O. K., Park, C. W., et al. (2003). Immunomodulatory effects of aqueous-extracted Astragali radix in methotrexate-treated mouse spleen cells. *Journal of Ethnopharmacology, 84,* 193–198.

Li, S. G. (2009). Yong-qi characterization and renal protective effect of a polysaccharide from Astragalus membranaceus. *Carbohydrate Polymers, 78,* 343–348.

Li, R., Chen, W. C., Wang, W. P., et al. (2009a). Extraction, characterization of astragalus polysaccharides and its immune modulating activities in rats with gastric cancer. *Carbohydrate Polymers, 78,* 738–742.

Li, S. P., Zhao, X. J., & Wang, J. Y. (2009b). Synergy of Astragalus polysaccharides and probiotics (Lactobacillus and Bacillus cereus) on immunity and intestinal microbiota in chicks. *Poultry Science, 88,* 519–525.

Lippman, S. M., Klein, E. A., Goodman, P. J., et al. (2009). Effect of selenium and vitamin E on risk of prostate cancer and other cancers: The Selenium and Vitamin E Cancer Prevention Trial (SELECT). *JAMA: The Journal of the American Medical Association, 301,* 39–51.

Liu, M., Wu, K., Mao, X., et al. (2010). Astragalus polysaccharide improves insulin sensitivity in KKAy mice: Regulation of PKB/GLUT4 signaling in skeletal muscle. *Journal of Ethnopharmacology, 127,* 32–37.

Lu, M. C., Yao, C. H., Wang, S. H., et al. (2010). Effect of Astragalus membranaceus in rats on peripheral nerve regeneration: in vitro and in vivo studies. *The Journal of Trauma, 68,* 434–440.

Luo, Z., Zhong, L., Han, X., et al. (2009). Astragalus membranaceus prevents daunorubicin-induced apoptosis of cultured neonatal cardiomyocytes: Role of free radical effect of Astragalus membranaceus on daunorubicin cardiotoxicity. *Phytotherapy Research: PTR, 23,* 761–767.

Ma, X. Q., Shi, Q., Duan, J. A., et al. (2002). Chemical analysis of Radix Astragali (Huangqi) in China: A comparison with its adulterants and seasonal variations. *Journal of Agricultural and Food Chemistry, 50,* 4861–4866.

Ma, W. H., Duan, K. N., Feng, M., et al. (2009). Aidi injection as an adjunct therapy for non-small cell lung cancer: A systematic review. *Zhong Xi Yi Jie He Xue Bao, 7,* 315–324.

Mak, D. H., Chiu, P. Y., Dong, T. T., et al. (2006). Dang-Gui Buxue Tang produces a more potent cardioprotective effect than its component herb extracts and enhances glutathione status in rat heart mitochondria and erythrocytes. *Phytotherapy Research: PTR, 20,* 561–567.

Mantovani, A., Garlanda, C., & Allavena P. (2010). Molecular pathways and targets in cancer-related inflammation. *Annals of Medicine, 42,* 161–170.

McCulloch, M., See, C., Shu, X. J., et al. (2006). Astragalus-based Chinese herbs and platinum-based chemotherapy for advanced non-small-cell lung cancer: Meta-analysis of randomized trials. *Journal of Clinical Oncology: Official Journal of the American Society of Clinical Oncology, 24,* 419–430.

Meijerman, I., Beijnen, J. H., & Schellens, J. H. (2006). Herb-drug interactions in oncology: Focus on mechanisms of induction. *Oncologist, 11,* 742–752.

Meyer, F., Bairati, I., Fortin, A., et al. (2008). Interaction between antioxidant vitamin supplementation and cigarette smoking during radiation therapy in relation to long-term effects on recurrence and mortality: A randomized trial among head and neck cancer patients. *International Journal of Cancer. Journal International Du Cancer, 122,* 1679–1683.

Motomura, K., Fujiwara, Y., Kiyota, N., et al. (2009). Astragalosides isolated from the root of astragalus radix inhibit the formation of advanced glycation end products. *Journal of Agricultural and Food Chemistry, 57,* 7666–7672.

Na, D., Liu, F. N., Miao, Z. F., et al. (2009). Astragalus extract inhibits destruction of gastric cancer cells to mesothelial cells by anti-apoptosis. *World Journal of Gastroenterology: WJG, 15,* 570–577.

Park, H. J., Kim, H. Y., Yoon, K. H., et al. (2009). The effects of Astragalus membranaceus on repeated restraint stress-induced biochemical and behavioral responses. *Korean Journal of Physiology Pharmacology, 13,* 315–319.

Pharmacopoeia, C. (2005). *Pharmacopoeia of the People's Republic of China*. PR China: People's Medical Publishing House.

Qiu, L. H., Xie, X. J., & Zhang, B. Q. (2010). Astragaloside IV improves homocysteine-induced acute phase endothelial dysfunction via antioxidation. *Biological & Pharmaceutical Bulletin, 33,* 641–646.

Qu, Y. Z., Li, M., Zhao, Y. L., et al. (2009). Astragaloside IV attenuates cerebral ischemia-reperfusion-induced increase in permeability of the blood-brain barrier in rats. *European Journal of Pharmacology, 606,* 137–141.

Qun, L., Luo, Q., Zhang, Z. Y., et al. (1999). Effects of astragalus on IL-2/IL-2R system in patients with maintained hemodialysis. *Clinical Nephrology, 52,* 333–334.

Ragupathi, G., Yeung, K. S., Leung, P. C., et al. (2008). Evaluation of widely consumed botanicals as immunological adjuvants. *Vaccine, 26,* 4860–4865.

Rittenhouse, J. R., Lui, P. D., & Lau, B. H. (1991). Chinese medicinal herbs reverse macrophage suppression induced by urological tumors. *The Journal of Urology, 146,* 486–490.

Ryu, M., Kim, E. H., Chun, M., et al. (2008). Astragali Radix elicits anti-inflammation via activation of MKP-1, concomitant with attenuation of p38 and Erk. *Journal of Ethnopharmacology, 115,* 184–193.

Shao, B. M., Xu, W., Dai, H., et al. (2004). A study on the immune receptors for polysaccharides from the roots of Astragalus membranaceus, a Chinese medicinal herb. *Biochemical and Biophysical Research Communications, 320,* 1103–1111.

Shen, H. H., Wang, K., Li, W., et al. (2008). Astragalus membranaceus prevents airway hyperreactivity in mice related to Th2 response inhibition. *Journal of Ethnopharmacology, 116,* 363–369.

Shu, X., McCulloch, M., Xiao, H., et al. (2005). Chinese herbal medicine and chemotherapy in the treatment of hepatocellular carcinoma: a meta-analysis of randomized controlled trials. *Integrative Cancer Therapies, 4,* 219–229.

Su, L., Mao, J. C., & Gu, J. H. (2007). Effect of intravenous drip infusion of cyclophosphamide with high-dose Astragalus injection in treating lupus nephritis. *Zhong Xi Yi Jie He Xue Bao, 5,* 272–275.

Sun, Y., Hersh, E. M., Talpaz, M., et al. (1983). Immune restoration and/or augmentation of local graft versus host reaction by traditional Chinese medicinal herbs. *Cancer, 52,* 70–73.

Sun, W. Y., Wei, W., Gui, S. Y., et al. (2008). Protective effect of extract from Paeonia lactiflora and Astragalus membranaceus against liver injury induced by bacillus Calmette-Guerin and lipopolysaccharide in mice. *Basic & Clinical Pharmacology & Toxicology, 103,* 143–149.

Taixiang, W., Munro, A. J., & Guanjian, L. (2005). Chinese medical herbs for chemotherapy side effects in colorectal cancer patients. *Cochrane Database of Systematic Reviews (Online),* CD004540.

Tanaka. K., Tamura, T., Fukuda, S., et al. (2008). Quality evaluation of Astragali Radix using a multivariate statistical approach. *Phytochemistry, 69,* 2081–2087.

The toxicology of the astragalus genus. Plants poisonous to livestock. http://www.ansci.cornell.edu/plants/locoweed/astrag1.html. Accessed 5 August 2010.

Tin, M. M., Cho, C. H., Chan, K., et al. (2007). Astragalus saponins induce growth inhibition and apoptosis in human colon cancer cells and tumor xenograft. *Carcinogenesis, 28,* 1347–1355.

Verhoef, M. J., Balneaves, L. G., Boon, H. S., et al. (2005). Reasons for and characteristics associated with complementary and alternative medicine use among adult cancer patients: A systematic review. *Integrative Cancer Therapies, 4,* 274–286.

Vickers, A. J. (2006). How to design a phase I trial of an anticancer botanical. *Journal of the Society for Integrative Oncology, 4,* 46–51.

Vickers, A., Goyal, N., Harland, R., et al. (1998). Do certain countries produce only positive results? A systematic review of controlled trials. *Controlled Clinical Trials, 19,* 159–166.

Virk-Baker, M. K., Nagy, T. R., & Barnes, S. (2010). Role of phytoestrogens in cancer therapy. *Planta Medica, 76,* 1132–1142.

Wei, H., Sun, R., Xiao, W., et al. (2003). Traditional Chinese medicine Astragalus reverses predominance of Th2 cytokines and their up-stream transcript factors in lung cancer patients. *Oncology Reports, 10,* 1507–1512.

Weng, X. S. (1995). Treatment of leucopenia with pure Astragalus preparation—An analysis of 115 leucopenic cases. *Zhongguo Zhong Xi Yi Jie He Za Zhi, 15,* 462–464.

Wojcikowski, K., Wohlmuth, H., Johnson, D. W., et al. (2010). Effect of Astragalus membranaceus and Angelica sinensis combined with Enalapril in rats with obstructive uropathy. *Phytotherapy Research: PTR, 24,* 875–884.

Yan, F., Zhang, Q. Y., Jiao, L., et al. (2009). Synergistic hepatoprotective effect of Schisandrae lignans with Astragalus polysaccharides on chronic liver injury in rats. *Phytomedicine, 16,* 805–813.

Yang, Z. G., Sun, H. X., & Fang, W. H. (2005). Haemolytic activities and adjuvant effect of Astragalus membranaceus saponins (AMS) on the immune responses to ovalbumin in mice. *Vaccine, 23,* 5196–5203.

Yang, M., Chan, G. C., Deng, R., et al. (2009). An herbal decoction of Radix astragali and Radix angelicae sinensis promotes hematopoiesis and thrombopoiesis. *Journal of Ethnopharmacology, 124,* 87–97.

Yang, M., Qian, X. H., Zhao, D. H., et al. (2010). Effects of Astragalus polysaccharide on the erythroid lineage and microarray analysis in K562 cells. *Journal of Ethnopharmacology, 127,* 242–250.

Yeung, K. S., & Gubili, J. (2007). Clinical guide to herb-drug interactions in oncology. *Journal of the Society for Integrative Oncology, 5,* 113–117.

Yip, P. Y., & Kwan, H. S. (2006). Molecular identification of Astragalus membranaceus at the species and locality levels. *Journal of Ethnopharmacology, 106,* 222–229.

Yu, D. H., Bao, Y. M., Wei, C. L., et al. (2005). Studies of chemical constituents and their antioxidant activities from Astragalus mongholicus Bunge. *Biomedical and Environmental Sciences: BES, 18,* 297–301.

Yu, S. Y., Ouyang, H. T., Yang, J. Y., et al. (2007). Subchronic toxicity studies of Radix Astragali extract in rats and dogs. *Journal of Ethnopharmacology, 110,* 352–355.

Zhang, Y. D., Shen, J. P., Zhu, S. H., et al. (1992). Effects of astragalus (ASI, SK) on experimental liver injury. *Yao Xue Xue Bao, 27,* 401–406.

Zhang, W. J., Hufnagl, P., Binder, B. R., et al. (2003). Antiinflammatory activity of astragaloside IV is mediated by inhibition of NF-kappaB activation and adhesion molecule expression. *Thrombosis and Haemostasis, 90,* 904–914.

Zhang, C. Z., Wang, S. X., Zhang, Y., et al. (2005). In vitro estrogenic activities of Chinese medicinal plants traditionally used for the management of menopausal symptoms. *Journal of Ethnopharmacology, 98,* 295–300.

Zhang, W. D., Chen, H., Zhang, C., et al. (2006a). Astragaloside IV from Astragalus membranaceus shows cardioprotection during myocardial ischemia in vivo and in vitro. *Planta Medica, 72,* 4–8.

Zhang, W. D., Zhang, C., Wang, X. H., et al. (2006b). Astragaloside IV dilates aortic vessels from normal and spontaneously hypertensive rats through endothelium-dependent and endothelium-independent ways. *Planta Medica, 72,* 621–626.

Zhang, Z. X., Qi. F., Zhou, D. J., et al. (2006c). Effect of 5-fluorouracil in combination with Astragalus membranaceus on amino acid metabolism in mice model of gastric carcinoma. *Zhonghua Wei Chang Wai Ke Za Zhi, 9,* 445–447.

Zhang, Y. W., Wu, C. Y., & Cheng, J. T. (2007). Merit of Astragalus polysaccharide in the improvement of early diabetic nephropathy with an effect on mRNA expressions of NF-kappaB

and IkappaB in renal cortex of streptozotoxin-induced diabetic rats. *Journal of Ethnopharmacology, 114,* 387–392.
Zhang, J., Xie, X., Li, C., et al. (2009). Systematic review of the renal protective effect of Astragalus membranaceus (root) on diabetic nephropathy in animal models. *Journal of Ethnopharmacology, 126,* 189–196.
Zou, Y. H., & Liu, X. M. (2003). Effect of astragalus injection combined with chemotherapy on quality of life in patients with advanced non-small cell lung cancer. *Zhongguo Zhong Xi Yi Jie He Za Zhi, 23,* 733–735.

Chapter 4
An Evidence-based Perspective of *Panax Ginseng* (Asian Ginseng) and *Panax Quinquefolius* (American Ginseng) as a Preventing or Supplementary Therapy for Cancer Patients

Lee Jia and Keduo Qian

Abstract Ginseng roots are taken orally as adaptogens and nourishing stimulants in traditional Chinese medicine. Along with the rapid advancement of modern research technologies, ginseng's effects in cancer treatment and prevention have been better elucidated both at molecular levels and on clinical aspects. Herein, we presented some techniques used for preparation for ginseng powders and separation of saponins, evidence from *in vitro* and *in vivo* tests and human studies to demonstrate effects of ginseng and its ingredients on cancer. Preparations of ginseng powders and saponins involve extensive use of extractions by methanol, water, hexane, n-BuOH, thin layer and high-performance liquid chromatography separations, depending on the constituents desired from the crudes. Ginseng is characterized by the presence of its active ingredients-ginsenosides, such as Rg3, Rg5, and Rh2, etc, which have been extensively studied *in vitro* and *in vivo*. Evidences from the *in vitro* studies demonstrated that ginsenosides can inhibit cancer cell invasion and metastasis, as well as induce apoptosis through different signaling pathways, which may substantiate the clinical use of ginseng as a preventing or supplementary therapy for cancer. *In vivo* animal studies further confirmed that Rg3 can improve the life quality and survival of the tumor-bearing mice and reduce their tumor weight. In addition, ginsenosides Rg3 and Rg5 showed statistically significant reduction in lung tumor incidence in mouse models, proving the chemo-preventive effects of red ginseng. In clinical studies, ginseng can improve the immune system activity of cancer patients and their appetites. Most importantly, regular ginseng consumption reduced the risk of development of all types of cancers, as observed in several randomized, double-blinded, placebo-controlled pilot trials. However, there is no conclusive evidence that ginseng itself can cure cancer. Its roles in cancer treatment should be viewed as a supplementary therapy to enhance host immune response to cancer and quality of patients' life. Nonetheless, the beneficial effects of ginseng

L. Jia (✉)
Developmental Therapeutics Program, DCTD, National Cancer Institute/NIH, Bethesda, MD, USA
e-mail: jiale@mail.nih.gov

W. C. S. Cho (ed.), *Evidence-based Anticancer Materia Medica,* Evidence-based Anticancer Complementary and Alternative Medicine,
DOI 10.1007/978-94-007-0526-5_4,

on cancer patients and high-risk populations (such as chronic immunosuppressive patients and the elders) have been well-demonstrated.

4.1 Introduction

The term ginseng refers to several species of the genus *Panax*. The two most commonly used species are Asian ginseng (*Panax ginseng* C.A. Meyer), which is almost extinct in its natural habitat but is still cultivated, and American ginseng (*Panax quinquefolius* L), which is both harvested from the wild and cultivated. Ginseng has long been known as a multifunctional tonic and anti-fatigue agent as described in the oldest Chinese materia medica book—Shennong's Classic of Materia Medica. The initial identification and description of ginseng can be dated back to about 5,500 years ago, when Shennong (a divine folk medical doctor) started to taste hundreds of herbs. Shennong's work was then recorded by Tao Hongjing into a book entitled 'Shennong's Classic of Materia Medica' during the Liang Dynasty (505–557 AD) (Zheng 1985).

The fundamentals of traditional Chinese medicine (TCM) stress the Yin-Yang balance. These theories apply the phenomena and laws of nature to the study of physiological activities and pathological changes of the human body and its interrelationships with other external elements. Yin and Yang are in conflict but at the same time mutually interdependent, with neither being able to exist in isolation. Under normal conditions, tumor and host immunosurveillance are kept within certain bounds. When the balance is broken, cancer grows. Chemo-therapeutic agents inhibit tumor growth in the tumor-bearing nude mice (Jia et al. 2008), but they often damage the host immunosurveillance simultaneously, resulting in a ruin of the physiological dynamic equilibrium. Such chemotherapy-derived imbalance could produce excess of Yin (cytotoxicity to all fast proliferative cells) and deficiency of Yang (quality of life and human's nature power to fight diseases for full recovery). Consequently, cancer is only delayed for months to finally take a life away which would otherwise have been saved if the patient's own recovery power was not damaged.

In general, the roles of ginseng in cancer treatment should be viewed as a supplementary therapy. Ginseng enhances host immune response (Scaglione et al. 1990), quality of patients' life, patients' appetite, and thus inhibiting further spreading of cancer. Generally, characteristics of the anticancer effects of ginseng may be summarized as follows: (1) it is observed only in slow-growing tumors such as Ehrlich and sarcoma 180 ascites tumors *in vivo*; (2) it is not observed in rapidly growing tumors such as L1210, P388 ascites tumors and Walker carcinosarcoma; and (3) there is no dose-response relationship and no cumulative effects (Yun et al. 2001a).

Although there is no conclusive evidence that ginseng itself can cure cancer, it makes sense that ginseng usage in cancer therapy should focus on synergistic combinations or palliative treatment. During active cancer therapy, ginseng should generally be administered in combination with chemotherapy and radiation. In this role, it acts as a biological response modifier and an adaptogen to synergistically enhance efficacy of the conventional therapy.

It should be mentioned that Chinese, Russian, Korean, and Japanese scientists did not explore the possibility of ginseng's preventive effects on cancer until 1990 (Yun et al. 2001a). Before that, ginseng research was heavily concentrated on its general beneficial effects, including its aphrodisiac effect, cognitive effect, tonic effect, and cardiovascular effect (Jia and Zhao 2009; Jia et al. 2009).

4.2 Preparations of Ginseng Powders and Saponins

Isolation and separation of ginseng saponins start with extraction of these ingredients from ginseng with methanol. The methanolic suspension is then subjected to column chromatographic separation (column styles: amberlite XAD-2, Diaion MCI Gel HP20, or Kogel BG4600). After removing saccharides and amino acids with water, elution of the columns with methanol is applied to obtain a saponin fraction. Those ginseng saponins, ginsenoside Rx (Table 4.1), are further separated on thin layer and high-performance liquid chromatography. A water-containing silica gel column and the water-containing solvent systems (Pak Aquasil) yield a good separation of ginsenosides (Kaizuka and Takahashi 1983).

Ginsenosides are generally labile under acidic conditions: ordinary acidic hydrolysis is always accompanied by many side reactions, such as cyclization of side chains, glycosyl elimination, and epimerization of carbone-20 by SN1 reaction. Therefore, the chemical transformations of secondary metabolites may occur during preparation processes.

Table 4.1 The average yields (%) of saponins from different species of ginseng. (Reproduced from Tanaka et al. 1986)

Ginsenosides		White ginseng	Red ginseng	American ginseng
20(s) protopanaxadiol-type	Rb1	0.5	0.4	1.8
	Rb2	0.2	0.2	0.03
	Rc	0.3	0.1	0.3
	Rd	0.2	0.036	0.5
	Rg3	0.0003	0.014(20R) 0.015(20S)	–
	Rh2	–	0.001	–
20(s) protopanaxatriol-type	Re	0.2	0.2	1.0
	Rf	0.05	0.07	–
	Rg1	0.2	0.3	1.9
	Rg2	0.014	0.01(20R) 0.02(20S)	0.008
	Rh1	0.0015	0.007(20R) 0.006(20S)	–
Oleanane-type	R0	0.02	0.045	0.07

For separation of panaxadiol type and panaxatriol type saponins, powdered red ginseng (2 kg) of 6 years old could be extracted with water (2 l × 2) at 90°C and filtered. One-tenth of the combined filtrates were evaporated to give a "water extract" (104.4 g). The remaining combined filtrates were successively extracted with hexane (1 l × 3) and water saturated n-BuOH (700 ml × 3). The dried hexane fraction (1.2 g) was named panaxan A-U, consisting of glucose, arabinose, galatose, rhamnose, xylose, and/or uronic acid. The water layer was also evaporated to give water fraction (715.9 g). The n-BuOH fraction was chromatographed on silica gel column, and the gel was eluted with $CHCl_3$-MeOH-H_2O (10:3:1 → 7:3:1) system. Eluates were examined by thin layer chromatography together with authentic samples, and the panaxadiol type saponins (29.2 g) and panaxatriol type saponins (32.8 g) were obtained (Kasai et al. 1983; Yun et al. 2000).

For separation of the total saponins, fresh ginseng can be air-dried and powdered. The powdered fresh ginseng (1 kg) can be extracted with water (2 l × 2) at 90°C and filtered. One tenth of the combined filtrates can be evaporated to give a water extract (49.2 g), and nine-tenths of the combined filtrates are extracted with ethyl ether (1 l × 3) and water saturated n-BuOH (700 ml × 3), to give n-BuOH fraction. The combined n-BuOH fraction can be dried and evaporated under the reduced pressure to give total saponin (63.6 g).

For preparation of the ginsenoside Rg3 and Rg5 mixture, the ginsenoside Rb1 (10 g) obtained from Korean ginseng can be hydrolyzed with 50% acetic acid (500 ml) at 70°C for 3 hours. The reaction mixture, concentrated to appropriate volume, is left at 4°C for 1 day and filtered. The filtrate is diluted with water (500 ml) and extracted with n-BuOH (250 ml × 3). The combined n-BuOH fractions are washed with saturated $NaHCO_3$ solution and evaporated under the reduced pressure. The residue is chromatographed on silica gel column, using $CHCl_3$-MeOH-H_2O (9:3:1) as eluting solvent, to obtain ginsenoside Rg3 and Rg5 mixture. Ginsenoside Rg3 and Rg5 mixtures can be subjected to high-performance liquid chromatography using acetonitrile-water (60:40) as mobile phase to analyze the ratio of ginsenoside Rg3 to Rg5 (2.6 g) (Kim et al. 1996).

4.3 Evidences of Ginseng as a Preventing or Supplementary Therapy for Cancer

4.3.1 In Vitro Experimental Evidences

Kitagawa et al. (1995) demonstrated that ginsenosides (Fig. 4.1), especially 20(*R*)-ginsenoside Rg3, can specifically inhibit cancer cell invasion and metastasis *in vitro*. Kitagawa's group developed an *in vitro* model of tumor cell invasion and metastasis. Tumor cells were seeded on a monolayer of mesothelial or endothelial host cells. The number of tumor cells that penetrated through the monolayer indicated their capacity of invasion. The capacity of the penetration of tumor cells *in vitro* was in parallel with

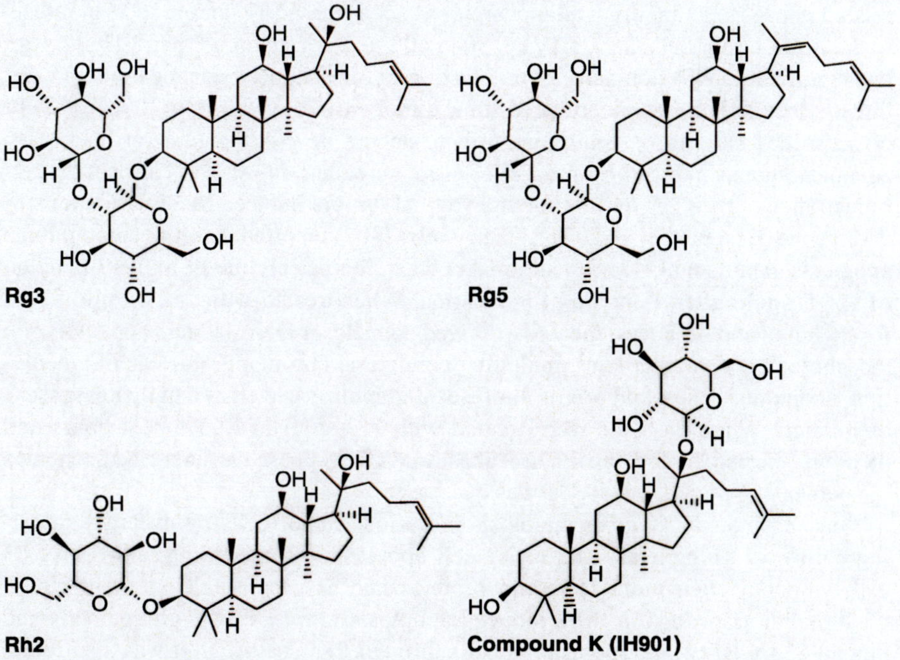

Fig. 4.1 Examples of active ginsenosides in the *in vitro, in vivo*, and clinical anticancer or cancer prevention evaluations

that of the *in vivo* implantation experiment. In the *in vitro* invasion system, 10% fetal calf serum was essentially required for the culture medium. It was found that fetal calf serum could be replaced with 1-oleoyl-lysophosphatidic acid. By using this model, more than 10 kinds of ginsenosides Rx were tested for their inhibition activity of tumor cell invasion and metastasis, and ginsenoside Rg3 was found to be the most effective in preventing cancer cell invasion. In addition, Azuma and Mochizuki (1994) found that ginsenoside Rb2 inhibited tumor angiogenesis, and Kikuchi et al. (1991) reported that ginsenoside Rh2 inhibited the human ovarian cancer growth in nude mice model. Recently, ginsenoside Rg3 was produced as an anti-angiogenic anticancer drug in China.

Rg3 is one of the most effective cytostasis ginsenosides isolated from ginseng. Rg3 inhibits human prostate cancer cells and other androgen dependent cells from proliferating (Liu et al. 2000). The mechanisms of action of Rg3 include: (1) decreasing genetic expression of 5α-reductase; (2) inhibiting cell cycle evolution genes such as proliferating cell nuclear antigen gene and cell cycle protease D1 gene that would stop cells from proliferating; (3) increasing cyclinase suppressor genes such as, *p21* and *p27*, to arrest cells at G1 stage; (4) down-regulating *Bcl-2* (the anti-induction apoptosis gene); and (5) activating caspase-3 (the induction apoptosis gene) to induce cell death. Rg3 was discovered to inhibit tumor cell proliferation and induce cell apoptosis in mice with induced liver cancer (Li et al. 2005). In addition, Rg3 can affect the differential expression of cell signaling genes and other related genes in human lung adenocarcinoma cell line A549, and induce apoptosis in the A549 tumor cells and HUVEC 304 cell lines. It was found that Rg3 and Rg5 had significant inhibition activity on benzo[a]pyrene-induced adenocarcinoma and dimethylbenz[a]anthracene-induced lung tumor in mice (Yun et al. 2001a). Both ginsenosides had strong inhibitory effects on the development of rat mammary adenocarcinoma induced by methyl-N-nitrosourea and N-ethyl-N-nitrosourea administration, as well as on uterine and vaginal tumors induced by dimethylbenz[a] anthracene (Bespalov et al. 2001). Chen et al. (2007) reported that Rg3 could induce apoptosis in human bladder transitional cell carcinoma cell line EJ at the IC_{50} value of 125.5 μg/ml after 48 hours of incubation. When treated with 150 μg/ml of Rg3 for 24 hours and 48 hours, the cells showed significant DNA ladders and apoptotic morphological characteristics, including condensed chromatin, nuclear fragmentation, apoptotic bodies, and bright fluorescent granules, as well as a higher caspase-3 expression. When the cells were treated with 75 μg/ml of Rg3 for 24 hours and 48 hours, the percentage of cells in S phase and G2/M phase was increased, whereas the percentage of cells in G0/G1 was decreased.

Ginsenoside Rh2 induces apoptosis in various tumor cells by different signaling pathways. In general, Rh2 cause cell apoptosis by interfering with Bcl-2 (B cell lymphocyte/leukemia-2) family proteins and caspase-3/PKC (protein kinase C) signaling transduction. Rh2 induced apoptosis in rat C6Bu-1 glioma cells and human SK-N-BE(2) neuroblastoma cells through PKC pathway. It was also found to induce apoptosis in human malignant melanoma, which was partially dependent on caspase-8 and caspase-3 (Fei et al. 2002). Cheng et al. (2005) demonstrated that ginsenoside Rh2 can induce apoptosis and inhibit cell growth in C6 glioma cells, human lung adenocarcinoma A549 cells, and various ovarian cancer cell lines. Mediating G1 growth arrest and apoptosis in human lung adenocarcinoma A549 cells appeared to be the molecular mechanisms of Rh2 in this research. Rh2 could also inhibit human hepatoma Bel-7404 cell lines *via* arresting cell cycle, up-regulating Bax protein expression, and down-regulating mutated p53 protein expression. In addition, Rh2 inhibited the growth of MCF-7 cells by inducing p21 protein expression and reducing cyclin D levels. As a result, cyclin/Cdk complex kinase activity, pRb phosphorylation, and E2F release could be inhibited.

Another ginsenoside, Compound K (or IH901), was shown to induce apoptosis in human hepatoblastoma HepG2 cells and KMS-11 cells through a cytochrome C-mediated activation of caspase-3 and caspase-8 proteases and inhibition of the FGFR3 (fibroblast growth factor receptor 3) expression (Oh and Lee 2004). Incubation of leukemic cells HL260 with Compound K produced apoptosis in the cells in a concentration- and time-dependent manner with morphologic changes in chromatic agglutination, atrophy, and nuclear fragmentation (Suda et al. 2000). Compound K suppressed melanoma cell proliferation in BI6-BL6 mice by activating PKC and releasing cytochrome C in mitochondria into cytoplasm. Western blot test revealed that Compound K could elevate p27Kipl expression, degrade expression of c-Myc and cyclin D1, and induce cell LLC apoptosis through activating caspase-3 protein kinase at the same time. Recently, IH-901 has been reported to induce both G1 arrest and apoptosis, and the apoptosis could be inhibited by COX-2 induction (Yim et al. 2005).

The combinational use of ginseng extracts or active ginsenosides with clinical anticancer drugs has also been studied. For example, the effect of panaxytriol (isolated from Asian ginseng) on the cytotoxicity of mitomycin C (MMC) against a human gastric carcinoma cell line, MK-1, was investigated, and a synergistic effect was observed when MK-1 cells were treated with the mixture of MMC and panaxytriol or treated with MMC followed by panaxytriol. Further research suggested that these synergistic effects may be induced by acceleration of the effect of MMC on cellular accumulation by panaxytriol (Matsunaga et al. 1994). In another research, quasipanaxatriol, ginsenoside Rh2, and compound K was found to greatly enhance the cytotoxicity of anticancer drugs daunomycin (DAU) and vinblastine in multidrug-resistant P388 leukemia cells (P388/ADM). The extent of enhancement ranged from 4- to 46-fold in DAU cytotoxicity, and 2- to 37-fold in vinblastine cytotoxicity. The reversal of DAU resistance in P388/ADM by quasipanaxatriol could be explained by the effective accumulation of the drugs mediated by the DAU-efflux blockage (Hasegawa et al. 1995).

4.3.2 *In Vivo Animal Test Evidences*

Using athymic mice transplanted with ovarian SKOV-3 cancer cells, Xu et al. (2007) showed that intraperitoneal injection of ginsenoside Rg3 alone, or Rg3 combined with cyclophosphamide, to the mice for 10 days improved the life quality and survival of the mice and reduced the tumor weight of the mice in comparison to that of the untreated control mice.

The inhibitory effect of ginsenoside Rh2 alone or together with cisplatin on the growth of human ovarian cancer cells (HRA) in nude mouse was studied, and it was discovered that po administration of Rh2 has a dose-dependent inhibitory effect on the tumor growth, while ip administration showed little activity. In addition, Rh2 significantly potentiated the inhibitory effect of cisplatin against ovarian cancer cell growth in the mice model, and these mice survived longer when given the ginsenoside preparation (Nakata et al. 1998).

Korean investigators carried out extensive long-term anti-carcinogenicity experiments with 2,000 newborn mice to investigate whether Asian ginseng could inhibit carcinogenesis induced by several chemical carcinogens. By using the so-called Yun's model, they confirmed significant anti-carcinogenic effects of powders and extracts of the 6-year-old dried fresh ginseng, 5- and 6-year old white ginsengs, and 4-, 5-, and 6-year old red ginseng. They also demonstrated that the anti-carcinogenicity of ginseng was more prominent in aged or heat treated extracts of ginseng and in red ginseng made by steaming (Yun et al. 2001b). The Yun's 9–12 weeks medium-term anti-carcinogenicity test mouse model was created as follows: The N:GP(S) newborn mice less than 24 hours old are subcutaneously injected once in the scapular region with 0.02 ml of benzo[a]pyrene (0.5 mg suspension of benzo[a] pyrene in aqueous gelatin) (Yun et al. 1995). After weaning, anti-carcinogenicity test materials are administered for 6 weeks through drinking water or diets. All mice are usually sacrificed at the 9th week after birth. The procedures to score the index of lung tumor incidence are then conducted.

To investigate the active components for cancer prevention, several fractions of 6-year old fresh ginseng and red ginseng, four semi-synthetic ginsenoside Rh1, Rh2, Rg3, and Rg5, major saponin components in red ginseng, were prepared. Among the ginsenosides, Rg3 and Rg5 showed statistically significant reduction of lung tumor incidence and Rh2 had a tendency of decreasing the incidence. Ginsenoside Rg3, Rg5, and Rh2 were found to be active anti-carcinogenic compounds. Rg3, Rg5, and Rh2 are active components in red ginseng, and they prevent cancer either singularly or synergistically (Yun et al. 2001a).

4.3.3 Evidence-based on Human Studies

Ginseng can improve immune system activity of cancer patients and their appetite, and function as a supplementary agent of chemotherapy. For instance, in Scaglione's human studies (1990), they examined effects of ginseng extract on following parameters of healthy volunteers: chemotaxis, phagocytosis index, phagocytosis fraction, intracellular killing, total lymphocytes (T3), T helper (T4) subset, suppressor cells (T8) subset, blastogenesis of circulating lymphocytes, natural killer cell activity. Each group ($n=20$) was given 100 mg of the ginseng extract every 12 hours daily for up to 8 weeks. Blood samples were withdrawn before beginning the treatment, at the 4th week and at the 8th week. They found that the activity of these immune biomarkers was significantly enhanced at the end of the 8-week study in comparison with the placebo group ($P<0.05$ or 0.001).

Xing et al. (2001) treated 35 rectal cancer patients with retention enema containing 85% ginsenoside for 4–6 hours per day for 6–8 consecutive days before surgical operation. The control group ($n=15$) received retention enema containing saline in the same way. They reported that after ginsenoside treatment, symptoms such as frequent defecation, hematochezia, and tenesmus were palliated in most patients (25 out of 35), and abdominal pain was relieved in 7 patients with incomplete intestinal obstruction. Electron microscopic examination showed apoptosis in 23 treated patients. In comparison, the above-mentioned changes were not observed in the control group. Pre-clinical studies have also showed some immune-stimulating activity of ginseng and ginsenosides (Xing et al. 2001; Block and Mead 2003).

Suh et al. (2002) reported that the red ginseng powder from Asian ginseng inhibits the recurrence of AJCC stage III gastric cancer. The CD4 and CD3 levels in patients during postoperative chemotherapy were restored to the initial preoperative values after red ginseng powder ingestion demonstrated by the flow cytometric analyses for peripheral T lymphocyte subsets, suggesting some immunomodulatory properties of ginseng in patients with advanced gastric cancer during postoperative chemotherapy. This study further demonstrated that the 5-year disease free survival and overall survival rate in patients taking the red ginseng powder during postoperative chemotherapy was significantly higher *versus* control

(68.2% *versus* 33.3%, 76.4% *versus* 38.5%, respectively, $P<0.05$). In spite of the limitation of a small number of patients ($n=42$), these findings suggest that red ginseng powder may help to improve postoperative survival in these gastric cancer patients. In another study, a commercial combination product that contained ginseng was administered to 126 cachectic cancer patients. Improvements were also observed in fatigue, pain tolerance, mental concentration, physical activity, and appetite (Chang et al. 2003).

Ginseng consumption reduced the risk of development of all types of cancer. Yun proposed that all non-toxic chemo-preventive agents should be classified into three categories: organ-specific, multiorgan-specific, and non-organ-specific agents (Yun 2001). He suggested that the first priority should be clinical studies of non-organ-specific cancer preventives. This approach would be justified in terms of time and cost-effectiveness and raise the possibility of providing a real chemopreventive medicine for general use in cancer. In a case-control study, Yun and Choi (1990, 1995) performed epidemiological studies among Korean people and the results demonstrated the non-organ-specific cancer-preventive activity of ginseng extracts. In their case-control study on the cancer-preventive effect of ginseng, the number of the subjects was extended from the original 905 pairs to later 1,987 pairs. In both the epidemiological studies, odds ratios (OR) of white ginseng powder intake were 0.44 (for 905 pairs) and 0.30 (for 1,987 pairs), and odds ratios of red ginseng extracts intake were 0.45 (for 905 pairs) and 0.20 (for 1,987 pairs). The OR indicates the possibility of being falsely significant. The smaller the number is, the lesser the possibility of being falsely significant is. Yun's studies stated that the preventive effects of ginseng on cancers were non-organ-specific. The consumption of fresh ginseng slices, fresh ginseng juice, and white ginseng tea did not decrease the risk for cancer. However, as the odds ratios show, the risk for cancer was rather unexpectedly low in the cases of intake of 1–3 times/year (OR=0.62), 4–11 times/year (OR=0.48), and 1 time/month or more (OR=0.31). Overall, the risk decreased as the frequency and duration of ginseng consumption increased. With respect to the site of cancer, the ORs for cancers of the lip, oral cavity, pharynx, esophagus, stomach, colorectum, liver, pancreas, larynx, lung, and ovary were significantly reduced by ginseng usage. Smokers with ginseng intake showed lower ORs for cancers of lung, lip, oral cavity, pharynx, and liver than those without ginseng intake.

The cancer preventive effect was also observed in a prospective study conducted on 4,634 people living in the ginseng cultivation area (Yun and Choi 1998). Subjects born before 1947 (over 40 years) were selected. A cohort of 4,634 persons over 40 years of age residing in Kangwha-eup (the ginseng cultivation area) was interviewed and examined between August 1987 and December 1989. Each study subject was interviewed by means of a standard questionnaire about demographic characteristics, life-long occupation, smoking and drinking habits, and past history of diseases, etc. In an attempt to obtain detailed information on ginseng intake, they used the same questionnaire as used in the previous two case-control studies (Yun and Choi 1990, 1995). The interviewers had been instructed and trained beforehand to ensure uniformity in the method of inquiry. The follow-up studies were carried out on all cohort members to document the development of cancer and other illnesses and to update exposure information. The length of the follow-up was calculated for each individual in the study as the number of days elapsed since completion of the questionnaire until death from cancer or other diseases. Deaths among the cohort from August 1987 to December 1997 were traced by population registration cards with no follow-up loss. A cohort member was classified as a cancer case if they had any disease code of cancer in hospital records, death certificates of the provincial government, privileged data of Korea Medical Insurance Corporation, etc. In the 5-year follow-up cohort study cancer risk significantly decreased among consumers of fresh ginseng extract, alone or together with other ginseng products. Among 24 red ginseng consumers, no cancer death occurred during the follow-up period. The risk for stomach and lung cancers was significantly reduced by ginseng intake, showing a statistically significant dose-response relationship in each follow-up year. These results suggest that ginseng exerted a preventive effect against the development of cancers of all organs, i.e. non-organ-specific.

4.4 Perspectives and Challenges

Ginseng is used in traditional Chinese medicine and is of interests to cancer patients in many other countries chiefly for its reputed anti-fatigue properties. However, ginseng is chemically complex, with more than 30 ginsenosides, polysaccharides or glycans, enzymes, organic acids and ester, amino acids, sugars, essential oils, minerals, etc. The active ingredients are believed to be the ginsenosides, and the contents of individual ginsenosides are more important than the total amount, since the individual efficacy or/and potency of each ginsenoside differs and will affect the pharmacological effects overall. However, as other botanic products, the percentage of the active ginsenosides, such as Rg3, Rg5, Rb1, and Rh2, varies from species to species and from extraction preparation methods, making the quality control difficult and challenging.

Numerous studies have examined cytotoxicity of ginseng and its fractions *in vitro* against a wide variety of cancer cell lines or *in vivo* in neoplasm models, with mixed results, probably due to the difference in constituents. Several of the isolated ginsenosides have demonstrated interesting anticancer activities, including induction of apoptosis, inhibition of cell cycle progression, and anti-angiogenic activity. Yet, the presences of these active ingredients are in trace amount in the crude extracts, which further necessitate the quality control guideline of ginseng products.

Studies on ginseng and cancer therapy in humans are rare. More reports focus on cancer prevention, which showed beneficial preventive effects of ginseng towards the risk of cancers. However, more carefully designed clinical studies are still needed to further address issues concerning the immunomodulation, anti-stress, anticancer, and anti-fatigue properties of ginseng, which are not statistically conclusive based on current random studies.

References

Azuma, I., & Mochizuki, M. (1994). Inhibition of tumor cell invasion and metastasis by ginsenosides. *Ginseng Review, 18,* 37–39.

Bespalov, V. G., Alexandrov, V. A., Limarenko, A. Y., et al. (2001). Chemoprevention of mammary, cervix and nervous system carcinogenesis in animals using cultured Panax ginseng drugs and preliminary clinical trials in patients with precancerous lesions of the esophagus and endometrium. *Journal of Korean Medical Science, 16*(Suppl), S42–S53.

Block, K. I, & Mead, M. N. (2003). Immune system effects of echinacea, ginseng, and astragalus: A review. *Integrative Cancer Therapies, 2,* 247–267.

Chang, Y. S., Seo, E. K., Gyllenhaal, C., et al. (2003). Panax ginseng: A role in cancer therapy? *Integrative Cancer Therapies, 2,* 13–33.

Chen, J. X., Peng, H. M., Pu, S. P., et al. (2007). Inducement effect of ginsenoside Rg3 on apoptosis of human bladder transitional cell carcinoma cell line EJ. *Zhongguo Zhong Yao Za Zhi, 32,* 1680–1684.

Cheng, C. C., Yang, S. M., Huang, C. Y., et al. (2005). Molecular mechanisms of ginsenoside Rh2-mediated G1 growth arrest and apoptosis in human lung adenocarcinoma A549 cells. *Cancer Chemotherapy and Pharmacology, 55,* 531–540.

Fei, X. F., Wang, B. X., Tashiro, S., et al. (2002). Apoptotic effects of ginsenoside Rh2 on human malignant melanoma A375-S2 cells. *Acta Pharmacologica Sinica, 23,* 315–322.

Hasegawa, H., Sung, J. H., Matsumiya, S., et al. (1995). Reversal of daunomycin and vinblastine resistance in multidrug-resistant P388 leukemia in vitro through enhanced cytotoxicity by triterpenoids. *Planta Medica, 61,* 409–413.

Jia, L., & Zhao, Y. (2009). Current evaluation of the millennium phytomedicine—Ginseng (I): Etymology, pharmacognosy, phytochemistry, market and regulations. *Current Medicinal Chemistry, 16,* 2475–2484.

Jia, L., Noker, P. E., Piazza, G. A., et al. (2008). Pharmacokinetics and pharmacodynamics of Phor21-betaCG(ala), a lytic peptide conjugate. *The Journal of Pharmacy and Pharmacology, 60,* 1441–1448.

Jia, L., Zhao, Y., & Liang, X. J. (2009). Current evaluation of the millennium phytomedicine-ginseng (II): Collected chemical entities, modern pharmacology, and clinical applications emanated from traditional Chinese medicine. *Current Medicinal Chemistry, 16,* 2924–2942.

Kaizuka, H., & Takahashi, K. (1983). High-performance liquid chromatographic system for a wide range of naturally occurring glycosides. *Journal of Chromatography, 258,* 135–146.

Kasai, R., Bess, H., Tanaka, O., et al. (1983). Saponin of red ginseng. *Chemical Pharmaceutical Bulletin, 31,* 2120–2125.

Kikuchi, Y., Sasa, H., Kita, T., et al. (1991). Inhibition of human ovarian cancer cell proliferation in vitro by ginsenoside Rh2 and adjuvant effects to cisplatin in vivo. *Anti-Cancer Drugs, 2,* 63–67.

Kim, S. I., Park, J. H., Ryu, J. H., et al. (1996). Ginsenoside Rg5, a genuine dammarane glycoside from Korean red ginseng. *Archives of Pharmacal Research, 19,* 551–553.

Kitagawa, I., Kobayashi, M., Akedo, H., et al. (1995). Inhibition of tumor cell invasion and metastasis by ginsenoside Rg-3. *Ginseng Review, 20,* 41–46.

Li, X., Guan, Y. S., Zhou, X. P., et al. (2005). Anticarcinogenic effect of 20(R)-ginsenoside Rg3 on induced hepatocellular carcinoma in rats. *Sichuan Da Xue Xue Bao Yi Xue Ban, 36,* 217–220.

Liu, W. K., Xu, S. X., & Che, C. T. (2000). Anti-proliferative effect of ginseng saponins on human prostate cancer cell line. *Life Sciences, 67,* 1297–1306.

Matsunaga, H., Katano, M., Saita, T., et al. (1994). Potentiation of cytotoxicity of mitomycin C by a polyacetylenic alcohol, panaxytriol. *Cancer Chemotherapy and Pharmacology, 33,* 291–297.

Nakata, H., Kikuchi, Y., Tode, T., et al. (1998). Inhibitory effects of ginsenoside Rh2 on tumor growth in nude mice bearing human ovarian cancer cells. *Japanese Journal of Cancer Research, 89,* 733–740.

Oh, S., & Lee, B. H. (2004). A ginseng saponin metabolite-induced apoptosis in HepG2 cells involves a mitochondria-mediated pathway and its downstream caspase-8 activation and Bid cleavage. *Toxicology and Applied Pharmacology, 194,* 221–229.

Scaglione, F., Ferrara, F., Dugnani, S., et al. (1990). Immunomodulatory effects of two extracts of Panax ginseng C.A. Meyer. *Drugs Under Experimental and Clinical Research, 16,* 537–542.

Suda, K., Murakami, K., Hasegawa, H., et al. (2000). Induction of apoptosis in Lewis lung carcinoma cells by an intestinal bacterial metabolite produced from orally administered ginseng protopanaxadiol saponins. *Journal of Traditional Medicine, 17,* 236–244.

Suh, S. O., Kroh, M., Kim, N. R., et al. (2002). Effects of red ginseng upon postoperative immunity and survival in patients with stage III gastric cancer. *The American Journal of Chinese Medicine, 30,* 483–494.

Tanaka, O., Kasai, R., & Morita, T. (1986). Chemistry of Ginseng and related plants: Recent advances. *Chinese University of Hong Kong: Abstract of Chinese Medicines, 1,* 130–152.

Xing, J. H., Chen, Y. Q., & Ji, M. X. (2001). Clinical study on effect of ginsenoside in inducing rectal cancer cell apoptosis. *Zhongguo Zhong Xi Yi Jie He Za Zhi, 21,* 260–261.

Xu, T. M., Xin, Y., Cui, M. H., et al. (2007). Inhibitory effect of ginsenoside Rg3 combined with cyclophosphamide on growth and angiogenesis of ovarian cancer. *Chinese Medical Journal, 120,* 584–588.

Yim, H. W., Jong, H. S., Kim, T. Y., et al. (2005). Cyclooxygenase-2 inhibits novel ginseng metabolite-mediated apoptosis. *Cancer Research, 65,* 1952–1960.

Yun, T. K. (2001). Panax ginseng—A non-organ-specific cancer preventive? *The Lancet Oncology, 2,* 49–55.

Yun, T. K., & Choi, S. Y. (1990). A case-control study of ginseng intake and cancer. *International Journal of Epidemiology, 19,* 871–876.

Yun, T. K., & Choi, S. Y. (1995). Preventive effect of ginseng intake against various human cancers: A case-control study on 1987 pairs. *Cancer Epidemiology Biomarkers Prevention, 4,* 401–408.

Yun, T. K., & Choi, S. Y. (1998). Non-organ specific cancer prevention of ginseng: A prospective study in Korea. *International Journal of Epidemiology, 27,* 359–364.

Yun, T. K., Kim, S. H., & Lee, Y. S. (1995). Trial of a new medium-term model using benzo(a) pyrene induced lung tumor in newborn mice. *Anticancer Research, 15,* 839–845.

Yun, T. K., Lee, Y. S., Choi, K. J., et al. (2000). Anticarcinogenicity of various ginseng fractions and components in red ginseng using Yun's anticarcinogenicity test model. *Journal of the Korean Association Cancer Prevention, 5,* 186–192.

Yun, T. K., Lee, Y. S., Lee, Y. H., et al. (2001a). Anticarcinogenic effect of Panax ginseng C.A. Meyer and identification of active compounds. *Journal of Korean Medical Science, 16*(Suppl), S6–S18.

Yun, T. K., Lee, Y. S., Lee, Y. H., et al. (2001b). Cancer chemopreventive compounds of red ginseng produced from Panax ginseng C.A. Meyer. *Journal of Ginseng Research, 25,* 107–111.

Zheng, B. C. (1985). Shennong's herbal—One of the world's earliest pharmacopoeia. *Journal of Traditional Chinese Medicine, 5,* 236.

Chapter 5
Panax Quinquefolius (American Ginseng) and *Panax Notoginseng* (Notoginseng) in Cancer Chemoprevention

Chong-Zhi Wang and Chun-Su Yuan

Abstract The clinical management of cancer invariably involves diverse conventional modalities, including surgery, radiation, and chemotherapy. However, the complexity of human cancer requires some alternative management to improve the therapeutic efficacy of conventional treatment and/or the quality of life of cancer patients. Medicinal botanicals have recently gained more attention for cancer management. Numerous effective anticancer drugs have been developed from botanicals, and identifying new herbal sources to develop ideal chemoprevention remains an essential step in advancing the treatment of cancer. In this chapter, potential roles of ginseng herbs, especially American ginseng and notoginseng, in cancer chemoprevention are presented. The major pharmacologically active constituents of ginsengs are ginsenosides, which can be mainly classified into protopanaxadiol and protopanaxatriol groups. The recognized active anticancer compounds from American ginseng and notoginseng are ginsenosides Rg3, Rh2, and protopanaxadiol. The structure-activity relationship between their chemical structures and pharmacological activities is discussed. Sugar molecules within a ginsenoside have a high impact on cancer cells. Anticancer activities increase with the decrease of sugar number. In addition, various steaming temperatures and time treatments of the ginseng herbs can change their ginsenoside profiles and enhance their anticancer activities. This heat treatment process may increase the role and efficacy of American ginseng and notoginseng in cancer chemoprevention.

5.1 Introduction

Cancer is a leading cause of death worldwide. In the United States, a total of 1,529,560 new cancer cases and 569,490 deaths from cancer are projected to occur in 2010 (Jemal et al. 2010). Although early diagnosis with rigorous screening may

C.-Z. Wang (✉)
Tang Center for Herbal Medicine Research, University of Chicago, 5841 South Maryland Avenue, MC 4028, Chicago, IL 60637, USA
e-mail: czwang@dacc.uchicago.edu

W. C. S. Cho (ed.), *Evidence-based Anticancer Materia Medica,* Evidence-based Anticancer Complementary and Alternative Medicine,
DOI 10.1007/978-94-007-0526-5_5,

have reduced its incidence compared to a few years ago, the prognosis associated with metastatic disease remains bleak. Current cancer treatment generally employs surgical resection combined with chemotherapy, using cytotoxic drugs and radiation therapy. Because this therapy is only moderately successful for late stage cancers, novel approaches to the treatment of cancer are required. The data from several controlled clinical trials supports a multimodal and multidisciplinary approach, including combinations of treatments and schedules in which they are administered, for treating both early and advance stage cancers (Goldberg et al. 2004; Hurwitz et al. 2004; Ma and Adjei 2009). Studies also showed that patients with cancer often resort to complementary and alternative medicine for the treatment of cancer-related symptoms and/or to reduce the adverse effects of chemotherapy (Shumay et al. 2001; Ott 2002; Lee et al. 2006; Wu et al. 2007).

There is compelling evidence that patients in this country resort to supplements or substitute them for conventional pharmacotherapy. Several national surveys indicate that at least one third of American adults take some form of dietary supplement, and botanicals comprise approximately 25% of the supplement market (Barnes et al. 2004). Botanicals have also been the major source of therapy in many traditional medical systems and have been used clinically for the treatment of a variety of diseases (Mashour et al. 1998; Wang et al. 2007a, c; Wicks et al. 2007; Xie et al. 2006). Botanical ingredients in natural products contain bioactive constituents with medical benefits (Akerele 1993; Leung 2007; Zhou et al. 2007; Li and Zhang 2008). Furthermore, botanicals have contributed significantly to cancer therapy, and it is likely that extracts and active constituents from herbal medicine will continue to play an important role in cancer therapeutics (Liu and Jiang 2006; Ng et al. 2006; Shieh et al. 2006; Ozaslan et al. 2007). In this chapter, we will discuss the potential roles of ginseng herbs in the chemoprevention of cancer.

5.2 Medicinal Use of Botanicals in Ginseng Family

Panax L. is a small genus in the family Araliaceae. Nearly all species in the genus *Panax*, such as *Panax ginseng* C.A. Meyer (Asian ginseng), *Panax quinquefolius* L (American ginseng), and *Panax notoginseng* (Burk) F.H. Chen (notoginseng), are important herbs used for different medical conditions (Chen et al. 2001; Wang et al. 2007c). Asian ginseng and notoginseng are considered Chinese herbal medicines, and American ginseng is one of the most commonly used botanicals in the US (Wang et al. 1999, 2006c).

It is generally believed that the active compounds in Asian ginseng, American ginseng, and notoginseng are triterpene glycosides or dammarane saponins, commonly referred to as ginseng saponins (ginsenosides and notoginsenosides). These ginseng saponins are the major active ingredients in the herbs, and their levels can be used to develop quality controls for these herbs (Fuzzati 2004; Chao et al. 2006; Wang et al. 2006a). There are over 100 different known ginseng saponins,

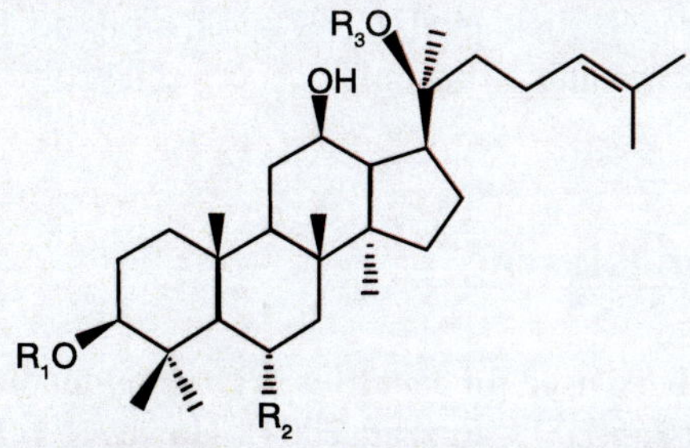

Compound	R_1	R_2	R_3
Notoginsenoside R1	-H	-O-glc^{2}–^{1}xyl	-glc
Ginsenoside Rb1	-glc^{2}–^{1}glc	-H	-glc^{6}–^{1}glc
Ginsenoside Rb2	-glc^{2}–^{1}glc	-H	-glc^{6}–^{1}ara(pyr)
Ginsenoside Rb3	-glc^{2}–^{1}glc	-H	-glc^{6}–^{1}xyl
Ginsenoside Rc	-glc^{2}–^{1}glc	-H	-glc^{6}–^{1}ara(fur)
Ginsenoside Rd	-glc^{2}–^{1}glc	-H	-glc
Ginsenoside Re	-H	-O-glc^{2}–^{1}rha	-glc
Ginsenoside Rg1	-H	-O-glc	-glc
Ginsenoside Rg2	-H	-O-glc^{2}–^{1}rha	-H
Ginsenoside Rg3	-glc^{2}–^{1}glc	-H	-H

Fig. 5.1 Chemical structures of saponins from American ginseng and notoginseng

and they are characterized by a four *trans*-ring rigid steroid aglycone skeleton and attached sugar moieties. Based on the aglycone skeleton, all representative ginseng saponins can be divided into protopanaxadiol group and protopanaxatriol group, except for the ginsenoside Ro, which is derived from the oleanolic acid group (Fig. 5.1).

Ginseng has many reported health benefits (Attele et al. 1999; Liu et al. 2006; Yamakage et al. 2006; Yoo et al. 2006). Regarding its anticancer effects, a case-control study on over one thousand subjects in Korea showed that Asian or Korean ginseng intakers had a decreased risk for many different cancers compared with nontakers (Yun and Choi 1995, 1998). It also suggested that ginseng has a non-organ specific preventive effect against cancer (Yun 2003).

Regarding the responsible anticancer constituents from Asian ginseng, published studies showed that some saponins could reduce the proliferation of cancer cells and sensitize cancer cells to chemotherapeutic agents *in vitro* (Lee and Huemer 1971; Kim et al. 2007; Koo et al. 2007). Several investigators found antitumor properties and other pharmacological activities of ginseng, and ginsenosides Rg3 and Rh2 are recognized as active anticancer saponins (Helms 2004). Jia et al. (2004) noted that ginsenoside Rh2 inhibited proliferation and induced apoptosis in cancer cell lines, and sensitized drug-resistant breast cancer cells to paclitaxel. Kim et al. (2004) studied 11 ginsenosides and determined that Rg3 and Rh2 inhibited the proliferation of prostate cancer cells. Iishi et al. (1997) used a rat model to determine the effects of ginsenoside Rg3 on inhibiting colon cancer cell proliferation.

5.3 American Ginseng

Ginseng root has been used for centuries in Oriental medicine as a panacea that promotes longevity (Yun 2003; Fuzzati 2004). However, relatively few studies focus on American ginseng, which is a popular herbal supplement in the US with consumers and patients (Attele et al. 1999; Helms 2004).

American ginseng is an obligate shade perennial plant native to eastern North America. The most commonly used part of the plant is the root, which is harvested after several years' cultivation. The largest ginseng growing area in the US is in Wisconsin. The bioactive constituents of American ginseng are ginsenosides, which are present in the root, leaf, stem, and berry of the plant. More than 60 ginsenosides, such as Rb1, Rb2, Rb3, Rc, Rd, Re, Rg1, Rg2, and Rg3 have been identified (Wang et al. 1999, 2006b; Assinewe et al. 2003; Qi et al. 2010) in American ginseng (Fig. 5.1). Previous studies of American ginseng focused on its activities on the cardiovascular system, such as its anti-ischemic, anti-arrhythmic, and antihypertensive effects (Yuan and Dey 2001; Kim and Park 2003). These pharmacological effects are, to a significant extent, considered to be linked to the antioxidant properties of the herb (Kitts et al. 2000; Wang et al. 2007c).

American ginseng extracts were found to inhibit the growth of breast cancer cells (Corbit et al. 2006). We investigated the effects of several herbal extracts on reducing chemotherapeutic side effects and found that American ginseng and one of its major constituents, ginsenoside Re, can attenuate cisplatin-induced nausea and vomiting in a rat model, while not affecting its anticancer properties in human cancer cells (Aung et al. 2007; Mehendale et al. 2005). In addition, the extract from American ginseng enhanced the anti-proliferation effect of cisplatin on human breast cancer cells, suggesting that it possesses its own anticancer activity (Aung et al. 2007). Our group also showed that after steaming American ginseng, its anti-proliferative effects improved significantly, possibly due to the altered ginsenoside profile (Wang et al. 2006c, 2007a, 2008).

To explore the mechanisms involved in cancer cell inhibition, we observed the effects of American ginseng on the gene expression and apoptotic pathways. From the analysis of microarray hybridization, we found that the anticancer mechanism of American ginseng extract and its representative compound, ginsenoside Rg3, have many of the same characteristics and the alterations in gene expression level imply important information for exploring this mechanism. The two recognized genes regulated by the extract and Rg3 (AKAPA8L and PITPNA) suggest that American ginseng takes effect through the regulation of cell mitosis and an intracellular singling pathway (Luo et al. 2008). In a separate study, the observed expression profiling of the selected pathways revealed various apoptotic related genes that inhibited growth in human colorectal cancer cells treated with American ginseng. The mitochondrial apoptotic pathway was a key target in cancer chemoprevention by steamed American ginseng extract (Wang et al. 2009a; Li et al. 2010). These expression analyses may lead to the identification of markers that predict the responsiveness of human cancer cells to American ginseng treatment.

Chronic inflammation is associated with increased cancer risk (Aggarwal et al. 2009). Ginseng has been observed to play a role in reducing inflammation, and suppressing colitis through p53-mediated apoptosis of inflammatory cells (Hofseth and Wargovich 2007; Jin et al. 2010). Ginsenoside compound K [20-O-beta-D-glucopyranosyl-20(S)-protopanaxadiol] exerts immunomodulatory and anti-inflammatory effects by deactivating the inflammatory response through the inhibition of COX-2 (cyclooxygenase-2) expression (Kimura et al. 2006; Hofseth and Wargovich 2007). Ginsenoside Rg3 attenuates COX-2 expression, NF-κB (nuclear factor-κB) activation, and activator protein-1 transcription factors (Lee et al. 2005). PPD inhibits inducible nitric oxide synthase and COX-2 expression through the inactivation of NF-κB (Kim et al. 2010).

For the clinical study, to investigate whether American ginseng might help cancer-related fatigue, in a randomized, double-blind study, 290 cancer patients received American ginseng in doses of 750, 1,000, or 2,000 mg/day or a placebo given twice daily over 8 weeks. Overall, this study suggested that American ginseng, at a dose of 750 mg/day, did not provide any benefit over that seen with a placebo. However, the two highest doses of ginseng (1,000 and 2,000 mg/day) did appear to decrease fatigue more than a placebo, as measured by various scales of fatigue, vitality, and well being. Data suggested that the higher doses studied may be helpful in cancer-related fatigue (Barton et al. 2010).

5.4 Notoginseng

Notoginseng is a Chinese herbal medicine that has a long history of use in China and other Asian countries. This herb is distributed in southwest China, Burma, and Nepal. Notoginseng is cultivated commercially in southwest China, especially in the Yunnan Province. The root is commonly used in remedies and it is dug up after the fruit has ripened.

The earliest scientific description of notoginseng was in Compendium of Materia Medica, a dictionary of Chinese herbs, written by Li Shi Zhen (1518–1593 AD). In Compendium of Materia Medica, notoginseng was described as "more valuable than gold," indicating the significance of this herb in traditional Chinese medicines. Notoginseng is regarded as the emperor herb in the treatment of different types of wounds because it is favored for the treatment of both internal and external hemorrhage (Ng et al. 2006; Wang et al. 2006c).

Modern pharmacological research on notoginseng has found that notoginseng exerts various effects on the cardiovascular system, central nervous system, endocrine system, and the inflammation response (Sun et al. 2005; Wang et al. 2006a, c).

Consistent with the hemostatic effect of notoginseng reported in ancient China, recent studies showed that the alcohol extract of notoginseng resulted in a reduction of the extent of bleeding and provides better hemostatic effects than no treatment, placebo treatment, or treatment with hydrophilic or lipophilic extracts (White et al. 2001). Notoginseng can also decrease blood pressure, improve blood supply, protect against shock, and protect the cardiovascular system and brain vasculature. Its protective mechanism could work by conferring protection against damage by oxygen free radicals, and also by binding to the estrogen receptor, since ginsenosides share many of the protective actions of estrogen in various body systems. Pharmacokinetic and pharmacodynamic studies have shown that the intranasal preparation of notoginseng saponins is a promising development and may be beneficial for the treatment of Alzheimer's disease. Notoginseng extracts were also found to possess the capacity to adjust energy metabolism and treat diabetes (Ng 2006).

Notoginseng has a very distinct saponin profile compared to that of American ginseng (Chen et al. 2001; Sun et al. 2005). The main bioactive compounds in notoginseng are dammarane saponins. Oleanane-type saponins, present in Asian ginseng and American ginseng, are not found in notoginseng. To date, 56 saponins have been isolated from the notoginseng plant. 35 of these notoginseng saponins belong to the protopanaxadiol group, while 21 of them belong to the protopanaxatriol group (Sun et al. 2005; Wang et al. 2006a). Ginsenosides Rb1, Rg1, Rd, and notoginsenoside R1 are the main saponins in the notoginseng root (Fig. 5.1).

Some studies have shown that notoginseng has antitumor effects (Chen et al. 2001; Wang et al. 2006c). We observed that the notoginseng root extract and its constituents have significant antiproliferative effects on human colorectal cancer cells (Wang et al. 2007a). Other plant parts of notoginseng also displayed potential antiproliferative effects on colorectal cancer cells (Wang et al. 2009b). The flower extract's most potent cancer cell growth inhibitory effects were shown within special chemical compositions (Wang et al. 2009b).

Ginsenosides Rb1 and Rd are major constituents in American ginseng and notoginseng. After oral administration, protopanaxadiol-type ginsenosides such as Rb1 and Rd are mostly metabolized by intestinal bacteria to a protopanaxadiol monoglucoside, compound K. In humans, compound K is detected in plasma 7 hours after the intake of ginsenosides and in urine 12 hours after the intake, indicating that compound K is the final metabolite of this type of ginsenoside (Ren et al. 2008). Ginsenoside Rb1 was not detectable in serum for 24 hours, indicating a major intestinal bacterial metabolism (Wakabayashi et al. 1997). Compound K has been recognized as an potential anticancer compound (Jeong et al. 2010).

Recently, we performed a steaming treatment on notoginseng root. We observed that there was a significant difference in the ginsenoside content after steaming the notoginseng root. No differences were observed in total ginsenoside content and antiproliferative effect between steaming the root for 4 hours or 6 hours. After steaming, ginsenoside Rg3 content increased significantly, an increase that was partially responsible for the increase in anticancer activity. On the other hand, ginsenoside Rh2 content increased only slightly after steaming. Thus, some other active anticancer components, in addition to Rg3 and Rh2, may form in the notoginseng root extract after the steaming process (Sun et al. 2010).

Our group also found that the notoginseng extract can increase the effects of cancer chemotherapy. Using the HCT-116 human colorectal cancer cell line, the antiproliferative effect of notoginseng extract combined with 5-FU was investigated. Compared to the control, when cells were treated with 5-FU or notoginseng separately, cell proliferation was reduced by 31% and 25%, respectively. The combination of 5-FU and notoginseng reduced cell proliferation by 59%, suggesting that combining notoginseng with 5-FU can reduce the dose of 5-FU, while significantly increasing the overall anti-proliferation effect on the cancer cells. Since it is well known that 5-FU has cytotoxic effects on primary cells, this synergistic effect between notoginseng and 5-FU makes it possible to reduce the dose of 5-FU in combination with notoginseng and thereby further decrease dose-related toxicity (Wang et al. 2007b).

In another study, notoginseng's potential to enhance the effects of irinotecan without affecting irinotecan's activity was observed. It appears that irinotecan-induced toxicity can be reduced by using notoginseng as a chemo-adjuvant (Wang et al. 2007a). When notoginseng potentiates the tumoricidal effects of chemotherapeutic agents, smaller chemotherapy doses can be used. Data obtained from our studies will have the potential to advance treatment regimens and improve the quality of life for patients suffering from cancer.

5.5 Saponin Structure-activity Observation and Heat-treatment of Ginsengs

Ginseng saponins belong to a family of triterpene glycosides or triterpene saponins. Ginseng saponins (except ginsenoside Ro) possess the four *trans*-ring rigid steroid skeleton, with a modified side chain at C-20. Sugar residues are attached to the -OH of the aglycon. As mentioned above, ginsenosides can be mainly classified into protopanaxadiol and protopanaxatriol groups. For the protopanaxadiol group, sugar residues are attached to the β-OH at C-3 and another -OH at C-20 of the aglycon, e.g. ginsenosides Rb1, Rb2, Rc, Rd, Rg3, and Rh2. For the protopanaxatriol group, sugar residues are attached to the α-OH at C-6 and another -OH at C-20 of the aglycon, e.g. ginsenosides Re, Rg1, Rh1, and notoginsenoside R1 (Fig. 5.1).

The structure-activity relationship elucidates the relations between chemical structure and the pharmacological activity for a series of compounds (Ooi et al. 2006; Benjamin et al. 2008). The anticancer activities of ginseng saponins are related to their aglycons and sugar residues (Helms 2004; Wang et al. 2007d). Sugar molecules within a ginsenoside have a high impact on cancer cells. Anticancer activities increase with a decrease in sugar number. The main anticancer saponins so far identified are from the protopanaxadiol group. The three most potent compounds in this group are Rg3, Rh2, and their aglycon, protopanaxadiol, and the latter two may have stronger effects (Popovich and Kitts 2002; Wang et al. 2007d).

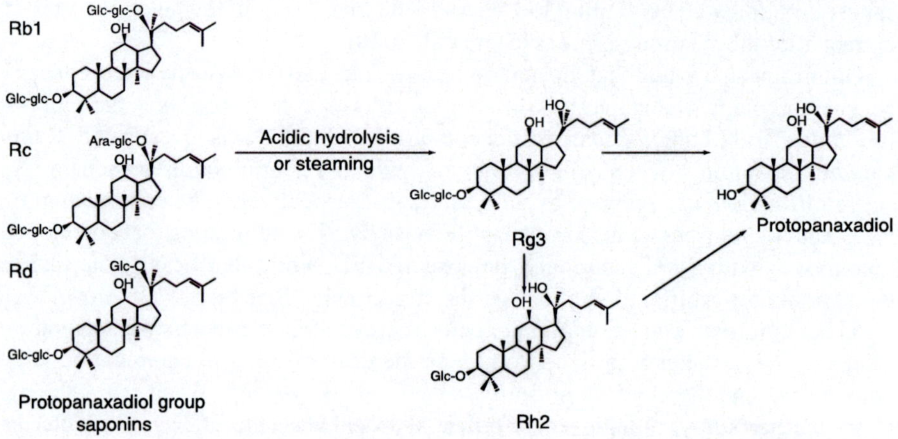

Fig. 5.2 Chemical conversions starting from protopanaxadiol group saponins using acidic hydrolysis or steaming process

Other compounds in the protopanaxadiol group showed less or no anticancer activities, probably due to the fact that their sugar residues are attached to the -OH at C-20 (Wang et al. 2006c).

Ginsenoside Rg3 was isolated from Asian ginseng, American ginseng, and notoginseng (Chen et al. 2002; Xu et al. 1987). However, Rg3 is only a trace saponin in different species of the genus *Panax* (Fuzzati 2004). Rg3 can also be obtained from a mild acidic hydrolysis of protopanaxadiol group saponins, such as Rb1, Rb2, and Rc (Fig. 5.2). Since Rg3 was found to effectively inhibit the growth of cancer cells (Mochizuki et al. 1995), studies of Rg3 sources were emphasized. In 2003, Rg3 was approved as a new anticancer drug in China (Lu et al. 2008). Although this saponin can be obtained by biological transformation and chemical synthesis, the process is complicated, the yield is limited and thus, the cost of the product is high. As shown in Fig. 5.2, Rh2 and protopanaxadiol are also derived from the protopanaxadiol group saponins. In Asia, Asian ginseng root can be prepared as (1) air-dried to white ginseng, or (2) steamed at approximately 100°C to red ginseng. Red ginseng has stronger anticancer activities than white ginseng (Yun et al. 2001) due to its relatively high Rg3 content. It seems likely that the steaming process or heat-treatment of ginseng is a good approach to transform inactive ginsenosides to active anticancer compounds, such as Rg3, Rh2, and protopanaxadiol.

Our laboratory treated the American ginseng berry at various temperatures and heating times to observe the changes in ginsenoside content and anticancer activities on human colorectal cancer cells. We found that steaming the American ginseng berry extract very significantly augmented the content of Rg3. When human colorectal cancer cells were treated with steamed berry extract (120°C, 2 hours), the anti-proliferation effects were 98% for HCT-116 and 99% for SW-480 cells. At the same treatment concentration, the effects of unsteamed extract were 34% for HCT-116 and 5% for SW-480 cells. This suggests that the steamed American ginseng berry augmented Rg3 content and anticancer activity significantly (Wang et al. 2006b). We also steamed American ginseng root and found a comparable change to its chemical constituent and antiproliferative activities (Wang et al. 2007a). Recent studies suggested that increasing the steaming time resulted in additional chemical changes and an increase in cancer cell growth inhibitory effects (Wang et al. 2009a).

Constituent changes of notoginseng after steaming treatment have also been reported (Lau et al. 2004). After the treatment, the content of Rb1, Rg1, Rd, and notoginsenoside R1 decreased, while Rg3 had increase, and this trend is similar to what we observed after the steaming treatment of American ginseng. Recently, we performed steaming treatment on the notoginseng root. After the treatment, the content of Rg3 was found to have increased remarkably, and the antiproliferative effects on cancer cells significantly increased (Sun et al. 2010).

After assaying the chemical structure-functional relationship of ginsenosides, we propose that the number of sugar molecules, structure of hydroxyl groups, and stereoselectivity in ginsenosides affect their anticancer activity. An understanding of this relationship is a prerequisite for purposeful modifications to produce novel agents for use in medical oncology (Qi et al. 2010).

5.6 Perspectives and Challenges

Previous studies suggested that American ginseng and notoginseng possess anticancer activities. We recently observed that after using a special heat-preparation or steaming process, the content of Rg3, a previously identified anticancer ginsenoside, increased significantly and became the main constituent in the steamed American ginseng. As expected, using the steamed extract increased anticancer activity significantly. Notoginseng has a very distinct saponin profile compared to that of American ginseng. Steaming notoginseng also significantly increased its anticancer effect.

The next logical step would be to characterize the effects of the two ginseng herbs (unsteamed and steamed) and their active constituents on cancer, and their mechanisms of action. Data obtained from future studies will help develop useful products for complementary and alternative therapies in oncology and expand our understanding about the biological mechanism behind the antitumor activity of ginseng and its active compounds.

Although ginseng plants have been extensively studied, much more knowledge is required to answer the questions regarding the observed effects of American ginseng and notoginseng in cancer chemoprevention. Thus, looking to the future, the widespread research of American ginseng and notoginseng seems certain to ensure continued interest in the development of this herb. With the trend of interdisciplinary research and the development of modern combinatorial techniques, it appears promising that new insights will be gained into novel cancer chemopreventive agents from ginseng in drug discovery.

Acknowledgments This work was supported in part by the NIH/NCCAM grants AT004418 and AT005362.

References

Aggarwal, B. B., Van Kuiken, M. E., Iyer, L. H., et al. (2009). Molecular targets of nutraceuticals derived from dietary spices: Potential role in suppression of inflammation and tumorigenesis. *Experimental Biology and Medicine (Maywood), 234,* 825–849.

Akerele, O. (1993). Nature's medicinal bounty: Don't throw it away. *World Health Forum, 14,* 390–395.

Assinewe, V. A., Baum, B. R., Gagnon, D., et al. (2003). Phytochemistry of wild populations of *Panax quinquefolius* L. (North American ginseng). *Journal of Agricultural and Food Chemistry, 51,* 4549–4553.

Attele, A. S., Wu, J. A., & Yuan, C. S. (1999). Ginseng pharmacology: Multiple constituents and multiple actions. *Biochemical Pharmacology, 58,* 1685–1693.

Aung, H. H., Mehendale, S. R., Wang, C. Z., et al. (2007). Cisplatin's tumoricidal effect on human breast carcinoma MCF-7 cells was not attenuated by American ginseng. *Cancer Chemotherapy and Pharmacology, 59,* 369–374.

Barnes, P. M., Powell-Griner, E., McFann, K., et al. (2004). Complementary and alternative medicine use among adults: United States, 2002. *Advance Data, 22,* 1–19.

Barton, D. L., Soori, G. S., Bauer, B. A., et al. (2010). Pilot study of *Panax quinquefolius* (American ginseng) to improve cancer-related fatigue: A randomized, double-blind, dose-finding evaluation: NCCTG trial N03CA. *Supportive Care in Cancer: Official Journal of the Multinational Association of Supportive Care in Cancer, 18,* 179–187.

Benjamin, M., Kaiser, E., & Milz, S. (2008). Structure-function relationships in tendons: A review. *Journal of Anatomy, 212,* 211–228.

Chao, Z., Shoyama, Y., & Tanaka, H. (2006). Pharmacokinetic study of ginsenosides Rb1 and Rg1 in rat by ELISA using anti-ginsenosides Rb1 and Rg1 monoclonal antibodies. *The American Journal of Chinese Medicine, 34,* 1069–1081.

Chen, F. D., Wu, M. C., Wang, H. E., et al. (2001). Sensitization of a tumor, but not normal tissue, to the cytotoxic effect of ionizing radiation using *Panax notoginseng* extract. *The American Journal of Chinese Medicine, 29,* 517–524.

Chen, Y., Zhan, E., Chen, H., et al. (2002). Saponins with low sugar chain from the leaves of *Panax notoginseng* (Burk) F. H. Chen. *Zhong Yao Cai Zhongyaocai (Journal of Chinese Medicinal Materials), 25,* 176–178.

Corbit, R., Ebbs, S., King, M. L., et al. (2006). The influence of lead and arsenite on the inhibition of human breast cancer MCF-7 cell proliferation by American ginseng root (*Panax quinquefolius* L.). *Life Sciences, 78,* 1336–1340.

Fuzzati, N. (2004). Analysis methods of ginsenosides. *Journal of Chromatography. B: Analytical Technologies in the Biomedical and Life Sciences, 812,* 119–133.

Goldberg, R. M., Sargent, D. J., Morton, R. F., et al. (2004). A randomized controlled trial of fluorouracil plus leucovorin, irinotecan, and oxaliplatin combinations in patients with previously untreated metastatic colorectal cancer. *Journal of Clinical Oncology: Official Journal of the American Society of Clinical Oncology, 22,* 23–30.

Helms, S. (2004). Cancer prevention and therapeutics: *Panax ginseng. Alternative Medicine Review: A Journal of Clinical Therapeutic, 9,* 259–274.

Hofseth, L. J., & Wargovich, M. J. (2007). Inflammation, cancer, and targets of ginseng. *The Journal of Nutrition, 137,* 183S–185S.

Hurwitz, H., Fehrenbacher, L., Novotny, W., et al. (2004). Bevacizumab plus irinotecan, fluorouracil, and leucovorin for metastatic colorectal cancer. *The New England Journal of Medicine, 350,* 2335–2342.

Iishi, H., Tatsuta, M., Baba, M., et al. (1997). Inhibition by ginsenoside Rg3 of bombesin-enhanced peritoneal metastasis of intestinal adenocarcinomas induced by azoxymethane in Wistar rats. *Clinical & Experimental Metastasis, 15,* 603–611.

Jemal, A., Siegel, R., Xu, J., et al. (2010) Cancer statistics, 2010. *CA: A Cancer Journal For Clinicians, 60,* 277–300.

Jeong, A., Lee, H. J., Jeong, S. J., et al. (2010). Compound K inhibits basic fibroblast growth factor-induced angiogenesis via regulation of p38 mitogen activated protein kinase and AKT in human umbilical vein endothelial cells. *Biological & Pharmaceutical Bulletin, 33,* 945–950.

Jia, W. W., Bu, X., Philips, D., et al. (2004). Rh2, a compound extracted from ginseng, hypersensitizes multidrug-resistant tumor cells to chemotherapy. *Canadian Journal of Physiology and Pharmacology, 82,* 431–437.

Jin, Y., Hofseth, A. B., Cui, X., et al. (2010). American ginseng suppresses colitis through p53-mediated apoptosis of inflammatory cells. *Cancer Prevention Research (Philadelphia, Pa.), 3,* 339–347.

Kim, S. H., & Park, K. S. (2003). Effects of *Panax ginseng* extract on lipid metabolism in humans. *Pharmacological Research: The Official Journal of the Italian Pharmacological Society, 48,* 511–513.

Kim, H. S., Lee, E. H., Ko, S. R., et al. (2004). Effects of ginsenosides Rg3 and Rh2 on the proliferation of prostate cancer cells. *Archives of Pharmacal Research, 27,* 429–435.

Kim, S. J., Jeong, H. J., Yi, B. J., et al. (2007). Transgenic *Panax ginseng* inhibits the production of TNF-alpha, IL-6, and IL-8 as well as COX-2 expression in human mast cells. *The American Journal of Chinese medicine, 35,* 329–339.

Kim, S. M., Lee, S. Y., Cho, J. S., et al. (2010). Combination of ginsenoside Rg3 with docetaxel enhances the susceptibility of prostate cancer cells via inhibition of NF-kappaB. *European Journal of Pharmacology, 631,* 1–9.

Kimura, Y., Sumiyoshi, M., Kawahira, K., et al. (2006). Effects of ginseng saponins isolated from red ginseng roots on burn wound healing in mice. *British Journal of Pharmacology, 148,* 860–870.

Kitts, D. D., Wijewickreme, A. N., & Hu, C. (2000). Antioxidant properties of a North American ginseng extract. *Molecular and Cellular Biochemistry, 203,* 1–10.

Koo, H. N., Jeong, H. J., Choi, I. Y., et al. (2007). Mountain grown ginseng induces apoptosis in HL-60 cells and its mechanism have little relation with TNF-alpha production. *The American Journal of Chinese Medicine, 35,* 169–182.

Lau, A. J., Seo, B. H., Woo, S. O., et al. (2004). High-performance liquid chromatographic method with quantitative comparisons of whole chromatograms of raw and steamed *Panax notoginseng*. *Journal of Chromatography, 1057,* 141–149.

Lee, K. D., & Huemer, R. P. (1971). Antitumoral activity of *Panax ginseng* extracts. *Japanese Journal of Pharmacology, 21,* 299–302.

Lee, J. Y., Shin, J. W., Chun, K. S., et al. (2005). Antitumor promotional effects of a novel intestinal bacterial metabolite (IH-901) derived from the protopanaxadiol-type ginsenosides in mouse skin. *Carcinogenesis, 26,* 359–367.

Lee, T. I., Chen, H. H., & Yeh, M. L. (2006). Effects of chan-chuang qigong on improving symptom and psychological distress in chemotherapy patients. *The American Journal of Chinese Medicine, 34,* 37–46.

Leung, P. C. (2007). The efficacy of Chinese medicine for SARS: A review of Chinese publications after the crisis. *The American Journal of Chinese Medicine, 35,* 575–581.

Li, X. J., & Zhang, H. Y. (2008). Western-medicine-validated anti-tumor agents and traditional Chinese medicine. *Trends in Molecular Medicine, 14,* 1–2.

Li, B., Wang, C. Z., He, T. C., et al. (2010). Antioxidants potentiate American ginseng induced killing of colorectal cancer cells. *Cancer Letters, 289,* 62–70.

Liu, X. S., & Jiang, J. (2006). Molecular mechanism of matrine-induced apoptosis in leukemia K562 cells. *The American Journal of Chinese Medicine, 34,* 1095–1103.

Liu, R., Xing, D., Lu, H., et al. (2006). Pharmacokinetics of puerarin and ginsenoside Rg1 of CBN injection and the relation with platelet aggregation in rats. *The American Journal of Chinese Medicine, 34,* 1037–1045.

Lu, P., Su, W., Miao, Z. H., et al. (2008). Effect and mechanism of ginsenoside Rg3 on postoperative life span of patients with non-small cell lung cancer. *Chinese Journal of Integrative Medicine, 14*(1), 33–36.

Luo, X., Wang, C. Z., Chen, J., et al. (2008). Characterization of gene expression regulated by American ginseng and ginsenoside Rg3 in human colorectal cancer cells. *International Journal of Oncology, 32,* 975–983.

Ma, W. W., & Adjei, A. A. (2009). Novel agents on the horizon for cancer therapy. *CA: A Cancer Journal for Clinicians, 59,* 111–137.

Mashour, N. H., Lin, G. I., & Frishman, W. H. (1998). Herbal medicine for the treatment of cardiovascular disease: Clinical considerations. *Archives of Internal Medicine, 158,* 2225–2234.

Mehendale, S., Aung, H., Wang, A., et al. (2005). American ginseng berry extract and ginsenoside Re attenuate cisplatin-induced kaolin intake in rats. *Cancer Chemotherapy and Pharmacology, 56,* 63–69.

Mochizuki, M., Yoo, Y. C., Matsuzawa, K., et al. (1995). Inhibitory effect of tumor metastasis in mice by saponins, ginsenoside-Rb2, 20(*R*)- and 20(*S*)-ginsenoside-Rg3, of red ginseng. *Biological & Pharmaceutical Bulletin, 18,* 1197–1202.

Ng, T. B. (2006). Pharmacological activity of sanchi ginseng (*Panax notoginseng*). *The Journal of Pharmacy and Pharmacology, 58,* 1007–1019.

Ng, L. T., Chiang, L. C, Lin, Y. T., et al. (2006). Antiproliferative and apoptotic effects of tetrandrine on different human hepatoma cell lines. *The American Journal of Chinese Medicine, 34,* 125–135.

Ooi, L. S., Li, Y., Kam, S. L., et al. (2006). Antimicrobial activities of cinnamon oil and cinnamaldehyde from the Chinese medicinal herb *Cinnamomum cassia* blume. *The American Journal of Chinese Medicine, 34,* 511–522.

Ott, M. J. (2002). Complementary and alternative therapies in cancer symptom management. *Cancer Practice, 10,* 162–166.

Ozaslan, M., Didem Karagoz, I., Kalender, M. E., et al. (2007). In vivo antitumoral effect of *Plantago major* L. extract on Balb/C mouse with Ehrlich ascites tumor. *The American Journal of Chinese Medicine, 35,* 841–851.

Popovich, D. G., & Kitts, D. D. (2002). Structure-function relationship exists for ginsenosides in reducing cell proliferation and inducing apoptosis in the human leukemia (THP-1) cell line. *Archives of Biochemistry and Biophysics, 406,* 1–8.

Qi, L. W., Wang, C. Z., & Yuan, C. S. (2010). American ginseng: Potential structure-function relationship in cancer chemoprevention. *Biochemical Pharmacology, 80,* 947–954.

Ren, H. C., Sun, J. G., Wang, G. J., et al. (2008). Sensitive determination of 20(*S*)-protopanaxadiol in rat plasma using HPLC-APCI-MS: Application of pharmacokinetic study in rats. *Journal of Pharmaceutical and Biomedical Analysis, 48,* 1476–1480.

Shieh, D. E., Cheng, H. Y., Yen, M. H., et al. (2006). Baicalin-induced apoptosis is mediated by Bcl-2-dependent, but not p53-dependent, pathway in human leukemia cell lines. *The American Journal of Chinese Medicine, 34,* 245–261.

Shumay, D. M., Maskarinec, G., Kakai, H., et al. (2001). Why some cancer patients choose complementary and alternative medicine instead of conventional treatment. *The Journal of Family Practice, 50,* 1067.

Sun, H., Ye, Y., & Pan, Y. (2005). Immunological-adjuvant saponins from the roots of *Panax notoginseng*. *Chemistry & Biodiversity, 2,* 510–515.

Sun, S., Wang, C. Z., Tong, R., et al. (2010). Effects of steaming the root of *Panax notoginseng* on chemical composition and anticancer activities. *Food Chemistry, 118,* 307–314.

Wakabayashi, C., Hasegawa, H., Murata, J., et al. (1997). In vivo antimetastatic action of ginseng protopanaxadiol saponins is based on their intestinal bacterial metabolites after oral administration. *Oncology Research, 9,* 411–417.

Wang, C. Z., McEntee, E., Wicks, S., et al. (2006a). Phytochemical and analytical studies of *Panax notoginseng* (Burk.) F.H. Chen. *Journal of Natural Medicines, 60,* 97–106.

Wang, C. Z., Wu, J. A., McEntee, E., et al. (2006b). Saponins composition in American ginseng leaf and berry assayed by high-performance liquid chromatography. *Journal of Agricultural and Food Chemistry, 54,* 2261–2266.

Wang, C. Z., Zhang, B., Song, W. X., et al. (2006c). Steamed American ginseng berry: Ginsenoside analyses and anticancer activities. *Journal of Agricultural and Food Chemistry, 54,* 9936–9942.

Wang, C. Z., Aung, H. H., Ni, M., et al. (2007a). Red American ginseng: Ginsenoside constituents and antiproliferative activities of heat-processed *Panax quinquefolius* roots. *Planta Medica, 73,* 669–674.

Wang, C. Z., Luo, X., Zhang, B., et al. (2007b). Notoginseng enhances anti-cancer effect of 5-fluorouracil on human colorectal cancer cells. *Cancer Chemotherapy and Pharmacology, 60,* 69–79.

Wang, C. Z., Mehendale, S. R., & Yuan, C. S. (2007c). Commonly used antioxidant botanicals: Active constituents and their potential role in cardiovascular illness. *The American Journal of Chinese Medicine, 35,* 543–558.

Wang, W., Zhao, Y., Rayburn, E. R., et al. (2007d). In vitro anti-cancer activity and structure-activity relationships of natural products isolated from fruits of *Panax ginseng*. *Cancer Chemotherapy and Pharmacology, 59,* 589–601.

Wang, C. Z., Aung, H. H., Zhang, B., et al. (2008). Chemopreventive effects of heat-processed *Panax quinquefolius* root on human breast cancer cells. *Anticancer Research, 28,* 2545–2551.

Wang, C. Z., Li, X. L., Wang, Q. F., et al. (2009a). The mitochondrial pathway is involved in American ginseng-induced apoptosis of SW-480 colon cancer cells. *Oncology Reports, 21,* 577–584.

Wang, C. Z., Xie, J. T., Fishbein, A., et al. (2009b). Antiproliferative effects of different plant parts of *Panax notoginseng* on SW480 human colorectal cancer cells. *Phytotherapy Research: PTR, 23,* 6–13.

Wang, X., Sakuma, T., Asafu-Adjaye, E., et al. (1999). Determination of ginsenosides in plant extracts from *Panax ginseng* and *Panax quinquefolius* L. by LC/MS/MS. *Analytical Chemistry, 71,* 1579–1584.

White, C. M., Fan, C., Song, J., et al. (2001). An evaluation of the hemostatic effects of hydrophilic, alcohol, and lipophilic extracts of notoginseng. *Pharmacotherapy, 21,* 773–777.

Wicks, S. M., Tong, R., Wang, C. Z., et al. (2007). Safety and tolerability of *Ganoderma lucidum* in healthy subjects: A double-blind randomized placebo-controlled trial. *The American Journal of Chinese Medicine, 35,* 407–414.

Wu, H., Yang, F., Cui, S., et al. (2007). Hematopoietic effect of fractions from the enzyme-digested colla corii asini on mice with 5-fluorouracil induced anemia. *The American Journal of Chinese Medicine, 35,* 853–866.

Xie, J. T., Chang, W. T., Wang, C. Z., et al. (2006). Curry leaf (*Murraya koenigii* Spreng.) reduces blood cholesterol and glucose levels in ob/ob mice. *The American Journal of Chinese Medicine, 34,* 279–284.

Xu, S. X., Chen, Y. J., Cai, Z. Q., et al. (1987). Studies on the chemical constituents of *Panax quinquefolius* Linn. *Yao Xue Xue Bao, 22,* 750–755.

Yamakage, M., Hattori, J., Satoh, J., et al. (2006). Effects of the Chinese herbal medicines Bupleuri radix, Ginseng radix, and Zingiberis rhizoma on lymphatic vessel activity in rats. *The American Journal of Chinese Medicine, 34,* 1063–1068.

Yoo, H. H., Yokozawa, T., Satoh, A., et al. (2006). Effects of ginseng on the proliferation of human lung fibroblasts. *The American Journal of Chinese Medicine, 34,* 137–146.

Yuan, C. S., & Dey, L. (2001). Multiple effects of American ginseng in clinical medicine. *The American Journal of Chinese Medicine, 29,* 567–569.

Yun, T. K. (2003). Experimental and epidemiological evidence on non-organ specific cancer preventive effect of Korean ginseng and identification of active compounds. *Mutation Research, 523/524,* 63–74.

Yun, T. K., & Choi, S. Y. (1995). Preventive effect of ginseng intake against various human cancers: A case-control study on 1987 pairs. *Cancer Epidemiology, Biomarkers & Prevention: A Publication of the American Association for Cancer Research, Cosponsored by the American Society of Preventive Oncology, 4,* 401–408.

Yun, T. K., & Choi, S. Y. (1998). Non-organ specific cancer prevention of ginseng: A prospective study in Korea. *International Journal of Epidemiology, 27,* 359–364.

Yun, T. K., Lee, Y. S., Lee, Y. H., et al. (2001). Anticarcinogenic effect of *Panax ginseng* C.A. Meyer and identification of active compounds. *Journal of Korean Medical Science, 16*(Suppl), S6–S18.

Zhou, X., Lin, J., Yin, Y., et al. (2007). Ganodermataceae: Natural products and their related pharmacological functions. *The American Journal of Chinese Medicine, 35,* 559–574.

Chapter 6
An Evidence-based Perspective of *Coptis Chinensis* (Chinese Goldthread) for Cancer Patients

Faqing Tang, Wenhua Mei, Daofa Tian and Damao Huang

Abstract *Coptis chinensis* (Chinese goldthread) and herbal complexes containing Chinese goldthread have been proven to possess anticancers effect *in vitro* and *in vivo*. Some herbal complexes containing Chinese goldthread, YiQiJueDu Granule (YQJDG), San-Huang-Xie-Xin Decoction, and ZuoJin Pill possess an inhibitory effect on nasopharyngeal carcinoma (NPC), HepG2, or S180 tumor cells, respectively. YQJDG's inhibitory effects on NPC growth and nasopharyngeal carcinogenesis are associated with reducing-telomerase besides down-regulating cell cycle and inducing apoptosis genes expression. Specially, YQJDG also has a therapeutic effect on the population at high risk for NPC. As a main component of YQJDG, Chinese goldthread has an anticancer effect on NPC, leukemia, melanoma, epidermoid carcinoma, hepatoma, oral carcinoma, glioblastoma, prostate carcinoma, and gastric carcinoma. Berberine and coptisine are two major components of Chinese goldthread. Berberine exerts anticancer effects through modulating Mcl-1, Bcl-xL, COX-2, MDR, TNF-α, IL-6, iNOS, IL-12, intercellular adhesion molecule-1, ELAM-1, MCP-1, CINC-1, cyclin D1, AP-1, HIF-1α, PPAR-γ, topoisomerase II, and inhibiting stress-induced mitogen-activated protein kinase activation. Its mechanisms include inducing cell cycle arrest, apoptosis, antiangiogenic, and anti-metastasis. Berberine also has a therapeutic effect-enhancing and toxicity-reducing effect on other antitumor therapies and preventive effect on carcinogenesis. However, berberine possesses some potential toxicity and adverse effects. Coptisine exerts anticancer effect mainly *via* inhibiting vascular smooth muscle cell proliferation.

F. Q. Tang (✉)
The Third Affiliated Hospital of Jinan University, 79 Kangning Road, 519000 Zhuhai, Guangdong, People's Republic of China; Xiangya Hospital, Central South University, 41008 Changsha, Hunan, People's Republic of China
e-mail: tangfaqing33@hotmail.com

W. C. S. Cho (ed.), *Evidence-based Anticancer Materia Medica,* Evidence-based Anticancer Complementary and Alternative Medicine,
DOI 10.1007/978-94-007-0526-5_6,

6.1 Introduction

Herbal medications are currently promoted more widely for clinical use in cancer therapy. Many claims of efficacy for these medications are based on anecdotal evidence in traditional Chinese medicine. However, certain Chinese herbs could presumably have potent anti-cancer properties. *Coptis chinensis* (Chinese goldthread or Huanglian) has been widely used in China for several thousand years and is prepared as a herbal tea from its roots. Chinese goldthread has been used for treating inflammatory conditions accompanied by high fever, including pneumonia and infections of the head and face, and is also routinely used in China for the treatment of gastroenteritis. Furthermore, it has been reported to inhibit the growth of *Heliobacter pylori* and the intestinal parasite *Blastocystis hominis in vitro* (Franzblau and Cross 1986; Yang et al. 1996; Zhang et al. 1997).

Chinese goldthread comprises various alkaloids, including berberine, palmaline, jatrorrhizine, epiberberine, magnoflorine, and coptisine (Sun et al. 2006). It exerts diverse activities, including anti-hypertensive (Ko et al. 2000), anti-diabetic (Tang et al. 2006; Jung et al. 2008), anti-inflammatory (Kuo et al. 2004), hypolipidemic (Doggrell 2005), and antioxidant effects (Rackova et al. 2004; Yokozawa et al. 2005; Hsieh et al. 2007; Hung et al. 2007). In particular, Chinese goldthread and its isolated alkaloids have been shown to suppress the growth of various tumor cells, including leukemia (Lin et al. 2006b), melanoma (Letasiova et al. 2006), epidermoid carcinoma (Kettmann et al. 2004), hepatoma (Hwang et al. 2006), oral carcinoma (Kuo et al. 2005), glioblastoma (Sanders et al. 1998), prostate carcinoma (Mantena et al. 2006), and gastric carcinoma (Lin et al. 2006c).

Animal studies have revealed that berberine can suppress chemically inducing carcinogenesis (Anis et al. 2001), tumor promotion (Nishino et al. 1986), and tumor invasion (Peng et al. 2006). Berberine is also a radiosensitizer of tumor cells, but not normal cells (Yount et al. 2004). Berberine modulates Mcl-1 (myeloid leukemia cell differentiation protein) (Kuo et al. 2005), Bcl-xL (Hwang et al. 2006), cyclooxygenase (COX)-2 (Kuo et al. 2005), MDR (multi-drug resistance) (Lin et al. 1999), tumor necrosis factor (TNF)-α and interleukin (IL)-6 (Choi et al. 2006), iNOS (inducible nitric oxide synthase) (Tan et al. 2005), IL-12 (Kang et al. 2002), intercellular adhesion molecule-1 and ELAM-1 (endothelial-leukocyte adhesion molecule-1) (Peng et al. 2006), MCP-1 (monocyte chemotactic protein-1) and CINC-1 (cytokine-induced neutrophil chemoattractant-1) (Cui et al. 2007), cyclin D1 (Mantena et al. 2006), AP-1 (activator protein 1) (Kuo et al. 2004), hypoxia-inducible factor (HIF)-1α (Lin et al. 2004b), peroxisome proliferator-activated receptor (PPAR)-γ (Huang et al. 2006), and topoisomerase II (Kang and Chung 2002). Using yeast mutants, berberine was found to bind and inhibit stress-induced, mitogen-activated protein kinase activation (Kang et al. 2002). Apoptotic, carcinogenic, and inflammatory effects, as well as various gene products (such as TNF-α, IL-6, COX-2, adhesion molecules, cyclin D1, and MDR) modulated by berberine are regulated by the transcription factor NF-κB (nuclear factor-κB). Berberine and coptisine constitute the major components of Chinese goldthread. Herein, evidences supporting the use of Chinese goldthread are summarized for cancer patients based on the herb itself, compounds containing the herb, and the extracts.

6.2 Anticancer Effects of Herbal Complexes Containing Chinese Goldthread

6.2.1 *Effect of YiQiJueDu Granule (YQJDG) on Anti-nasopharyngeal Carcinoma (NPC)*

6.2.1.1 Inhibitory Effect of YQJDG on the Proliferation of NPC Cells

YQJDG comprises Chinese goldthread, *Astragalus membranaceus*, *Codonopsis pilosula*, and *Glycyrrhiza uralensis* Fisch (Tang et al. 2001b). Cell dynamics studies have demonstrated that YQJDG inhibits the proliferation of NPC cells time- and dose-dependently (Tang and Tian 1995). YQJDG induces NPC cell apoptosis at low concentrations (<40 μM), possesses a cytotoxic effect at high concentrations (>50 μM) (Tang and Tian 1996), and inhibits NPC cell telomerase activity (Tang et al. 2001a). Proteomic studies have revealed that YQJDG down-regulates expression of aldehyde dehydrogenase, heat shock protein (HSP)27, matrix metalloproteinase (MMP)-2, CD27L receptor, IL-12 alpha chain, DNA-PKcs (DNA-dependent protein kinase catalytic subunit), MPRI (mannose 6-phosphate receptor), and follistatin *in vitro* (Wang et al. 2003). *In vitro*, YQJDG can inhibit NPC cell proliferation and induce apoptosis, and these effects may be associated with telomerase and DNA-dependent protein kinase.

6.2.1.2 Inhibitory Effect of YQJDG on the Growth of Implant Tumors of NPC Cells

The antitumor effect of YQJDG was investigated in nude mice carrying the NPC cancer cell line CNE1. YQJDG can inhibit the implant tumor growth when orally administered. Genomic studies (Tang et al. 2004a) have demonstrated that YQJDG down-regulates expression of genes encoding growth factor receptors [NGF (nerve growth factor), pleiotrophin, interferon-α receptor, and transforming growth factor (TGF)-β receptor 1], protein kinases (STK2, FER, and cGMP-dependent kinase type 1 α), cell cycle regulators (cdc2-related protein kinase CHED, protein kinase CHK1, and CDKN2C), cell signal transduction regulators (TRAF5, MAD3, and p68 kinase), and gene expression regulators (Ets transcription factor, zinc finger protein, and transcriptional activator). Conversely, YQJDG up-regulates expression of genes encoding the apoptosis regulators *TIEG/EGRα* and *H731*. Proteomic studies (Tang et al. 2004b) have revealed that YQJDG up-regulates expression of HKR2, phosphoribosyl pyrophosphate synthetase, TNFR superfamily members, BAX, mucin 5B precursor, and GTP:RNA guanylyltransferase, and down-regulates expression of fibulin-3, zinc finger protein 266, and carboxyl terminus of HSP70-interacting protein *in vivo*, which are associated with cell division, cell cycle, DNA repair, and apoptosis signal transduction. These indicate that YQJDG can inhibit NPC growth *in vivo*, and this inhibition is associated with growth factor, transforming growth factor, cell cycle, and cell division.

6.2.1.3 Inhibitory Effect of YQJDG on Nasopharyngeal Tumorigenesis Induced by N,N′-dinitrosopiperazine

Rat NPC induced by N,N′-dinitrosopiperazine was served as a nasopharyngeal tumorigenesis animal model to investigate the anti-NPC effect of YQJDG. In this model, telomerase was activated and involved in nasopharyngeal tumorigenesis (Tang et al. 2001a). Treatment with YQJDG inhibited nasopharyngeal tumorigenesis and telomerase activation (Tang et al. 2001a). Genomic studies have indicated that *PDGFB*, *M6P/IGFR*$_2$, *ErbB*$_3$*EGF*, *CCNE*, *CCND*$_3$, *NF*-κB, *JNK*, prothymosin-α, Cu-Zn SOD1, and *MAPK1* are involved in the inhibition of nasopharyngeal tumorigenesis by YQJDG. In addition, proteomic studies have revealed that YQJDG decreased expression of E_3 isozyme, mitochondrial processing peptidase, hydroxymeth-methylglutary 1-coalyase, CD27L receptor, MPIR I, HSP27, MMP2, and metalloproteinase inhibitor in inhibition of nasopharyngeal tumorigenesis. YQJDG may have some preventive effect on nasopharyngeal carcinogenesis by chemicals.

6.2.1.4 Therapeutic Effect of YQJDG on Patients with Nasopharyngeal Pre-cancerous Lesions

237 patients with nasopharyngeal dysplasia (pre-cancerous lesion) were randomized to trial group and control group, and either received YQJDG orally ($n = 112$) or did not ($n = 125$). In total, 96 g of YQJDG was boiled twice with water and reconstituted into 100 ml of YQJDG decoction each time. Subsequently, 100 ml of YQJDG decoction was administered orally twice daily in the trial group for 3 months as one therapeutic course repeated once each year. These patients were followed up after 2 years, at which time their nasopharyngeal tissues were pathologically detected using a comparison determined before therapy. The results demonstrate that YQJDG could markedly block the development of nasopharyngeal tumorigenesis in the trial group at a total effective rate of 88.4% (99/112) in the short-term, significantly higher than that in the control group ($P < 0.05$), which was only 45.5% (57/125) (Ouyang et al. 1999; Tao et al. 2001). YQJDG can block nasopharyngeal carcinogenesis.

6.2.1.5 Therapeutic Effect of YQJDG on Individuals at High Risk for NPC

Epstein-Barr virus (EBV) is associated with nasopharyngeal carcinoma, and individuals harboring the EBV-associated antibodies VCA/IgA and EA/IgA are considered at high risk for nasopharyngeal carcinoma. With YQJDG treatment, it could reduce EBV potential infection and incidence rate of nasopharyngeal carcinoma (Tian et al. 2000). YQJDG may have some preventive effect on the population with high risk for NPC through reducing EBV infection.

6.2.1.6 YQJDG Enhances Therapeutic Effect and Reduces Toxicity of Radiationtherapy on Patients with NPC

Radiotherapy has been a main therapeutic strategy on patients with NPC for decades, but some patients discontinue therapy due to the adverse effects of radiotherapy. 70 patients with NPC were divided into trial (TG) and control (CG) groups. TG received the routine radiotherapy and YQJDG (n=35), 100 ml of YQJDG was administered orally twice daily for 3 months. CG only received the radiotherapy (n=35). Rate of tumor reduction was determined using solid tumor WHO objective curative effect standards, and CD3, CD4, and CD8 were detected. Results showed that the rate of tumor reduction, CD3, and CD4 in TG were higher than that in CG, and 3 years survival rate of TG (91.43%) was higher than CG (70%). The adverse-toxicity of radiation in TG was lower than that in CG. These data indicated that YQJDG could effectively reduce the adverse effects of radiotherapy and enhance therapeutic effect (Wang and Tian 2006).

6.2.2 *Anti-liver Cancer Effect of San-Huang-Xie-Xin Decoction*

San-Huang-Xie-Xin Decoction (SHXXD) consists of *Rheum officinale* rhizomes, *Scutellaria baicalensis* roots, and Chinese goldthread rhizomes. HepG2 cells were treated with extracts of SHXXD and gene expression profiles were analyzed by DNA microarray. Gene set enrichment analysis indicated that SHXXD displayed a unique anti-proliferation pattern *via* p53-signaling, p53-activated, and DNA damage-signaling pathways in HepG2 cells. Network analysis confirmed that most genes were regulated by one molecule, p53. In addition, hierarchical clustering analysis showed that *Rhizoma Coptidis* shares a similar gene expression profile with SHXXD (Cheng et al. 2008).

6.2.3 *Anticancer Effect of ZuoJin Pill*

ZuoJin Pill (ZJP) is a well-known classic formula in traditional Chinese medicine, comprising Chinese goldthread and *Evodia rutaecarpa* (Juss.) Benth at a ratio of 6:1 (w/w). Anticancer activity experiments were performed by inhibiting the growth of S180 tumors *in vivo*. Tumor growth inhibition rate, spleen index, lymphocyte proliferation activity, apoptosis index, TNF-α level, serum tumor marker activity, increase in life span, histopathology, and gene expression were tested. The results indicated that ZJP could significantly induce apoptosis of cancer cells. The inhibition ratio, increase in life span, and TNF-α level of mice treated with ZJP alone were 50.54%, 64.91%, and 1.04 ng/ml, respectively, much higher than in mice treated with Chinese goldthread or *Evodia rutaecarpa* alone. In addition, acid and alkaline phosphatase activities were significantly increased, aldolase and lactate dehydrogenase activities were reduced in serum, and the expression of Bax and wild-type p53 proteins were much higher for mice treated with ZJP alone compared with those treated with Chinese goldthread or *Evodia rutaecarpa* alone. A clear synergistic effect on anticancer activity was observed with ZJP, and the mechanism of antitumor growth may be due to an effect on gene expression and activities of serum tumor markers (Wang et al. 2009).

6.3 Anticancer Effects of the Single Herb Chinese Goldthread

Chinese goldthread exhibited the strongest activity against human liver and leukemia cell lines when used alone. The IC_{50} values for Chinese goldthread in HepG2, Hep3B, and PLC/PRF/5 cell lines were 20, 55, and 35 mg/ml, respectively. The IC_{50} values for Chinese goldthread in K562, U937, and P3H1 cell lines were 29, 29, and 31 mg/ml, respectively (Lin et al. 2004a). Chinese goldthread is a potential anti-carcinogenic agent for treating hepatocellular carcinoma by inducing cell cycle arrest and promoting apoptosis, with *NAG-1* (NSAID-activated gene) as the molecular target. Inhibition of cell proliferation, induction of apoptosis, and cell cycle arrest at the G2/M phase was observed in HepG2 cells treated with Chinese goldthread. The pro-apoptotic effects of Chinese goldthread were associated with a corresponding down-regulation of Bcl-2, activation of procaspase-3 and -9 as well as cleavage of poly (ADP-ribose) polymerase. Further studies demonstrated the involvement of *NAG-1* in the pro-apoptotic events following prior activation of its upstream transcriptional factor *Egr-1* (early growth response gene-1) (Auyeung and Ko 2009). Considering the traditional use, the anticancer effects of Chinese goldthread can be ascribed to its CM trait by removing damp-heat, fire, and toxicity. The Chinese goldthread molecular mechanisms involve cell-cycle arrest, apoptosis induction, and anti-inflammation. Although berberine is an essential anticancer compound in Chinese goldthread, the latter is shown more effective and beneficial than the use of berberine alone in some studies. The clinical application of Chinese goldthread as a novel cancer therapeutic agent requires *in vivo* validation and further investigation of its anticancer mechanisms (Tang et al. 2009b).

6.4 Anticancer Effects of Chinese Goldthread Extracts

Chinese goldthread contains diverse alkaloids, including berberine, palmaline, jatrorrhizine, epiberberine, magnoflorine, and coptisine. Berberine and coptisine are the major components of Chinese goldthread.

6.4.1 *Anticancer Effects of Berberine*

Berberine could modulate Mcl-1 (Kuo et al. 2005), Bcl-xL (Hwang et al. 2006), COX-2 (Kuo et al. 2005), MDR (Lin et al. 1999), TNF-α and IL-6 (Choi et al. 2006), iNOS (Tan et al. 2005), IL-12 (Kang et al. 2002), intercellular adhesion molecule-1 and ELAM-1 (Peng et al. 2006), MCP-1 and CINC-1 (Cui et al. 2007), cyclin D1 (Mantena et al. 2006), AP-1(Kuo et al. 2004), HIF-1α (Lin et al. 2004b), PPAR-γ (Huang et al. 2006), and topoisomerase II (Kang and Chung 2002). Berberine was also found to bind and inhibit stress-induced mitogen-activated protein kinase kinase activation (Kang et al. 2002).

6.4.1.1 Inhibitory Effect of Berberine on the Proliferating Activity of Tumor Cells

Berberine (Natural Yellow 18, 5,6-dihydro-9,10-dimethoxybenzo-(g)-1,3-benzodio-xolo(5,6-a) quinolizinium) is a major component of Chinese goldthread used in traditional Chinese herbal medicine. The chemical structure of berberine chloride, which has a molecular weight of 371.8, is shown in Fig. 6.1. Because of its ability to cause cell cycle arrest and apoptosis in several malignant cell lines, berberine has received much attention as a potential anticancer therapeutic agent. Recently, berberine has been examined for its anticancer activity following evidence of anti-neoplastic properties (Nishino et al. 1986).

Berberine induced G2/M arrest in leukemia cells *via* the inhibition of cyclin B1 and the promotion of Wee1 (Lin et al. 2006b). In addition, treatment with berberine induced cell cycle arrest at G0/G1 in the anoikis-resistant MCF-7 and MDA-MB-231 breast cancer cells (Kim et al. 2010). Furthermore, berberine effectively induced mitotic arrest of HONE1 cells (Tsang et al. 2009), greatly decreased G0/G1 phase-associated cyclin and cyclin-dependent kinase (cyclin D1, cyclin E, Cdk2, and Cdk4) expression, and increased apoptotic gene expression and activation of caspase-3 in SK-N-SH cells (Choi et al. 2008). After treatment with berberine, the proportion of pulmonary giant cell carcinoma (PG) cells at the G0/G1 phase increased, while cells at the S and G2/M phases decreased. Berberine at low doses (12.5–50 μM) was concentrated in mitochondria and promoted G1 arrest. In contrast, higher doses (>50 μM) resulted in cytoplasmic and nuclear berberine accumulation, as well as G2 arrest. DNA synthesis was not markedly affected by low doses

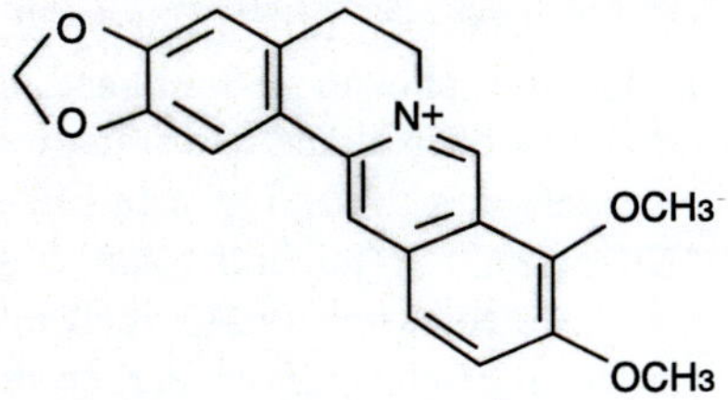

Fig. 6.1 Chemical structure of berberine

of berberine, but 100 μM was strongly inhibitory. Berberine displayed multiphasic effects in these malignant cell lines, which correlated with the concentration and intracellular distribution of this alkaloid (Serafim et al. 2008). Berberine-induced inhibition of proliferation of DU145, PC-3, and LNCaP cells was associated with G1 arrest, which in DU145 cells was associated with inhibition of expression of cyclins (D1, D2, and E) and cyclin-dependent kinases (Cdks 2, 4, and 6), increased expression of Cdk inhibitory proteins (Cip1/p21 and Kip1/p27), and enhanced binding of Cdk inhibitors to Cdk (Mantena et al. 2006). In SNU-5 cells treated with 25–200 μmol/l berberine, G2/M cell cycle arrest was observed, which was associated with a marked increase in the expression of p53, Wee1, and Cdk1 proteins and decreased cyclin B. A concentration-dependent decrease of cells in G0/G1 phase and an increase in G2/M phase were detected (Lin et al. 2006c).

Berberine increased apoptosis through reduction of MMP, release of cytochrome c, and activation of caspase-3 (Lin et al. 2007a). Berberine also suppressed the expression of NF-κB-regulated gene products involved in anti-apoptosis (Bcl-xL, survivin, IAP1, IAP2, and cFLIP). Suppression of anti-apoptotic gene products correlated with enhancement of apoptosis induced by TNF-α and chemotherapeutic agents, as well as with inhibition of TNF-induced cellular invasion (Pandey et al. 2008). Berberine inhibited the levels of anti-apoptotic protein Bcl-2 but increased the levels of pro-apoptotic protein Bax before leading to decreased mitochondrial membrane potential (DeltaPsim), cytochrome *c* release, and activation of capase-9 and -3 for an apoptotic response. Caspase-8, -9, and -3 were activated by berberine based on the substrate solution (PhiPhiLux-G1D1, CaspaLux 8-L1D2, and CaspaLux 9-M1D2 for caspase-3, -8, and -9, respectively). Each inhibitor of caspase-8, -9, and -3 led to an increase in the percentage of viable C6 cells after exposure to berberine (Lin et al. 2006a; Mantena et al. 2006; Eom et al. 2008; Meeran et al. 2008; Chen et al. 2009; Ho et al. 2009a). Berberine enhanced the apoptosis of CaSki cells through the induction of a higher ratio of p53 and Bax/Bcl-2 proteins, increased levels of reactive oxygen species (ROS) and Ca^{2+}, disruption of the mitochondrial membrane potential, and promotion of caspase-3 activity. In CaSki cells pre-treated with the pan-caspase inhibitor zVAD-fmk, berberine-induced caspase-3 activity and apoptosis were significantly blocked as confirmed by flow cytometric analysis. Expression of GADD153, a transcription factor involved in apoptosis, was induced by berberine. Thus, berberine increased ROS levels, leading to endoplasmic reticulum stress based on the increase in GADD153 and Ca^{2+} release from the endoplasmic reticulum (Lin et al. 2007b). Berberine-induced, dose-dependent induction of apoptosis was accompanied by sustained phosphorylation of JNK and p38 MAPK, as well as generation of ROS. In addition, the induction of apoptosis was alleviated by inhibitors specific for JNK and p38. Furthermore, there was an increase in the cellular levels of phospho-c-Jun, FasL, and t-BID in berberine-induced apoptosis *via* the activation of JNK and p38 signaling pathways. Administration of N-acetylcysteine, a scavenger of ROS, reversed berberine-induced apoptotic effects *via* inhibition of JNK and p38, c-jun activation, FasL t-BID expression (Hsu et al. 2007). Apoptosis, detected as the sub-G0 cell population in cell cycle measurements, was confirmed in 25–200 μmol/l berberine-treated cells by monitoring the apoptotic pathway. Apoptosis was identified by the sub-G0 cell population, up-regulation of Bax, down-regulation of Bcl-2, release of Ca^{2+}, and decreased mitochondrial membrane potential leading to the release of mitochondrial cytochrome *c* into the cytoplasm, activation of caspase-3, and concomitant apoptosis (Lin et al. 2006c).

Apoptosis mediated by berberine is associated with p53 expression. Treatment of human lung cancer A549 cells, which express wild-type p53, and human lung cancer H1299 cells, which are p53-deficient, with berberine resulted in inhibition of cell proliferation and an increase in apoptotic cell death. However, A549 cells were more sensitive to berberine-induced cytotoxic effects than H1299 cells. In addition, treatment of A549 cells with pifithrin-α, a specific inhibitor of p53, or transfection of A549 cells with a *p53* antisense oligodeoxynucleotide resulted in a reduction in the berberine-induced inhibition of cell proliferation and apoptosis. Berberine-induced apoptosis of both A549 and H1299 human lung cancer cells was associated with disruption of the mitochondrial membrane potential, reduction in the levels of Bcl-2 and Bcl-xL, an increase in Bax and Bak, and activation of caspase-3. Furthermore, berberine administration by oral gavage inhibited the growth of subcutaneous A549 and H1299 lung tumor xenografts in athymic nude mice. However, the growth of tumor xenografts with H1299 cells was faster than with A549 cells in mice and the chemotherapeutic effect of berberine was more pronounced in the p53-positive A549 tumor xenograft than in the p53-deficient H1299 tumor xenograft (Katiyar et al. 2009). Moreover, expression profiling and Ingenuity pathway analysis results showed that the cytotoxicity of berberine was accompanied by activation of BRCA1-mediated DNA damage response, G1/S and G2/M cell cycle checkpoint regulation, and p53 signaling pathways. The activation of these signaling pathways may be caused by berberine intercalating DNA and inducing DNA strand breaks through the inhibition of topoisomerases and induction of DNA lesions (Pinto-Garcia et al. 2010). The p53-expressing SK-N-SH cells were more susceptible to berberine ($IC_{50}=37$ μM) than the p53-deficient SK-N-MC cells ($IC_{50}\geq 100$ μM) without cytotoxic effects on cortical neuronal cells.

6.4.1.2 Anti-angiogenic Effect of Berberine on Tumor Tissue

Tumor-induced angiogenesis is a prerequisite for excessive tumor growth. Blood vessels invade the tumor tissue after degradation of the extracellular matrix scaffold by matrix metalloproteinases (MMPs). Therefore, inhibition of MMPs may be a useful tool to abolish neoangiogenesis of solid tumors. Berberine significantly inhibited angiogenesis in embryoid bodies and decreased intracellular ROS levels (Wartenberg et al. 2003). In addition, berberine could directly inhibit *in vitro* human umbilical vein endothelial cell tube formation and migration. Based on berberine's antiangiogenic property and its clinical potential, it may be an inhibitor of tumor angiogenesis. Furthermore, berberine prevented hypoxic SC-M1 cultures from expressing VEGF (vascular endothelial growth factor) and HIF-1α two key factors in mediating tumor angiogenesis. However, overexpression of HIF-1α in SC-M1 cells dramatically reversed the inhibitory effect of berberine on SC-M1-induced *in vitro* human umbilical vein endothelial cell migration. HIF-1α repression is a critical step in the inhibitory effect of berberine on tumor-induced angiogenesis. Berberine did not down-regulate *HIF-1α* mRNA but destabilized HIF-1α protein. In addition, berberine -induced HIF-1α degradation was blocked by a 26S proteasome inhibitor and berberine increased lysine-acetylated HIF-1α in hypoxic SC-M1 cultures (Lin et al. 2004b).

6.4.1.3 Inhibitory Effect of Berberine on the Metastatic Potential of Cancer Cells

In cancer cell migration and invasion processes, matrix-degrading proteinases are required. In particular, invasion of cancer cells induced by MMP-9 is a pivotal step in cancer metastasis. Berberine induced down-regulation of MMP-1, -2, and -9, but not MMP-7. Berberine appears to exert its anti-cancer properties by inducing ROS production and preventing cell migration *via* inhibition of the gene expression of MMP-1, -2, and -9 in a human gastric cancer cell line SNU-5 (Kim et al. 2008; Lin et al. 2008a). Berberine inhibited migration and invasion of human tongue squamous carcinoma SCC-4 cells. These processes were mediated by the p-JNK, p-ERK, p-p38, IκK, and NF-κB signaling pathways, resulting in inhibition of MMP-2 and -9 in these cells. Berberine down-regulated the expression of u-PA (urokinase-plasminogen activator), MMP-2, and MMP-9 in SCC-4 cells through the FAK, IKK, and NF-κB mediated pathways and a novel function of berberine was to inhibit the invasive capacity of malignant cells (Ho et al. 2009b). A549 cells treated with berberine demonstrated reduced levels of ECM (extracellular matrix) proteinases, including MMP-2 and u-PA, by gelatin and casein zymography analyses. This inhibitory effect likely occurs at the transcriptional level, since reduction in the transcript levels corresponds to that in the protein levels. Moreover, berberine exerted its action *via* regulation of TIMP-2 (tissue inhibitor of metalloproteinase-2) and u-PAI (urokinase-plasminogen activator inhibitor). The upstream mediators of the effect involved c-jun, c-fos, and NF-κB, as revealed by reduced phosphorylation of the proteins. These findings suggest that berberine possesses an anti-metastatic effect in non-small lung cancer cells and, therefore, may be helpful in clinical treatment (Peng et al. 2006).

Berberine suppressed the presence of phosphorylated ezrin, and this inhibitory effect was dependent on the suppression of Rho kinase activity. Reduction of ezrin phosphorylation at Thr567 by berberine was associated with its inhibitory effect on filopodia formation. In addition, reduction of Rho kinase-mediated ezrin phosphorylation may be involved in the anti-metastatic effect of berberine on NPC (Tang et al. 2009a). Berberine suppressed the activation of Rho GTPases, including RhoA, Cdc42, and Rac1, indicating a novel function of berberine in the suppression of Rho GTPase signaling to mediate its inhibitory action on cell migration and motility. The SDF-1/CXCR4 axis involves in the migration process of leukemic cells and berberine partially inhibited SDF-1-induced AML cells and the migration of leukemia stem cells. Berberine reduced the levels of SDF-1 secreted by bone marrow-derived stromal cells in the microenvironment but did not affect CXCR4 expression on the HL-60 cell membrane. Therefore, berberine might be a potentially effective agent for the prevention of leukemia (Li et al. 2008).

6.4.1.4 Anticancer Effect of Berberine in Animal Experiments

Retroviral infection with Friend murine leukemia virus (FMuLv) has been used as a model for identifying potential anti-viral medicinal preparations or establishing new treatment strategies. In BALB/c mice, treatment with berberine was found to extend the life span of leukemia-harboring animals by more than 60 days, decrease the anemic condition prevalent in the FMuLv alone-treated group, inhibit the massive leukemic cell infiltrations to sinusoidal spaces in the spleen, and decrease the expression of Bcl-2, Raf-1, Erk-1, IFN-γ receptor, and erythropoietin. Berberine was able to suppress the progression of leukemia induced by FMuLv and further support its chemopreventive potential against virally induced cancers (Harikumar et al. 2008). The effect of berberine on WEHI-3 leukemia cells *in vivo* was studied, revealing the reduction of Mac-3 and CD11b markers and indicating differentiation inhibition of macrophage and granulocyte precursors. No effect was observed concerning the CD14 marker but the CD19 marker demonstrated the promotion of differentiation of B cell precursors. The weights of spleen samples from mice treated with berberine were found to be lower (Yu et al. 2007). In *in vivo* studies, a significant reduction in tumor volume was observed on the 16th day with berberine administration at 5 and 10 mg/kg. The 1 mg/kg-dose stimulated the tumor mass, but the other tested doses, 5 and 10 mg/kg, reduced the tumor weight (Letasiova et al. 2005). In addition, berberine reduced tumor weights and volumes accompanied by apoptotic cell death and increased expression of apoptotic cell death proteins (Choi et al. 2009).

SCC-4 tumor cells were implanted into mice and groups of mice were treated with berberine (10 mg/kg of body weight). Treatment with berberine resulted in a reduction in tumor incidence. Tumor size in xenograft mice treated with 10 mg/kg berberine was significantly smaller than that in the control group (Ho et al. 2009c). Oral administration of berberine for 14 days significantly inhibited spontaneous mediastinal lymph node metastasis produced by orthotopic implantation of Lewis lung carcinoma into the lung parenchyma in a dose-dependent manner, but did not affect tumor growth at the lung implantation site. Combined treatment with berberine and an anti-cancer drug, CPT-11, resulted in marked inhibition of tumor growth at the implantation site and lymphatic metastasis (Mitani et al. 2001). Chinese rhizome supplement significantly attenuated the weight loss of tumor-bearing mice without a change in food or water intake (Iizuka et al. 2000).

6.4.1.5 Effect of Berberine on Prevention of Carcinogenesis

Berberine inhibited neoplastic transformation by the induction of an antioxidant defense system and the ability to induce apoptotic-like changes, thus clarifying its anticancer role (Thirupurasundari et al. 2009). Oral administration of berberine inhibited hepatocyte proliferation and iNOS expression, as well as inhibited the activities of CYP2E1 and CYP1A2 in diethylnitrosamine (DEN) plus phenobarbital (PB) DEN-plus-PB-treated rats *in vivo*. Moreover, berberine inhibited the activities of CYP2E1 and CYP1A2 in microsomes isolated from DEN-plus-PB-treated rats *in vitro*, suggesting that the anti-hepatocarcinogenic potential of berberine might be due to inhibiting the oxidative metabolic activities of CYP2E1 and CYP1A2 in rats (Zhao et al. 2008). Berberine has been found to significantly inhibit carcinogenesis induced by 20-methylcholanthrene or N-nitrosodiethylamine in a dose-dependent manner in small animals. Administration of berberine could significantly reduce the tumor incidence in animals after injection of 20-methylcholanthrene and increase their life span compared with the control (Anis et al. 2001). COX-2 is abundantly expressed in colon cancer cells and plays a key role in colon tumorigenesis. Compounds inhibiting *COX-2* transcriptional activity could potentially have a chemopreventive property against colon tumor formation. Berberine effectively inhibited *COX-2* transcriptional activity in colon cancer cells in a dose- and time-dependent manner. Thus, these findings may indicate that berberine has antitumor promoting effects (Fukuda et al. 1999). AP-1 (activator protein 1) is a transcription factor that plays a critical role in carcinogenesis. Berberine was shown to inhibit AP-1 activity in a dose- and time-dependent manner. The inhibitory effect on AP-1 activity in cancer cells may further explain the anti-tumor activity of berberine (Fukuda et al. 1999).

6.4.1.6 Berberine Combining with Other Antitumor Therapies Enhances Therapeutic Effect and Reduce Toxicity

The combination of berberine and gamma-radiation enhanced the anticancer effects through the p38 MAPK pathway and ROS generation (Hur et al. 2009). In addition, combined treatment of ER antagonists and berberine conferred synergistic growth inhibitory effects on MCF-7 cells (ER^+), but not MDA-MB-231 cells (ER^-). Similar results were observed using the combined treatment of berberine and fulvestrant, a specific aromatase antagonist. Analysis of the expression of breast cancer-related genes indicated that *EGFR*, *HER2*, *bcl-2*, and *COX-2* were significantly down-regulated, while *IFN-β* and *p21* were remarkably up-regulated by berberine. This suggests that Chinese goldthread extracts could be promising adjuvants to ER antagonists in ER-positive breast cancer treatment through regulation of expression of multiple genes (Liu et al. 2009). Berberine enhanced As_2O_3-mediated inhibition of glioma cell growth. The formation of confluent cell layer was inhibited after incubation with As_2O_3; and this effect was even more pronounced in the presence of berberine. In addition, As_2O_3-mediated reduction in motility and invasion of glioma cells were enhanced upon co-treatment with berberine. Furthermore, it has been reported that PKC isoforms influence the morphology of the actin cytoskeleton as well as the activation of metalloproteinases MT1-MMP and MMP-2, which were involved in cancer cell migration. Treatment of glioma cells with As_2O_3 and berberine significantly decreased the activation of PKC-α and led to actin cytoskeleton rearrangements. The levels of two downstream transcription factors, myc and jun, as well as MT1-MMP and MMP-2 were also significantly reduced. After co-treatment of glioma cells with As_2O_3 and berberine, cancer cell metastasis can be significantly inhibited, most likely due to blocking of the PKC-mediated signaling pathway involved in cancer cell migration. The abovementioned study is potentially interesting for development of novel chemo-therapeutic approaches in the treatment of malignant gliomas, as well as cancer in general (Lin et al. 2008b).

Radiotherapy is the most efficacious strategy for lung cancer. Radiation-enhancing effects and the underlying mechanisms of berberine have been investigated. The cellular ultrastructure revealed the presence of an autophagosome and an increased proportion of acridine orange stained-positive cells, demonstrating that berberine enhanced radio-sensitivity *via* autophagy. An animal tumor model verified the synergistic cytotoxic effect of berberine and irradiation that resulted in a substantial shrinkage of tumor volume. Supplementation with berberine enhanced the cytotoxicity of radiation in both *in vivo* and *in vitro* lung cancer models. The mechanisms underlying this synergistic effect involved the induction of autophagy, suggesting that berberine could be used as an adjuvant therapy to treat lung cancer (Peng et al. 2008). In experiments measuring clonogenic survival, treatment with a non-toxic dose of berberine rendered glioblastoma multiforme (GBM) cells more sensitive than vehicle-treated control cells to X-rays. Berberine could be integrated with post-operative radiotherapy to selectively promote residual GBM tumor cell death (Yount et al. 2004).

6.4.1.7 Potential Toxicity and Adverse Effects of Berberine

Electrospray ionization mass spectrometric (ESI-MS) studies indicated that berberine shows a binding stoichiometries with d(AAGAATTCTT), d(AAGGATCCTT) and d(AAGCATGCTT). Their relative binding affinities toward these three double-stranded DNA were semi-quantitatively evaluated by measuring the ratios of the complex signals ([ds+alkaloid-5H](4−)+[ds+2alkaloid-6H](4−)) to those of the duplexes ([ds-4H](4−)) and also by ESI-MS competitive binding experiments (Chen et al. 2005). This binding may block DNA replication in normal cells. Berberine was selectively accumulated by mitochondria, resulting in arrest of cell proliferation, mitochondrial fragmentation and depolarization, oxidative stress, and a decrease in ATP levels. Electron microscopy of berberine-treated cells showed a reduction in mitochondria-like structures, accompanied by a decrease in mitochondrial DNA copy number. Isolated mitochondrial fractions treated with berberine exhibited slower mitochondrial respiration, particularly when complex I substrates were used, and increased complex I-dependent oxidative stress. These results demonstrate that many previously unknown alterations of mitochondrial physiology may be induced by berberine. Pereira et al. suggested that high doses of berberine should not be used without a proper toxicology assessment (Pereira et al. 2007). Berberine up-regulated multidrug- resistance transporter (pgp-170) expression in two oral (KB, OC2), two gastric (SC-M1, NUGC-3), and two colon (COLO 205, CT 26) cancer cell lines, and decreased the retention of rhodamine 123. Treatment of cells with berberine before paclitaxel treatment resulted in increased viability. In addition, pre-treatment with berberine blocked paclitaxel-induced cell cycle responses and morphological changes. Berberine modulated the expression and function of pgp-170, leading to a reduced response to paclitaxel in digestive tract cancer cells (Lin et al. 1999).

6.4.2 *Effect of Coptisine*

6.4.2.1 Anticancer Effect of Coptisine

Acceleration of vascular smooth muscle cell (VSMC) proliferation is closely linked to the pathogenesis of vascular diseases. Coptisine (Fig. 6.2), an alkaloid of the Chinese goldthread, was the active ingredient in Chinese goldthread, selectively preventing VSMC proliferation with a GI_{50} of 3.3 μM (1.2 μg/ml). Coptisine selectively prevented VSMC proliferation at lower concentrations (Tanabe et al. 2006). In addition, coptisine was cytotoxic in LoVo and HT29, but less potent in L1210 cells, and it was partially crossresistant in the human colon tumor cell line (Colombo et al. 2001). Furthermore, coptisine exhibited a strong inhibitory effect on the proliferation of hepatoma and leukemia cell lines, with IC_{50} values varying from 1.4 to 15.2 μg/ml and from 0.6 to 14.1 μg/ml, respectively. Thus, coptisine possesses anti-hepatoma and anti-leukemia activities (Lin et al. 2004a).

Coptisine displayed antiproliferative action against VSMCs by blocking the cell cycle at G1 and G2/M phases. The G1 block was shown by inhibition of ^{3}H-thymidine incorporation into VSMCs at coptisine concentrations higher than 15 μM. The mechanism underlying the G1 arrest involved a decrease in cyclin D1 protein, although cyclin E, A, and B were not affected by coptisine treatment. The selective reduction in cyclin D1 protein was mainly attributable to accelerated proteolysis *via* proteasome-dependent pathway, since it was inhibited by a proteasome inhibitor, N-carbobenzoxy-L-leucinyl -L-leucinyl-L-norleucinal (MG132) and further the mRNA level of *cyclin D1*, protein synthesis, and MAPK (mitogen-activated protein kinase) activity remained unaltered. The mechanism underlying the G2/M arrest involved partial inhibition of tubulin polymerization, which was apparent at coptisine concentration of 3 μM. Berberine arrested the cell cycle at G1 phase *via* a mechanism identical with coptisine, but did not cause block at G2/M phase (Tanabe et al. 2005).

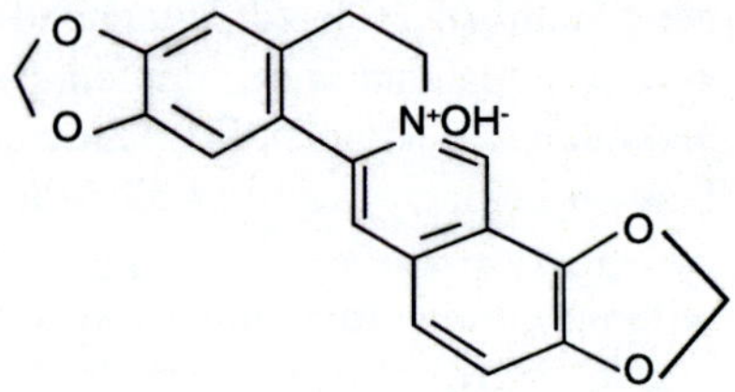

Fig. 6.2 Chemical structure of coptisine

6.4.2.2 Potential Toxicity and Adverse Effects of Coptisine

ESI-MS spectrometric studies indicated that coptisine shows binding stoichiometries with d(AAGAATTCTT), d(AAGGATCCTT), and d(AAGCATGCTT). Their relative binding affinities toward these three double-stranded DNA were semi-quantitatively evaluated by measuring the ratios of the complex signals ([ds+alkaloid-5H](4−)+[ds+2alkaloid-6H](4−)) to those of the duplexes ([ds-4H](4−)) and also by ESI-MS competitive binding experiments (Chen et al. 2005). This effect may block DNA replication in normal cells. In A10 cells (a rat VSMC line), coptisine up-regulated the mRNAs of *Mdr1a* and *Mdr1b*. Coptisine also induced *Mdr1a/1b* protein expression and enhanced the efflux of rhodamine 123 from A10 cells (Suzuki et al. 2010). This implies that coptisine may up-regulate *Mdr* expression in tumors. Coptisine showed an inhibitory effect on MAO-A (type A monoamine oxidase) activity in a concentration-dependent manner using a substrate kynuramine. Coptisine is also a potent reversible inhibitor of MAO-A, and that coptisine functions to regulate the catecholamine content (Ro et al. 2001). Coptisine may possess some potential toxicity and adverse effect.

6.5 Perspectives and Challenges

Chinese goldthread and herbal complexes containing Chinese goldthread have been used in tumor therapy in traditional Chinese medicine. Chinese goldthread has been proved to have an antitumor effect on many cancers, and this antitumor effect distributes to its main components, berberine and coptisine. It possesses both tumor therapeutic and preventive effects. Chinese goldthread or its extract beberine may be a potential cancer therapeutic agent and it may be widely used in clinical therapy, especially at cancer prevention.

Acknowledgments This work was supported in part by the National Natural Science Foundation of China (30973400, 81071718).

References

Anis, K. V., Rajeshkumar, N. V., & Kuttan, R. (2001). Inhibition of chemical carcinogenesis by berberine in rats and mice. *The Journal of Pharmacy and Pharmacology, 53,* 763–768.

Auyeung, K. K., & Ko, J. K. (2009). Coptis chinensis inhibits hepatocellular carcinoma cell growth through nonsteroidal anti-inflammatory drug-activated gene activation. *International Journal of Molecular Medicine, 24,* 571–577.

Chen, W. H., Qin, Y., Cai, Z., et al. (2005). Spectrometric studies of cytotoxic protoberberine alkaloids binding to double-stranded DNA. *Bioorganic & Medicinal Chemistry, 13,* 1859–1866.

Chen, T. C., Lai, K. C., Yang, J. S., et al. (2009). Involvement of reactive oxygen species and caspase-dependent pathway in berberine-induced cell cycle arrest and apoptosis in C6 rat glioma cells. *International Journal of Oncology, 34,* 1681–1690.

Cheng, W. Y., Wu, S. L., Hsiang, C. Y., et al. (2008). Relationship between San-Huang-Xie-Xin-Tang and its herbal components on the gene expression profiles in HepG2 cells. *The American Journal of Chinese Medicine, 36,* 783–797.

Choi, B. H., Ahn, I. S., Kim, Y. H., et al. (2006). Berberine reduces the expression of adipogenic enzymes and inflammatory molecules of 3T3-L1 adipocyte. *Experimental & Molecular Medicine, 38,* 599–605.

Choi, M. S., Yuk, D. Y., Oh, J. H., et al. (2008). Berberine inhibits human neuroblastoma cell growth through induction of p53-dependent apoptosis. *Anticancer Research, 28,* 3777–3784.

Choi, M. S., Oh, J. H., Kim, S. M., et al. (2009). Berberine inhibits p53-dependent cell growth through induction of apoptosis of prostate cancer cells. *International Journal of Oncology, 34,* 1221–1230.

Colombo, M. L., Bugatti, C., Mossa, A., et al. (2001). Cytotoxicity evaluation of natural coptisine and synthesis of coptisine from berberine. *Farmaco, 56,* 403–409.

Cui, H. S., Hayasaka, S., Zheng, L. S., et al. (2007). Effect of berberine on monocyte chemotactic protein-1 and cytokine-induced neutrophil chemoattractant-1 expression in rat lipopolysaccharide-induced uveitis. *Ophthalmic Research, 39,* 32–39.

Doggrell, S. A. (2005). Berberine—A novel approach to cholesterol lowering. *Expert Opinion on Investigational Drugs, 14,* 683–685.

Eom, K. S., Hong, J. M., Youn, M. J., et al. (2008). Berberine induces G1 arrest and apoptosis in human glioblastoma T98G cells through mitochondrial/caspases pathway. *Biological & Pharmaceutical Bulletin, 31,* 558–562.

Franzblau, S. G., & Cross, C. (1986). Comparative in vitro antimicrobial activity of Chinese medicinal herbs. *Journal of Ethnopharmacology, 15,* 279–288.

Fukuda, K., Hibiya, Y., Mutoh, M., et al. (1999). Inhibition by berberine of cyclooxygenase-2 transcriptional activity in human colon cancer cells. *Journal of Ethnopharmacology, 66,* 227–233.

Harikumar, K. B., Kuttan, G., & Kuttan, R. (2008). Inhibition of progression of erythroleukemia induced by Friend virus in BALB/c mice by natural products—Berberine, curcumin and picroliv. *Journal of Experimental Therapeutics & Oncology, 7,* 275–284.

Ho, Y. T., Lu, C. C., Yang, J. S., et al. (2009a). Berberine induced apoptosis via promoting the expression of caspase-8, -9 and -3, apoptosis-inducing factor and endonuclease G in SCC-4 human tongue squamous carcinoma cancer cells. *Anticancer Research, 29,* 4063–4070.

Ho, Y. T., Yang, J. S., Li, T. C., et al. (2009b). Berberine suppresses in vitro migration and invasion of human SCC-4 tongue squamous cancer cells through the inhibitions of FAK, IKK, NF-kappaB, u-PA and MMP-2 and -9. *Cancer Letters, 279,* 155–162.

Ho, Y. T., Yang, J. S., Lu, C. C., et al. (2009c). Berberine inhibits human tongue squamous carcinoma cancer tumor growth in a murine xenograft model. *Phytomedicine, 16,* 887–890.

Hsieh, Y. S., Kuo, W. H., Lin, T. W., et al. (2007). Protective effects of berberine against low-density lipoprotein (LDL) oxidation and oxidized LDL-induced cytotoxicity on endothelial cells. *Journal of Agricultural and Food Chemistry, 55,* 10437–10445.

Hsu, W. H., Hsieh, Y. S., Kuo, H. C., et al. (2007). Berberine induces apoptosis in SW620 human colonic carcinoma cells through generation of reactive oxygen species and activation of JNK/p38 MAPK and FasL. *Archives of Toxicology, 81,* 719–728.

Huang, C., Zhang, Y., Gong, Z., et al. (2006). Berberine inhibits 3T3-L1 adipocyte differentiation through the PPARgamma pathway. *Biochemical and Biophysical Research Communications, 348,* 571–578.

Hung, T. M., Lee, J. P., Min, B. S., et al. (2007). Magnoflorine from Coptidis rhizoma protects high density lipoprotein during oxidant stress. *Biological & Pharmaceutical Bulletin, 30,* 1157–1160.

Hur, J. M., Hyun, M. S., Lim, S. Y., et al. (2009). The combination of berberine and irradiation enhances anti-cancer effects via activation of p38 MAPK pathway and ROS generation in human hepatoma cells. *Journal of Cellular Biochemistry, 107,* 955–964.

Hwang, J. M., Kuo, H. C., Tseng, T. H., et al. (2006). Berberine induces apoptosis through a mitochondria/caspases pathway in human hepatoma cells. *Archives of Toxicology, 80,* 62–73.

Iizuka, N., Miyamoto, K., Hazama, S., et al. (2000). Anticachectic effects of Coptidis rhizoma, an anti-inflammatory herb, on esophageal cancer cells that produce interleukin 6. *Cancer Letters, 158,* 35–41.

Jung, H. A., Yoon, N. Y., Bae, H. J., et al. (2008). Inhibitory activities of the alkaloids from Coptidis rhizoma against aldose reductase. *Archives of Pharmacal Research, 31,* 1405–1412.

Kang, M. R., & Chung, I. K. (2002). Down-regulation of DNA topoisomerase IIalpha in human colorectal carcinoma cells resistant to a protoberberine alkaloid, berberrubine. *Molecular Pharmacology, 61,* 879–884.

Kang, B. Y., Chung, S. W., Cho, D., et al. (2002). Involvement of p38 mitogen-activated protein kinase in the induction of interleukin-12 p40 production in mouse macrophages by berberine, a benzodioxoloquinolizine alkaloid. *Biochemical Pharmacology, 63,* 1901–1910.

Katiyar, S. K., Meeran, S. M., Katiyar, N., et al. (2009). p53 Cooperates berberine-induced growth inhibition and apoptosis of non-small cell human lung cancer cells in vitro and tumor xenograft growth in vivo. *Molecular Carcinogenesis, 48,* 24–37.

Kettmann, V., Kosfalova, D., Jantova, S., et al. (2004). In vitro cytotoxicity of berberine against HeLa and L1210 cancer cell lines. *Pharmazie, 59,* 548–551.

Kim, S., Choi, J. H., Kim, J. B., et al. (2008). Berberine suppresses TNF-alpha-induced MMP-9 and cell invasion through inhibition of AP-1 activity in MDA-MB-231 human breast cancer cells. *Molecules, 13,* 2975–2985.

Kim, J. B., Yu, J. H., Ko, E., et al. (2010). The alkaloid Berberine inhibits the growth of Anoikis-resistant MCF-7 and MDA-MB-231 breast cancer cell lines by inducing cell cycle arrest. *Phytomedicine, 17,* 436–440.

Ko, W. H., Yao, X. Q., Lau, C. W., et al. (2000). Vasorelaxant and antiproliferative effects of berberine. *European Journal of Pharmacology, 399,* 187–196.

Kuo, C. L., Chi, C. W., & Liu, T. Y. (2004). The anti-inflammatory potential of berberine in vitro and in vivo. *Cancer Letters, 203,* 127–137.

Kuo, C. L., Chi, C. W., & Liu, T. Y. (2005). Modulation of apoptosis by berberine through inhibition of cyclooxygenase-2 and Mcl-1 expression in oral cancer cells. *In Vivo, 19,* 247–252.

Letasiova, S., Jantova, S., & Muckova, M. (2005). Antiproliferative activity of berberine in vitro and in vivo. *Biomedical Papers of the Medical Faculty of the University Palacký, Olomouc, Czechoslovakia Republic, 149,* 461–463.

Letasiova, S., Jantova, S., Cipak, L., et al. (2006). Berberine-antiproliferative activity in vitro and induction of apoptosis/necrosis of the U937 and B16 cells. *Cancer Letters, 239,* 254–262.

Li, H., Guo, L., Jie, S., et al. (2008). Berberine inhibits SDF-1-induced AML cells and leukemic stem cells migration via regulation of SDF-1 level in bone marrow stromal cells. *Biomedical Pharmacotherapy, 62,* 573–578.

Lin, H. L., Liu, T. Y., Wu, C. W., et al. (1999). Berberine modulates expression of mdr1 gene product and the responses of digestive track cancer cells to Paclitaxel. *British Journal of Cancer, 81,* 416–422.

Lin, C. C., Ng, L. T., Hsu, F. F., et al. (2004a). Cytotoxic effects of Coptis chinensis and Epimedium sagittatum extracts and their major constituents (berberine, coptisine and icariin) on hepatoma and leukaemia cell growth. *Clinical and Experimental Pharmacology & Physiology, 31,* 65–69.

Lin, S., Tsai, S. C., Lee, C. C., et al. (2004b). Berberine inhibits HIF-1alpha expression via enhanced proteolysis. *Molecular Pharmacology, 66,* 612–619.

Lin, C. C., Kao, S. T., Chen, G. W., et al. (2006a). Apoptosis of human leukemia HL-60 cells and murine leukemia WEHI-3 cells induced by berberine through the activation of caspase-3. *Anticancer Reseaerch, 26,* 227–242.

Lin, C. C., Lin, S. Y., Chung, J. G., et al. (2006b). Down-regulation of cyclin B1 and up-regulation of Wee1 by berberine promotes entry of leukemia cells into the G2/M-phase of the cell cycle. *Anticancer Research, 26,* 1097–1104.

Lin, J. P., Yang, J. S., Lee, J. H., et al. (2006c). Berberine induces cell cycle arrest and apoptosis in human gastric carcinoma SNU-5 cell line. *World Journal of Gastroenterology, 12,* 21–28.

Lin, C. C., Yang, J. S., Chen, J. T., et al. (2007a). Berberine induces apoptosis in human HSC-3 oral cancer cells via simultaneous activation of the death receptor-mediated and mitochondrial pathway. *Anticancer Research, 27,* 3371–3378.
Lin, J. P., Yang, J. S., Chang, N. W., et al. (2007b). GADD153 mediates berberine-induced apoptosis in human cervical cancer Ca ski cells. *Anticancer Research, 27,* 3379–3386.
Lin, J. P., Yang, J. S., Wu, C. C., et al. (2008a). Berberine induced down-regulation of matrix metalloproteinase-1, -2 and -9 in human gastric cancer cells (SNU-5) in vitro. *In Vivo, 22,* 223–230.
Lin, T. H., Kuo, H. C., Chou, F. P., et al. (2008b). Berberine enhances inhibition of glioma tumor cell migration and invasiveness mediated by arsenic trioxide. *BMC Cancer, 8,* 58.
Liu, J., He, C., Zhou, K., et al. (2009). Coptis extracts enhance the anticancer effect of estrogen receptor antagonists on human breast cancer cells. *Biochemical and Biophysical Research Communications, 378,* 174–178.
Mantena, S. K., Sharma, S. D., & Katiyar, S. K. (2006). Berberine, a natural product, induces G1-phase cell cycle arrest and caspase-3-dependent apoptosis in human prostate carcinoma cells. *Molecular Cancer Therapeutics, 5,* 296–308.
Meeran, S. M., Katiyar, S., & Katiyar, S. K. (2008). Berberine-induced apoptosis in human prostate cancer cells is initiated by reactive oxygen species generation. *Toxicology and Applied Pharmacology, 229,* 33–43.
Mitani, N., Murakami, K., Yamaura, T., et al. (2001). Inhibitory effect of berberine on the mediastinal lymph node metastasis produced by orthotopic implantation of Lewis lung carcinoma. *Cancer Letters, 165,* 35–42.
Nishino, H., Kitagawa, K., Fujiki, H., et al. (1986). Berberine sulfate inhibits tumor-promoting activity of teleocidin in two-stage carcinogenesis on mouse skin. *Oncology, 43,* 131–134.
Ouyang, Z., Li, Z. Y., & Yuan, W. H. (1999). An observation on therapeutic effect of Copitis chenesis and Scutellaria baicalensis on EBV-VCA/IgA of patients with nasopharyngeal carcinoma precancer. *Chinese Journal of Integrated Traditional and Western Medicine, 7,* 167–168.
Pandey, M. K., Sung, B., Kunnumakkara, A. B., et al. (2008). Berberine modifies cysteine 179 of IkappaBalpha kinase, suppresses nuclear factor-kappaB-regulated antiapoptotic gene products, and potentiates apoptosis. *Cancer Research, 68,* 5370–5379.
Peng, P. L., Hsieh, Y. S., Wang, C. J., et al. (2006). Inhibitory effect of berberine on the invasion of human lung cancer cells via decreased productions of urokinase-plasminogen activator and matrix metalloproteinase-2. *Toxicology and Applied Pharmacology, 214,* 8–15.
Peng, P. L., Kuo, W. H., Tseng, H. C., et al. (2008). Synergistic tumor-killing effect of radiation and berberine combined treatment in lung cancer: The contribution of autophagic cell death. *International Journal of Radiation Oncology, Biology, Physics, 70,* 529–542.
Pereira, G. C., Branco, A. F., Matos, J. A., et al. (2007). Mitochondrially targeted effects of berberine [Natural Yellow 18, 5,6-dihydro-9, 10-dimethoxybenzo(g)-1,3-benzodioxolo(5,6-a) quinolizinium] on K1735-M2 mouse melanoma cells: Comparison with direct effects on isolated mitochondrial fractions. *The Journal of Pharmacology and Experimental Therapeutics, 323,* 636–649.
Pinto-Garcia, L., Efferth, T., Torres, A., et al. (2010). Berberine inhibits cell growth and mediates caspase-independent cell death in human pancreatic cancer cells. *Planta Medica, 76,* 1155–1161.
Rackova, L., Majekova, M., Kost'alova, D., et al. (2004). Antiradical and antioxidant activities of alkaloids isolated from Mahonia aquifolium. Structural aspects. *Bioorganic & Medicinal Chemistry, 12,* 4709–4715.
Ro, J. S., Lee, S. S., Lee, K. S., et al. (2001). Inhibition of type A monoamine oxidase by coptisine in mouse brain. *Life Science, 70,* 639–645.
Sanders, M. M., Liu, A. A., Li, T. K., et al. (1998). Selective cytotoxicity of topoisomerase-directed protoberberines against glioblastoma cells. *Biochemical Pharmacology, 56,* 1157–1166.
Serafim, T. L., Oliveira, P. J., Sardao, V. A., et al. (2008). Different concentrations of berberine result in distinct cellular localization patterns and cell cycle effects in a melanoma cell line. *Cancer Chemotherapy and Pharmacology, 61,* 1007–1018.

Sun, J., Ma, J. S., Jin, J., et al. (2006). Qualitative and quantitative determination of the main components of huanglianjiedu decoction by HPLC-UV/MS. *Yao Xue Xue Bao (Acta pharmaceutica Sinica), 41,* 380–384.

Suzuki, H., Tanabe, H., Mizukami, H., et al. (2010). Selective regulation of multidrug resistance protein in vascular smooth muscle cells by the isoquinoline alkaloid coptisine. *Biological & Pharmaceutical Bulletin, 33,* 677–682.

Tan, Y., Ming, Z., Tang, Q., et al. (2005). Effect of berberine on the mRNA expression of nitric oxide synthase (NOS) in rat corpus cavernosum. *Journal of Huazhong University of Science and Technology. Medical Sciences, 25,* 127–130.

Tanabe, H., Suzuki, H., Mizukami, H., et al. (2005). Double blockade of cell cycle progression by coptisine in vascular smooth muscle cells. *Biochemical Pharmacology, 70,* 1176–1184.

Tanabe, H., Suzuki, H., Nagatsu, A., et al. (2006). Selective inhibition of vascular smooth muscle cell proliferation by coptisine isolated from Coptis rhizoma, one of the crude drugs composing Kampo medicines Unsei-in. *Phytomedicine, 13,* 334–342.

Tang, F. Q., & Tian, D. F. (1995). Study on dynamics of Zhizoma Coptidis and its complex to kill human nasopharyngeal carcinoma cell. *Journal of Hunan College of Traditional Chinese Medicine, 15,* 41–44.

Tang, F. Q., & Tian, D. F. (1996). Morphological observation on Rhizoma Copidis and its complex to kill nasopharyngeal carcinoma cells. *Journal of Hunan College of Traditional Chinese Medicine, 16,* 46–48.

Tang, F. Q., Jiang, H. Y., Duan, C. J., et al. (2001a). Profile of telomerase and telomerase RNA expression in nasopharyngeal carcinogenesis of rats induced by N, N'dinitrosopiperazine (DNP). *Chinese Journal of Pathology, 30,* 125–128.

Tang, F. Q., Tian, D. F., Chen, C. H., et al. (2001b). Inhibitory effect of YiQiJieDu power to telomerase and telomerase RNA in nasopharyngeal tumorigenesis. *Bulletin of Hunan Medical University, 26,* 301–304.

Tang, F. Q., Gong, Z. J., Zhou, H., et al. (2004a). Gene expression analysis of the implanted-tumor of nasopharngeal carcinoma treated by YiQiJieDu granule using human cancer cDNA array. *Practical Preventive Medicine, 11,* 637–640.

Tang, F. Q., Tian, D. F., Deng, F. L., et al. (2004b). Effect of YiQiJieDu granule on the implanted-tumor protein expression in nasopharyngeal carcinoma. *Bulletin of Hunan Medical University, 29,* 577–582.

Tang, L. Q., Wei, W., Chen, L. M., et al. (2006). Effects of berberine on diabetes induced by alloxan and a high-fat/high-cholesterol diet in rats. *Journal of Ethnopharmacology, 108,* 109–115.

Tang, F. Q., Wang, D. S., Duan, C. J., et al. (2009a). Berberine inhibits metastasis of nasopharyngeal carcinoma 5-8F cells by targeting Rho kinase-mediated Ezrin phosphorylation at threonine 567. *The Journal of Biological Chemistry, 284,* 27456–27466.

Tang, J., Feng, Y., Tsao, S., et al. (2009b). Berberine and Coptidis rhizoma as novel antineoplastic agents: A review of traditional use and biomedical investigations. *Journal of Ethnopharmacology, 126,* 5–17.

Tao, Z. D., & Tian, D. F. (2001). Observation of therapeutic effect in prerenting and treating pharyngeal mucous reaction dueto radiotherapy with YiQiYangYin Fang. *Journal of Hunan College of Traditional Chinese Medicine, 13,* 20–21.

Thirupurasundari, C. J., Padmini, R., & Devaraj, S. N. (2009). Effect of berberine on the antioxidant status, ultrastructural modifications and protein bound carbohydrates in azoxymethane-induced colon cancer in rats. *Chemico-Biological Interactions, 177,* 190–195.

Tian, D. F., Tang, F. Q., Chen, X. Y., et al. (2000). A clinical observation of YIJIEDU granule's inhibitory effect on the EBV activity in population at nasopharyngeal carcinoma risk. *Journal of Hunan College of Traditional Chinese Medicine, 20,* 47–49.

Tsang, C. M., Lau, E. P., Di, K., et al. (2009). Berberine inhibits Rho GTPases and cell migration at low doses but induces G2 arrest and apoptosis at high doses in human cancer cells. *International Journal of Molecular Medicine, 24,* 131–138.

Wang, D. H., & Tian, D. F. (2006). A clinical observation of radiotherapy and YIQIJIEDU granule treatment on nasopharyngeal carcinoma. *Journal of Hunan College of Traditional Chinese Medicine, 26,* 36–38.

Wang, G. P., Tang, F. Q., & Zhou, J. P. (2003). Effect of Coptis Chinensis compound on the gene expression in transplanted tumor tissue in nasopharyngeal carcinoma cell line of CNE1 by cDNA microarray. *Hunan Yi Ke Da Xue Xue Bao, 28,* 347–352.

Wang, X. N., Xu, L. N., Peng, J. Y., et al. (2009). In vivo inhibition of S180 tumors by the synergistic effect of the Chinese medicinal herbs Coptis chinensis and Evodia rutaecarpa. *Planta Medica, 75,* 1215–1220.

Wartenberg, M., Budde, P., De Marees, M., et al. (2003). Inhibition of tumor-induced angiogenesis and matrix-metalloproteinase expression in confrontation cultures of embryoid bodies and tumor spheroids by plant ingredients used in traditional chinese medicine. *Laboratory Investigation, 83,* 87–98.

Yang, L. Q., Singh, M., Yap, E. H., et al. (1996). In vitro response of Blastocystis hominis against traditional Chinese medicine. *Journal of Ethnopharmacology, 55,* 35–42.

Yokozawa, T., Satoh, A., Cho, E. J., et al. (2005). Protective role of Coptidis Rhizoma alkaloids against peroxynitrite-induced damage to renal tubular epithelial cells. *The Journal of Pharmacy Pharmacology, 57,* 367–374.

Yount, G., Qian, Y., Moore, D., et al. (2004). Berberine sensitizes human glioma cells, but not normal glial cells, to ionizing radiation in vitro. *Journal of Experimental Therapeutics & Oncology, 4,* 137–143.

Yu, F. S., Yang, J. S., Lin, H. J., et al. (2007). Berberine inhibits WEHI-3 leukemia cells in vivo. *In Vivo, 21,* 407–412.

Zhang, L., Yang, L., & Zheng, X. (1997). A study of Helicobacterium pylori and prevention and treatment of chronic atrophic gastritis. *Journal of Traditional Chinese Medicine, 17,* 3–9.

Zhao, X., Zhang, J. J., Wang, X., et al. (2008). Effect of berberine on hepatocyte proliferation, inducible nitric oxide synthase expression, cytochrome P450 2E1 and 1A2 activities in diethylnitrosamine- and phenobarbital- treated rats. *Biomedicine & Pharmacotherapy, 62,* 567–572.

Chapter 7
An Evidence-based Perspective of *Angelica Sinensis* (Chinese Angelica) for Cancer Patients

Po-Cheng Lin, Tzyy-Wen Chiou and Horng-Jyh Harn

Abstract The main compounds in *Angelica sinensis* (Chinese angelica) acetone extract AS-C are ferulic acid, ligustilide, brefeldin A, butylidenephthalide, as well as polysaccharides. Polysaccharide have been determined their effects on various human cancer cells. Subsequently, the active component of AS-C, butylidenephthalide (BP), has been investigated for its antitumor effects on glioblastoma multiforme (GBM) brain tumors and colon cancer. *In vitro*, GBM cells were treated with BP, and the effects on proliferation, cell cycle, and apoptosis were determined. *In vivo*, the human GBM tumor, DBTRG-05MG and RG2, the rat GBM tumor, were injected subcutaneously or intracerebrally with BP. BP increased the expression of cyclin kinase inhibitor, including p21 and p27, to decrease the phosphorylation of the Rb proteins, and down-regulated the cell cycle regulators, resulting in cell arrest and apoptosis at the G0/G1 phase. We also examined BP-induced changes in gene expression by microarray screening using human GBM brain tumor cells. Among these genes, *Nur77* is particularly interesting because it plays an important role in the apoptotic processes in various tumor cell lines. BP was able to increase *Nur77* mRNA and protein expression in a time-dependent manner. After the GBM 8401 cells were treated with BP, Nur77 translocated from the nucleus to the cytoplasm while the cytochrome c was released from the mitochondria, and caspase-3 became activated. Since BP has difficulty passing through the blood-brain barrier, we developed a local release system that incorporates BP into a biodegradable polyanhydride material, p(CPP-SA) (BP/Wafer), and investigated its antitumor effects. We used two xenograft animal models, F344 rats (for rat GBM) and nude mice (for human GBM), which were injected with RG2 and DBTRG-05MG cells, respectively, for tumor formation and subsequently treated subcutaneously with BP Wafers. In addition, to

T.-W. Chiou (✉)
Department of Life Science and Graduate Institute of Biotechnology, National Dong Hwa University, No 1, Sec 2, Da Hsueh Road, Shoufeng, Hualien, Taiwan
e-mail: twchiou@mail.ndhu.edu.tw

H.-J. Harn (✉)
Department of Pathology, China Medical University and Hospital, 2 Yuh-Der Road, Taichung 40447, Taiwan
e-mail: duke_harn @yahoo.com.tw

W. C. S. Cho (ed.), *Evidence-based Anticancer Materia Medica,* Evidence-based Anticancer Complementary and Alternative Medicine,
DOI 10.1007/978-94-007-0526-5_7,

study the effect of the interstitial administration of BP against cranial brain tumors, BP/Wafers were surgically placed in FGF-SV40 transgenic mice. Our BP/Wafer significantly reduced tumor size and prolonged survival in a dose-dependent manner.

7.1 *Angelica Sinensis* (Chinese Angelica or Dong Quai) as a Traditional Chinese Medicine

Many natural product sources have been considered for anticancer drugs, especially plants, microorganisms, and marine life (Schwartsmann et al. 2002). Herbal medicines from the Chinese Pharmacopoeia have long been prescribed for many diseases and are drawing increased scientific attention (Vickers 2002). In a herb having different components, synergistic activities, or buffering toxic effects, mixtures or extracts of herbs may have more therapeutic or preventive activity than the isolated herb (Li et al. 2000; Vickers 2002). Several studies have clarified that extracts from several herbal medicines or mixtures have possible anticancer effects either *in vitro* or *in vivo* (Yano et al. 1994; Kao et al. 2001; Bonham et al. 2002; Hu et al. 2002; Lee et al. 2002). Hu et al. (2002) demonstrated that a *Ganoderma lucidum* alcohol extract could promote apoptosis in MCF-7 human breast cancer cells, and Lee et al. (2002) showed that a water extract of *Paeoniae lactiflora* could promote apoptosis in HepG2 and Hep3B hepatoma cells. Kao et al. reported that a water extract of Bu-Zhong-Yi-Qi Decoction (a mixture of ten herbs) could promote apoptosis in hepatoma cells (Kao et al. 2001). Yano et al. (1994) clarified that the water-soluble elements of Sho-Saiko-To (mixture of seven herbs) suppressed the proliferation of KIM-1 human hepatoma cells and KMC-1 cholangiocarcinoma cells. PE-SPES (a mixture of eight herbs) was developed for the clinical treatment of prostate cancer (Bonham et al. 2002).

7.2 Other Anticancer Studies on Chinese Angelica

The main compounds in Chinese angelica are ferulic acid, ligustilide, brefeldin A, butylidenephthalide, and polysaccharides (Zhao et al. 2000). The ferulic acid and lingustilide are usually used as chemical markers for the quality control of Chinese angelica roots (Zhao et al. 2000). As regards to butylidenephthalide, it is used as a standard profile for ligusticum Chuanxing identification. Ferulic acid is able to inhibit platelet aggregation while ligustilide has anti-asthmatic activity (Liu et al. 2000).

In 1994, Choy et al. (1994) identified a low molecular weight polysaccharide from the rhizome of Chinese angelica. Furthermore, it shows strong antitumor and immunostimulating activities both *in vitro* and *in vivo*. In addition, one investigator recently isolated several kinds of polysaccharides from Chinese angelica, and determined their effects on cancer cells. Interestingly, a novel polysaccharide, APS-1, with a backbone composed of (1,4)-α-D-glucopyranosyl residues, and branches composed of (1,6)-α-D-Glcp residues with a terminal β-L-arabofuranose residue showed significant antitumor effects *in vitro,* especially in human cervical cancer

HeLa cells (Cao et al. 2006). Additionally, Cao et al. (2010) have further determined three acidic polysaccharides (APS-3a, APS-3b, and APS-3c) obtained from Chinese angelica. Diels. They displayed different structural features and antitumor activities. This consequence is mediated by an increase in *IFN-gamma, IL-2,* and *IL-6* mRNA expressions in splenocytes and an increase in NO and TNF-gamma production in macrophage (Cao et al. 2010).

In terms of anti-brain tumor study, acetone extract AS-AC or chloroform extract AS-C significantly inhibited the proliferative activity of against glioblastoma (GBM) 8401 cultured cells by 30–50%, as well as the expression of cathepsin B and vascular endothelial growth factor. For the *in vivo* study in nude mice, the growth of the tumor was inhibited by 30% (AS-AC) to 60% (AS-C). S-AC and AS-C also significantly inhibited microvessel formation in the tumors of nude mice (Lee et al. 2006). Subsequently, bio-based assays for extracts of Chinese angelica showed that the acetone extract AS-C had dose-dependent anti-proliferative effect on A549, HT29, DBTRG-05MG, and J5 human cancer cells. Their results also demonstrate that the AS-C could induce cell cycle arrest and activate the mechanism of apoptosis in human cancer cells (Cheng et al. 2004). Finally, the active component of butylidenephthalide from acetone extract of Chinese angelica has been identified for anti-brain effect (Tsai et al. 2006). Butylidenephthalide belongs to one of the phthalide compounds. There are many phthalide identified from other nature components that have been reported for antitumor effect (Mullady et al. 2004; Yoshikawa et al. 2010). Later, butylidenephthalide showed to have anticancer potential on colon cancer. Moreover, in combination with other ingredients (senkyunolide A and z-ligustilide) it displays synergy antitumor effect (Kan et al. 2008). Actually, cytotoxicity against L1210 and K562 tumor cell line by z-ligustilide also has been reported (Chen et al. 2007).

Additionally, in irradiated lung tissue, Chinese angelica down-regulation of TNF-α and TGF-β1 results in inhibition of the progress of radiation-induced pulmonary fibrosis (Han et al. 2006a; Xie et al. 2006; Zhong et al. 2007). These data suggest that Chinese angelica may be useful in preventing and treating radiation-induced pulmonary fibrosis in the clinic.

Interestingly, phthalide like butylidenephthalide not only have antitumor effect but chemoprevention of benzo[a]pyrene-induced forestomach cancer in mice (Zheng et al. 1993). A synergistic antiproliferative effect was also observed when butylidenephthalide was combined with the chemotherapeutic drug 1,3-bis(2-chloroethyl)-1-nitrosourea (BCNU or carmustine) (Yu et al. 2010). This synergistic effect is mediated by down-regulation of the DNA repair gene *MGMT* (O6-methylguanine methyltransferase) expression (Yu et al. 2010).

7.3 Chinese Angelica on Clinical Trial of Cancer Patient

Only a few clinical trials have been conducted of Chinese angelica, but so far clinical effect of Chinese angelica injection in treating acute cerebral infarction have been reported (Liu et al. 2004). Treatment of menopausal symptom has also been studied (Kupfersztain et al. 2003). In addition, Chinese angelica can significantly inhibit platelet activation, relieve vascular endothelial cell injury, and improve microcirculation in ulcerative colitis (Dong et al. 2004). Besides, Chinese angelica can improve pulmonary hemodynamic *via* effect of the metabolism of ET-1, AT-II, EDF, and increase PaO_2 in the body (Xu and Li 2000). Regarding adjuvant chemotherapy in clinical trial, Chinese angelica can prevent or ameliorate these sequelae caused by chemotherapy, e.g. hot flashes due to premature menopause (Rock and DeMichele 2003). It could also inhibit and delay the thrombosis of rabbit and inhibit platelet aggregation (Zhao et al. 1994).

7.4 The Root of Chinese Angelica

The root of Chinese angelica, also known as "Dong quai", is a popular herbal medicine that has been used throughout China for over a thousand years to treat gynecological conditions. Several extracts or compounds purified from these herbs have been found to increase myocardial blood blow, reduce radiation damage, and purify blood quality (Yim et al. 2000; Wang et al. 2001; Xie et al. 2001; Kim et al. 2002). Chinese angelica has been determined to consist mainly of polysaccharides, and has been shown to have a protective effect against gastrointestinal damage and hepatic injury (Ye et al. 2001a, b, c, 2003). In conclusion, different components of *A. sinensis* may be involved in different pharmacological activities.

Our materials were supplied by Chung-Yuan Co (Taipei, Taiwan), and were identified by Professor Han-Ching Lin. A voucher herbarium specimen was deposited at the School of Pharmacy, National Defense Medical Center (Taiwan). The dried and powdered roots of Chinese angelica (2 kg) were extracted sequentially using acetone (5 L, 3 times), methanol (5 L, 3 times), and water (2 L). The extracts were concentrated under reduced pressure to yield a chloroform extract AS-C, methanol extract AS-M, and water-soluble extract AS-W (Fig. 7.1). These extracts were dissolved in dimethyl sulfoxide and incubated by shaking at 37°C for 1 hour. The extracts were stored at 4°C before each experiment. Six major compounds have been isolated from Chinese angelica: (E)-ligustilide, (Z)-ligustilide, (Z)-n-butylidenephthalide, palmitic acid, beta-sitosterol, and ferulic acid (Wang et al. 1998). Butylidenephthalide (BP, BdPh, or K1; molecular weight approximately 188.22) derived from the Chinese angelica chloroform extract AS-C was the major component (over 30% of AS-C crude) and was therefore chosen for experiments on its antitumor activity. BP has several medicinal properties, such as an anti-platelet effect, as well as an inhibitory effect on cyclooxygenase (Teng et al. 1987). It also abates angina (Ko et al. 1998, 2002; Bardon et al. 2002). In addition, it has been suggested that the anti-proliferation effect of synthetic BP-42 (3-butylidene-4,5-dihydroxyphthalide), modified by adding the hydroxyl molecules in the phthalide group of BP, could supply anti-atherosclerotic therapeutics (Mimura et al. 1995). Yet, the antitumor activities of BP have not yet been determined.

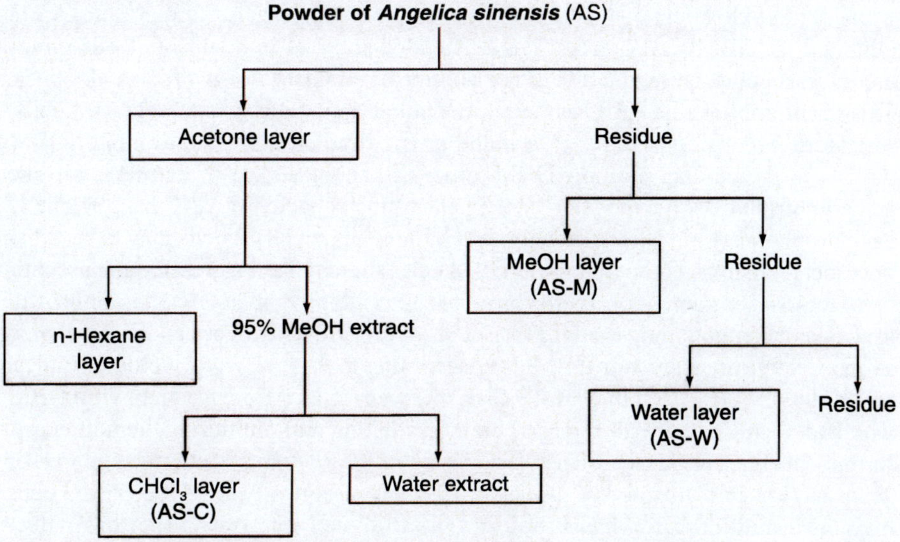

Fig. 7.1 Flow chart of the extraction of the root of *Angelica sinensis* (AS). *Angelica sinensis* extracts were concentrated under reduced pressure to yield a chloroform extract AS-C, methanol extract AS-M, and water-soluble extract AS-W

7.5 Chloroform Extract AS-C of Chinese Angelica: Antitumor Effect *In Vitro* and *In Vivo*

Based on its long record in traditional Chinese medicine, Chinese angelica has displayed a gamut of biological activities, e.g. regulation of the immune system, alleviation of menopausal symptoms, and recovery in myocardial blood flow (Yano et al. 1994; Kao et al. 2001; Bonham et al. 2002; Hu et al. 2002; Lee et al. 2002; Giese et al. 2003). There has been very little advancement in drugs for malignant brain tumors, in part because of the difficulties drugs encounter when passing through the blood-brain barrier. This present study is the first to verify that AS-C demonstrates strong activity against GBM *in vitro* and *in vivo*. Using the AS-C extract, the cytotoxic assays showed not only brain tumor cells, but also other tumor cells *in vitro*; these same experiments proved that regular fibroblast cells were resistant to AS-C. No indication of AS-C induced cytotoxic response was found in either the liver or the kidney after a single injection of 500 mg/kg (either intraperitoneal or subcutaneous). In this study, the anticancer efficacy of AS-C was better than that of BCNU. Furthermore, the cytotoxicity of AS-C to normal cells was lower than that of Taxol (Tsai et al. 2005). In clinical studies, several antitumor drugs like Temozolomide extended the survival time for patients with GBM only slightly (Blacklock et al. 1986; Shapiro and Green 1987; Mahaley et al. 1989; Cobb et al. 1996; Elliott et al. 1996; Santarius et al. 1997; Ross et al. 1998; Bello et al. 2001; Bardon et al. 2002; Giese et al. 2003). Based on the *in vitro* cytotoxic results, AS-C had a proportionately greater cytotoxic effect on GBM cells than on other tumor cells. Furthermore, AS-C inhibited rat GBM tumor growth and extended survival in a subcutaneous tumor model, particularly reducing the tumor volume of rat GBM *in situ* (Tsai et al. 2005). This result implies that AS-C can cross the blood-brain barrier to affect GBM cells. Therefore, according to both *in vitro* and *in vivo* studies, AS-C could inhibit GBM tumor cell growth and promote GBM tumor cell apoptosis. AS-C extract is oil-like and arduous to dissolve in water, as the penetrating capacity of AS-C may be due to its hydrophobic (lipid-like) character. In addition, human GBM cells may have a distinct metabolism as compared to rat GBM cells, thereby having a different susceptibility to AS-C extract. AS-C therapeutics can also inhibit human GBM tumor growth in a subcutaneous tumor model (Tsai et al. 2005). AS-C extract can be developed as an effective and safe anti-GBM drug according to these *in vivo* cytotoxic studies.

The human GBM cell line DBTRG-05MG was isolated from a patient with GBM who had been treated with regional brain irradiation and multidrug chemotherapy. In this cell line, no losses of heterozygosity in the *p53* and *Rb* tumor suppressor genes have been confirmed (Kruse et al. 1992). Consequently, DBTRG-05MG cells may have multidrug resistance ability. The RG2 cell line (rat GBM) has defective p53 expression and homozygous deletions of the *p16/Cdkn2a/Ink4a* gene locus (Schlegel et al. 1999; Bowers et al. 2003). Several studies have verified that both brain tumor cell lines are resistant to BCNU, an alkylating agent among the main chemotherapeutic agents for brain tumors (Kornblith and Walker 1988; Belanich et al. 1996; Tsai et al. 2005). By comparison, AS-C impressively inhibited growth of the two GBM tumors *in vitro* and *in vivo*. It is reasonable to suggest that AS-C could promote GBM cell apoptosis by both p53-dependent and -independent pathways and may offer a significantly better treatment effect in tumors than BCNU, for particular tumors. In addition, *in vitro* and *in vivo* cytotoxic experimental evidence showed that there were no significant side effects under either high doses or high frequency of AS-C treatment (Tsai et al. 2005), implying that AS-C is promising as an antitumor drug for malignant tumors with multidrug resistance.

To investigate the mechanisms that might account for the effects of AS-C on GBM tumors, cell cycles were monitored. The results showed suppressed tumor growth due to cell cycle arrest at the G0/G1 phase (>90%), with induction of multiple apoptosis molecules resulting in apoptotic cell death after the AS-C treatment. AS-C prompted high levels of expression of p21 and p16 and reduced the phosphorylation of Rb protein 6 hours after therapy (Tsai et al. 2005), indicating that AS-C arrests the cell cycle at the G0/G1 phase by moderating the gene expression, including the cyclin/CDK (cyclin-dependent kinase)/cyclin kinase inhibitor system. Both p16 and p21 are CDK inhibitors that bind to and negatively manage CDK or cyclin/CDK complexes (Sherr and Roberts 1995). The p16 protein, a member of the INK4 family, binds to CDK4 or CDK6 to suppress kinase activity at mid-G1 phase (Reed et al. 1994). The binding of p21 and cyclin/CDK complexes cause the cell cycle G1 arrest through RB phosphorylation (Dobashi et al. 2003). AS-C, therefore, could up-regulate the CDK inhibitor and result in a decrease in the phosphorylated Rb proteins.

These effects, AS-C anticancer compounds triggering cell cycle G1 arrest and leading to increased p21 expression, are also observed in monoterpene activity, an essential oil in plants (Bardon et al. 2002). On the other hand, a compound isolated from marine sponges (aragusterol A) has been reported to be a potent anticancer marine steroid causing decreased Rb phosphorylation and triggering of the cell cycle G1 arrest (Fukuoka et al. 2000).

7.5.1 The p53-Dependent Apoptosis Pathway

In Tsai's report (2005), they found AS-C could promote tumor cell apoptosis and the supposition effects could be the p53-dependent apoptosis pathway. *p53* is a tumor suppression gene and, in their experiment, the brain tumor cells' p53 was up-regulated and the phosphorylation of p53 increased after AS-C treatment in a time-dependent manner. Once the NH2-terminal region of p53 has been phosphorylated, *p53* will depart from *mdm2* (murine double minute 2) gene. Following this process, the tumor cells Bax (pre-apoptosis protein) and Bcl-2 (anti-apoptosis protein) become imbalanced. The up-regulated Bax will form a homodimer and trigger caspase-9 activation. Caspase-9 activity will then trigger caspase-3 activity to execute the apoptosis process. In some reports, DNA damage can activate p53-dependent apoptosis by induced phosphorylation of the p53 at the Ser15 site and increased expression of p16 (Shapiro and Green 1987).

7.5.2 The Fas-FasL-Induced Apoptosis Pathway

In addition to the p53-dependent apoptosis, AS-C treatment also increases Fas expression in human GBM tumor cells (Tsai et al. 2005). Fas can be the receptor for death signaling from the cell membrane and will then activate pro-caspase-8, subsequently promoting pro-caspase-3 activation, which results in apoptosis. In the GBM tumor cells, AS-C induced apoptosis might be mediated, at least partially, *via* Fas and mitochondrial apoptosis pathways (Tsai et al. 2005). In order to verify this point, we applied AS-C on the RG2 cells which had impaired p53 expression and homozygous deletions of the *p16/Cdkn2a/Ink4a* gene locus (Schlegel et al. 1999; Bowers et al. 2003). AS-C induced apoptosis in the RG2 cells was observed. Yet, while the RG2 cells were impaired for p53 and p16, caspase-9 and p21 expressions induced by AC-S were still displayed in these cells. This result indicates that AS-C entered into the RG2 cells through the p53-independent apoptosis pathway (Fas-induced apoptosis pathway), but not the p53-dependent apoptosis pathway. Thus, AS-C induced cell apoptosis and cell growth inhibition also involve the p53-independent pathway.

In Shang's reports, AP-0 (extracts from Chinese angelica), have invasion inhibiting and metastasis effects on hepatocellular carcinoma (Shang et al. 2003). However, AP-0 is very different in nature from the active component in Tsai's AS-C extract (Tsai et al. 2005). In Tsai's study, the AS-C has two properties including oil-like consistency (non-polar) and smaller molecular weight. In clinical brain tumor therapy, these properties allow the drugs to pass through the blood-brain barrier and decrease the tumor volume in the brain. From Tsai's *in vitro* and *in vivo* evidence, AS-C has potent anticancer effects including tumor cell cycle arrest and apoptosis. If the specific active AS-C compounds can be isolated from AS-C and the antitumor mechanisms investigated, AS-C has the potential to provide a new choice for future clinical brain tumor therapy.

7.6 BP Is AS-C's Major Antitumor Component Which Can Suppress Telomerase Activity

Since AS-C produces antitumor activity in human glioblastomas (Tsai et al. 2005), we were encouraged to isolate these antitumor components from this crude extraction of Chinese angelica. In AS-C, two major components (codename: BP and K2) have been identified. Their molecular weights are 188.22 (BP) and 190.23 (K2). Whether one or both of these components is an active antitumor component has yet to be determined. After separating the compounds out of the AS-C, the major anticancer component (over 30% of the crude AS-C), has been identified as BP (Tsai et al. 2006). In our laboratory, BP has been shown to arrest the cell cycles of glioblastoma and hepatocellular carcinomas in the G0/G1 phase, as well as promote apoptosis (Tsai et al. 2006; Chen et al. 2008; Lin et al. 2008). *In vitro*, BP promotes apoptosis through both the p53-dependent and p53-independent pathways. In the *in situ* rat's GBM tumor, BP not only suppresses tumor growth, but also reduces the GBM volume in situ. These findings provide evidence to suggest that BP is the potential anticancer compound from Chinese angelica (Tsai et al. 2006; Chen et al. 2008; Lin et al. 2008).

Chinese angelica is a candidate for telomerase suppression (Tsai et al. 2006; Chen et al. 2008; Lin et al. 2008). Since the BP isolated from the chloroform extract of Chinese angelica displayed dramatic antitumor effect both *in vitro* and *in vivo*, the BP may be targeting the telomerase. To clarify this hypothesis, BP's ability to suppress telomerase activity was tested in an experiment on GBM. According to our study, in brain cancer cell lines DBTRG-05MG and GBM 8401, BP can suppress telomerase mRNA transcription, reducing telomerase protein expression. However, the RNA component of telomerase hTR RNA expression in the GBM cells was not affected after the BP treatment. Decreased telomerase expression arose from transcriptional *hTERT* (human telomerase reverse transcriptase) inhibition and correlated with reduced telomerase activity. These results determined that BP's down-regulation of telomerase expression and decreasing effect on telomerase activity could be potential antitumor mechanisms on GBM.

In the telomerase promoter regulatory region, there are several potential transcription factor binding sites to regulate telomerase mRNA transcription. Among them, the Myc/Mad binding sites E-boxes play a critical role in telomerase mRNA transcription (Flores et al. 2006). In the tumor cell, c-myc protein can bind to the

E-boxes site and activate telomerase mRNA transcription (Wu et al. 1999). In addition, Kyo's report (2000) indicates that there are five Sp1 binding sites in the telomerase core promoter region. In most cancer cells, Sp1 proteins are up-regulated and telomerase expression is much higher in cancer cells than in normal cells. When the telomerase promoter Sp1 sites mutate, telomerase mRNA transcription is substantially reduced (Kyo et al. 2000). Analysis of the GBM cells' mRNA and protein expression supported the fact that BP suppresses telomerase expression, but the *c-myc* mRNA and protein expression were unaffected after the BP treatment. However, the other key transcription factor, Sp1, decreased in protein level. The electrophoretic mobility shift assay analysis indicated that the Sp1 protein expression declined after the BP treatment and then decreased Sp1 binding activity at the telomerase promoter. Thus, the protein level of Sp1 expression may be a critical determinant of telomerase transcription.

Telomerase activities are tightly related to cell proliferation and senescence. BP demonstrates antitumor effects including suppression of cancer cell proliferation and promotion of apoptosis. In the anti-proliferation effect, the possible mechanism is BP function as a telomerase suppresser. Upon BP treatment, GBM cells stopped proliferation and became senescent. The evidence was observed in the transfection and overexpression of telomerase in the GBM cells. These cells can defend against BP's stop proliferation effect and resist senescence, indicating that BP suppresses cellular proliferation and inhibits telomerase activity by repressing telomerase transcriptional activity. The present study evaluated molecular mechanisms underlying the cell cycle G0/G1 arrest effect of BP in brain cancer cells.

7.7 Using Oligodeoxynucleotide-based Microarray Analysis to Determine BP's Effectors Against GBM

To determine the mechanism by which BP induces apoptosis, an oligodeoxynucleotide-based microarray was used to screen out BP-mediated changes in gene expression. Human GBM tumor cells were treated with BP (IC_{50} concentration) for 3 or 24 hours. After treatment, the mRNA was extracted by microarray screening and compared with the vehicle group (GBM cells treated with dimethyl sulfoxide). In all of the 422 BP up-regulated genes (49 in 3 hours and 373 in 24 hours), Lin et al. (2008) focused on those showing more than a two-fold increase as compared to the vehicle. Among them, one gene family was up-regulated by BP in a very short time. They are *NOR-1*, *Nurr1*, and *Nur77*.

NOR-1 (also called *NR4A3*), *Nurr1* (also called *NR4A2*), and *Nur77* (also called *NR4A1*) are early response genes which can induce, by serum, growth factors, receptor engagement, and apoptotic stimuli (Williams and Lau 1993; Woronicz et al. 1994; Li et al. 2000; Winoto and Littman 2002). These family members share similar protein structural features and have no known natural ligand yet (Li et al. 2006). Because their receptor ligand has not yet been identified, they are classified as orphan receptors (Kim et al. 2006). In this family, Nur77-mediated apoptosis has been extensively studied in T cells and several cancer cells (Williams and Lau 1993; Youn et al. 1999; Li et al. 2000; Wilson et al. 2003; Lin et al. 2004). In addition to *Nur77*, *NOR-1*, and *Nurr1* have also been reported to be involved in cell growth and apoptosis (Li et al. 2006). Nur77-mediated apoptosis has been reported *via* two mechanisms: (1) Nur77 responds to apoptotic stimuli and translocates to the mitochondria to interact with Bcl-2. This Nur77 and Bcl-2 interaction reverses the function of Bcl-2 from anti-apoptotic to pro-apoptotic. As a result, cytochrome c release from the mitochondrion is triggered and, subsequently, apoptosis occurs. These processes have been demonstrated in human prostate carcinoma and other cancer cells (Li et al. 2000; Lin et al. 2004). (2) Nur77 can function as a transcription factor. Nur77 appears to up-regulate genes responsible for promoting apoptosis, e.g. the Fas ligand, the tumor necrosis factor-related apoptosis-inducing ligand, the Nur77 downstream gene 1, and the Nur77 downstream gene 2 (Weih et al. 1996; Rajpal et al. 2003; Chintharlapalli et al. 2005). Nur77-mediated apoptosis is stimulated by many apoptosis inducers, e.g. retinoid-related 6-[3′-(1-adamantyl)-4′-hydroxyphenyl]-2-naphthalenecarboxylic acid (also called CD437), tetradecanoylphorbol-1, 3-acetate, etoposide VP-16, cisplatin, etoposide, and insulin-like growth factor binding protein-3 (Li et al. 2000; Bardon et al. 2002; Liu et al. 2002; Wilson et al. 2003). More recently, some reports highlight both activation of JNK (c-Jun NH2-terminal kinase) and inhibitions of Akt as involved in Nur77's translocation from nucleus to cytoplasm (Han et al. 2006b). Additionally, in T cells, the PKC (protein kinase C) activation pathway resulting in Nur77 expression plays a key role in T cell receptor signaling leading to thymocyte apoptosis (Kim et al. 2005).

7.8 BP Causes Nur77-Translocation Leading to Tumor Apoptosis

It is certain that *Nur77* plays a critical role in cancer cells (Li et al. 2000). Lin and co-workers (2008) used a fluorescence microscope and the fluorescent-linked Nur77-specific antibody to track and observe the Nur77 translocation by its immunofluorescence stain. After treatment with BP for 24 hours, they found Nur77 translocated from the nucleus to the cytoplasm. The fluorescence images showed that Nur77 was more abundant in the nucleus than in the cytosol. They separated the cytosolic and nucleus protein fractions and assayed the Nur77 spread by Western blot. Both results showed the Nur77 predominantly localized in the nucleus. However, when the GBM cells were treated with BP, the Nur77 was up-regulated and migrated to the cytoplasm. After translocating to the cytoplasm, the Nur77 functioned as a potent pro-apoptotic molecule by triggering cytochrome c release and activating caspase-9 and then caspase-3 and finally the process of apoptosis. To prove this finding, *Nur77* siRNA was used to silence the BP-induced *Nur77*. In this condition, BP no longer triggered GBM cell apoptosis. This study evaluated the molecular mechanisms underlying the apoptosis effect of BP in brain cancer cells.

7.9 BP's Inhibition of Telomerase Activity Can Cause Tumor Senescence and Contribute to Tumor Apoptosis

In Lin et al.'s study (2008), BP induced GBM cell apoptosis by *Nur77* up-regulation. However, the *Nur77* siRNA only increased cell viability by 40%, indicating that there is another mechanism decreasing the cell viability after the BP treatment. In our study, BP could down-regulate telomerase activity. Subsequently, we found that BP-inhibited GBM cell growth is partially due to the *hTERT* down-regulation. Cell viability is controlled by proliferative and apoptotic signals. In the GBM cell, BP suppressing telomerase induces a cytostatic effect. In our laboratory, after BP treatment, the doubling time of the DBTRG cell is increased according to the BP concentration. In the cytomegalovirus-driven *hTERT*-HA over-expression phTERT GBM cells, the cells are still under apoptosis. We found that cell viability was almost rescued after *hTERT* expression was restored and combined with *Nur77* siRNA treatment. This result indicates that *Nur77* and *hTERT* contribute to BP partially inhibiting GBM cell growth. The BP-induced cytotoxic and cytostatic effects are contributed by Nur77 over-expression and telomerase inhibition separately.

In order to study the trail of ectopic *hTERT* expression, GBM 8401 cells were transfected with the *hTERT*-HA construct phTERT. *hTERT*-HA has proven to be defective for increasing telomere length, despite displaying enzymatic activity on artificial substrates in Counter et al.'s report (1998). In our laboratory, in the *hTERT*-HA expressing 8401 cells, the chromosomes' telomeres are slightly longer as compared to the control cells, agreeing with Counter's report that the HA tag may not completely disrupt the *in vivo* function of the TERT protein. However, in our study, the exogenous *hTERT*-HA expression exterminated the BP-mediated cell cycle arrest. In addition, another possible mechanism might be that *hTERT* could regulate some growth-promoting factors (i.e. Akt, STAT3) and cell cycle regulatory components (i.e. cyclin D1) to promote cell growth besides telomere elongation (Jagadeesh et al. 2006).

Senescence is an irreversible cell cycle barrier. In senescent cells, increased activity of SA-β-gal (senescence-associated β-galactosidase) has been reported. Senescent cells remain viable but do not synthesize DNA (Schmitt 2007). As cells age, their telomeres become shorter until they reach a critical point when their structures change. The shortened telomeres trigger cell senescence (Herbig et al. 2004). Culture shock-induced senescence has been confirmed in cultured cells under extrinsic stress (Keith et al. 2007).

Finally, some DNA-damaging drugs, such as the topoisomerase I inhibitor camptothecin, the topoisomerase II inhibitor adriamycin, and the cross-linking agent cisplatin, could induce senescence *in vitro* (Chang et al. 1999; Schmitt 2007). SA-β-gal activity is widely recognized as a biomarker correlated with senescence (Dimri et al. 1995). In our studies, we found brain cancer cells treated with BP displayed intense SA-β-gal staining in association with decreased telomerase activity.

By using a colony formation assay, we observed the number of SA-β-gal positive colonies decrease (~87%) in the 8401 phTERT transformation cells.

Senescence depends on a lot of signaling pathways which result in a permanent and irreversible cell cycle blockade. In these pathways, p53, p21, and p16 are three major cell cycle regulatory and senescence proteins (Coates et al. 2007). After telomere shortening and chromosome instability, p53, a downstream signaling molecule, could be activated. Active p53 could arrest cell division and allow repair of damaged chromosomes and restore genomic stability (Liu 2000). In our previous study, we observed that BP treatment increases the p53 phosphorylation in the DBTRG-05MG tumor cells and increases the p53 (Tsai et al. 2006). In addition, senescence-associated markers p21 and p16 were also increased as a result of the BP treatment. Future experiments could confirm the mechanism of BP-induced tumor cell senescence.

7.10 Therapeutic Effects of the Chloroform Extract of Chinese Angelica on a Rat's *In Situ* GBM Tumor

F344 rats were implanted *via* intracutaneous injection with 5×10^4 RG2 cells and were treated with subcutaneous AS-C (500 mg/kg/day) on days 4–8 to investigate the AS-C antitumor effects on rat *in situ* GBM tumors. There were significant decreases in tumor volume in the treated group as compared to the untreated group ($P<0.05$) (Tsai et al. 2005). Average tumor volumes at days 14 and 16 were 70 ± 4.8 and 126.4 ± 11.1 mm^3 in the control group *vs* 46.2 ± 3.6 and 99.5 ± 9.5 mm^3 in the AS-C treated group. *In situ* tumor volumes in the AS-C treated group were smaller than in the control group according to the magnetic resonance imaging data (Tsai et al. 2005). Ki-67 proteins decreased and cleaved caspase-3 proteins increased in the immunohistochemistry results, indicating that the tumor cells were under apoptosis at day 16 after AS-C treatment as compared to the control group *in vivo* (Tsai et al. 2005). Similar results were found in the subcutaneous RG2 tumor model.

7.11 Therapeutic Effects of the Chloroform Extract of Chinese Angelica on Xenograft Tumor Growth

To investigate whether AS-C can suppress human GBM tumor growth, nude mice were inoculated with human DBTRG-05MG cells subcutaneously. After 5 days, the mice were treated with a single dose of AS-C (500 mg/kg). There was significant suppressive effect in the AS-C IP500 and AS-C SC500 treatment groups as compared to the untreated group ($P<0.005$) (Tsai et al. 2005). Average tumor sizes at day 38 were 849.9 ± 150.1 mm^3 in the control group and 295.5 ± 25.3 mm^3 in the AS-C IP500 treatment group, and 155.1 ± 56.4 mm^3 in the AS-C SC500 treatment group (Tsai et al. 2005).

7.12 Cytotoxic Activity of the Chloroform Extract of Chinese Angelica on a Human GBM Tumor *In Vivo*

To confirm whether AS-C can induce human GBM tumor cell death *in vivo*, we examined the cytotoxic activity of AS-C in human GBM tumors by using a xenograft nude mouse model. Histologic analysis revealed that a significant proportion of cells (30–50% in AS-C treated *versus* 0% in controls) in the AS-C treated tumor are dying. However, AS-C induced cytotoxic effects were not observed in the control group (Tsai et al. 2005). In Tsai et al.'s study (2005), AS-C treatment could decrease Ki-67 expression, increase cleaved caspase-3 protein expressions, and induce tumor cell apoptosis at day 10 as compared to the control group *in vivo*.

7.13 Antitumor Effects of BP on the Survival of Animals Bearing Subcutaneous GBM Tumors

In vivo animal experiments were used to examine the antitumor activity of BP. An inhibitory effect on RG2 tumor growth was observed in the BP-treated group, but not in the control group ($P<0.005$) (Tsai et al. 2006). In the control group, the average tumor size at day 26 was 20.7 ± 1.5 cm^3 *vs* only 9.6 ± 0.4 cm^3 in the BP-treated group. The survival time of the BP-treated rats was significantly prolonged (30 ± 2.1 days in the BP-treated group *vs* 41.5 ± 4.2 days in the control group, $P<0.0001$) (Tsai et al. 2006). In addition, there were no significant differences in the body weight between the control and BP-treated groups (Tsai et al. 2006). The Ki-67 expression was obviously decreased, indicating the anti-proliferation activity of BP *in vivo*. Cleaved caspase-3, an apoptotic marker, increased after the BP treatment and apoptosis of the tumor cells *in vivo* was induced (Tsai et al. 2006). *In vitro* and *in vivo*, BP could inhibit tumor cell growth and tumor cell apoptosis. In addition, no drug-related toxicity was observed by using BP at a dose of 500 mg/kg, as histological analysis of organs confirmed (data not shown).

7.14 Antitumor Effects of BP on the *In Situ* GBM Tumors of Rats

To confirm the antitumor effects of BP on *in situ* GBM tumors in rats, 5×10^4 RG2 cells were transplanted *via* intracutaneous injection into the brains of F344 rats, and these rats were then treated with subcutaneous BP (300 mg/kg/day) on days 4–8. MRI data indicated the tumor volume of the BP-treated group was much smaller as compared to the control group (Tsai et al. 2006).

7.15 Autologous Implantation of RG2 into a Rat Brain as a Model for Studying the BP Antitumor Effect

Since RG2 is a spontaneous rat GBM tumor, Tsai et al. (2006) implanted it into a rat brain to evaluate BP's antitumor effect. After administering BP subcutaneously for five days using a dose of 300 mg/kg/day, the MRI result showed a significant decrease in tumor size as compared to the control group. The tumor size was 70.0 ± 4.8 mm^3 and 126.4 ± 11.1 mm^3 in the control group at day 14 and 16, respectively. In contrast, in the group treated with BP, the tumor size was 46.4 ± 1.8 and 91.7 ± 8.3 mm^3 at days 14 and 16, respectively. By using an immunohistochemical stain, Tsai et al. (2006) found that the proliferation index of Ki-67 indicated it had decreased its protein expression while the apoptosis index of cleaved caspase-3 indicated that it had increased its protein expression in the BP treatment group relative to the control group. This result is similar to the subcutaneous RG2 tumor model which suggested that the antitumor effect of BP remained the same, even across the blood-brain barrier. Finally, the BP-treated group had a significantly prolonged survival rate as compared to the controls. At day 19 in the BP-treated group, the survival rate was much improved (50%, 3/6) as compared to the untreated group (16.7%, 1/6) ($P = 0.0016$).

7.16 Heterologous Implantation of Human GBM into Nude Mice as a Model to Study BP's Antitumor Effect

In order to investigate BP's antitumor effect in human GBM, DBTRG-05MG was implanted into nude mice. Five escalating doses of BP (70, 150, 300, 500, and 800 mg/kg) were administered subcutaneously from day 1 to day 5. Significant suppression of tumor growth as compared to the control group was observed in all five escalating doses. The tumor growth inhibition was dose-dependent. After 200 days, Tsai et al. (2006) found that the tumor growth rates were 16.7% (1/6), 33.3% (2/6), 50% (3/6), 80% (4/5), and 100% (6/6) in the treated BP-800 mg, BP-500 mg, BP-300 mg, BP-150 mg, and BP-70 mg groups, respectively. The average tumor size was larger than 1,000 mm^3 in the control (vector only) group and 171.7 mm^3 in the BP-800 mg treatment group. The survival rate at day 200 was significantly prolonged for the nude mice treated with BP as compared to the controls. The survival rate was 0% (0/6), 20% (1/5), 50% (3/6), and 83.3% (5/6) in the untreated, BP-70 mg, BP-300 mg, and BP-800 mg treated groups, respectively. In addition, the immunohistochemical stain of Ki-67 indicated a decrease in protein expression but an increase in the cleaved caspase protein expression, indicating that BP can inhibit tumor proliferation and induce tumor apoptosis (Fig. 7.2). Consequently, the evidence indicates that BP can decrease tumor size and increase mice survival rate. Finally, 200 days after the BP treatment, there were no body weight changes

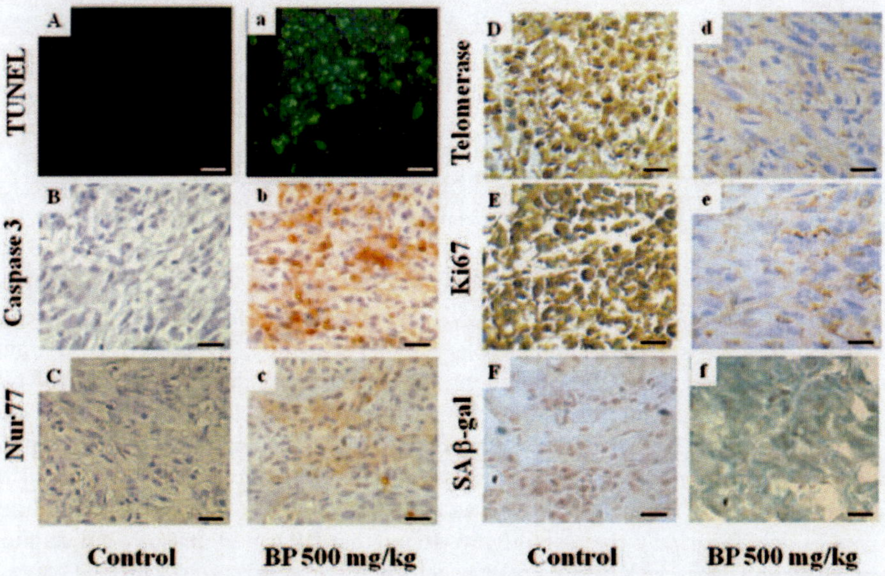

Fig. 7.2 Butylidenephthalide (BP)-induced *in vivo* growth inhibition of DBTRG-05MG xenografts in nude mice. Human brain cancer xenografts were established by subcutaneous injection of approximately 2×10^6 GFP-DBTRG-05MG cells into the dorsal subcutaneous tissue of nude mice. After tumors reached approximately 50 mm^3, the mice were given daily subcutaneous injections of vehicle alone (control group; $n=3$) or the indicated amount of BP (treatment group; $n=3$ each) for five successive days. Paraffin-embedded tumor tissue sections (5 μm) were subjected to immunohistochemical staining using TUNEL assay for DNA fragmentation (**A** and **a**), antibodies against Caspase-3 (**B** and **b**), Nur77 (**C** and **c**), telomerase (**D** and **d**) and Ki-67 (**E** and **e**). Additional tumor tissues were snap-frozen in OCT mounting medium and 12-μm sections were subjected to SA-β-gal staining (**F** and **f**). Representative images are shown; bar=50 μm

or histologically specific changes in the hollow organs, even with a high BP dose of 800 mg/kg for 5 days of continuing treatment, which supports the premise that BP has no toxic effect on acute and subacute toxicity.

7.17 Nur77 Is One of the Promising Effectors of BP Against Brain GBM

From the *in vitro* investigation, we learned that BP can up-regulate Nur77 expression and Nur77's translocation promotes tumor apoptosis, and we were prompted to examine whether this also happened *in vivo*. Using a xenograft nude mice model, we implanted 2×10^6 DBTRG cells into dorsal subcutaneous tissue and evaluated the antitumor effect of BP and its relation to Nur77 translocation. Western blot and immunohistochemical stain (Fig. 7.2) both showed that Nur77 overexpression after the BP treatment and Nur77's translocation were parallel with tumor apoptosis

(Lin et al. 2008). At present, Nur77 is one of the most promising effectors of BP against brain GBM.

7.18 BP's Use Against Brain GBM Mediates Tumor Senescence by Suppressing Telomerase

As mentioned previously, we determined the active component of the chloroform extract of Chinese angelica against GBM by using telomerase activity as the screening platform. This active component is BP. Subsequently, by using RT-PCR and the TRAP method, we demonstrated that BP can not only decrease *hTERT* mRNA but also decrease telomerase activity. Since telomerase is related to senescence and cell proliferation (Blagoev 2009), we proposed setting BP against the tumor apoptosis-mediated tumor senescence and the proliferation pathway. An immunohistochemical stain of X-gal and Ki-67 was applied to monitor these two possibilities respectively (Fig. 7.2). Interestingly, we found that BP can increase X-gal staining intensity and decrease Ki-67, as well as cell cyclin protein, e.g. p21 protein expression. This result is associated with BP concentration. This consequence implies that BP may mediate tumor senescence or cell proliferation leading to apoptosis. Restoring telomerase activity by transfection cytomegalovirus driving *hTERT* engineering vector can abolish this BP-induced tumor apoptosis effect. At present, we suggest telomerase is possibly another BP effector against brain tumors.

7.19 Helping BP Cross the Blood-brain Barrier

The biodistribution study of BP has been performed using labeled BP from a dermal application (Sekiya et al. 2000). Most labeled BP is present in the liver, gallbladder, bile duct, and kidneys. Unfortunately, there is only 0.029 μg/g of brain tissue which can detect labeled BP after 1 hour. On the other hand, only 3% of the total amount of BP can reach the whole brain, not to mention the tumor. As we have noted, BP is hydrophobic, making it difficult to administer orally and also difficult to dissolve in water. In terms of targeting the brain lesion, the blood-brain barrier is another obstacle preventing not only microorganisms, but also drugs like BP from crossing over. This helps to explain why only a few antitumor drugs, e.g. temozolomide or BCNU, have been applied to treat brain tumors. Developing a new dosage form against brain tumors is urgent and necessary.

One such control release wafer is characterized by local delivery of Carmustine using biodegradable polymers Gliadel. Unfortunately, Gliadel only prolongs survival for about 2 months, but some complications have been observed. These side effects include a higher incidence of wound infection and wound dehiscence (Tamargo et al. 2003). Since BP has a high LD_{50} 7.5 g/kg) and a low ED_{50} (0.8 g/kg), as well as a relatively safe therapeutic window (therapeutic index: $LD_{50}/ED_{50}=94$), we wanted to test the possibility of a BP-wafer for treating brain tumors.

In the *in vitro* releasing study, we determined that 50% of the BP can be released by the 6th day. This release can reach 90% by the 30th day. Fortunately, the BP released from the wafer retains antitumor properties similar to BP itself.

Regarding the *in vivo* investigation, three different rodent models have been used. As we mentioned earlier, RG2 is a spontaneous rat F344 GBM and subcutaneous implantation has been performed. The second model is human GBM, and subcutaneous DBTRG xenografts into nude mice have been performed. The third model is an FGF-SV40 transgenic mouse with a spontaneous brain tumor. In all the three *in vivo* studies, we observed a significant inhibitory effect on tumor growth with no significant adverse effects on the rodents. This interstitial administration of BP can prolong survival time and rate relative to the wafer only control group. In the rat F344 model, the survival rate is 0% in the wafer no drug group (0/6, wafer no drug), but improvement is seen notably in the BP groups: 67% (4/6, wafer + 3% BP) and 100% (3/3 wafer + 10% BP) with $P<0.001$ at the end of 30 days. These results are in contrast to the 3% BCNU wafer, where the survival rate was 50% (3/6, wafer + 3% BCNU).

In the FGF-SV40 transgenic mice, the survival rate was 0% in wafer no drug (0/3, wafer no drug), but improved in the BP groups: 33% (1/3, wafer + 3% BP), 100% (3/3 wafer + 15% BP) with $P<0.001$ at the end of 5 months.

Since BP is less toxic (LD_{50}, 7.5 g/kg) as compared to BCNU (LD_{50}, 20 mg/kg), it is reasonable to expect that there was no brain edema, no delay in wound healing, no CSF leakage, and no brain infection—all symptoms that have been observed for BCNU Wafers (Subach et al. 1999; McGovern et al. 2003).

Finally, based on previous studies, we showed that 300 mg/kg BP, when administered subcutaneously, reduced tumor volume by about 25% at day 15 (Tsai et al. 2006), whereas the 15% BP-Wafers (15 mg BP) implanted into the brains of FGF-SV40 transgenic mice decreased tumor size by about 64% at day 20. We therefore estimate that our BP-Wafer interstitial controlled-release device increases the antitumor BP effect to 50 times more potency than a subcutaneous injection.

7.20 Conclusion and Perspectives

We find BP to be one of a few potential target drugs for treating human brain tumors. The potential effectors are Nur77 and telomerase (Fig. 7.3). BP is also a possible compound against human GBM-mediated tumor senescence. Using BP-loaded wafers can increase the local concentration of BP and help cross the blood-brain barrier, resulting in 50 times more BP being directed against the brain tumor. Plus, BP is less toxic than Gliadel or temozolomide, the most popular drugs in current clinical use against GBM. A pre-IND package will be prepared in the near future.

Oral feeding of BP is the most common route of administration for this antitumor drug. However, BP is hydrophobic and poorly dissolved in water. Structure activity modification of BP is urgently required. In addition, though BP has a relatively high

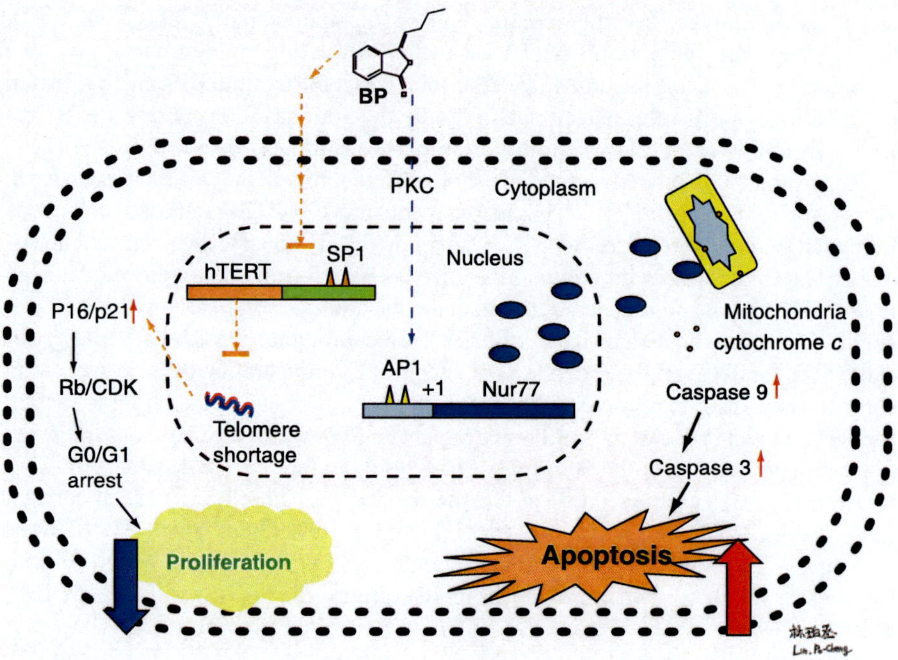

Fig. 7.3 The mechanism by which butylidenephthalide (BP) induces glioblastoma multiforme growth inhibition and apoptosis. In response to BP stimuli, glioblastoma multiforme cells undergo two pathways: (1) Telomerase activity is suppressed by BP which then induces cytostatic effect in glioblastoma multiforme cells. (2) Nur77 migrates from the nucleus to the cytoplasm, where it targets mitochondria to release cytochrome c and trigger apoptosis

therapeutic index ($LD_{50}/ED_{50}=94$), its efficacy still needs improvement. Since the target genes of BP are *Nur77* and *hTERT*, it is a potential target drug for use against brain tumors and is worth exploring for clinical application.

Acknowledgments We would like to thank Shinn-Zong Lin (Center for Neuropsychiatry, China Medical University and Hospital, Taichung; Department of Neurosurgery, China Medical University Beigan Hospital, Yunlin, Taiwan), Po-Yen Liu (Graduate Institute of Chinese Medical Science, China Medical University, Taichung, Taiwan), You Chorn Horn (Graduate Institute of Networking and Multimedia, National Taiwan University, Taipei, Taiwan), Li-Fu Chang (Department of Life Science and Graduate Institute of Biotechnology, National Dong Hwa University, Hualien, Taiwan), and Ivy I-Wei Lin (School of Medicine, China Medical University, Taichung, Taiwan) for their contributions.

References

Bardon, S., Foussard, V., Fournel, S., et al. (2002). Monoterpenes inhibit proliferation of human colon cancer cells by modulating cell cycle-related protein expression. *Cancer Letters, 181*, 187–194.

Belanich, M., Pastor, M., Randall, T., et al. (1996). Retrospective study of the correlation between the DNA repair protein alkyltransferase and survival of brain tumor patients treated with carmustine. *Cancer Research, 56,* 783–788.

Bello, L., Carrabba, G., Giussani, C., et al. (2001). Low-dose chemotherapy combined with an antiangiogenic drug reduces human glioma growth in vivo. *Cancer Research, 61,* 7501–7506.

Blacklock, J. B., Wright, D. C., Dedrick, R. L., et al. (1986). Drug streaming during intra-arterial chemotherapy. *Journal of Neurosurgery, 64,* 284–291.

Blagoev, K. B. (2009). Cell proliferation in the presence of telomerase. *PLoS One, 4,* e4622.

Bonham, M., Arnold, H., Montgomery, B., et al. (2002). Molecular effects of the herbal compound PC-SPES: Identification of activity pathways in prostate carcinoma. *Cancer Research, 62,* 3920–3924.

Bowers, G., He, J., Schulz, K., et al. (2003). Efficacy of adenoviral p53 delivery with SCH58500 in the intracranial 9 l and RG2 models. *Frontiers in Bioscience: A Journal and Virtual Library, 8,* a54–a61.

Cao, W., Li, X. Q., Liu, L., et al. (2006). Structural analysis of water-soluble glucans from the root of Angelica sinensis (Oliv.) Diels. *Carbohydrate Research, 341,* 1870–1877.

Cao, W., Li, X. Q., Wang, X., et al. (2010). Characterizations and anti-tumor activities of three acidic polysaccharides from Angelica sinensis (Oliv.) Diels. *International Journal of Biological Macromolecules, 46,* 115–122.

Chang, B. D., Broude, E. V., Dokmanovic, M., et al. (1999). A senescence-like phenotype distinguishes tumor cells that undergo terminal proliferation arrest after exposure to anticancer agents. *Cancer Research, 59,* 3761–3767.

Chen, Q. C., Lee, J., Jin, W., et al. (2007). Cytotoxic constituents from angelicae sinensis radix. *Archives of Pharmacal Research, 30,* 565–569.

Chen, Y. L., Jian, M. H., Lin, C. C., et al. (2008). The induction of orphan nuclear receptor Nur77 expression by n-butylidenephthalide has pharmaceuticals on hepatocellular carcinoma cell therapy. *Molecular Pharmacology, 74,* 1046–1058.

Cheng, Y. L., Chang, W. L., Lee, S. C., et al. (2004). Acetone extract of Angelica sinensis inhibits proliferation of human cancer cells via inducing cell cycle arrest and apoptosis. *Life Sciences, 75,* 1579–1594.

Chintharlapalli, S., Burghardt, R., Papineni, S., et al. (2005). Activation of Nur77 by selected 1,1-Bis(3′-indolyl)-1-(p-substituted phenyl) methanes induces apoptosis through nuclear pathways. *The Journal of Biological Chemistry, 280,* 24903–24914.

Choy, Y. M., Leung, K. N., Cho, C. S., et al. (1994). Immunopharmacological studies of low molecular weight polysaccharide from Angelica sinensis. *The American Journal of Chinese Medicine, 22,* 137–145.

Coates, S. S., Lehnert, B. E., Sharma, S., et al. (2007). Beryllium induces premature senescence in human fibroblasts. *The Journal of Pharmacology and Experimental Therapeutics, 322,* 70–79.

Cobb, M. A., Husain, M., Andersen, B. J., et al. (1996). Significance of proliferating cell nuclear antigen in predicting recurrence of intracranial meningioma. *Journal of Neurosurgery, 84,* 85–90.

Counter, C. M., Hahn, W. C., Wei, W., et al. (1998). Dissociation among in vitro telomerase activity, telomere maintenance, and cellular immortalization. *Proceedings of the National Academy of Sciences of the United States of America, 95,* 14723–14728.

Dimri, G. P., Lee, X., Basile, G., et al. (1995). A biomarker that identifies senescent human cells in culture and in aging skin in vivo. *Proceedings of the National Academy of Sciences of the United States of America, 92,* 9363–9367.

Dobashi, Y., Takehana, T., & Ooi, A. (2003). Perspectives on cancer therapy: Cell cycle blockers and perturbators. *Current Medicinal Chemistry, 10,* 2549–2558.

Dong, W. G., Liu, S. P., Zhu, H. H., et al. (2004). Abnormal function of platelets and role of angelica sinensis in patients with ulcerative colitis. *World Journal of Gastroenterology: WJG, 10,* 606–609.

Elliott, P. J., Hayward, N. J., Huff, M. R., et al. (1996). Unlocking the blood-brain barrier: A role for RMP-7 in brain tumor therapy. *Experimental Neurology, 141,* 214–224.

Flores, I., Benetti, R., & Blasco, M. A. (2006). Telomerase regulation and stem cell behaviour. *Current Opinion in Cell Biology, 18,* 254–260.

Fukuoka, K., Yamagishi, T., Ichihara, T., et al. (2000). Mechanism of action of aragusterol a (YTA0040), a potent anti-tumor marine steroid targeting the G(1) phase of the cell cycle. *International Journal of Cancer. Journal International du Cancer, 88,* 810–819.

Giese, A., Bjerkvig, R., Berens, M. E., et al. (2003). Cost of migration: Invasion of malignant gliomas and implications for treatment. *Journal of Clinical Oncology: Official Journal of the American Society of Clinical Oncology, 21,* 1624–1636.

Han, G., Zhou, Y. F., Zhang, M, S., et al. (2006a). Angelica sinensis down-regulates hydroxyproline and Tgfb1 and provides protection in mice with radiation-induced pulmonary fibrosis. *Radiation Research, 165,* 546–552.

Han, Y. H., Cao, X., Lin, B., et al. (2006b). Regulation of Nur77 nuclear export by c-Jun N-terminal kinase and Akt. *Oncogene, 25,* 2974–2986.

Herbig, U., Jobling, W. A., Chen, B. P., et al. (2004). Telomere shortening triggers senescence of human cells through a pathway involving ATM, p53, and p21 (CIP1), but not p16 (INK4a). *Molecular Cell, 14,* 501–513.

Hu, H., Ahn, N. S., Yang, X., et al. (2002). Ganoderma lucidum extract induces cell cycle arrest and apoptosis in MCF-7 human breast cancer cell. *International Journal of Cancer. Journal International du Cancer, 102,* 250–253.

Jagadeesh, S., Kyo, S., & Banerjee, P. P. (2006). Genistein represses telomerase activity via both transcriptional and posttranslational mechanisms in human prostate cancer cells. *Cancer Research, 66,* 2107–2115.

Kan, W. L., Cho, C. H., Rudd, J. A., et al. (2008). Study of the anti-proliferative effects and synergy of phthalides from Angelica sinensis on colon cancer cells. *Journal of Ethnopharmacology, 120,* 36–43.

Kao, S. T., Yeh, C. C., Hsieh, C. C., et al. (2001). The Chinese medicine Bu-Zhong-Yi-Qi-Tang inhibited proliferation of hepatoma cell lines by inducing apoptosis via G0/G1 arrest. *Life Sciences, 691,* 485–496.

Keith, W. N., Thomson, C. M., Howcroft, J., et al. (2007). Seeding drug discovery: Integrating telomerase cancer biology and cellular senescence to uncover new therapeutic opportunities in targeting cancer stem cells. *Drug Discovery Today, 12,* 611–621.

Kim, H., Lee, J. E., Kim, B. Y., et al. (2005). Menin represses JunD transcriptional activity in protein kinase C theta-mediated Nur77 expression. *Experimental & Molecular Medicine, 37,* 466–475.

Kim, H., Kim, B. Y., Soh, J. W., et al. (2006). A novel function of Nur77: Physical and functional association with protein kinase C. *Biochemical and Biophysical Research Communications, 348,* 950–956.

Kim, S. H., Lee, S. E., Oh, H., et al. (2002). The radioprotective effects of bu-zhong-yi-qi-tang: A prescription of traditional Chinese medicine. *The American Journal of Chinese Medicine, 30,* 127–137.

Ko, W. C., Sheu, J. R., Tzeng, S. H., et al. (1998). The selective antianginal effect without changing blood pressure of butylidenephthalide in conscious rats. *Planta Medica, 64,* 229–232.

Ko, W. C., Liao, C. C., Shih, C. H., et al. (2002). Relaxant effects of butylidenephthalide in isolated dog blood vessels. *Planta Medica, 68,* 1004–1009.

Kornblith, P. L., & Walker, M. (1998). Chemotherapy for malignant gliomas. *Journal of Neurosurgery, 68,* 1–17.

Kruse, C. A., Mitchell, D. H., Kleinschmidt-DeMasters, B. K., et al. (1992). Characterization of a continuous human glioma cell line DBTRG-05MG: Growth kinetics, karyotype, receptor expression, and tumor suppressor gene analyses. *In vitro Cellular & Developmental Biology: Journal of the Tissue Culture Association, 28A,* 609–614.

Kupfersztain, C., Rotem, C., Fagot, R., et al. (2003). The immediate effect of natural plant extract, Angelica sinensis and Matricaria chamomilla (Climex) for the treatment of hot flushes during menopause. A preliminary report. *Clinical and Experimental Obstetrics & Gynecology, 30,* 203–206.

Kyo, S., Takakura, M., Taira, T., et al. (2000). Sp1 cooperates with c-Myc to activate transcription of the human telomerase reverse transcriptase gene (hTERT). *Nucleic Acids Research, 28,* 669–677.

Lee, S. M., Li, M. L., Tse, Y. C., et al. (2002). Paeoniae Radix, a Chinese herbal extract, inhibits hepatoma cells growth by inducing apoptosis in a p53 independent pathway. *Life Sciences, 71,* 2267–2277.

Lee, W. H., Jin, J. S., Tsai, W. C., et al. (2006). Biological inhibitory effects of the Chinese herb danggui on brain astrocytoma. *Pathobiology: Journal of Immunopathology, Molecular and Cellular Biology, 73,* 141–148.

Li, H., Kolluri, S. K., Gu, J., et al. (2000). Cytochrome c release and apoptosis induced by mitochondrial targeting of nuclear orphan receptor TR3. *Science, 289,* 1159–1164.

Li, Q. X., Ke, N., Sundaram, R., et al. (2006). NR4A1, 2, 3—An orphan nuclear hormone receptor family involved in cell apoptosis and carcinogenesis. *Histology and Histopathology, 21,* 533–540.

Lin, B., Kolluri, S. K., Lin, F., et al. (2004). Conversion of Bcl-2 from protector to killer by interaction with nuclear orphan receptor Nur77/TR3. *Cell, 116,* 527–540.

Lin, P. C., Chen, Y. L., Chiu, S. C., et al. (2008). Orphan nuclear receptor, Nurr-77 was a possible target gene of butylidenephthalide chemotherapy on glioblastoma multiform brain tumor. *Journal of Neurochemistry, 106,* 1017–1026.

Liu, C. X., Xiao, P. G., & Li, D. P. (2000). *Modern research and application of Chinese medicinal plants* (pp. 166–169). Hong Kong Medical: Hong Kong.

Liu, J. P. (2000). Telomerase: Not just black and white, but shades of gray. *Molecular Cell Biology Research Communications: MCBRC, 3,* 129–135.

Liu, S., Wu, Q., Ye, X. F., et al. (2002). Induction of apoptosis by TPA and VP-16 is through translocation of TR3. *World Journal of Gastroenterology: WJG, 8,* 446–450.

Liu, Y. M., Zhang, J. J., & Jiang, J. (2004). Observation on clinical effect of Angelica injection in treating acute cerebral infarction. *Zhongguo Zhong Xi Yi Jie He Za Zhi, 24,* 205–208.

Mahaley, M. S., Jr., Mettlin, C., Natarajan, N., et al. (1989). National survey of patterns of care for brain-tumor patients. *Journal of Neurosurgery, 71,* 826–836.

McGovern, P. C., Lautenbach, E., Brennan, P. J., et al. (2003). Risk factors for postcraniotomy surgical site infection after 1,3-bis (2-chloroethyl)-1-nitrosourea (Gliadel) wafer placement. *Clinical Infectious Diseases: An Official Publication of the Infectious Diseases Society of America, 36,* 759–765.

Mimura, Y., Kobayashi, S., Naitoh, T., et al. (1995). The structure-activity relationship between synthetic butylidenephthalide derivatives regarding the competence and progression of inhibition in primary cultures proliferation of mouse aorta smooth muscle cells. *Biological & Pharmaceutical Bulletin, 18,* 1203–1206.

Mullady, E. L., Millett, W. P., Yoo, H. D., et al. (2004). A phthalide with in vitro growth inhibitory activity from an oidiodendron strain. *Journal of Natural Products, 67,* 2086–2089.

Rajpal, A., Cho, Y. A., Yelent, B., et al. (2003). Transcriptional activation of known and novel apoptotic pathways by Nur77 orphan steroid receptor. *The EMBO Journal, 22,* 6526–6536.

Reed, S. I., Bailly, E., Dulic, V., et al. (1994). G1 control in mammalian cells. *Journal of Cell Science. Supplement, 18,* 69–73.

Rock, E., & DeMichele, A. (2003). Nutritional approaches to late toxicities of adjuvant chemotherapy in breast cancer survivors. *Clinical and Experimental Obstetrics & Gynecology, 30,* 203–206.

Ross, B. D., Zhao, Y. J., Neal, E. R., et al. (1998). Contributions of cell kill and posttreatment tumor growth rates to the repopulation of intracerebral 9L tumors after chemotherapy: An MRI study. *Proceedings of the National Academy of Sciences of the United States of America, 95,* 7012–7017.

Santarius, T., Kirsch, M., Rossi, M. L., et al. (1997). Molecular aspects of neuro-oncology. *Clinical Neurology and Neurosurgery, 99,* 184–195.

Schlegel, J., Piontek, G., Kersting, M., et al. (1999). The p16/Cdkn2a/Ink4a gene is frequently deleted in nitrosourea-induced rat glial tumors. *Pathobiology: Journal of Immunopathology, Molecular and Cellular Biology, 67,* 202–206.

Schmitt, C. A. (2007). Cellular senescence and cancer treatment. *Biochimica et Biophysica Acta, 1775,* 5–20.

Schwartsmann, G., Ratain, M. J., Cragg, G. M., et al. (2002). Anticancer drug discovery and development throughout the world. *Journal of Clinical Oncology: Official Journal of the American Society of Clinical Oncology, 20,* 47S–59S.

Sekiya, K., Tezuka, Y., Tanaka, K., et al. (2000). Distribution, metabolism and excretion of butylidenephthalide of Ligustici chuanxiong rhizoma in hairless mouse after dermal application. *Journal of Ethnopharmacology, 71,* 401–409.

Shang, P., Qian, A. R., Yang, T. H., et al. (2003). Experimental study of anti-tumor effects of polysaccharides from Angelica sinensis. *World Journal of Gastroenterology: WJG, 9,* 1963–1967.

Shapiro, W. R., & Green, S. B. (1987). Reevaluating the efficacy of intra-arterial BCNU. *Journal of Neurosurgery, 66,* 313–315.

Sherr, C. J., & Roberts, J. M. (1995). Inhibitors of mammalian G1 cyclin-dependent kinases. *Genes & Development, 9,* 1149–1163.

Subach, B. R., Witham, T. F., Kondziolka, D., et al. (1999). Morbidity and survival after 1,3-bis(2-chloroethyl)-1-nitrosourea wafer implantation for recurrent glioblastoma: A retrospective case-matched cohort series. *Neurosurgery, 45,* 17–22.

Tamargo, R. J., Myseros, J. S., Epstein, J. I., et al. (2003) Interstitial chemotherapy of the 9L gliosarcoma: Controlled release polymers for drug delivery in the brain. *Cancer Research, 53,* 329–333.

Teng, C. M., Chen, W. Y., Ko, W. C., et al. (1987). Antiplatelet effect of butylidenephthalide. *Biochimica et Biophysica Acta, 924,* 375–382.

Tsai, N. M., Lin, S. Z., Lee, C. C., et al. (2005). The antitumor effects of Angelica sinensis on malignant brain tumors in vitro and in vivo. *Clinical Cancer Research: An Official Journal of the American Association for Cancer Research, 11,* 3475–3484.

Tsai, N. M., Chen, Y. L., Lee, C. C., et al. (2006). The natural compound n-butylidenephthalide derived from Angelica sinensis inhibits malignant brain tumor growth in vitro and in vivo. *Journal of Neurochemistry, 99,* 1251–1262.

Vickers, A. (2002). Botanical medicines for the treatment of cancer: Rationale, overview of current data, and methodological considerations for phase I and II trials. *Cancer Investigation, 20,* 1069–1079.

Wang, H., Chen, R., & Xu, H. (1998). Chemical constituents of radix Angelicae Sinensis. *Zhongguo Zhong Yao Za Zhi, 23,* 167–168.

Wang, X., Wei, L., Ouyang, J. P., et al. (2001). Effects of an angelica extract on human erythrocyte aggregation, deformation and osmotic fragility. *Clinical Hemorheology and Microcirculation, 24,* 201–205.

Weih, F., Ryseck, R. P., Chen, L., et al. (1996). Apoptosis of nur77/N10-transgenic thymocytes involves the Fas/Fas ligand pathway. *Proceedings of the National Academy of Sciences of the United States of America, 93,* 5533–5538.

Williams, G. T., & Lau, L. F. (1993). Activation of the inducible orphan receptor gene nur77 by serum growth factors: Dissociation of immediate early and delayed-early responses. *Molecular and Cellular Biology, 13,* 6124–6136.

Wilson, A. J., Arango, D., Mariadason, J. M., et al. (2003). TR3/Nur77 in colon cancer cell apoptosis. *Cancer Research, 63,* 5401–5407.

Winoto, A., & Littman, D. R. (2002). Nuclear hormone receptors in T lymphocytes. *Cell, 109*(Suppl), S57–S66.

Woronicz, J. D., Calnan, B., Ngo, V., et al. (1994). Requirement for the orphan steroid receptor Nur77 in apoptosis of T-cell hybridomas. *Nature, 367,* 277–281.

Wu, K. J., Grandori, C., Amacker, M., et al. (1999). Direct activation of TERT transcription by c-MYC. *Nature Genetics, 21,* 220–224.

Xie, C. H., Zhang, M. S., Zhou, Y. F., et al. (2006). Chinese medicine Angelica sinensis suppresses radiation-induced expression of TNF-alpha and TGF-beta1 in mice. *Oncology Reports, 15,* 1429–1436.

Xie, F., Li, X., Sun, K., et al. (2001). An experimental study on drugs for improving blood circulation and removing blood stasis in treating mild chronic hepatic damage. *Journal of Traditional Chinese Medicine, 21,* 225–231.

Xu, J., & Li, G. (2000). Observation on short-term effects of Angelica injection on chronic obstructive pulmonary disease patients with pulmonary hypertension. *Zhongguo Zhong Xi Yi Jie He Za Zhi, 20,* 187–189.

Yano, H., Mizoguchi, A., Fukuda, K., et al. (1994). The herbal medicine sho-saiko-to inhibits proliferation of cancer cell lines by inducing apoptosis and arrest at the G0/G1 phase. *Cancer Research, 54,* 448–454.

Ye, Y. N., Koo, M. W., Li, Y., et al. (2001a). Angelica sinensis modulates migration and proliferation of gastric epithelial cells. *Life Sciences, 68,* 961–968.

Ye, Y. N., Liu, E. S., Li, Y., et al. (2001b). Protective effect of polysaccharides-enriched fraction from Angelica sinensis on hepatic injury. *Life Sciences, 69,* 637–646.

Ye, Y. N., Liu, E. S., Shin, V. Y., et al. (2001c). A mechanistic study of proliferation induced by Angelica sinensis in a normal gastric epithelial cell line. *Biochemical Pharmacology, 61,* 1439–1448.

Ye, Y. N., So, H. L., Liu, E. S., et al. (2003). Effect of polysaccharides from Angelica sinensis on gastric ulcer healing. *Life Sciences, 72,* 925–932.

Yim, T. K., Wu, W. K., Pak, W. F., et al. (2000). Myocardial protection against ischaemia-reperfusion injury by a Polygonum multiflorum extract supplemented 'Dang-Gui decoction for enriching blood,' a compound formulation, ex vivo. *Phytotherapy Research: PTR, 14,* 195–199.

Yoshikawa, K., Kokudo, N., Hashimoto, T., et al. (2010). Novel phthalide compounds from Sparassis crispa (Hanabiratake), Hanabiratakelide A-C, exhibiting anti-cancer related activity. *Biological & Pharmaceutical Bulletin, 33,* 1355–1359.

Youn, H. D., Sun, L., Prywes, R., et al. (1999). Apoptosis of T cells mediated by Ca2+-induced release of the transcription factor MEF2. *Science, 286,* 790–793.

Yu, Y. L., Yu, S. L., Su, K. J., et al. (2010). Extended O6-methylguanine methyltransferase promoter hypermethylation following n-butylidenephthalide combined with 1,3-bis(2-chloroethyl)-1-nitrosourea (BCNU) on inhibition of human hepatocellular carcinoma cell growth. *Journal of Agricultural and Food Chemistry, 58,* 1630–1638.

Zhao, K. J., Dong, T. T., Tu, P. F., et al. (2003). Molecular genetic and chemical assessment of radix Angelica (Danggui) in China. *Journal of Agricultural and Food Chemistry, 51,* 2576–2583.

Zhao, L., Zhang, Y., & Xu, Z. X. (1994). Clinical effect and experimental study of xijian tongshuan pill. *Zhongguo Zhong Xi Yi Jie He Za Zhi, 14,* 71–73.

Zheng, G. Q., Kenney, P. M., Zhang, J., et al. (1993). Chemoprevention of benzo[a]pyrene-induced forestomach cancer in mice by natural phthalides from celery seed oil. *Nutrition and Cancer, 19,* 77–86.

Zhong, Y. H., Han, G., Zhou, Y. F., et al. (2007). Protection of Angelica sinensis against radiation-induced pulmonary fibrosis in mice. *Zhonghua Yu Fang Yi Xue Za Zhi, 41,* 105–109.

Chapter 8
An Evidence-based Perspective of *Scutellaria Barbata* (Skullcap) for Cancer Patients

Brian Y. Y. Wong and Hannah H. L. Wong

Abstract *Scutellaria barbata* (skullcap) has been used in traditional Chinese and Korean medicine for treating various inflammation illnesses and cancers. *In vitro* studies have demonstrated the anti-mutagenesis (chemo-preventive) effect of skullcap *via* modulating the metabolism of mutagenic compounds such as aflatoxin B_1 and benzo[a]pyrene to reduce their DNA binding efficiency. *In vitro* and *in vivo* studies using flavonoid compound of skullcap in both human and animal models have shown promising anticancer effects. The aqueous extract of skullcap has shown to have the most effective anticancer chemical constituents. *In vitro* studies indicated that skullcap might be effective against all three stages of carcinogenesis (initiation, promotion, and progression). It has also been found to exert anticancer mechanisms such as anti-inflammation, anti-proliferation, induction of apoptosis against numerous cancer cell lines including digestive (liver and colorectal), respiratory (lung and nasopharyngeal), and lymphatic (leukemia) systems, and induction of sex hormone-specific glycolytic necrosis, especially those of the reproductive system (breast, uterine, and prostate) while inactive on normal human mammary epithelial cells. *In vivo* murine model showed aqueous extract of skullcap may enhance macrophage cell line activity leading to inhibition of tumor growth. It has also been shown to inhibit aberrant crypt formation in colon, delay prostate cancer development and progression in transgenic adenocarcinoma of mouse prostate mice, and reduced solid ascites tumor proliferation in the breast of mice. Aqueous extracts of skullcap have been found to have a favorable toxicity profile when used as an oral feeding in Phase 1 and 1B clinical studies in metastatic breast cancer patients. Based on the studies reported on skullcap, the best prospective therapeutic application of skullcap would be in breast, prostate, liver, colorectal, uterine, and lung cancer patients.

B. Y. Y. Wong (✉)
Department of Biology, Pacific Union College, 1 Angwin Avenue, Angwin, CA 94508, USA
e-mail: bwong@puc.edu

W. C. S. Cho (ed.), *Evidence-based Anticancer Materia Medica,* Evidence-based Anticancer Complementary and Alternative Medicine,
DOI 10.1007/978-94-007-0526-5_8,

Fig. 8.1 Picture of *Scutellaria barbata* D Don (skullcap or Ban-Zhi-Lian). (http://healingpastures.com/2010/08/02/cancer-fighting-mint-plant, Accessed 26 August 2010)

8.1 Introduction

Scutellaria barbata D. Don (skullcap), Chinese name *Ban-Zhi-Lian* of the Lamiaceae family, is a 12–35 cm tall perennial herb, found along fields and ditches in southern China and Korea (Fig. 8.1). It is commonly known as skullcap, a mint with a mildly bitter taste commonly used in traditional Chinese and Korean medicine for the treatment of a variety of ailments such as appendicitis, hepatitis, snake bites, lung, liver and rectal cancer (Jiangsu New Medical College 1977; Chong and Lee 1988; Zhu 1998; Huang 1999). In China, skullcap has been used clinically in treating breast cancer, lung cancer, liver cancer, digestive tract cancers, and chorioepithelioma. Skullcap has been used in combination with *Oldenlandia diffusa* in patients with liver, rectal and lung cancer. However, it was stated that skullcap alone or in combination with other herbs has not been successful in complete cure of cancer (complete remission) but rather provides symptomatic relief (Chong and Lee 1988). Literature review also shows very limited *in vitro* and *in vivo* research on the inhibitory or anticancer properties and mechanisms of skullcap following the ethnopharmacological approach in the early 1900s. The purpose of this chapter is to describe the early chemopreventive studies of skullcap and follow its development into the isolation and understanding of the various chemical constituents as well as their *in vitro* and *in vivo* effect or mechanism based on research literature.

8.2 Early Studies

8.2.1 Anti-mutagenesis and Anti-carcinogenesis Chemo-prevention Properties

Our early studies found that aqueous extracts (whole) of skullcap possess anti-mutagenic and chemopreventive properties (Wong et al. 1992a, b, c, 1993a, b). These studies indicate that skullcap contains phytochemicals which inhibited mutagenesis, DNA binding, and metabolism of the pro-carcinogens aflatoxin B_1 (AFB_1) and benzo[a]pyrene bioactivated by Arcoclor 1254-induced rat S9. Inhibition effects of skullcap were also shown in the mutagenicity of AFB_1 on mutant bacteria *Salmonella typhimurium* TA100 using dexamethansone (DXM)-induced rat hepatic S9, on cytochrome P450-linked aminopyrine N-demethylase (APND) activity in DXM-induced hepatic microsomes, and on the metabolism of AFB_1 by DXM-induced S9. Aqeuous extract of skullcap consistently inhibited the mutagenicity of AFB_1 bioactivated by either non-induced or DXM-induced hepatic S9. The effects correlated with the inhibition of cytochrome P450-linked APND activity in DXM-induced S9 mediated metabolism of $[^3H]AFB_1$. These findings suggest that skullcap contains antimutagenic and antitumorigenic property on AFB_1 *via* inhibition of cytochrome p450 isozyme (CYP3)-mediated metabolism of the carcinogens. This data suggests that skullcap is effective against all three stages of carcinogenesis (initiation, promotion, and progression).

8.2.2 Immune Enhancing and Anti-inflammation Effect of Skullcap

A later study showed that oral feeding with the aqueous herbal extract inhibited the growth of transplanted murine renal cell carcinoma (RenCa) in Balb/c mice significantly (Wong et al. 1996). *In vitro* data of the same study revealed that it enhanced the phagocytic oxidative burst in murine macrophage, J774 cells. This suggests its ability to inhibit tumor growth observed *in vivo*. An aqueous extract of skullcap was also shown to have greater antimutagenic effect in inhibiting DNA damage of peripheral lymphocytes when injected into cigarettes (Han et al. 1997). Our later study revealed skullcap treatment could elevate H_2O_2 and hydroxyl radical production in the macrophage-like RAW 264.7 mouse peritoneal cell line. It showed that skullcap modulated COX-2 (cyclooxygenase-2) and inducible NO (nitric oxide) synthase protein expression, as well as the activity of prostaglandin E_2 and stable oxidation products of NO *in vitro*. This may be related to the formation of reactive oxygen species (Harris et al. 2003). Together these early studies aroused and led to continuous research interest in the chemical isolation and specific anticancer effect and mechanism of skullcap since the early 2000s.

8.3 Chemical Constitution

Skullcap has been reported to contain several bioactive flavonone compounds such as scutellarein, scutellarin, carthamidin, isocarthamidin, wogonin, apigenin, luteolon, pheophoride-*a*, and various clerodane diterpeonoids (Zhu 1998; Kim et al. 2005; Chan et al. 2006; Dai et al. 2006a, b; 2007a, b, 2008, 2009a, b, 2010). Tables 8.1 and 8.2 list more detail chemical compounds and extracts of skullcap with their respective *in vitro* and *in vivo* anticancer properties (effect and mechanism) classified

Table 8.1 Anticancer properties of chemical constitution of *Scutellaria barbata* (skullcap) on murine *in vitro* and *in vivo* studies according to organ system

Organ system (Cancer cell line/Tissue/Animal model) Chemical constitution	Effect/Mechanism	Reference
Integumentary:		
Skin		
Aqueous extract	Inhibition of tumorigenesis (DMBA-induced) in mouse skin cancer model	Suh et al. 2007
Lymphatic:		
Macrophage-like		
(RAW 264.7) Aqueous extract (whole)	Elevation of H_2O_2 and hydroxyl radical in macrophage-like RAW 264.7 mouse peritoneal cell line	Harris et al. 2003
	Modulation of cyclooxygenase-2 (COX-2) and inducible nitric oxide (NO) synthase protein expression, and prostaglandin E_2 and stable oxidation products of NO	
Macrophage		
(J744) Aqueous extract (whole)	Enhancing J744 macrophage cell oxidative burst	Wong et al. 1996
Respiratory:		
Lung		
(LLC) Luteolin	Induction of apoptosis in Lewis lung carcinoma LLC cells *via* caspase activation and extracellular signal-regulated kinase ERK/Akt inhibition	Kim et al. 2007b
	Reduction of proliferation cell nuclear antigen (PCNA)	
	Enhancement of Annexin-V-positive cell growth and G1 DNA	
	Activation of caspase-3 and -9, and PARP cleavage	
	Increasing the ratio of Bax/Bcl-2	
	Reduction of mitochondrial membrane potential	
	Inhibition of growth of LLC cells implanted on the flank of mice	
(LLC) Aqueous extract	Inhibition of growth	Shoemaker et al. 2005
Digestive:		
Colon		
Aqueous extract (whole)	Inhibition of azoxymethane (AOM)-induced aberrant crypt foci (ACF) formation in C57BL/6 mice	Wong et al. 2010
Liver		
(H22) Aqueous extract (ESB)	Inhibition of proliferation and induction of apoptosis of mouse hepatoma cells through loss of mitochondrial transmembrane potential, release of cytochrome c, and activation of caspase-3	Dai et al. 2008a

Table 8.1 (continued)

Organ system (Cancer cell line/Tissue/Animal model) Chemical constitution	Effect/Mechanism	Reference
(H22) Aqueous extract (ESB)	Induction of apoptosis *via* reduction of mitochondrial transmembrane potential of mouse hepatoma cells by serum containing ESB	Dai et al. 2008b
(H22) Aqueous extract (ESB)	Inhibition of proliferation and induction of apoptosis of hepatoma cells	Dai et al. 2008c
(Hep3B) Pheophorbide-a $C_{35}H_{36}N_4O_5$	Reduction of tumor size (57%) in nude mice model	Tang et al. 2006
Urinary:		
Kidney		
(RenCa) Aqueous extract (whole)	Inhibition of tumor growth of transplanted RenCa cancer cells in Balb/c mice	Wong et al. 1996
Reproductive (female):		
Mammary adenocarinoma		
(EAC- Ehrlich ascites tumor cells) Chloroform fraction: Phytol, wogonin, luteolin, hispidulin	Inhibition of solid Ascites tumor proliferation and prolonging life span in mice	Yu et al. 2007
Mammary gland tissue Aqueous extract	Inhibition of preneoplastic lesions development in carcinogen (DMBA)-induced mouse mammary glands in culture	Suh et al. 2007
Breast		
(MCNeuA) Aqueous extract	Inhibition of murine breast carcinoma MCNeuA cell growth (while relatively inactive on normal human mammary epithelial huMEC cells)	Shoemaker et al. 2005
(MCNeuA) Aqueous extract	Inhibition of cell growth Apoptosis	Campbell et al. 2002
Reproductive (male):		
Prostate		
(TRAMP-C1) Aqueous extract (whole)	Induction of apoptosis *via* DNA fragmentation	Wong et al. 2009
Prostate		
Aqueous extract (whole)	Delaying prostate cancer development & progression in TRAMP mice Activation of caspase-3 in TRAMP mice prostate tissue	Wong et al. 2009

Table 8.2 Anticancer properties of chemical constitution of *Scutellaria barbata* (skullcap) on human *in vitro* and *in vivo* studies classified according to organ system

Organ system (Cancer cell line/Tissue/Clinical study) Chemical constitution	Effect/Mechanism	Reference
Circulatory & Lymphatic:		
Leukemia		
(HL-60) Neo-clerodane diterpenoids: barbatellarines A (1), B (2)	Weak cytotoxic against human leukemia HL-60 cells by barbatellarine B	Lee et al. 2010
Flavonoids: Apigenin, luteolin	Induction of G1 arrest and apoptosis in promyeloctyic leukemia HL-60 cell line Inhibition of Cyclin A, D1, D2, D3, E; CDK 2, 4, 6 Up-regulation of p21, CDK inhibitors Down-regulation of retinoblastoma protein phosphorylation	Kim et al. 2007a
(KG-1) Aqueous extract (whole)	Anti-proliferation and induction of apoptosis in this acute myelogenous leukemia cell line	Chui et al. 2005
(U937) Methylene Chloride fraction	Induction of apoptosis (activating caspase-3 and -9, cleaving PARP, increasing the ratio of Bax/Bcl-2, and releasing cytochrome c *via* the mitochondrial signaling pathways)	Cha et al. 2004
(Red blood cell)		
Polysaccharide B3-PS1 (1,700KDa) Gal 4.3:Glc 1.6:Man 1.1:Ara 1.0+Rha, Fuc, Xyl	Anti-complementary in hemolytic assay (C1q, C1r, C1s, C2, C3, C4, C5, C9)	Wu and Chen 2009
(Peripheral lymphocytes)		
Aqueous extract	Inhibition of damage of DNA in lymphocytes caused by the total particle material extract of cigarette tar and greater antimutagenesis effect when the herb was injected into cigarettes ahead	Han et al. 1997
Respiratory:		
Lung		
(A549) Flavonoid (wogonin)	Inhibition of phorbol 12-myristate 13-acetate (PMA)-induced *COX-2* gene expression *via* inhibition of *c-Jun* expression and *AP-1* activation in this epithelial cancer cell line	Chen et al. 2008

Table 8.2 (continued)

Organ system (Cancer cell line/Tissue/ Clinical study) Chemical constitution	Effect/Mechanism	Reference
Aqueous extract (whole)	Inhibition of cell growth	Shoemaker et al. 2005
Aqueous extract (whole)	Anti-proliferation and induction of apoptosis	Chui et al. 2005
Ethanol extract (30%)	Cytotoxic and Induction of apoptosis	Yin et al. 2004
	Induction of gene expression (5-fold change in 10 genes involved in DNA damage, cell cycle control, nucleic acid binding and protein phosphorylation) by *c*DNA microarray analysis	
(SPC-A-1)		
Ethanol extract (75%)	Induction of apoptosis by up-regulation of caspase-3 expression and down-regulation of survival genes (*survinin*) expression	Wei et al. 2007
Nasopharyngeal		
(HONE-1)		
Neo-clerodane diterpenoids : Scutehenanine H (1) & 6-(2,3-epoxy-2-isopropyl-n-propoxyl) barbatin C (2)	Cytotoxic against the cell line	Dai et al. 2010
Scutehenanines A-D (1,4,5,6), 6-O-acetylscutehenanine A (2), 6-O-(2-carbonyl-3-methylbutanoyl)scutehenanine A (3)		Dai et al. 2009a
Scutebarbatine O (1), 6-O-nicotinoylscutebarbatine G (2)		Dai et al. 2009b
6,7-dibenzoyloxybarbatin C (1, barbatin D), 6-(2-acetoxy-3-methylbutanoloxy)-7-(2-carbonyl-3-methylbutanoyloxy) barbatin C (2, barbatin E)		Dai et al. 2008
Ent-clerodane diterpenoids (1–4)		Dai et al. 2007a
Scutebarbatines I-L (1–4)		Dai et al. 2007b

Table 8.2 (continued)

Organ system (Cancer cell line/Tissue/Clinical study) Chemical constitution	Effect/Mechanism	Reference
Barbatins A-C (1–3), Scutebarbatine B (4)		Dai et al. 2006
Scutebarbatine C-F (1–4)		Dai et al. 2006b
Digestive:		
Oral epidermoid carcinoma		
(KB)		
Colorectal carcinoma		
(HT29)		
[Chemicals same as those of nasopharyngeal above]	Cytotoxic against these two cancer cell lines	[References same as above]
Stomach		
(AGS) Ethanol (30%) extract	Inhibition of growth of gastric AGS cancer cell line	Yin et al. 2004
Pancreas		
(Panc-1) Aqueous extract	Inhibition of growth of pancreatic carcinoma cells (while relatively inactive on normal human mammary epithelial huMEC cells)	Shoemaker et al. 2005
Colon		
(LoVo) Methanol extract (chemically standardized with: scutellarein, sutellarin, carthamidin, isocarthamidin, wogonin)	Induction of cell death (increase in sub G1 phase) in colon adenocarinoma LoVo cells using a functional proteomic approach	Goh et al. 2005
Liver		
(Hep-G2) Pheophorbide-a $C_{35}H_{36}N_4O_5$	Reversal of P-glycoprotein multidrug resistance mediated (reduction of expression at both transcriptional and translational level) in hepatoma Hep-G2 cell line Induction of cell cycle arrest at G2/M phase and inhibition of cylcin-A1 and cdc2	Tang et al. 2007
(Hep-G2) Ethanol extract	Inhibition of cell proliferation, blockage of cell cycle induction (decreasing S phase and increasing G0/G1phase), and induction of apoptosis (up-regulation of *Fas* expression related to the activation of *FNFR* superfamily) of hepatocellular Hep-G2 cell	Lin et al. 2006a

Table 8.2 (continued)

Organ system (Cancer cell line/Tissue/ Clinical study) Chemical constitution	Effect/Mechanism	Reference
(Hep-G2) Ethanol extract (30%)	Inhibition of growth of liver Hep-G2	Yin et al. 2004
(Hep3B) Pheophorbide-*a* $C_{35}H_{36}N_4O_5$	Inhibition of cell proliferation and induction of apoptosis (*via* the releasing of cytochrome c mitochondria and decreasing pro-caspase-3 and -9) in hepatocellular carcinoma Hep3B cells	Tang et al. 2006
(Hep3B) Pheophorbide- *a* $C_{35}H_{36}N_4O_5$	Anti-proliferation (*via* sub-G1 cell cycle arrest) and induction of apoptosis (*via* the suppression of Bcl-2, releasing in Hep3B of cytochrome c to cytosol, and activation of pro-caspase-3 and -9) (viral-induced) cells	Chan et al. 2006
(Hep3B) Aqueous extract (whole)	Anti-proliferation and induction of apoptosis in Hep3B hepatocellular carcinoma and Hep-G2 hepatoblastoma	Chui et al. 2005
(Hep-G2, BEL-7402) Chloroform fraction: Phytol, wogonin, luteolin, hispidulin	Cytotoxic to both cell lines Induction of apoptosis in hepatocellular carcinoma BEL-7402 cells with a lower cytotoxic effect on normal liver cell line (L-O2) Activation of caspase-9 and cytochrome c Decrease *S* phase content	Yu et al. 2007
(QGY-7701) Aqueous extract (ESB)	Inhibition of cell proliferation, activation of anti-oncogene *Bcl-2*, induction of apoptosis and inhibition of G to S phase cell cycle progression in hepatocellular carinoma QGY-7701 cells	Lin et al. 2006b
Urinary		
Kidney		
(ACHN) *n*-hexane, ethyl acetate, & chloroform (most effective) fractions	Cytotoxic against the renal adenocarcinoma cells	Yu et al. 2007
Reproductive (female):		
Breast		
(Metastatic breast cancer) Aqueous extract (BZL101)	Oral feeding (40 g/day) was safe, well-tolerated in phase 1B does escalation trial of human metastatic breast cancer	Perez et al. 2010

Table 8.2 (continued)

Organ system (Cancer cell line/Tissue/ Clinical study) Chemical constitution	Effect/Mechanism	Reference
(Metastatic breast cancer) Aqueous extract (BZL101)	Favorable toxicity profile & promising efficacy in Phase I clinical trial of advanced breast cancer	Rugo et al. 2007
Breast		
(MCF-7, MDA-MB-231) Aqueous extract (BZL101)	Phenotype specific anti-proliferative gene expression responses in human breast (early stage estrogen sensitive MCF-7 *vs* late estrogen insensitive MDA-MB-231) reproductive cancer cells Induction of G(1) cell cycle arrest & ablated expression of regulators Cyclin D1, CDK2, CDK4, growth factor stimulatory pathways & estrogen receptor-α expression in estrogen sensitive MCF-7 breast cancer cells (ablation of promoter activities)	Marconett et al. 2010
(MDA-MB-231) Pheophorbide-a (photo-activated) $C_{35}H_{36}N_4O_5$	Activation of mitochondria-mediated apoptosis through the suppression of extracellular signal-regulated kinase (ERK)-mediated autophagy in estrogen receptor-negative human breast adenocarcinoma cells MDA-MB-231 by pheophorbide-a based photodynamic therapy (Pa-PDT) Triggering mitogen-activated protein kinase (MAPK) pathway Activation of *c-June N*-terminal kinase (JNK) Inhibition of ERK	Bui-Xuan et al. 2010
(MDA-MB-231) Brucea javanica extract *(Scutellaria barbata + Gleditsia sinensis + Radix Sophorae Tonkinesis)*	Induction of apoptosis (*via* mitochrondrial dependent pathway associated with caspase-3 activity) in human breast carcinoma MDA-MB231 cell	Lau et al. 2005
(MDA-MB-231) Aqueous extract (whole)	Anti-proliferation and induction of apoptosis	Chui et al. 2005
(SK-BR-3) Neoclerodane diterpenoids: Scutebatas A-G (1–7)	Weak cytotoxicity against human mammary cancer cells SK-BR-3 by Scutebata A	Zhu et al. 2010

Table 8.2 (continued)

Organ system (Cancer cell line/Tissue/ Clinical study) Chemical constitution	Effect/Mechanism	Reference
(SK-BR-3, BT474) Aqueous extract (BZL101)	Selectively induction of cell death in human breast cancer SK-BR-3 and BT474 cells but not in non-transformed mammary epithelial cells (MCF10A) *via* induction of reactive oxygen species (ROS) Inhibition of glycolysis selectively in tumor cells (decrease in the enzymatic activities within the glycolytic pathway and the inhibition of lactate production) Induction of oxidative damage leads to hyperactivation of poly(ADP-ribose)polymerase (PARP), followed by a sustained decrease in levels of NAD and depletion of ATP in cancer cells Induction of expression of genes involved in oxidative response, DNA damage & cell death	Fong et al. 2008
(SK-BR-3, BT474, MDA-MB-231, MDA-MB-361, MDA-MB-453, MDA-MB-468, Du4475) Aqueous extract (BZL101)	Induction of strong growth inhibition and DNA fragmentation (while relatively inactive on normal human fibroblast IMR90 cell lines)	Rugo et al. 2007
(MCF-7, MDA-MB-435S) *n*-hexane, ethyl acetate, & chloroform (most effective) fractions	Cytotoxic against both cell lines	Yu et al. 2007
(MCF-7) Aqueous extract	Inhibition of breast carcinoma cell growth (while relatively inactive on normal human mammary epithelial huMEC cells)	Shoemaker et al. 2005
(MCF-7, MDA-MB-231, SK-BR-3, BT474) Aqueous extract	Inhibition of human breast cancer cells growth	Campbell et al. 2002
Ovarian		
(HOC) Aqueous extract	Function as anti-mutagen, anti-inflammation and anti-proliferation agent. Inhibition of COX-2 and hydroperoxidase expression	Suh et al. 2007
(SKOV-3, CAOV3) Aqueous extract	Induction of apoptosis	Powell et al. 2003

Table 8.2 (continued)

Organ system (Cancer cell line/Tissue/ Clinical study) Chemical constitution	Effect/Mechanism	Reference
Uterine		
(Leiomyoma cells)	Inhibition of leiomyoma growth	Kim et al. 2008
Aqueous extract (evaporated by ethanol)	Down-regulation of the expression of Bcl-2 protein in uterine leiomyoma cells	
Aqueous extract	Reduction of tumor volume in leiomyomal smooth muscle cells *via* regulation of IGF-I production and induction of increased rate of apoptosis	Kim et al. 2005
Aqueous extract	Anti-proliferation (*via* inhibition of expression of proliferating cell nuclear antigen, Cyclin E, cdc2) of HCG-promoted leiomyomal cells	Lee et al. 2004c
Aqueous extract	Inhibition of aromatase activity of leiomyomal cells	Lee et al. 2004b
Aqueous extract	Induction of *c-fos* gene expression in uterine leiomyomal cells *via* increases in adenosine-3',5', cAMP mediated by β2-adrenergic receptors (β2-ARs) which in turn activated the cAMP/protein kinase (PKA) pathway	Lee et al. 2004a
Aqueous extract	Inhibition of uterine leiomyomal smooth muscle cell proliferation *via* the induction of alpha-smooth muscle actin (SMA), calponin h1 and p27	Lee et al. 2004d
Cervix		
(HeLa)		
n-hexane, ethyl acetate, & chloroform (most effective) fractions	Cytotoxic to the cervix carcinoma cell line	Yu et al. 2007
(HeLa)	Function as anti-mutagen, anti-inflammation and anti-proliferation agent.	Suh et al. 2007
Aqueous extract	Inhibition of COX-2 and hydroperoxidase expression	
Reproductive (male):		
Prostate		
(LNCaP, PC-3)	Phenotype specific anti-proliferative gene expression responses in human prostate (androgen sensitive LNCaP *vs* late androgen insensitive PC-3) reproductive cancer cells	Marconett et al. 2010
Aqueous extract (BZL101)	Arresting early stage androgen sensitive LNCaP in G(2)/M phase with corresponding decreases in Cyclin B1, CDK1 & androgen receptor expression	
	Inducing S phase arrest with corresponding ablations in Cyclin A2 & CDK2 expression	

Table 8.2 (continued)

Organ system (Cancer cell line/Tissue/ Clinical study) Chemical constitution	Effect/Mechanism	Reference
(LNCaP, PC-3) Aqueous extract	Inhibition of prostate cancer cell growth (while relatively inactive on normal human mammary epithelial huMEC cells)	Shoemaker et al. 2005
(LNCaP, DU-145) Aqueous extract (whole)	Regulation of caspase-8 activation in LNCaP (androgen dependent) and DU-145 (androgen independent) prostate cancer cells Activation of apoptosis through an upstream androgen-receptor pathway Promotion of apoptosis, anti-inflammation, & anti-proliferation by inhibiting Akt pathway Binding to membrane androgen receptors (mARs) and activation of secondary pathways further lead to apoptosis and inhibit tumor progression	Nguyen et al. 2008
(LNCaP) Aqueous extract (whole)	Induction of apoptosis in LNCaP hormone dependent prostate cancer cells Up-regulation of apoptotic pathway (enhance Bax, p53, Akt, JNK production) and down-regulation of survival pathway	Wong et al. 2009
(LNCaP) Aqueous extract (whole)	Activation of caspase-3, -8, and -9 in LNCaP prostate cancer cell	Wong et al. 2007

according to organ system in animal and human respectively. These chemical constitutions of skullcap include various types of extracts (aqueous, chloroform, ethanol, ethyl acetate, methanol, methylene chloride, and *n*-hexane) and compounds. Among these, clerodane diterpeonoids are the most abundantly isolated compounds (>28) and are potential candidates for further studies of *in vitro* and *in vivo* mechanisms; while aqueous skullcap extract is the most dominant form of extraction with greater and more detailed effect and mechanism. One of such total aqueous extracts, BZL101 is suggested to have greater activity in *in vitro* assays compared to enriched chromatographically isolated pure compounds (Perez et al. 2010). For example, 9 flavonoid phytochemicals showed cytotoxic activity *in vitro*, but only at concentrations far higher than those found in the total aqueous extract. Also, the whole extract is more cytotoxic than any combination of the purified flavonoids. To learn more about the various chemical compounds of skullcap, readers may refer to Tables 8.1 and 8.2 for their specific chemical nature and specific anticancer effects or mechanisms respectively. The following sections will focus on the discussion of some of the anticancer properties and mechanisms of these chemical constituents on animal and human *in vitro* and *in vivo*, classified according to organ system.

8.4 Anticancer *In Vitro* and *In Vivo* Animal Studies

8.4.1 In Vitro Murine Cancer Cell Line Studies

Guided by ethnopharmacological approach, *in vitro* cell line data in Table 8.1 reveals studies done on the specific effect and mechanism of various chemical constituents of skullcap classified according to organ system. There are 6 murine cancer cell lines, 1 murine tissue (mammary glands in culture), and 4 organ systems studied. These include the lymphatic (macrophage—RAW 264.7 and J744 cancer cells), respiratory (lung—LLC), digestive (liver—H22), and reproductive [(breast—MC-NeuA, and prostate—transgenic adenocarcinoma of mouse prostate (TRAMP)-C1]. Major anticancer mechanisms include anti-inflammation, induction of phagocytosis, induction of apoptosis, and anti-proliferation while inactive on normal human mammary epithelial huMEC cells relatively. Some of the molecular mechanisms include elevation of H_2O_2 and hydroxyl radical, modulation of COX-2, enhancement of macrophage oxidative burst, activation of caspase-3 and -9), activation of poly(ADP-ribose) polymerase (PARP) cleavage, reduction of mitochondrial membrane potential, and release of cytochrome *c*. See Table 8.1 for detail mechanism and references for each cell line respectively.

8.4.2 In Vivo Murine Models

On the other hand, *in vivo* studies show that five mouse organ systems were affected by various chemical constituents of skullcap, including the integumentary (inhibition of DMBA-induced skin cancer), digestive (inhibition of AOM-induced ACF colon cancer and inhibition of tumor volume in nude mouse transplanted liver Hep3B cancer cells), urinary (inhibition of tumor growth with transplanted kidney RenCa cancer cells), and reproductive (inhibition of solid Ehrlich ascites tumor and prolonging mice life span; delay of prostate cancer development and progression in TRAMP mice, and activation of caspase-3 in the prostate tissue of TRAMP mice). Again, the majority of the effects came from aqueous extract of skullcap (Table 8.1).

The TRAMP model is a spontaneous autochthonous transgenic mouse model. It mimics heterogenic tumor progression in human prostate cancer, providing a relevant pre-clinical model for identifying important pathways in tumorigenesis, androgen independence, and metastasis of prostate cancer (Gingrich et al. 1996, 1997; Gupta et al. 2000). Our *in vivo* data shows a delayed in tumor onset and development in the TRAMP mice. Palpable tumor development in 50% of the mice happened at 25 weeks in the placebo group, 29 weeks in the low-dose (8 mg skullcap daily) and mid-dose (16 mg) treatment groups, and 33 weeks in the high-dose (32 mg) group (log rank, $P=0.0211$). Hematoxylin and eosin histopathological dorsal prostate tissue also reveals delay of prostate tumor progression and the activation of caspase-3 in the prostate tissue of the skullcap-treated mouse. These findings further suggest the potential efficacy of skullcap as a chemo-preventive and plausible treatment agent in prostate cancer (Wong et al. 2009).

8.5 Anticancer *In Vitro* Human Cell Lines and Clinical Studies

In this section, relevant *in vitro* human cell line and *in vivo* human data on the anticancer effect and mechanism of skullcap as classified by organ system, deserve more detailed description. For most of the systems, especially for reproductive system (breast, prostate, ovarian, and uterine), aqueous extract of skullcap is the most effective and dominant anticancer constituent in both *in vitro* and *in vivo* studies (Table 8.2).

8.5.1 In Vitro Cell Line Studies

Table 8.2 summarizes *in vitro* cell line data in showing the specific effects and mechanisms of the various chemical constituents of skullcap classified according to organ system. There are 34 human cancer cell lines (from 5 different organ systems) and 2 hematopoietic cell lines (red blood cell and peripheral lymphocytes). These include the lymphatic (leukemia—HL-60, KG-1, U937), respiratory (lung—A549, SPC-A-1; nasopharyngeal—HONE-1), digestive (oral—KB; stomach—AGS; pancreas—Panc-1; liver—Hep-G2, Hep3B, BEL-7402, QGY-7701); colon—LoVo; colorectal—HT29), urinary (kidney—ACHN), and reproductive (breast—MCF-7,

MDA-MB-231, MDA-MB-361, MDA-MB-435S, MDA-MB-468, SK-BR-3, BT474; ovarian—HOC, SKOV-3, CAOV3; uterine—leiomyoma cells; cervix—HeLa; prostate—LNCaP, PC-3, DU-145). Major anticancer mechanisms include cytotoxic, anti-inflammation, induction of phagocytosis, induction of apoptosis, anti-proliferation while inactive on normal human mammary epithelial huMEC, and inhibition of glycolysis selectively in tumor cells but not in non-transformed MCF10A cells. Detailed mechanism of each cell line is listed in Table 8.2. BZL101 was reported to induce cell death *via* caspase activation in breast cancer cells but not in non-transformed (benign) mammary epithelial cells MCF10A and normal human fibroblasts IMR90. Hyperactivation of PARP and inhibition of glycolysis are likely the key mechanisms resulting in the energetic collapse and necrotic death that of breast cancer cells (Rugo et al. 2007; Fong et al. 2008).

Some of the molecular mechanisms include but not limited to induction of G1 and G2/M arrest; inhibition of Cyclins (A, D1, D2, D3, E) and CDKs (2, 4, 6); inhibition of COX-2; enhancement of macrophage oxidative burst; activation of caspase-3, -8, and -9; up-regulation of Bax (Bcl-2-associated X protein), p53 (tumor suppressor protein), Akt, JNK, *Fas,* MAPK (mitogen-activated protein kinase); activation of PARP cleavage; reduction of mitochondrial membrane potential; releasing of cytochrome c *via* the mitochondrial signaling pathways; down-regulation of *Bcl-2* (B cell lymphoma 2), *ERK*; induction of *c-fos* gene expression; induction of expression of genes involved in oxidative response (*GCLM*, *CBS*, *TRAF3*, etc), DNA damage (*TIPARP*, *CADD45a*, etc), cell death (*A20*, *TNF*, etc), xenobiotic response (*CYP1A1*, *CYP1B1*, *HSP70*, etc), and NF-κB pathway (*TNF*, *ICAM1*, *IL-8,* etc). Detailed mechanism of each cell line is listed in Table 8.2. Recent *in vitro* study of breast cancer cells (early stage estrogen sensitive MCF-7 *versus* late estrogen insensitive MDA-MB-231) and prostate cancer cells (androgen sensitive LNCaP *versus* late androgen insensitive PC3) revealed phenotype specific anti-proliferative gene expression responses in these cancer cells (Marconett et al. 2010). Induction of G1 cell cycle arrest and ablated expression of regulators Cyclin D1, CDK2, CDK4, growth factor stimulatory pathways, and estrogen receptor-α expression in estrogen sensitive MCF-7 breast cancer cells (ablation of promoter activities) were observed. The skullcap extract also arrested early stage androgen sensitive LNCaP in G2/M phase with corresponding decreases in Cyclin B1, CDK1, and androgen receptor expression. It also induced S phase arrest with corresponding ablations in Cyclin A2 and CDK2 expression.

8.5.2 Clinical Studies

There were two clinical studies on patients with metastatic breast cancer (Rugo et al. 2007; Perez et al. 2010). An aqueous extract of skullcap BZL101 was found to have favorable toxicity profile and promising efficacy in a Phase I clinical trial in the treatment of advance breast cancer (Rugo et al. 2007). Oral feeding of BZL (40 g/day) was demonstrated to be safe and well-tolerated in Phase IB dose escalation trial of metastatic breast cancer (Perez et al. 2010). Preliminary success in the Phase I clinical trial on 21 metastatic breast cancer patients demonstrated that oral intake of BZL101, an aqueous extract of skullcap (up to 12 g in 350 ml solution per day) had a favorable toxicity profile and encouraging clinical activity in heavily chemotherapy pretreated patients (Rugo et al. 2007). In this study, the mean age of patients was 54 years and the mean number of prior treatments for metastatic disease was 3.9. There was no grade III or IV adverse events (AEs). The most frequently reported skullcap-related grade I and II AEs were: nausea (38%), diarrhea (24%), headache (19%), flatulence (14%), fatigue (10), constipation (10%), and vomiting (10%). At the conclusion of this Phase I clinical study, 16 patients were available for evaluation. Among them, 4 had stable disease (SD) for >90 days (25%), 3/16 had SD for >180 days (19%), 5 had objective tumor regression (1 of that was 1 mm short of a partial remission using the Response Evaluation Criteria in Solid Tumors criteria. A follow-up study using oral feeding on BZL (40 g/day) was demonstrated to be safe and well-tolerated in a Phase IB dose escalation trial of metastatic breast cancer (Perez et al. 2010). In this open-label, Phase IB, multicenter, dose escalation study, all 27 women had histologically confirmed breast cancer and measurable stage IV disease. These patients had a median of 2 prior chemotherapy treatments for metastatic disease, and were treated in four different dose cohorts. At the end, grade 3 and 4 AEs were uncommon. The following dose-limiting toxicities were observed: grade 3 diarrhea, fatigue, rib pain, and grade 4 AST elevation (aspartate aminotransferase liver enzyme and metastases). Among 14 evaluable patients according to the Response Evaluation Criteria in Solid Tumors criteria, 3 were classified with stable disease for >120 days (21%), 1 remains stable for 700+days after 449 days BZL101 treatment. Three patients with objective tumor regression (>0% and <30%) were identified by independent radiology review. The maximum tolerated dose of BZL101 was not reached and it was defined as 40 g/day.

8.6 Discussion

8.6.1 Prospective Therapeutic Application and Direction for Cancer Patients

8.6.1.1 Breast Cancer

Breast cancer is the most common cancer among females in the United States (Jemal et al. 2010). The recent Phase I and Phase IB clinical studies involving the use of skullcap in advanced breast cancer patients indicate a promising therapeutic application of skullcap (Rugo et al. 2007; Perez et al. 2010). This data in collaboration with the known mechanisms of skullcap understood through *in vitro* studies (as described in Sect. 8.5.1 and Table 8.2), along with the two positive preliminary clinical results suggest that skullcap aqueous extract may have a promising application in hormone related female cancers like ovarian, uterine, cervical and breast cancer, despite originally believed to be less effective in treatment of reproductive organ cancers (Chong and Lee 1988; Zhu 1998; Huang 1999).

8.6.1.2 Prostate Cancer

Prostate cancer is the most common form of cancer and the second leading cause of cancer death in American men (Jemal et al. 2010). Data from of our *in vitro* and *in vivo* animal studies suggested a few possible cancer prevention and inhibition mechanism of skullcap on prostate cancer (Wong et al. 2009). Phytochemicals in the whole aqueous extract of skullcap might work together to induce programmed cell death in prostate cancer cells through the expression of p53 and Bax, which activating the apoptosis pathway. Skullcap might also affect cancer progression through the regulation of Akt, phosphorylated Akt, and JNK to suppress the survival pathway *in vitro*.

Prostate cancer is an ideal subject for clinical trials of cancer prevention due to its prevalence, long natural history, relative ease of prostate gland biopsies, and the relative availability of surrogate tumor markers. Patients with early prostate-specific antigen elevation ('prostate-specific antigen-only' disease progression) and with early disease should be ideal candidates for novel investigational therapies such as skullcap (Smith and Kantoff 2001). The preliminary success of Phase I and IB clinical trials on the safety and efficacy of high dosage (40 g/day) aqueous extracts of skullcap (Rugo et al. 2007; Perez et al. 2010) suggest that a Phase I clinical study of the efficacy skullcap on prostate cancer would be rewarding. Prostate cancer like breast cancer is a hormonally-related tumor, and thus skullcap may also be effective. There is currently still no cure in Western medicine for advanced prostate cancer leading many patients to seek alternative medicines, such as traditional Chinese medicine (Tang and Eisenbrand 1992; The Pharmacopoeia Commission of PRC 2000). Further translational studies of skullcap and its selective cytotoxicity in prostate cancer cells and non-transformed prostate epithelial cells may reveal other plausible mechanisms for skullcap, such as induction of reactive oxygen species and inhibition of glycolysis in cancer cells would give more support for the clinical trial of skullcap in prostate cancer patients.

8.6.1.3 Liver, Colorectal, Uterine, and Lung Cancers

Scientific literature reviews yield another group of promising candidates for skullcap treatment including liver cancer (9 *in vitro* studies with 1 murine and 3 human cancer cell lines), colorectal cancer (9 *in vitro* studies with 2 human cell lines and 1 *in vivo* murine ACF model), uterine (6 *in vitro* studies with leiomyoma cells), and lung cancer (7 *in vitro* studies with 1 murine and 2 human cell lines) (Table 8.2). These studies have shown a favorable anticancer effect. Besides aqueous extract, phenorbide-*a* ($C_{35}H_{36}N_4O_5$), ethanol extract, and chloroform extract (phytol, wogonin, luteolin, and hispidulin) are especially effective (cytotoxic, anti-proliferation, induction of apoptosis and expression of Bcl-2) against liver cancer cells but with a low cytotoxic effect on normal liver L-O2 cell lines (Chan et al. 2006; Lin et al. 2006a, b; Tang et al. 2006, 2007; Yu et al. 2007). Following ethnopharmacological approach and with further positive translational studies plus animal studies, skullcap could be a promising therapeutic anticancer alternative for these types of patients.

8.6.2 Prospective and Challenges

According to Newman and Cragg (2007), about 30% of all new chemical compounds discovered in the last 20 years are derived from natural products and a further 20% are derivatives of these natural products. Additionally, over 60% of drugs approved for cancer treatment were derived from natural products (Newman et al. 2003). Since the approval of the first botanical drug in 2006, the Food and Drug Administration in USA has received over 350 botanical investigational new drug applications and the aqueous extract of skullcap, BZL101 was one of the earliest botanical investigational new drugs issued (Perez et al. 2010). With the present understanding of skullcap's *in vitro* anticancer mechanisms (Tables 8.1 and 8.2) and the success of second Phase I clinical trial revealed that BZL101, a Phase II clinical trial for women with MBC is planned (Perez et al. 2010). Further positive clinical trial results would certainly facilitate the bringing of skullcap from bench to clinical use for breast cancer patients.

The Food and Drug Administration in USA defines dose by the total mass of the extract skullcap, BZL101, and not by the cumulative mass of the active compounds, and the aqueous extract of skullcap (BZL101 is the one of the well studied example) is the most effective type of extract. The challenge for future studies would be defining the relationship of response of pharmacokinetic profiles with the known active chemical components of skullcap. The drug development pathway of BZL101 is different from traditional pharmacognosy (identifying a single active chemical with significant enhanced activity per extract mass) in that the biological response appears to be dependent on simultaneous cytotoxic activity by a group of compounds rather than by just one (Perez et al. 2010). Since the mechanism for the biological response of skullcap has been studied and potential clinical response of BZL101 has been observed, studies on biomarkers for responses, or for patient selection, can be carried out to potentially replace traditional pharmacological analyses aiding clinicians to better understand the therapeutic index of BZL101. These would aid in the similar study and development of other aqueous extracts of skullcap.

References

Bui-Xuan, N. H., Tang, P. M., Wong, C. K., et al. (2010). Photo-activated pheophorbide-a, an active component of Scutellaria barbata, enhances apoptosis via the suppression of ERK-mediated autophagy in the estrogen receptor-negative human breast adenocarcinoma cells MDA-MB-231. *Journal of Ethnopharmacology, 13,* 95–103.

Campbell, M. J., Hamilton, B., Shoemaker, M., et al. (2002). Antiproliferative activity of Chinese medicinal herbs on breast cancer cells in vitro. *Anticancer Research, 22,* 3843–3852.

Cha, Y., Lee, E., Lee, H., et al. (2004). Methylene chloride fraction of Scutellaria barbata induces apoptosis in human U937 leukemia cells via the mitochondrial signaling pathways. *Clinica Chimica Acta; International Journal of Clinical Chemistry, 348,* 41–48.

Chan, J. Y., Tang, P. M., Hon, P. M., et al. (2006). Pheophorbide-a, a major anti-cancer component purified from Scutellaria barbata, induces apoptosis in human hepatocellular carcinoma. *Planta Medica, 72,* 28–33.

Chen, L. G., Hung, L. Y., Tsai, K. W., et al. (2008). Wogonin, a bioactive flavonoid in herbal teainhibits inflammatory cyclooxygenase-2 gene expression in human lung epithelial cancer cells. *Molecular Nutrition & Food Research, 52,* 1349–1357.

Chong, S. C., & Lee H. (1988). Chinese medicinal herbs of Hong Kong. Hong Kong: Commercial Press.

Chui, C. H., Lau, F. Y., Tang, J. C., et al. (2005). Activities of fresh juice of Scutellaria barbata and warmed water extract of Radix Sophorae Tonkinensis on anti-proliferation and apoptosis of human cancer cell lines. *International Journal of Molecular Medicine, 16,* 337–341.

Dai, S. J., Chen, M., Liu, K., et al. (2006a). Four new neo-clerodane diterpenoid alkaloids from Scutellaria barbata with cytotoxic activities. *Chemical & Pharmaceutical Bulletin (Tokyo), 54,* 869–872.

Dai, S. J., Tao, J. Y., Liu, K., et al. (2006b). Neo-clerodane diterpenoids from Scutellaria barbata with cytotoxic activities. *Phytochemistry, 67,* 1326–1330.

Dai, S. J., Sun, J. Y., Ren, Y., et al. (2007a). Bioactive ent-clerodane diterpenoids from Scutellaria barbata. *Planta Medica, 73,* 1217–1220.

Dai, S. J., Wang, G. F., Chen, M., et al. (2007b). Five new neo-clerodane diterpenoid alkaloids from Scutellaria barbata with cytotoxic activities. *Chemical & Pharmaceutical Bulletin (Tokyo), 55,* 1218–1221.

Dai, S. J., Shen, L., & Ren, Y. (2008). Two new neo-clerodane diterpenoids from Scutellaria barbata. *Journal of Integrrative Plant Biology, 50,* 699–702.

Dai, Z. J., Liu, X. X., Xue, Q., et al. (2008a). Anti-proliferative and apoptosis-inducing activity of Scutellaria barbata drug-containing serum on mouse's hepatoma H22 cells. *Zhong Yao Cai, 31,* 550–553.

Dai, Z. J., Wang, X. J., Xue, Q., et al. (2008b). Effects of Scutellaria barbata drug-containing serum on apoptosis and mitochondrial transmembrane potential of hepatoma H22 cells. *Zhong Xi Yi Jie He Xue Bao, 6,* 821–826.

Dai, Z. J., Wang, Z. J., Li, Z. F., et al. (2008c). Scutellaria barbata extract induces apoptosis of hepatoma H22 cells via the mitochondrial pathway involving caspase-3. *World Journal of Gastroenterology: WJG, 14,* 7321–7328.

Dai, S. J., Pheng, W. B., Shen, L., et al. (2009a). Two new neo-clerodane diterpenoid alkaloids from Scutellaria barbata with cytotoxic activities. *Journal of Asian Natural Products Research, 11,* 451–456.

Dai, S. J., Pheng, W. B., Zhang, D. W., et al. (2009b). Cytotoxic neo-clerodane diterpenoids alkaloids from Scutellaria barbata. *Journal of Natural Products, 72,* 1793–1797.

Dai, S. J., Qu, G. W., Yu, Q. Y., et al. (2010). New neo-clerodane diterpenoids from Scutellaria barbata with cytotoxic activities. *Fitoterapia, 81,* 737–741.

Fong, S., Shoemaker, M., Cadaoas, J., et al. (2008). Molecular mechanisms underlying selective cytotoxic activity of BZL101, an extract of Scutellaria barbata, towards breast cancer cells. *Cancer Biology & Therapy, 7,* 577–586.

Gingrich, J. R., Barrios, J. R., Morton, R. A., et al. (1996). Metastatic prostate cancer in a transgenic mouse. *Cancer Research, 56,* 4096–4102.

Gingrich, J. R., Barrios, J. R., Kattan, M. W., et al. (1997). Androgen-independent prostate cancer progression in the TRAMP model. *Cancer Research, 57,* 4687–4691.

Goh, D., Lee, Y. H., & Ong, E. S. (2005). Inhibitory effects of a chemically standardized extract from Scutellaria barbata in human colon cancer cell lines, LoVo. *Journal of Agricultural and Food Chemistry, 53,* 8197–8204.

Gupta, S., Ahmad, N., Marengo, S. R., et al. (2000). Chemoprevention of prostate carcinogenesis by x-difluoromethylornithine in TRAMP mice. *Cancer Research, 60,* 5125–5133.

Han, F., Hu, J., & Xu, H. (1997). Effects of some Chinese herbal medicine and green tea antagonizing mutagenesis caused by cigarette tar. *Zhonghua yu fang yi xue za zhi* [*Chinese Journal of Preventive Medicine*], *31,* 71–74.

Harris, G. K., Sharra, D. C., Leonard, S. S., et al. (2003). Effects of the Chinese medicinal herb Scutellaria barbata on cyclooxygenase-2 and inducible nitric oxide synthase expression on LPS-induced RAW 264.7 cells. In: Proceedings of the 94th Annual Meeting American Association of Cancer Research, 2003 April, Washington, DC. Philadelphia, PA: AACR, 44, 1419.

Huang, K. C. (1999). The pharmacology of Chinese herbs. Boca Raton: CRC Press.

Jemal, A., Siegel, R., Ward, E., et al. (2010). Cancer statistics, 2010. *C CA: A Cancer Journal for Clinicians, 60,* 277–300.

Jiangsu New Medical College. (1977). Dictionary of Chinese Materia Medica. Shanghai: Science and Technology Press of Shanghai.

Kim, D., Lee, T., Lim, I., et al. (2005). Regulation of IGF-I production and proliferation of human leiomyomal smooth muscle cells by Scutellaria barbata D. Don. in vitro: Isolation of flavonoids of apigenin and luteolon as acting compounds. *Toxicology and Applied Pharmacology, 205,* 213–224.

Kim, E. K., Kwon, K. B., Han, M. J., et al. (2007a). Induction of G1 arrest and apoptosis by Scutellaria barbata in the human promyeloctyic leukemia HL-60 cell line. *International Journal of Molecular Medicine, 20,* 123–128.

Kim, J. H., Lee, E. O., Lee, H. J., et al. (2007b). Caspase activation and extracellular signal-regulated kinase/Akt inhibition were involved in luteolin-induced apoptosis in Lewis lung carcinoma cells. *Annals of the New York Academy of Sciences, 1095,* 598–611.

Kim, K. W., Jin, U. H., Kim, D. I., et al. (2008). Antiproliferation effect of Scutellaria barbata D. Don. On cultured human uterine leiomyoma cells by down-regulation of the expression of Bcl-2 protein. *Phytotherapy Research: PTR, 22,* 583–590.

Lau, F. Y., Chui, G. H., Gambari, R., et al. (2005). Antiproliferative and apoptosis-inducing activity of Brucea javanica extract on human carcinoma cell. *International Journal of Molecular Medicine, 16,* 1157–1162.

Lee, T. K., Cho, H. L., Kim, D. I., et al. (2004a). Scutellaria barbata D. Don induces c-fos gene expression in human uterine leiomyomal cells by activating beta2-adrenergic receptors. *International Journal of Gynecological Cancer: Official Journal of the International Gynecological Cancer Society, 14,* 526–531.

Lee, T. K., Kim, D. I., Han, J. Y., et al. (2004b). Inhibitory effects of Scutellaria barbata D. Don and Euonymus alatus Sieb aromatase activity of human leiomyomal cells. Cycle analysis. *Immunopharmacology and Immunotoxicology, 26*, 315–327.

Lee, T. K., Kim, D. I., Song, Y. L., et al. (2004c). Differential inhibition of Scutellaria barbata D. Don (Laminaceae) on HCG-promoted proliferation of cultured uterine leiomyomal and myometrial smooth muscle cells. *Immunopharmacology and Immunotoxicology*, *26,* 329–342.

Lee, T. K., Lee, D. K., Kim, D. I., et al. (2004d). Inhibitory effects of Scutellaria barbata D. Don on human uterine leiomyomal proliferation through cell cycle analysis. *International Immunonopharmacology, 4,* 447–454.

Lee, H., Kim, Y., Choi, I., et al. (2010). Two novel neo-clerodane diterpenoids from Scutellaria barbata. *Bioorganic & Medicinal Chemistry Letters*, *20,* 288–290.

Lin, J. M., Liu Y, & Luo R. C. (2006a). Effect of Scutellaria barbata extract against human hepatocellular Hep-G2 cell proliferation and its mechanism. *Nan Fang Yi Ke Da Xue Xue Bao, 26,* 975–977.

Lin, J. M., Liu, Y., & Luo, R. C. (2006b). Inhibitory effect of Scutellaria barbata extracts against human hepatocellular carcinoma cells. *Nan Fang Yi Ke Da Xue Xue Bao, 26,* 591–593.

Marconett, C. N., Morgenstern, T. J., San Roman, A. K., et al. (2010). BZL 101, a phytochemcial extract form the Scutellaria barbata plant, disrupts proliferation of human breast and prostate cancer cells through distinct mechanisms dependent on the cancer cell phenotype. *Cancer Biology & Therapy, 10,* 397–405.

Newman, D. J., & Cragg, G. M. (2007). Natural products as sources of new drugs over the last 25 years. *Journal of Natural Products, 70,* 461–477.

Newman, D. J., Cragg, G. M., & Snader, K. M. (2003). Natural products as sources of new drugs over the period 1981–2002. *Journal of Natural Products, 66,* 1022–1037.

Nguyen, D. L., Devadhason, R. R., & Wong, B. Y. (2008). Dosage-dependent regulation of caspase 8 activation in LNCaP and DU145 human prostate cancer cell lines by the Chinese medicinal herb Scutellaria barbata. 7th Annual AACR International Conference of Frontiers in Cancer Prevention Research, A39.

Perez, A. T., Arun, B., Tripathy, D., et al. (2010). A phase 1B dose escalation trial of Scutellaria barbata (BZL101) for patients with metastatic breast cancer. *Breast Cancer Research and Treatment, 120,* 111–118.

Powell, C. B., Fung, P., Jackson, J., et al. (2003). Aqueous extract of herba Scutellaria barbatae, a Chinese herb used for ovarian cancer, induces apoptosis of ovarian cancer cell lines. *Gynecologic Oncology, 91,* 332–340.

Rugo, H., Shtivelman, E., Perez, A., et al. (2007). Phase I trial and antitumor effects of BZL101 for patients with advance breast cancer. *Breast Cancer Research: BCR, 105,* 17–28.

Shoemaker, M., Hamilton, B., Dairkee, S. H., et al. (2005). In vitro anticancer activity of twelve Chinese medicinal herbs. *Phytotherapy Research: PTR, 19,* 649–651.

Smith, M. R., & Kantoff, P. W. (2001). Molecular biology of prostate cancer. In J. Mendelsohn, P. M. Howley, M. A. Israel, & L. A. Liotta (Eds.), The molecular basis of cancer (2nd ed., pp. 343–360). Philadelphia: WB Saunders Company.

Suh, S. J., Yoon, J. W., Lee, T. K., et al. (2007). Chemoprevention of Scutellaria barbata on human cancer cells and tumorigenesis in skin cancer. *Phytotherapy Research: PTR, 21,* 135–141.

Tang, P. M., Chan, J. Y., Au, S. W., et al. (2006). Pheophorbide a, an active component in Scutellaria barbata, possesses photodynamic activities by inducing apoptosis in human hepatocellular carcinoma. *Cancer Biology & Therapy, 5,* 1111–1116.

Tang, P. M., Chan, J. Y., Zhang, D. M., et al. (2007). Pheophorbide a, an active component in Scutellaria barbata, reverses P-glycoprotein-mediated multidrug resistance on a human hepatoma cell line R-HepG2. *Cancer Biology & Therapy, 6,* 504–509.

Tang, W., & Eisenbrand G. (1992). *Chinese drugs of plant origin: Chemistry, pharmacology and use in traditional and modern medicine.* Berlin: Springer.

The Pharmacopoeia Commission of PRC. (2000). *Pharmacopoeia of the People's Republic of China* (English Edition). Beijing: People's Republic of China.

Wei, P. Y., Pu, H. Q, Wei, X., et al. (2007). Apoptosis-induced effect of Scutellaria barbata extract on human lung cancer SPC-A-1 cells and the expression of apoptosis associated genes. *Zhong Yao Cai, 30,* 1270–1273.

Wong, B. Y., Lau, B. H., & Teel, R. W. (1992a). Chinese medicinal herbs modulate mutagenesis, DNA binding and metabolism of benzo[a]pyrene. *Phytotherapy Research: PTR, 6,* 10–14.

Wong, B. Y., Lau, B. H., & Teel, R. W. (1992b). Chinese medicinal herbs modulate mutagenesis, DNA binding and metabolism of benzo[a]pyrene 7,8-dihydrodiol and benzo[a]pyrene 7,8-dihydrodiol-9,10-epoxide. *Cancer letters, 62,* 123–131.

Wong B. Y., Lau, B. H., Tadi, P. P., et al. (1992c). Chinese medicinal herbs modulate mutagenesis, DNA binding and metabolism of aflatoxin B_1. *Mutation Research, 279,* 209–216.

Wong, B. Y., Lau, B. H., Yamasaki, T., et al. (1993a). Inhibition of dexamethasone-induced cytochrome P450-mediated mutagenicity and metabolism of aflatoxin B_1 by Chinese medicinal herbs. *European Journal of Cancer Prevention, 2,* 351–356.

Wong, B. Y., Lau, B. H., Yamasaki, T., et al. (1993b). Modulation of cytochrome P450IA1-mediated mutagenicity, DNA binding and metabolism of benzo[a]pyrene by Chinese medicinal herbs. *Cancer Letters, 68,* 75–82.

Wong, B. Y., Jia, T. Y., Wan, C. P., et al. (1996). Oldenlandia diffusa and Scutellaria barbata augment macrophage oxidative burst and inhibit tumor growth. *Cancer Biotherapy & Radiopharmaceuticals, 11,* 51–56.

Wong, B. Y., Nguyen, D. L., Cabrera, I. B., et al. (2007). Modulation of apoptosis via the induction of caspase 3, 8, and 9 in LNCaP cells by two Chinese medicinal herbs Scutellaria barbata and Oldenlandia diffusa. 6th Annual AACR International Conference of Frontiers in Cancer Prevention Research, 112, A135.

Wong, B. Y., Nguyen, D. L., Lin, T., et al. (2009). Chinese medicinal herb Scutellaria barbata modulates apoptosis and cell survival in murine and human prostate cancer cells and tumor development in TRAMP mice. *European journal of cancer prevention: The Official Journal of the European Cancer Prevention Organisation (ECP), 18,* 331–341.

Wong, B. Y., Kim, L. Y., Kim, B. Y., et al. (2010). Inhibition of azoxymethane-induced aberrant crypt foci in C57BL/6 mice by the Chinese medicinal herbs Scutellaria barbata. 101st Annual Meeting American Association of Cancer Research 2010, A5677.

Wu, Y, & Chen, D. F. (2009). Anti-complementary effect of polysaccharide B3-PS1 in Herba Scutellariae barbatae (Scutelllaria barbata). *Immunopharmacology and Immunotoxicology, 30,* 696–701.

Yin, X., Zhou, X., Jie, C., et al. (2004). Anti-cancer activity and mechanism of Scutellaria barbata extract on human cancer cell lines A549. *Life Sciences, 75,* 2233–2244.

Yu, J., Liu, H., Lei, J., et al. (2007). Antitumor activity of chloroform fraction of Scutellaria barbata and its active constituent. *Phytotherapy Research: PTR, 21,* 817–822.

Zhu, F., Di, Y. T., Liu, L. L., et al. (2010). Cytotoxic neoclerodane diterpenoids from Scutellaria barbata. *Journal of Natural Products*, *73,* 233–236.

Zhu, Y. P. (1998). Chinese materia medica: Chemistry, pharmacology and applications. Amsterdam, Netherlands: Harwood Academic, 1998.

Chapter 9
An Evidence-based Perspective of *Hedyotis Diffusa* or *Oldenlandia Diffusa* (Spreading Hedyotis) for Cancer Patients

Jun Shao, Guoqing Gong and Louis Trombetta

Abstract *Hedyotis diffusa* or *Oldenlandia diffusa* (spreading hedyotis) is one of the most commonly used anticancer herbs. Its clinical use has a history more than several thousand years. It contains flavones, anthraquinones, polysaccharides, and other compounds possessing anticancer activities. In most cases, it is used together with other herbs. About 15% of the anticancer herbal formulas used in China contain this herb. Both pre-clinical and clinical studies have established the efficacy and safety of spreading hedyotis in treating various cancers including stomach cancer, liver cancer, lung cancer, esophagus cancer, and leukemia. It can directly inhibit the growth of various cancer cells and induce apoptosis both *in vitro* and *in vivo*. It shows selective cytotoxicity against cancerous cells. It can suppress some oncogenes and up-regulate anti-oncogenes. It also has immune modulation functions against cancer. It enhances the activities of natural killer cells and macrophages, promotes the proliferation of spleen cells, and up-regulates interleukin-2 and tumor necrosis factor-alpha. Clinical outcomes have demonstrated that it can enhance the efficacies and reduce the adverse effects (i.e. white blood cell decrease, nausea/vomit) by the conventional chemotherapies. It is also effective in relieving cancerous pain and fever. The commonly used clinical doses of 30–60 g/day usually do not cause any considerable adverse effects.

9.1 Introduction

Hedyotis diffusa Willd. or *Oldenlandia diffusa* (Willd.) Roxb (spreading hedyotis) is a commonly used herb for various diseases, especially for cancer. In Chinese, it is called Bai Hua She She Cao. Spreading hedyotis has been recorded in many Chinese medical literatures. It is usually 15–50 cm in height, and produces small white flowers during the summer. The whole plant is used for medical purposes. It is collected during the

J. Shao (✉)
Department of Pharmaceutical Sciences, College of Pharmacy and Allied Health Professions, St. John's University, 8000 Utopia Parkway, Queens, NY 11439, USA
e-mail: shaoj@stjohns.edu

W. C. S. Cho (ed.), *Evidence-based Anticancer Materia Medica,* Evidence-based Anticancer Complementary and Alternative Medicine,
DOI 10.1007/978-94-007-0526-5_9,

Fig. 9.1 *Hedyotis diffusa or Oldenlandia diffusa* (spreading hedyotis) modified from Luo et al. 1991

summer or autumn, washed and dried. It can also be used fresh. This herb grows in southern China. Fig. 9.1 is the drawing of the herb.

According to the traditional Chinese medicine (TCM), spreading hedyotis has functions of cleansing heat, detoxifying, promoting blood circulation, dispersing static blood, removing dampness, and facilitating urination. It is commonly used in clinical practice for the treatment of various cancer, hepatitis, gastrointestinal inflammation, urinary infections, pneumonia, etc. It is probably the most widely used herb for cancer treatment. It is generally used in combination with the other herbs. The book *Comprehensive Collection of Single- and Multiple-Ingredient Formulas for Tumor* (Chen et al. 1998) collects about 1,700 TCM formulas for various cancer treatments, and 15% of these formulas contain spreading hedyotis as one of the main ingredients. The anticancer activities of this herb have been demonstrated through many pre-clinical and clinical studies.

9.2 Pre-clinical Studies

9.2.1 Chemical Components

The chemical components of spreading hedyotis are numerous, mainly being anthraquinones such as 2-methyl-3-hydroyanthraquinone, 2-methyl-3-methoxyanthraquinone (Tai et al. 1979), flavones such as quercetin, quercetin-3-O-β-D-glucopyranoside,

Table 9.1 Representative chemical constituents in *Hedyotis diffusa* or *Oldenlandia diffusa* (spreading hedyotis) and their pharmacological activities

Constituents	Biological functions	References
Flavones	Down-regulating *pim-1, rel, fos* and *Bcl-2*; up-regulating TNF-α and IFN-γ; increasing spleen lymphocyte transformation	Zhang et al. 2007b
Methylanthraquinone	Inducing apoptosis *via* Ca^{2+}/calpain/caspase-4 pathway	Liu et al. 2010
Oleanolic acid	Inducing apoptosis; anti-invasion	Zhang et al. 2003
Polysaccharides	Activating p53, inhibiting *Bcl-xl*; inducing apoptosis; inhibiting tumor grow *in vivo*	Meng et al. 2008; Yang et al. 2010b
Quercetin	Anti-proliferation and inducing apoptosis	Ke et al. 2008
Stigmasterol	Up-regulating MAP2K6; inducing apoptosis; increasing G0-G1 proportion; reducing G2/M cells	Zhang et al. 2008
Ursolic acid	Down-regulating VEGF and MVD; inhibiting angiogenesis; up-regulating the expression of CDF15 and p21; inducing apoptosis	Li et al. 2007; Wang et al. 2008; Yang et al. 2010b; Ganbold et al. 2010

and eldrin (Lu et al. 2000; Zhou et al. 2007), terpenes such as ursolic acid and oleanolic acid (Cai et al. 1964; Ganbold et al. 2010), sterols such as stigmasterol, β-stigmasterol, and β-sitosterol (Fu et al. 1963; Cai et al. 1964; Zhang et al. 2008), polysaccharides, and essential oils (Ling et al. 2005; Si et al. 2008; Yu et al. 2008; Huang et al. 2008, 2009; Shi et al. 2010).

Aqueous extract of spreading hedyotis was fractioned by HPLC into 11 fractions. It was found that the fraction which was most effective in inducing apoptosis contained ursolic acid and its enantiomer oleanolic acid (Ganbold et al. 2010).

Listed in Table 9.1 are some representative chemical constituents and their pharmacological activities related to cancer treatment.

9.2.2 *In Vitro and In Vivo Anticancer Effects and Mechanisms*

9.2.2.1 Anticancer Effects

Gupta et al. (2004) studied the *in vitro* and *in vivo* anticancer activities of the aqueous extract of spreading hedyotis. The results demonstrate that the extract could significantly induce apoptosis, with 50% growth inhibition concentrations (IC_{50}) against seven human cancer cell lines in the range of 7–24 mg raw material/ml after 48-hour drug exposure. Human prostate cancer line Ln-Cap was the most sensitive cell line with an IC_{50} of 7 mg/ml. The apoptosis was confirmed by microscopic examination and DNA fragmentation assay. Rounding of the cells, cytoplasmic shrinkage, nucleus segregation and chromatin condensation were observed under the microscope after the cells being exposed to the herbal extract. A significant lad-

Fig. 9.2 Anticancer activities of *Hedyotis diffusa* or *Oldenlandia diffusa* (spreading hedyotis) in C57BL mice with B16-F10 lung metastases. The animals were iv inoculated through the tail vein with 0.2×10^6 live B16-F10 on day 1. The control group each was given 100 µl phosphate-buffered saline while the treated group each was given 100 µl of the herbal extract (5 g raw material/kg/day) on days 3–12 by oral gavage. The animals were sacrificed on day 14. The upper eight lungs were from the control group and the lower eight lungs from the treated group

der pattern of the DNA fragments was obtained after 24-hour exposure to the herbal extract. Oral administration of the herbal extract effectively reduced B16-F10 murine melanoma cell metastasis in the lungs of C57BL/J mice with a 70% ($P<0.001$) reduction in lung metastasis colonies at the daily dose of 5 g raw material/kg body weight (Fig. 9.2).

Intraperitoneal (ip) injection of the extract of spreading hedyotis at the daily dose of 15–60 mg/kg for 10 consecutive days inhibited the transplanted H22 liver cancer in Kunming mice. The reduction in tumor weight was 35–60% ($P<0.05$), respectively, corresponding to the given doses. The study also found that treatment with the herbal extract could increase the G0-G1 cell portion and reduce the G2/M portion (Zhang et al. 2008). Another study showed that the aqueous extract of spreading hedyotis could induce apoptosis of H22 cells *in vitro*. This was shown to be related with the up-regulation of HSP 70 (heat shock protein 70) and anti-oncogene protein P16 (Hu et al. 2009).

Spreading hedyotis is effective in treating cervical cancer. The growth of U14 cervical cancer in BALB/c mice was significantly inhibited by 28–49% by the oral administration of the aqueous extract at the dose of 30–120 mg/kg/day for 10 days. The treatment also significantly reduced the telomerase activities of the cancer cells, considered to be an important mechanism for its anticancer effect (Gao et al. 2007).

Total flavonoids extracted from spreading hedyotis inhibited the proliferation of human hepatic cancer cell line SMMC-7721 *in vitro*. The oncogenes such as *pim-1, rel, ras, fos, myc*, and *met* were down-regulated (Zhang et al. 2007a). The treatment of the flavonoids on Kunming mice transplanted with H22 mouse hepatic cancer cells showed inhibition of tumor growth, increase in G0/G1, increase in the plasma levels of tumor necrosis factor-alpha (TNF-α) and IFN-γ ($P<0.01$) as compared to the controls. The treatment also significantly ($P<0.05$) enhanced the spleen lymphocyte transformation rate as compared to the control (Zhang et al. 2007b).

Studies showed that the polysaccharides of spreading hedyotis were effective in inhibiting tumor growth *in vivo*. One hundred grams of the dried herb was extracted with 12 times of water. The extraction was repeated three times. After further purification, total 2.1 g of polysaccharides with molecular weight less than 10 K were obtained. Oral administration of the polysaccharides at the dose of 10–30 mg/kg/day for 10 consecutive days inhibited the growth of the transplanted S180 sarcoma in Kunming mice by 18–36%. Although the positive control with the treatment of cyclophosphamide with 30 mg/kg/day for 3 consecutive days through ip injection achieved 61% tumor growth inhibition, it showed 28% decrease in body weight. In contrast, the treatment with the polysaccharides of spreading hedyotis only resulted in negligible decrease in body weight (Yang et al. 2010b). The same group also studied the efficacy of the polysaccharides on the transplanted H22 hepatoma in Kunming mice. Oral administration of the polysaccharides at the dose of 10–30 mg/kg/day for 10 consecutive days achieved 27–50% tumor growth inhibition.

Treatment with stigmasterol from spreading hedyotis (15–60 mg/kg/day, ip injection for 10 consecutive days) on Kunming mice transplanted with H22 mouse hepatic tumor suppressed the tumor growth by 35–60% ($P<0.05$), increased the G0-G1 portion, and decreased the G2/M portion. (Zhang et al. 2008) In this study, 5-FU was used as the positive control at the dose of 30 mg/kg/day through ip injection for 10 consecutive days. This treatment resulted in 67% inhibition in tumor growth ($P<0.05$), slightly higher than the best efficacy achieved by spreading hedyotis stigmasterol (60% inhibition by the highest dose). However, treatment of 5-FU caused a significant decrease in body weight ($P<0.01$). The body weight of the positive control group was only 64% of the model control group receiving the treatment of normal saline, while the body weight of the groups treated with spreading hedyotis stigmasterol showed negligible differences from the model control group. These results demonstrated spreading hedyotis stigmasterol was as effective as 5-FU in treating H22 tumor but had much less adverse effects than 5-FU.

Ursolic acid was able to inhibit the proliferation of doxorubicin-resistant hepatoma cell line *in vitro*. It activated *Bak* but not *Bax*. Apoptosis was mainly through the caspase-independent apoptosis-inducing factor signaling pathway. Ursolic acid was also effective in inhibiting doxorubicin-resistant hepatoma cell line growth in athymic mice without any serious adverse effect on body weight, liver, heart and spleen (Yang et al. 2010a).

Oleanolic acid was found to be able to inhibit the growth of *ras* oncogene transformed Rat 6 fibroblast while not exhibiting any cytotoxic effect to the normal fibroblast cells. The results also suggest that oleanolic acid might induce the normal cells to secrete some inhibitory factor(s) against the transformed cells (Wu et al. 2009).

Spreading hedyotis also shows synergistic effect with chemo-therapeutic agents. Co-administration of the aqueous extract of spreading hedyotis (oral, 20 g raw material/kg/day for 10 consecutive days) with cisplatin (ip, 2.0 mg/kg/day for 2 days) achieved 60% inhibition of H22 tumor weight in Kunming mice in contrast to 49% and 40% inhibition by cisplatin and the herbal extract alone. This increase in efficacy was statistically significant ($P<0.01$). Interestingly, the treatment by the herb alone increased survival time by 43% in contrast to 34% and 21% by the co-administration and cisplatin alone. The differences in the survival time among the three groups were significant ($P<0.05$). Treatment by the herb alone could increase the body weight by 19% while the treatment by cisplatin alone decreased the body weight by 10%. Co-administration did not cause more body weight decrease than the treatment by cisplatin alone (Li et al. 2009).

9.2.2.2 Immune Modulation

Spreading hedyotis can increase immune functions. Studies showed that it could significantly increase the proliferation of spleen cells induced by Con A and lipopolysaccharides, and increase the killing effect of T cells (Qin 1990). It stimulated macrophages to produce interleukin (IL)-6 and tumor necrosis factor (Yoshida et al. 1997). Studies on the mice transplanted with S180 tumor cells indicated that intravenous injection of spreading hedyotis extract at the dose of 3.75 mg raw material/kg could significantly increase the natural killer (NK) cell and macrophage activities, promote the proliferation of spleen cells and enhance the biological activity of IL-2 (Liu et al. 2008). Oral administration of the aqueous extract of spreading hedyotis at the dose of 25 g raw material/kg/day for 12 days increased CD_4^+ and CD_8^+ T cells in H22-bearing Kunming mice (Hu et al. 2007).

In vitro studies found that spreading hedyotis could augment oxidative burst of murine macrophage cell line J774, indicating the enhancement of the macrophage functions. The results showed a dose-dependent increase in oxidative burst (Wong et al. 1996).

Studies also showed that spreading hedyotis reduced the adverse effects of cyclophosphamide in terms of white blood cell decrease. Oral administration of aqueous extract of spreading hedyotis at the doses of 2.74 and 5.48 g raw material/kg/day for 5 days counteracted the adverse effect of cyclophosphamide (ip 100 mg/kg/day for consecutive 3 days) on white blood cells in Kunming mice. Compared to the control group without any treatment, cyclophosphamide caused 63% decrease in white blood cells, while the treatment of low and high doses of spreading hedyotis reduced the decrease of white blood cell to 25% and 16% ($P<0.05$), respectively. The high dose treatment also significantly increased the mouse bone marrow karyocytes by 85% compared to the control group ($P<0.05$) (Su and Zhao 2007).

9.2.3 Toxicity and Side Effects

According to TCM practice, spreading hedyotis is generally considered as a relatively safe herb. Most TCM medical textbooks suggest not using the herb during pregnancy, which is the only caution listed. Acute toxicity study on the concentrated aqueous extract indicates that the LD_{50} for mice was 104 g/kg (raw material/body weight) through ip injection (Bureau of Traditional Chinese Medicine 1999).

Cell culture studies demonstrate about 10% growth inhibition on normal human pancreatic cells at the concentration of 50 mg raw material/ml while the IC_{50} against several human cancer cell lines fell in the range of 7–25 mg raw material/ml after 48 drug exposure (Gupta et al. 2004).

In vitro cell culture studies found that spreading hedyotis significantly inhibited the growth and induced apoptosis of leukemic cells HL60 while it did not induce apoptosis in human blood lymphocytes. However, the presence of spreading hedyotis prevented the progression of the stimulated human blood lymphocytes through the cell cycle, suggesting certain cytotoxicity against normal cells. In general, spreading hedyotis demonstrated selective cytotoxicity toward cancerous cells (Willimott et al. 2007).

Intraperitoneal injection to the Kunming mice at a daily dose of 60 mg extract/kg for 10 consecutive days did not cause significant change in body weight, while the same treatment with 5-FU at the dose of 30 mg/kg significantly reduced the body weight by 30% ($P<0.01$) (Zhang et al. 2008).

Oral administration of the aqueous extract of spreading hedyotis at the dose of 20 g raw material/kg/day for 10 consecutive days did not cause significant side effect in H22 hepatoma-bearing Kunming mice. However, 40 g raw material/kg/day for consecutive 10 days could cause 5% decrease in body weight, which was less than the 10% decrease caused by cisplatin (ip 2.0 mg/kg/day for 2 days) (Li et al. 2009).

Oral administration of the polysaccharides of spreading hedyotis at the dose of 30 mg/kg/day for 10 days did not cause any significant adverse effect on the weight of spleen and thymus in Kunming mice (Yang et al. 2010b).

9.3 Clinical Studies and Applications

Spreading hedyotis has been used clinically for a long time in China. It is mainly used for various cancer treatments in addition to other disease treatments. Clinical studies have demonstrated the effectiveness of this herb in treating many types of cancer including gastro-intestinal cancer, liver cancer, lung cancer, and leukemia. It can also improve the efficacy and reduce the adverse effects of the conventional chemotherapies.

9.3.1 Esophagus Cancer

Tang et al. (2003) studied the efficacy of spreading hedyotis in treating 106 cases of esophagus cancer of middle-late stage. The patients were given spreading hedyotis extract through iv infusion (40–60 drops/minute) at the dose of 24–60 g raw material/day. Each treatment cycle was consisted of 26 days: the treatment was given on days 1–5, 8–12, 15–19, and 22–26. The treatment lasted for total 4 cycles. There was a rest of 2 weeks between the treatment cycles. The treatment achieved 18% complete relief, 41% partial relief, 26% stabilization, and 16% progression. The treatment also resulted in complete disappearance of pain in 36/52 patients, partial relief of pain in 16/52 patients. The treatment was also effective in reducing cancerous fever—12/12 patients experienced complete relief of fever. There were 4 patients with lung metastases having chest edema and 2 with liver metastases having abdominal edema. After the treatment, all the edemas disappeared. No bone marrow suppression or renal toxicity was observed. The only adverse reaction observed was gastrointestinal reactions which occurred when the iv infusion rate was high.

9.3.2 Leukemia

Huang et al. (2001) studied the co-administration of spreading hedyotis with chemotherapeutic agents in treating acute non-lymphocytic leukemia. The control group of 21 patients received only chemotherapy of daunorubicin (45–60 mg/m^2/day for 3 days) and cytosine-1-β-D-arabinofuranoside (100–200 mg/m^2/day for 7 days); and the treated group of 19 patients received the same chemotherapy together with intravenous infusion of spreading hedyotis extract injection at the dose of 30 ml/day for 21 days (the injection concentration was not reported in the original paper; it was probably 1 g raw material/ml since this is the concentration of the commercially available products). After 2–3 terms of treatment, the treated group achieved 74% complete reliefs compared to 57% in the control group. The overall effective rate in the treated group was 95% compared to 71% in the control group ($P<0.01$).

9.3.3 Non-small Cell Lung Cancer

A clinical study on the treatment of late-stage non-small lung cancer shows that the addition of spreading hedyotis extract injection could significantly improve the efficacy and reduce the adverse effect of the chemotherapy. Thirty-three patients (38–69 years old with an average age of 55.5 years) in the control group were given only chemotherapy of cyclophosphamide (600–800 mg/day on day 1 and day 8), doxorubicin (40–60 mg on day 1), and cisplatin (50–60 mg/day on day 1 and day 2). Fifty-three patients (41–82 years old with an average age of 54.8 years old) in the treated group were given the same chemotherapy together with spreading hedyotis extract injection at the dose of 30 ml, twice daily, for 21 days (the injection concentration was not reported in the original paper; it was probably 1 g raw material/ml since this is the concentration of the commercially available products). The treatment efficacy on tumor was evaluated according to international objective efficacy evaluation standards. After 2–3 terms of treatment, in the treated group, 0/53 experienced complete relief, 23/53 partial relief, 18/53 no change, and 12/53 progression; in the control group, 0/33 experienced complete relief, 11/33 partial relief, 6/33 no change, and 16/33 progression. There was no significant difference in partial relief rate between the 2 groups. However, the rate of partial relief and stabilization was 77% in the treated group compared to 51% in the control group, showing significant difference ($P < 0.05$). The clinical symptoms including cough, chest tightness, chest pain, short breath, fever, cold sweat, and fatigue were monitored before and after the treatment. In the treated group, 13/53 experienced complete relief of the symptoms, 24/53 partial relief, and 16/53 no change or progression. In the control group, 6/33 experienced complete relief, 7/33 partial relief, and 20/33 no change or progression. There was a statistical significant difference in the improvement of clinical symptoms between the 2 groups ($P < 0.05$). The treated group had less adverse effects in terms of white blood cell decrease and nausea/vomiting than the control group ($P < 0.05$). In the treated group, 12/53 had degree I white blood cell decrease, 10/53 degree II white blood cell decrease, and 4/53 degree III white blood cell decrease; in the control group, 13/33 had degree I white blood cell decrease, 11/33 degree II white blood cell decrease, and 6/33 degree III white blood cell decrease. For nausea/vomiting, in the treated groups, there were 14/53 experiencing degree I, 11/53 degree II, and 4/53 degree III; in the control group, there were 17/33 experiencing degree I, 10/33 degree II, and 5/33 degree III (Li and Huang 2000).

9.3.4 *Liver Cancer*

Liu et al. (2004) studied the treatment of late-stage primary liver cancer with spreading hedyotis. Forty patients were given spreading hedyotis extract injection at the dose of 4 ml, 3 times daily, for 56 days (the injection concentration was not reported in the original paper). The treatment achieved 18% effective rate (complete and partial relief), and 63% stabilization rate (complete and partial relief and no change). The half-year and one-year survival rate was 60% and 28%, respectively. No adverse effect was observed during the treatment. The treatment also improved the conditions of liver pain, poor appetite, abdominal edema, fever, and fatigue. The improving rates were in the range of 58–76%. Total 75% of the patients experienced improved or stabilized life quality, which was an important aspect of the efficacies of cancer treatment. The authors also examined the effect of the treatment on the transformation rate of T cells and the subpopulations of T cells. The results showed that the treatment significantly ($P < 0.05$) increased the CD3 T cells by 11% and CD4/CD8 by 15%, respectively. T cell was also increased by 5%.

9.3.5 Various Cancers

Zhao and Zhang (2007) reported the efficacy of spreading hedyotis together with chemotherapies in treating various cancers of stages III and IV (Table 9.2). In general, the addition of spreading hedyotis extract injection at the dose of 4 ml, twice daily, for 19 days/term, total 2–4 terms (the injection concentration was not reported in the original paper) to the standard chemotherapies could significantly increase

Table 9.2 Efficacy and adverse effects of *Hedyotis diffusa* or *Oldenlandia diffusa* (spreading hedyotis) injection (intramuscular, 4 ml, twice daily) in combination with chemotherapy

			Spreading Hedyotis + Chemotherapy	Chemotherapy
N			78	76
Baseline KPS score	>60		61	58
	<60		17	18
Cancer stage	III		33	31
	IV		45	45
Cancer type	Lung		18	17
	Breast		19	18
	Liver		16	15
	Stomach		15	14
	Other		10	12
Efficacy[a]	CR		3 (4%)	1 (1%)
	PR		25 (32%)	14 (18%)
	NC		29 (37%)	36 (47%)
	PD		21 (27%)	25 (33%)
	KPS ↑ or no change		45 (58%)	30 (40%)
	KPS ↓		33 (42%)	46 (61%)
Adverse effects[b]	WBC decrease	0	5	1
		I	36	19
		II	30	36
		III	6	15
		IV	1	5
	Nausea	0	29	1
		I	11	17
		II	23	23
		III	13	26
		IV	2	9
	Vomiting	0	6	1
		I	33	23
		II	25	18
		III	10	20
		IV	4	14

[a] According to WHO efficacy categories of solid tumor treatment. CR: complete response, KPS: Karnofsky performance status, NC: no change, PD: progressive disease, PR: partial response

[b] According to WHO adverse event categories of anticancer drugs

the efficacy rate (complete and partial relief rate) from 20% to 60% ($P<0.05$), and significantly reduce the white blood cell decrease ($P<0.01$) and gastrointestinal adverse effect ($P<0.001$) caused by the chemotherapies. It could also achieve significantly more pain relief than the chemotherapy alone ($P<0.01$), and provide better life quality than the latter ($P<0.05$). Similar results were also observed by Luo et al. (2002) in the treatment of late-stage digestive and lung cancer.

9.3.6 Clinical Applications and Representative Formulas

Spreading hedyotis is widely used in various cancer treatments, in combination with other herbs or chemo-therapeutics. In the following, several representative formulas are listed.

Formula 1: fresh spreading hedyotis 70 g, *Scutellaria barbata* 20 g, *Cortex phellodendri* 10 g, *Coptis chinensis* 10 g, *Achyranthis bidentatae* 10 g, *Rhizoma atractylodis* 10 g, *Atractylodis macrocephalae* 10 g, *Panax notoginseng* 10 g, *Cortex moutan* 10 g, *Paeoniae rubra* 10 g, *Semen coicis* 10 g, *Rhizoma sparganii* 10 g, *Curcumae aeruginosae* 10 g, *Panax ginseng* 10 g, *Astragalus membranaceus* 10, *Radix bupleuri* 10, *Rehmanniae preparata* 10 g. Take one dose per day for the treatment of late-stage ovarian cancer. Extract with boiling water and divide the extract into 3 portions for drinking (Wu 2009).

Formula 2: spreading hedyotis 30 g, *Dioscoreae bulbiferae* 10 g, *Paeoniae rubra* 10 g, *Rhizoma homalomenae* 15 g, *Radix aristolochiae* 7 g, *Fructus trichosanthis* 30 g, *Herba prunellae* 10 g, *Bulbus fritillariae thunbergii* 10 g, *Concha ostreae* 15 g, *Alga sargassi fusiformis* 12 g, *Radix scrophulariae* 15 g, *Radix glehniae* 30 g, *Radix ophiopogonis* 10 g. Take one dose per day for the treatment of late-stage thyroid cancer. Extract with boiling water and divide the extract into 3 portions for drinking (Wang 2009).

Formula 3: *Radix adenophorae* 15 g, *Radix polygonatae* 15 g, *Radix asparagi* 15 g, *Flos inulae* 3 g (wrapped separately with cloth), *Rhizoma dioscoreae* 24 g, *Rhizoma imperatae* 60 g, spreading hedyotis 120 g. Take one dose per day for the treatment of esophagus cancer. Extract with boiling water, mix with suitable amount of honey, and divide the extract into 3 portions for drinking (Chang 1987).

Although these three formulas were used for the specific cancer in the original publications, the authors believe, based on our experience and knowledge, that indications of these formulas can be expanded to other cancers as well.

9.3.7 Dosage Forms and Doses

Spreading hedyotis is available in the form of a dried herb. In most cases, the dried herb is extracted with water and taken orally. Injections made of its extract are also available commercially for intramuscular and intravenous injection. The common daily dose is around 30 g, and 60 g is also used in many formulas. The highest daily dose recorded is 120 g (Chang 1987).

9.4 Concluding Remarks and Perspectives

Spreading hedyotis is one of the most widely used herbs for cancer treatment in TCM practice. Pre-clinical and clinical studies have established its effectiveness and safety in treating various cancers. It has been proven that the combination of spreading hedyotis with chemotherapies can result in higher efficacy and less adverse effects than the chemotherapies alone. Compared to the chemotherapies, spreading hedyotis provides better life quality to patients and usually does not cause any noticeable side effects based on currently commonly used doses. It seems that there is a possibility to increase the dose above the commonly used level (30–60 g/day) in order to achieve higher efficacy. In another word, there is a need to establish the appropriate doses through well-controlled clinical trials in order to render the best efficacy with the balance of acceptable adverse effects. In conclusion, spreading hedyotis is an effective anticancer herb with minimum adverse effects.

References

Bureau of Traditional Chinese Medicine. (1999). *Zhong Hua Ben Cao* (pp. 433–436). Shanghai: Shanghai Science and Technology Press.

Cai, C. C., Qian, X. L., & Jiang, D. Q. (1964). Analysis of the chemical constituents of Hedyotis diffusa I. *Acta Pharmaceutica Sinica, 11,* 809–814.

Chang, M. Y. (1987). Selection of anticancer formulas from contemporary famous doctors. *Fujian Traditional Chinese Medicine, 2,* 48–49.

Chen, Y., Cong, Z., Zhao, E. H., et al. (1998). *Comprehensive Collection of Single- and Multiple-Ingredient Formulas for Tumor*. Beijing: China Traditional Chinese Medicine Press.

Fu, F. Y., Xu, Z. P., Li, M. D., et al. (1963). Study on chemical constituents of Hedyotis diffusa. *Acta Pharmaceutica Sinica, 10,* 618–621.

Ganbold, M., Barker, J., Ma, R., et al. (2010). Cytotoxicity and bioavailability studies on a decoction of Oldenlandia diffusa and its fractions separated by HPLC. *Journal of Ethnopharmacology, 131,* 396–403.

Gao, C., Liu, Y., Cai, X. M., et al. (2007). Experimental research of chemotherapy on mice U14 cervical cancer by Hedyotis diffusa. *The Practice Journal of Cancer, 22,* 557–559.

Gupta, S., Zhang, D., Yi, J. Z., et al. (2004). Anticancer activities of Oldenlandia diffusa. *Journal of Herbal Pharmacotherapy, 4,* 21–33.

Hu, L., Luo, X. Y., Xie, Y. H., et al. (2009). Effect of heat shock protein 70 expression induced by herba Hedyotis diffusa on apoptosis of H22 hepatoma cells. *Traditional Chinese Drug Research Clinical Pharmacology, 20,* 536–539.

Hu, L., Wang, H. Q., Cui, N. J., et al. (2007). Effect of herba Hedyotis diffusa on living murine ascites hepatoma H22 cells and T lymphocytes. *Journal of Guangzhou University of Traditional Chinese Medicine, 24,* 313–316.

Huang, J. Y., Wang, X. L., & Gao, P. (2001). Clinical observation on the treatment of acute non-lymphocyte leukemia with co-administration of spreading hedyotis Injectable and chemotherapy. *Journal of Henan Traditional Chinese Medicine, 16,* 38.

Huang, W. H., Li, Y. B., & Jiang, J. Q. (2008). Chemical constituents from Hehyotis diffusa. *China Journal of Chinese Materia Medica, 33,* 524–526.

Huang, W. H., Li, Y. B., & Jiang, J. Q. (2009). Chemical constituents from Hehyotis diffusa. *China Journal of Chinese Materia Medica, 34,* 712–714.

Ke, Z. J., Ding, X. X., Dong, & W. Q., et al. (2008). The effect of quercetin on the proliferation and apoptosis of human bladder cancer cell BIU-87. *The Practical Journal of Cancer, 23,* 116–118.

Li, H., Zhang, D. Z., & Ma, M. F. (2009). Studies on the antitumor effect of Hedyotis diffusa on H22 hepatoma in mice. *China Journal of Traditional Chinese Medicine Science and Technology, 16,* 28–29.

Li, X. B., Sun, X. F., Yin, Q. W., et al. (2007). Ursolic acid inhibits the angiogenesis of transplanted human lung adenocarcinoma cell line A549 in nude mice. *Journal of Shandong University (Health Science), 45,* 188–190.

Li, Y., & Huang, J. Y. (2000). Treatment of late-stage non-small cell lung cancer with spreading hedyotis extract injection and chemotherapy. *Journal of Henan Traditional Chinese Medicine, 15,* 45.

Ling, Y. Z. (2005). Extraction, separation and assay on polysaccharides from Hedyotis diffusa willd. *Biology and Technology, 15,* 48–50.

Liu, J. B., & Yao, Z. W. (2004). The clinical effect of spreading hedyotis extract injection on primary liver caner. *Journal of Medicine Forum, 25,* 37–39.

Liu, J. X., Li, H. Y., Miao, G. X., et al. (2008). Experimental study on antitumor activities and immunity regulation of Hedyotis diffusa Willd. Injection. *Chinese Journal of Information on Traditional Chinese Medicine, 15*(Suppl.), 22–23.

Liu, Z., Liu, M., Liu, M., et al. (2010). Methylanthraquinone from Hedyotis diffusa WILLD induces Ca(2+)-mediated apoptosis in human breast cancer cells. *Toxicology in Vitro, 24,* 142–148.

Lu, C. M., Yang, J. J., Wang, P. Y., et al. (2000). A new acylated flavonol glycoside and antioxidant effects of Hedyotis diffusa. *Planta Medica, 66,* 374–378.

Luo, W. J., Wu, W. T., Li, S. J., et al. (2002). Efficacy analysis on the treatment of late-stage cancer by Hedyotis diffusa Injectable with chemotherapy. *Practice of Clinical Combined Traditional Chinese Medicine and Western Medicine, 2,* 1–2.

Luo, X. R., Huang, S. M., Wu, D. L., et al. (1991). *Shi Yong Zhong Cao Yao Cai Se Tu Pu* (Vol. 1, p. 37). Guangzhou: Guangdong Science and Technology.

Meng, Q. Y., Lv, X. F, Ren, F. Y., et al. (2008). Effect of polysaccharides from Hedyotis diffusa willd. On cell line Bel7402. *Chinese Journal of Gerontology, 7,* 1271–1272.

Qin, F. H. (1990). The strengthening effect of Hedyotis diffusa on mouse immune functions. *Shanghai Immunology Journal, 10,* 32.

Shi, F., Jia, X. B., Jia, D. S., et al. (2010). Research of material base of hedyotis diffusa Willd. In preventing lung cancer. *China Journal of Traditional Chinese Medicine and Pharmacy, 25,* 403–408.

Si, J. Y., Chen, D. H., Pan, R. L., et al. (2008). Studies on the glucosides from spreading hedyotis. *Chinese Traditional and Herbal Drugs, 39,* 507–509.

Su, C. Y., & Zhao, H. R. (2007). Study on efficacy of water extracts from Hedyotis diffusa and its formula with Radix Astragali for raising the leukocyte. *Pharmacology Clinical Research, 15,* 25–27.

Tai, D. F., Lin, Y. M., & Chen, F. C. (1979). Components of Hedyotis diffusa Willd. *Hua Hsueh, Chemistry* (Taipei), *3,* 60–61.

Tang, L. M., Wang, J. H., & Zhou, H. J. (2003). Clinical observation on the treatment of 106 cases of middle-late stage esophagus cancer with spreading hedyotis intravenous infusion. *HaiNan Medicine, 14,* 75.

Wang, J., Wang, W. P., Gu, Z. L., et al. (2008). Mechanisms of apoptosis induced by ursolic acid in human adenocarcinoma of lung SPC-A-1 cells. *Suzhou University Journal of Medical Science, 28,* 207–209.

Wang, Z. M. (2009). Treatment of late-stage thyroid cancer with spreading hedyotis. *Journal of Traditional Chinese Medicine, 50*(Suppl.), 79–80.

Willimott, S., Barker, J., Jones, L. A., et al. (2007). Apoptotic effect of Oldenlandia diffusa on the leukaemic cell line HL60 and human lymphocytes. *Journal of Ethnopharmacology, 114,* 290–299.

Wong, B. Y., Lau, B. H., Jia, T. Y., et al. (1996). Oldenlandia diffusa and Scutellaria barbata augment macrophage oxidative burst and inhibit tumor growth. *Cancer Biotherapy & Radiopharmaceuticals, 11,* 51–56.

Wu, J. X. (2009). Treatment of late-stage ovarian cancer with fresh spreading hedyotis. *Journal of Traditional Chinese Medicine, 50,* 629.

Wu, P. K., Chi Shing Tai, W., Liang, Z. T., et al. (2009). Oleanolic acid isolated from Oldenlandia diffusa exhibits a unique growth inhibitory effect against *ras*-transformed fibroblasts. *Life Sciences, 85,* 113–121.

Yang, L., Liu, X., Lu, Z., et al. (2010a). Ursolic acid induces doxorubicin-resistant HepG2 cell death via the release of apoptosis-inducing factor. *Cancer Letters, 298,* 128–138.

Yang, P. M., Dai, L., Liu, Y., et al. (2010b). Antitumor effect of Hedyotis diffusa polysaccharides on S180 and H22 experimental tumor. *North-west Pharmacy, J, 25,* 33–34.

Yoshida, Y., Wang, M. Q., Shan, B. E, et al. (1997). Immunomodulating activity of Chinese medicinal herbs and Oldenlandia diffusa in particular. *International Journal of Immunopharmacology, 19,* 359–370.

Yu, L., Li, J. M., Jiang, Z., et al. (2008). A new anthraquinone from Hedyotis diffusa. *Chinese Journal of Medicinal Chemistry, 18,* 298–302.

Zhang, D. F., Huang, W., Huang, J. Q., et al. (2003). Study on proliferation inhibition and anti-invasion and apoptotic induction of oleanolic acid in human lung cancer cell line. *Cancer Research Prevention Treatment, 30,* 180–182.

Zhang, S., Song, Y. Q., Zhou, S., et al. (2007a). Target gene regulation involved in the inhibitory effect of flavonoids from Hedyotis diffusa Willd. On human hepanoma cell line SMMC-7721. *World Chinese Journal of Digestology, 15,* 1060–1066.

Zhang, S., Song, Y. Q., Zhou, S., et al. (2007b). Inhibitory effect of flavonoids from Hedyotis diffusa willd. On hepatoma cells in vitro & vivo and its influence on transplanted H22 tumor cells' proliferation cycle, apoptosis and immune circumstances in mice. *World Chinese Journal of Digestology, 15,* 1347–1352.

Zhao, S. L., & Zhang, X. H. (2007). Treatment of cancer by spreading hedyotis extract injection with chemotherapy. *Journal of Medicine Forum, 28,* 74–75.

Zhang, S., Yue, W., Wang, L., et al. (2008). Inhibitive effect of stigmasterol from Hedyotis diffusa Willd. On hepatoma cells in vitro & vivo and its influence on transplanted H22 tumor cells multiplication cycle and apoptosis. *Progress in Modern Biomedicine, 8,* 2016–2017.

Zhou, Y. J., Wu, K. S., & Zeng, G. Y. (2007). Studies on constituents of Oldenlandia diffusa. *China Journal of Chinese Materia Medica, 32,* 590–593.

Chapter 10
An Evidence-based Perspective of *Allium Sativum* (Garlic) for Cancer Patients

Anna Herman-Antosiewicz, Anna Kawiak and Jędrzej Antosiewicz

Abstract *Allium sativum* (garlic) has been used since prehistoric times in various cultures as a spice as well as a medicine to combat microbial and fungal infections, help in cardiovascular problems, stimulate immunological system or stop tumor growth. Epidemiological studies indicate that increased consumption of garlic is inversely correlated with the risk of different types of cancer in various human populations. Garlic preparations inhibit chemically induced cancers in animals. This chemo-preventive activity is attributed to organosulfur compounds which modulate Phase I and II detoxification enzymes, thus inhibit pro-carcinogen activation and/or enhance carcinogen neutralization and removal. Laboratory studies also indicate that garlic compounds suppress cancer development at post-initiation phases inducing cell cycle block and apoptosis as well as inhibiting angiogenesis and metastasis. Results of the *in vitro* studies explain the mechanisms of action of garlic organosulfurs at the molecular level, which is a necessary step before their clinical use for cancer patients. This chapter reviews the evidence on garlic chemo-preventive activities in human populations, animal models and limited clinical trials. It also summarizes the current knowledge on molecular mechanisms of its anti-proliferative activity toward cancer cells, possible interactions with drugs and impact on immune system—factors that should be considered before use of garlic compounds in cancer therapy.

10.1 Introduction

Allium sativum (garlic) is a member of the *Liliaceae* family, which also includes onions, leeks, scallions or chives. Garlic is rich in sulfur-containing compounds, which contribute to its characteristic odor, taste and beneficial health effects. Thus, it has been used as a spice as well as a medicine since prehistoric times

A. Herman-Antosiewicz (✉)
Department of Molecular Biology, University of Gdańsk, Kładki 24, 80-822 Gdańsk, Poland
e-mail: wuzel@biotech.ug.gda.pl

W. C. S. Cho (ed.), *Evidence-based Anticancer Materia Medica,* Evidence-based Anticancer Complementary and Alternative Medicine,
DOI 10.1007/978-94-007-0526-5_10,

in various cultures (Rivlin 2001). Ancient Egyptians, Jews, Greeks and Romans consumed garlic to increase their strength, courage in battle and enhance work capacity. The *Codex Ebers*, an Egiptian medical text dating to 1500 BC, mentions garlic as a remedy for skin diseases, poisoning, heart problems, and abnormal growths (tumors). Hippocrates prescribed garlic for protecting the skin against toxins or treating abdominal tumors. In ancient China and Japan garlic was thought to provide energy, lift depression, and improve male potency, aid respiration and digestion. In India, 2,000 years ago, garlic was used to treat arthritis and heart disease.

In the modern era health benefits of garlic are well recognized and include antimicrobial, anti-fungal, lipid and glucose lowering, anti-thrombotic and immunostimulatory properties, as well as anticancer activity. The evidence for the chemopreventive and therapeutic application of garlic preparations is reviewed below.

10.2 Bioactive Compounds Derived from Garlic

Garlic bulbs contain approximately 65% water, 28% carbohydrates, 2.3% organosulfur compounds, 2% protein, 1.2% free amino acids and 1.5% fiber (Block 1985). The biological activity of garlic is attributed to organosulfur compounds (OSCs). The primary sulfur-containing compounds in whole and intact garlic are γ-glutamyl-*S*-alk(en)yl-*L*-cysteins. They are precursors of odorless alliin (*S*-alk(en)yl-*L*-cysteine sulfoxide). Processing of garlic bulbs (cutting, chewing or crushing) releases a vacuolar enzyme, alliinase, which acting upon alliin gives rise to allicin (diallyl thiosulphate), the principal active substance of fresh garlic extract, discovered by Cavallito in 1944 (Cavallito and Bailey 1944). Allicin is unstable and breaks down readily to produce odorous oil-soluble sulfur compounds, including diallyl sulfide (DAS), diallyl disulfide (DADS), diallyl trisulfide (DATS), allyl methyl trisulfide, dithiins and ajoenes (Fig. 10.1). Water-soluble garlic sulfur compounds, such as S-allylcysteine (SAC) or S-allylmercaptocysteine (SAMC), are products of the bioconversion of γ-glutamyl-*S*-alk(en)yl-*L*-cysteins which takes place during natural aging of plants (Fig. 10.1).

The composition and thus biological activity of garlic extracts depends on the mode of preparation. For example, rehydratated standardized powder of crushed garlic is almost devoid of alliin because rehydratation activates alliinase. It contains allicin (13.5 mg/g DW), allyl sulfides (0.15 mg/g DW, including DATS- 56–87% and DADS- 9–31%), γ-glutamyl-*S*-allylcysteine (5.9 mg/g DW), γ-glutamyl-*S*-*trans*-1-propenylcysteine (5.8 mg/g DW), S-allylcysteine (0.28 mg/g DW), S-allylmercaptocysteine (<0.02 mg/g DW) (Lawson and Wang 2005). Garlic oil obtained by steam distillation contains diallyl- (57%, including DADS and DATS), methyl allyl- (37%) and dimethyl oligosulphides (6%) (Lawson et al. 1991). On the other hand, oil macerated garlic, produced by homogenizing chopped garlic in vegetable oil, contains primarily vinyldithiins (70%), dialk(en)yl sulfides (18%) and ajoenes (12%) (Lawson et al. 1991). Aged garlic extract (AGE) which results from pro-

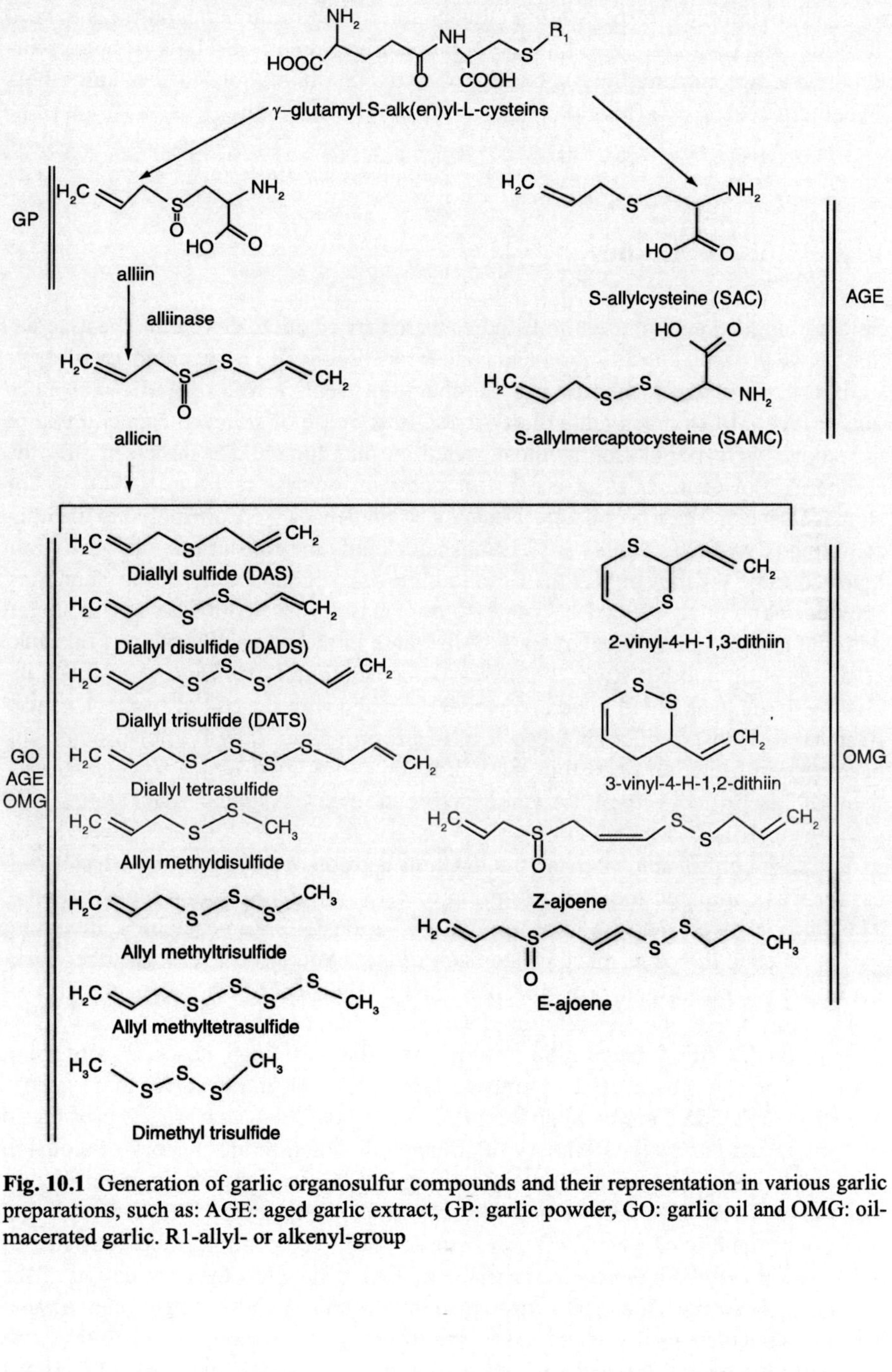

Fig. 10.1 Generation of garlic organosulfur compounds and their representation in various garlic preparations, such as: AGE: aged garlic extract, GP: garlic powder, GO: garlic oil and OMG: oil-macerated garlic. R1-allyl- or alkenyl-group

longed extraction of fresh garlic in ethanol solution at room temperature is devoid of allicin and contains water-soluble organosulfurs, such as S-allylcysteine, S-allylmercaptocysteine and lipid-soluble DAS, DADS, diallyl polysulfides and others (Weinberg et al. 1993) (Fig. 10.1).

10.3 Human Studies

Epidemiological and intervention studies were carried out to determine the association between garlic consumption and cancer risk (Table 10.1). The majority of these studies were related to addressing the chemo-preventive role of garlic consumption in regard to stomach and colon cancer. In the case of stomach cancer, most of the studies were population studies carried out in China and regarded mainly the comparison of death rates or cancer incidents in low-risk and high-risk areas for stomach cancer. The lower death rates/cancer incidences were attributed to the consumption of garlic. In the low-risk Cangshan County the consumption of garlic was reported to be 9 times higher than in the high-risk Linqu County of the Shandong Province (You et al. 1998, 1999). A 3-fold increase in garlic intake was associated with the low cancer incidence in a low-risk area in the Jiangsu Province (Takezaki et al. 1999). In another study the increased consumption of *Allium* vegetables, including garlic, was found to be inversely associated with the risk of stomach cancer (Gao et al. 1999). Although these results are promising they do not provide the necessary information regarding the validation of the food-frequency assessment, therefore according to the FDA evidence-based review system, are not statistically significant (Kim and Kwon 2009).

In a case-control study carried out in China a group of 564 patients with stomach cancer was compared to a control group of 1,131 residents from a high-risk area. The study showed that the consumption of various *Allium* vegetables, including garlic, reduced the occurrence of stomach cancer (You et al. 1989). Another case-control study performed in Korea indicated a significant decrease in the risk of gastric cancer with the increased intake of garlic (Kim et al. 2002).

In a double-blind intervention study the effects of high doses of allitridum (DATS) and low doses of selenium were examined on the occurrence of gastric cancer in the Qixia County, Shandong Province. Case subjects were selected based on criteria such as medical history of stomach disorder, family history of stomach cancer, alcohol consumption and/or smoking. 2,526 case subjects were enrolled in the trail and were given allitridum (200 mg/day) and selenium (100 µg every other day) for 1 month each year for 3 consecutive years. The control group consisted of 2,507 subjects, which were given a placebo. Five years after the termination of the study the occurrence of cancers was registered. 35 and 51 cases of malignant tumors were documented in the intervention group and control group, respectively, and the morbidity rates of all cancers in the intervention group declined by 22%. In the case of gastric cancer 10 and 19 cases were recorded in the intervention and control

Table 10.1 Studies on the evaluation of garlic intake and cancer risk

Study type	Cases/controls	Garlic preparation	Effects	References
Stomach cancer				
Case-control	564/1,131	Garlic (kg/year) 0.1–1.5	OR 0.8 (S)	You et al. 1989
		>1.5	OR 0.7 (S)	
Case-control	338/669	Garlic, frequency of intake >2 times/month	OR 0.89 (NS)	Hansson et al. 1993
Cohort	139/3123	Garlic supplement any/day/1 year (*vs* other supplement)	RR 0.93 (NS)	Dorant et al. 1996b
Case-control	153/234	Garlic 1–3 servings/month	OR 0.4 (S)	Gao et al. 1999
		>4 servings/month	OR 0.3 (S)	
Case-control	136/136	Garlic (raw), highest quartile	OR 0.53 (S)	Kim et al. 2002
Intervention	2526/2507	Allitridum (DATS) 200 mg/1 month/year/3 yrs	RR 0.36 (S)	Li et al. 2004
Intervention	3365 cases	200 mg AGE and 1 mg steam distilled garlic oil—2/2 × day/7.3 years	RR 1.06 (NS)	You et al. 2006
Colorectal cancer				
Cohort	212/35,004	Garlic >1 servings/week	Distal colon RR 0.52 (S) Total colon RR 0.68 (NS)	Steinmetz et al. 1994
Cohort	205/47,949	Garlic >2 servings/week	Colon RR 0.77 (NS) Distal colon RR 0.63 (S)	Giovannucci et al. 1994
Case-cohort	443/3,123	Garlic supplement any/day/1 year (*vs* other supplement)	RR 0.93 (NS)	Dorant et al. 1996a
Case-control	488/488	Garlic >3 servings/week	OR 0.66 (S)	Witte et al. 1996
Case-control	1192/1192	Garlic, highest tertile	OR 0.8 (S)	Le Marchand et al. 1997
Case-control	223/491	Garlic, highest tertile	OR 0.39 (S)	Levi et al. 1999
Intervention	37 cases	AGE, 2.4 ml/day/12 months	RR 0.71 (NS)	Tanaka et al. 2004
Case-control	2280/4765	Garlic, high intake	OR 0.74 (S)	Galeone et al. 2006

Table 10.1 (continued)

Study type	Cases/controls	Garlic preparation	Effects	References
Prostate cancer				
Case-control	328/328	Garlic, frequency of intake > 2 times/week	OR 0.64 (NS)	Key et al. 1997
Case-control	238/471	Garlic > 2.14 g/day	OR 0.47 (S)	Hsing et al. 2002
Case-control	1294/1451	Garlic, high intake	OR 0.81 (NS)	Galeone et al. 2006
Cohort	1338/29,361	Garlic > 1 serving/week	RR 0.88 (NS)	Kirsh et al. 2007
Breast cancer				
Cohort	469/1713	Garlic supplement any/day/1 year (*vs* other supplement)	RR 0.87 (NS)	Dorant et al. 1995
Case-control	345/345	Garlic, frequency of intake		Challier et al. 1998
		7–10 times/week	OR 0.52 (S)	
		11–12 times/week	OR 0.25 (S)	
Case-control	2900/3122	Garlic, high intake	OR 0.9 (NS)	Galeone et al. 2006
Lung cancer				
Cohort	484/3,123	Garlic supplement any/day/1 year	RR 1.78 (NS)	Dorant et al. 1994
Laryngeal cancer				
Case-control	201/414	Garlic, highest tertile	OR 0.5 (NS)	Zheng et al. 1992a
Case-control	527/1297	Garlic, high intake	OR 0.56 (S)	Galeone et al. 2006
Nasal cancer				
Case-control	60/414	Garlic daily	OR 0.6 (NS)	Zheng et al. 1992b
Esophageal cancer				
Case-control	196/392	Raw/cooked (kg/year) 0.6–2.0	OR 0.6 (S)	Hu et al. 1994
		> 2.0	OR 1.0 (NS)	
Case-control	395/1066	Garlic, high intake	OR 0.43 (S)	Galeone et al. 2006

NS: not significant, OR: odds ratio—highest *vs* lowest consumption, RR: relative risk, S: significant groups, respectively, and the morbidity rate declined by 47.3% in the intervention group (Li et al. 2004).

In a latter study carried out in the Linqu County of the Shandong Province, a long-term trial was conducted to evaluate, among others, the effects of AGE and steam distilled garlic oil on the development of advanced precancerous gastric lesions. The study was a factorial-design trial, in which 3,365 case subjects were given a one-time antibiotic treatment for *Helicobacter pylori* infection and/or 7.3 years of oral vitamin (vitamin C, E and selenium) supplementation and/or garlic supplementation. The results showed no association between long-term garlic supplementation and precancerous gastric lesions (You et al. 2006). The results of a cohort study carried out in the Netherlands and a case-control study carried out in Sweden showed that intake of garlic supplements did not correlate with the risk of stomach cancer (Dorant et al. 1996b; Hansson et al. 1993).

In the United States two large cohort studies were conducted to determine the correlation between garlic intake and colorectal cancer. In the first study, The Iowa Women's Health Study, a group of 41,837 women were interviewed for their dietary habits and were then monitored for cancer incidence over a period of five years. During this time 212 cases of colon cancer were documented. The results of the study showed that the consumption of more than 1 serving of garlic per week significantly decreased the risk of colorectal cancer (Steinmetz et al. 1994). In another cohort study, 47,949 men were followed for approximately 6 years. 205 cases of colon cancer were identified during this period. The results, however, did not associate garlic intake with colon cancer (Giovannucci et al. 1994). These results were further confirmed by a case-cohort study carried out in the Netherlands. Dietary habits of 120,852 cases were evaluated with a food-item frequency questionnaire. After a 3.3-year follow-up, 443 case subjects diagnosed with colorectal cancer and 3,123 randomly selected healthy control cases were analyzed and no association between garlic intake and colon cancer risk was reported (Dorant et al. 1996a).

In a case-control study conducted in Switzerland, including 223 case patients and 491 control subjects, the correlation between garlic consumption and colorectal cancer was shown (Levi et al. 1999). These results were supported by another case-control study carried out in Italy and Switzerland. 1,394 colon cancer cases, 886 rectal cancer cases and 4,765 control subjects were examined. The specific amounts of garlic intake were not given since garlic intake was specified in terms of qualitative variables such as non-use, low, intermediate or high. It was showed that intermediate to high use of garlic is associated with the reduced risk of colorectal cancer (Galeone et al. 2006). Two other case-control studies carried out in the United States and Hawaii confirmed that garlic intake was inversely associated with the development of adenocarcinomas (Le Marchand et al. 1997; Witte et al. 1996). The results of these two studies, however, where not considered as significant, as assessed by the FDA evidence-based review system.

A double-blind, intervention trial carried out in Japan was performed to evaluate the effects of AGE on patients with colorectal adenomas. 51 patients diagnosed with colorectal cancer were divided into an intervention group (high dose AGE,

2.4 ml/day) and control group (low dose, 0.16 ml/day). The results, verified after 6 and 12 months, showed no effect of AGE on the incidence of polyps, however AGE was found to decrease the size of existing carcinomas. Due to the low number of subjects and short duration of the trial, the results can only suggest the preventative role of AGE on colorectal adenocarcinomas (Tanaka et al. 2004).

The association between garlic consumption and the reduced risk of prostate cancer was also examined. In a case-control study carried out in England, 328 men diagnosed with prostate cancer and the same number of population controls were examined regarding their dietary habits, which included the consumption of garlic. The results showed that the intake of two or more servings of garlic per week was not associated with the reduction of prostate cancer risk (Key et al. 1997). In another study carried out in Shanghai, the association between garlic consumption and reduced prostate cancer risk was much more pronounced. 238 men diagnosed with prostate cancer and 471 men, as population control subjects, were interviewed to determine the amounts of *Allium* vegetables they consumed. The results showed that men who consumed more than 10.0 g/day had a significantly lower risk of prostate cancer compared to those who consumed less than 2.2 g/day (Hsing et al. 2002). Two other studies did not report an association between garlic intake and prostate cancer risk (Galeone et al. 2006; Kirsh et al. 2007).

Garlic consumption was also found to be inversely related to the risk of breast cancer as shown in a European case-control study conducted in France. 345 women diagnosed with primary breast carcinoma and 345 healthy women were evaluated for their dietary habits. The results of the study showed that an increase in garlic intake (more than 7 servings per week) reduced the risk of breast cancer (Challier et al. 1998). A case-control study carried out in Italy and Switzerland did not report an association between garlic intake and risk of breast cancer (Galeone et al. 2006). Similarly, a cohort study carried out in the Netherlands also reported no correlation. A cohort of 469 patients with breast cancer and 1,713 control subjects were analyzed after a 3.3-year follow-up and no association of garlic intake with breast cancer risk was reported (Dorant et al. 1995). The associations between garlic consumption and other cancers are presented in Table 10.1.

On the basis of the presented studies it is clear that to some extent there is a correlation between garlic intake and some types of cancer. The published epidemiological studies suggest the protective role of garlic against colon and stomach cancer. However, these studies do have some limitations, which decrease their credibility. One of these limitations is the diverse efficacy of the different types of garlic preparations consumed by the case subjects. These studies do not provide any consistency as to the type of preparation used, therefore too many variables are introduced. Moreover, the chemical composition of the garlic preparation can be affected through the method of preparation, such as cooking, extracting, etc. Another limitation is providing the appropriate data as to the amounts of garlic consumed. These are often not precisely specified and are reported as frequency instead of amount. Therefore, in order for the results of the studies to be more consistent, clinical trials need to be carried out to provide more credible evidence as to the association between garlic consumption and cancer development.

10.4 Evidence of the Anticancer Activity of Garlic in Animal Models

The anti-carcinogenic properties of garlic have been examined in animal experimental models. Garlic in the form of various preparations, administered either intragastrically or percutaneously, has been shown to inhibit the carcinogenesis of the colon, skin, tongue, breast and liver, induced by a variety of carcinogens (Table 10.2). For example, the chemo-preventive effects of garlic on 1,2-dimethylhydrazine (DMH)-induced colon cancer were studied in rats. Garlic was fed to rats and DMH was injected subcutaneously for 20 weeks, once a week. The results of the experiments showed that garlic significantly reduced the incidence of tumor formation in the group of rats, in which garlic constituted 2.5% of their diet (Cheng et al. 1995).

Likewise, garlic-derived compounds have been studied for their chemo-preventive activities. DAS inhibited tumor formation of the esophagus, lung, skin, forestomach, colon and breast, whereas contradictory results were obtained in the case of liver carcinogenesis (Table 10.2). DADS was reported to provide protection against chemically induced carcinogenesis of the skin, colon, kidney, fore-stomach, prostate and breast. DATS inhibited forestomach-induced carcinogenesis, whereas SAC, colon-induced carcinogenesis. Contradictory results were obtained with SAC regarding *N*-methylnitrosourea (MNU)-induced breast cancer, the inconsistencies of the obtained results could be due to the differences in the administered doses of SAC (Table 10.2).

The anti-carcinogenic properties of OSCs were found to be associated with the number of sulfur atoms in their chemical structure. In a study evaluating the protective effects of OSCs on benzo[a]pyrene (BP)-induced neoplasia of the forestomach and lungs, allyl methyl trisulfide (AMT), allyl methyl disulfide (AMD), DAS and DATS inhibited forestomach carcinogenesis, whereas only DAS and AMD inhibited lung carcinogenesis. These results suggested that the number of sulfur atoms in the molecule could determine the organ sites at which protection against carcinogens takes place (Sparnins et al. 1988). The number of allyl groups has also been shown to influence the inhibitory capacities of OSCs. Studies showed that OSCs containing allyl groups are more potent inhibitors of carcinogenesis (Sparnins et al. 1986, 1988).

Apart from the chemo-preventive properties of garlic-derived OSCs, the anticancer activities of these compounds have also been determined in animal xenograft tumor models (Table 10.3). DADS was found to inhibit the growth of HCT-15 human colon tumor xenografts in nude mice. DADS was either intraperitoneally or intragastrically administered to mice, however, its efficacy was 2 times higher when injected intraperitoneally, reducing tumor volume by 69%. The efficacy of DADS was similar to that of 5-fluorouracil (5-FU), a recognized chemo-therapeutic agent used in advanced colon cancer therapy. The concurrent treatment of DADS with 5-FU reduced the toxicity of 5-FU, improving white blood cell counts and the decrease in spleen weight and elevated plasma urea (Sundaram and Milner 1996c). DADS also suppressed the growth of KPL-1 breast cancer xenografts in nude mice.

Table 10.2 Effects of garlic and its constituents on chemically-induced carcinogenesis in animal models

Garlic preparation	Organ	Animal, dose	Carcinogen	Effect on tumor	References
Garlic	Colon	Rats, diet 2.5% (4.76 g/m^2 body surface/day)	DMH	Inh	Cheng et al. 1995
	Tongue	Rats, 250 mg/kg, po	4NQO	Inh	Balasenthil et al. 2001
Garlic oil	Skin	Mice, 10% solution, pc	BaP-croton oil	Inh	Sadhana et al. 1988
Garlic extract	Skin	Mice, 5 mg DW, pc	DMBA-TPA	Inh	Nishino et al. 1989
Garlic powder	Mammary	Rats, 1–4% in diet	DMBA	Inh	Liu et al. 1992
AGE	Liver	Rats, 2.5 & 10 ml/kg, pc	DEN	Inh	Uda et al. 2006
DAS	Esophagus	Rats, 200 mg/kg, po	NMBA	Inh	Wargovich et al. 1988
	Lung	Mice, 200 mg/kg, po	NKK	Inh	Yang et al. 2001
	Liver	Rats, 250 mg/kg, po	DEN, 2-AAF	Inh	Singh et al. 2004
	Liver	Rats, 50–100 mg/kg, po	DMH	Inh	Hayes et al. 1987
	Skin	Mice, 250 µg/mouse, pc	DMBA, TPA	Inh	Singh and Shukla 1998
	Forestomach, lung	Mice, 0.02 mmol, po	BaP	Inh	Sparnins et al. 1988
	Colon	Mice, 200 mg/kg, po	DMH	Inh	Wargovich 1987
	Colon	Rats, 50–200 mg/kg, po	Combination	No	Takahashi et al. 1992
	Liver	Rats, 50–200 mg/kg, po	Combination	Increase	Takahashi et al. 1992
	Mammary	Rats, 1.8 mmol/kg, po	DMBA	Inh	Ip et al. 1992
DADS	Mammary	Rats, 200 ppm in diet	PhIP	Inh	Suzui et al. 1997
	Mammary	Rats, 1.8 mmol/kg, po	DMBA	Inh	Ip et al. 1992
	Skin	Mice, topical application 1 mg/100 µl/2 × week/25 weeks	DMBA/TPA	Inh	Dwivedi et al. 1992
	Mammary	Rats, 57 µmol/kg diet	MNU	Inh	Schaffer et al. 1996
	Colon	Rats, 200 ppm in diet	AOM	Inh	Reddy et al. 1993
	Colon	Rats, 50–200 mg/kg, po	Combination	Inh	Takahashi et al. 1992
	Kidney	Rats, 50–200 mg/kg, po	Combination	Inh	Takahashi et al. 1992
	Forestomach	Mice, 0.02 mmol, po	NDEA	Inh	Wattenberg et al. 1989
	Prostate	Rats, 150 mg/kg, po	MNU	Inh	Arunkumar et al. 2006b

Table 10.2 (continued)

Garlic preparation	Organ	Animal, dose	Carcinogen	Effect on tumor	References
DATS	Forestomach	Mice, 0.02 mmol, po	BaP	Inh	Sparnins et al. 1988
	Lung	Mice, 0.02 mmol, po	BaP	No	Sparnins et al. 1988
SAC	Colon	Mice, 50–100 mg/kg, po	DMH	Inh	Sumiyoshi and Wargovich 1990
	Mammary	Rats, 57 µmol/kg, po	MNU	Inh	Schaffer et al. 1996
	Mammary	Rats, 666 & 2,000 ppm in diet	MNU	No	Cohen et al. 1999

2-AAF: 2-acetyl-aminofluorene, AOM: azoxymethane, BaP: benzo[*a*]pyrene, DAS: diallyl sulfide, DADS: diallyl disulfide, DATS: diallyl trisulfide, DEN: diethylnitrosamine, DMBA: 7,12-dimethylbenz(*a*)anthracene, DMH: 1,2-dimethylhydrazine, DW: dry weight, Inh: inhibition, MNU: *N*-methyl-N-nitrosourea, NDEA: *N*-nitrosodiethylamin, NMBA: *N*-nitrosomethylbenzylamine, NNK: 4-(methylnitrosamino)-1-(3-pyridyl)-1-butanone, No: no effect, 4NQO: 4-nitroquinoline 1-oxide, pc: percutaneously, po: per os, PhIP: 2-amino-1-methyl-6-phenylimidazo[4,5-b]pyridine, SAC: S-allylcysteine, TPA: 12,O-tetradecanoylphorbol-13-acetate

Table 10.3 Anticancer effects of garlic constituents in mouse xenograft models

Compound	Xenograft	Dose	Effect	References
Ajoene	B16/BL6 mouse melanoma tumor	25 μg/g bw/before tumor injection, pc	↓ 70–90% primary tumor growth	Taylor et al. 2006
		5, 15, 25 μg/g bw/every 2 days/4 weeks, pc	↓ 90% lung metastases	
DADS	HCT-15 human colon cancer cells	1 mg/3 × week, ip	↓ 69% tumor volume	Sundaram and Milner 1996c
	KPL-1 human estrogen receptor (ER)-positive breast cancer cells	1–2 mg/3 × week/35 weeks, ip	↓ 45% tumor volume	Nakagawa et al. 2001
DATS	PC-3 human prostate cancer cells	6 μmol/3 × week/20 days, po	↓ 3×tumor volume	Xiao et al. 2006a
DATS in nanoparticles	HepG2 human hepatoma cells	1.5 mg/kg/every 2 days/14 days, iv	↓ 45% tumor volume	Zhang et al. 2007
SAC	CWR22R human androgen-independent prostate cancer cells	1 g/kg/day/8 weeks	↓ 62.4% tumor volume	Chu et al. 2007
SAMC	CWR22R androgen-independent primary prostate cancer cells	200 mg/kg/day/23 days, po + docetaxel 7.5 mg/kg/week/23 days, po	↑ 53% docetaxel sensitization, ↑ tumor suppression, and ↓ toxicity than docetaxel alone	Howard et al. 2008
	PC-3 androgen-independent metastatic prostate cancer cells	300 mg/kg/day/28 days, po	↓ 71% tumor volume and ↓85.5% lung and adrenal metastases	Howard et al. 2007
Thiacremonone	SW620 colon cancer	5, 10, 30 mg/kg/4 weeks, po	↓ 32–50% tumor volume	Ban et al. 2009
		1 mg/kg/10 days, po, + docetaxel 1 mg/kg/1 × week/4 weeks, ip	↓ tumor volume and ↑mouse life span	

DADS: diallyl disulfide, DATS: diallyl trisulfide, ip: intraperitoneally, iv: intravenously, pc: percutaneously, po: per os, SAC: S-allylcysteine, SAMC: S-allylmercaptocysteine

DADS (1 or 2 mg) was injected intraperitoneally 3 times per week from the day of tumor injection. After 35 weeks a 45% reduction in primary tumor weight was observed in DADS-treated mice, which was associated with reduced cell proliferation (Nakagawa et al. 2001). DADS also suppressed the growth of *H-ras* oncogene transformed tumor xenografts in nude mice. DADS was administered orally 3 times a week from the day of tumor injection and the results showed that the growth of tumors was significantly delayed in comparison with control mice (Singh et al. 1996).

DATS inhibited the growth of PC-3 human prostate cancer xenografts in athymic mice. DATS was administered orally (6 μmoL, thrice weekly) to nude mice with implanted PC-3 cells and after 20 days of therapy the average tumor volume in DATS-treated mice was three-fold smaller than in control mice. Moreover, the treatment did not cause weight loss or any other side effects (Xiao et al. 2006a).

S-allylcysteine was reported to inhibit the growth of human prostate CWR22R cancer xenografts in mice with no detectable side-effects. The effects of SAC were associated with the inhibition of tumor proliferation and invasion, as shown by E-cadherin restoration, and increased apoptosis (Chu et al. 2007).

Thiacremonone, a novel sulfur compound isolated from garlic, was reported to inhibit the growth of SW620 human colon tumor xenografts in mice. Furthermore, the co-treatment of thiacremonone with low doses of docetaxel (1 mg/kg) increased the susceptibility of tumor cells to docetaxel, most probably through the inactivation of NF-κB. This combined treatment reduced docetaxel toxicity and greatly increased the life span of mice from 45 to 110 days. Therefore, these results indicate that thiacremonone could be used in association with other chemo-therapeutical drugs, enabling lower dose chemo-therapuetical treatment (Ban et al. 2007, 2009).

Another OSC, SAMC, was also reported to increase the efficacy of docetaxel, as shown in a prostate CWR22R cancer xenograft model. The combination was 53% more potent in inhibiting tumor growth than docetaxel alone (Howard et al. 2008).

The anti-metastatic activities of garlic compounds have also been studied *in vivo*. The effects of SAMC were examined on the growth and tumor spread of prostate cancer with the use of a fluorescent orthotopic severe combined immunodeficient (SCID) mouse model. PC-3 cells, constitutively expressing GFP, were injected into the dorsal prostate of SCID mice and after 30 days, metastatic GFP-expressing tumors in lungs and adrenals were identified. The oral administration of SAMC (300 mg/kg/d) not only inhibited the growth of primary tumors by 71% but also reduced the number of lung and adrenal metastases by 85.5% and reduced the number of circulating tumor cells by 91%. No toxic effects or organ pathology were observed in SAMC-treated mice (Howard et al. 2007). Moreover, studies have shown that 200 mg/kg SAMC protects against toxin-induced liver damage with an efficacy similar to that of N-acetylcysteine, a hepatoprotective agent (Sumioka et al. 1998). Another garlic constituent, ajoene, was also found to inhibit lung metastasis of B16/BL6 melanoma cells in mice (Taylor et al. 2006).

The effects of OSCs have also been examined in targeted therapy with the use of nanoparticles. In these studies polybutylcyanoacrylate nanoparticles, filled with DATS (DATS-PBCA-NP) were targeted against orthotopically transplanted liver hepatocellular HepG2 cells in mice. DATS-PBCA-NP was found to suppress the proliferation and induce apoptosis in transplanted cells leading to a significant inhibition of tumor growth (Zhang et al. 2007).

The results of the *in vivo* experiments are consistent and indicate the chemo-preventive, anti-proliferative and anti-metastatic properties of garlic and its constituents.

10.5 Molecular Targets of Garlic Compounds—*In Vitro* Studies

A large body of research on the chemo-preventive and anticancer activities of garlic constituents has been performed on cultured cells, the best model for the investigation on the molecular mechanism of their action and important step for identification of mechanism-based biomarkers which might be useful in future clinical trials and therapies. It appears that compounds from garlic, besides prevention of initiation and promotion of carcinogenesis, act at later phases inhibiting cancer cell proliferation and viability.

Growth suppressive activity of garlic compounds, documented for cancer cells of different origin, relies on their ability to block cell cycle progression and/or induce programmed cell death (Table 10.4). Interestingly, most garlic-derived OSCs induce accumulation of cancer cells with 4 N DNA content which indicates a block in G2, M or both phases of the cell cycle.

Progression of cells from G2 into mitosis depends on the Cdk1-cyclin B1 complex, whose activity is controlled at multiple levels, such as production and association of components, their intracellular localization and reversible phosphorylation. Dephosphorylation of Cdk1 at Thr14/Tyr15 by Cdc25 family of dual-specificity phosphatases is crucial for the activity of the complex and Cdc25s activities are under control of DNA damage checkpoint kinases Chk1 and Chk2 (Peng et al. 1997; Sanchez et al. 1997). Garlic compounds negatively influence activation of the Cdk1-cyclin B1 complex. It has been demonstrated that DADS decreased Cdk1 kinase activity within 4 h of treatment in HCT-15 colon cancer cells (Knowles and Milner 1998). It correlated with reduced complex formation between Cdk1 and cyclin B1 despite the time-dependent accumulation of cyclin B1 as well as increased inactivating phosphorylation of Cdk1 and drop in Cdc25C protein level upon DADS treatment (Knowles and Milner 1998, 2000). The same compound induced inactivating phosphorylation of Cdk1 in HL-60 human leukemia cells (Tan et al. 2004) or decreased Cdk1 kinase level in PC-3 human prostate cancer cells in a dose-dependent manner (Arunkumar et al. 2006b). Similarly, DATS induced accumulation of cyclin B1 and impaired Cdk1/cyclin B1 activity in J5 liver cancer cells (Wu et al. 2004), decreased Cdc25C and Cdk1 levels and induced Tyr15 phosphorylation of Cdk1 in H358 non-small and H460 large cell lung cancer cell lines (Xiao et al. 2009).

Table 10.4 Evidence of anti-proliferative and apoptosis inducing activities of the most commonly studied garlic compounds in cancer cell lines. Molecular targets whose statuses change upon OSCs treatment are indicated

OSCs	Model	Effect	Molecules affected	References
DAS	Human neuroblastoma	Apoptosis	Ca^{2+}, Bax:Bcl2	Karmakar et al. 2007
	Human lung	Apoptosis	Bax:Bcl2	Hong et al. 2000
	Human glioblastoma	Apoptosis	Ca^{2+}	Das et al. 2007
DADS	Human colon	G2/M arrest, apoptosis	Cdk1, cyclin B1, Cdc25C, acetylo-H3, H4, Ca^{2+}, microtubule network, ROS, JNK, Bax:Bcl2	Sundaram and Milner 1996a, b; Knowles and Milner 1998, 2000; Park et al. 2002; Druesne et al. 2004a, b; Xiao et al. 2005b; Song et al. 2009; Yang et al. 2009
	Human cervical	Apoptosis	ROS, Ca^{2+}, Bax:Bcl2	Lin et al. 2008
	Human gastric	G2/M arrest	Cdc25C	Yuan et al. 2004
	Human bladder	G2/M arrest, apoptosis	ROS	Lu et al. 2004
	Human leukemia	G2/M arrest, apoptosis	Cdk1, cyclin B1, ROS, acetylo-H3, H4	Lea et al. 1999; Kwon et al. 2002; Tan et al. 2004
	Human lung	G2/M arrest, apoptosis	ROS, Bax:Bcl2	Hong et al. 2000; Wu et al. 2005
	Human prostate	G2/M arrest, apoptosis	Cdk1, cyclin B1, acetylo-H3, H4	Arunkumar et al. 2006a; 2007
	Human breast	Apoptosis	Bax:Bcl-xL, BimEL	Nakagawa et al. 2001; Lund et al. 2005
	Human glioblastoma	Apoptosis	Ca^{2+}	Das et al. 2007
	Human neuroblastoma	Apoptosis	ROS, Bcl2, JNK Ca^{2+}, Bax:Bcl2	Filomeni et al. 2003; Aquilano et al. 2007; Karmakar et al. 2007

Table 10.4 (continued)

OSCs	Model	Effect	Molecules affected	References
DATS	Human prostate	G2/M, prometaphase arrest, apoptosis	Cdk1, cyclin B1, Cdc25C, securin, ROS, JNK, ferritin, Akt, Bid, Bcl2, Bax:Bcl2	Xiao et al. 2004, 2005a, 2006b; Herman-Antosiewicz and Singh 2005; Antosiewicz et al. 2006; Xiao and Singh 2006; Herman-Antosiewicz et al. 2007, 2010
	Human liver	G2/M arrest	Cyclin B1	Wu et al. 2004
	Human gastric	G2/M arrest, apoptosis	Bax:Bcl2	Lan and Lu 2004; Ha et al. 2005
	Human colon	M arrest, G2/M arrest apoptosis	Microtubule network, Cdc25C, ROS	Hosono et al. 2005, 2008; Busch et al. 2010
	Human bladder	Apoptosis	Akt, Bcl2	Wang et al. 2010b
	Human breast	G2/M arrest, apoptosis	Cyclin B1, Bax:Bcl2	Malki et al. 2009
	Human skin	G2/M arrest, apoptosis	ROS, Ca^{2+}	Wang et al. 2010a
	Human glioblastoma	Apoptosis	Ca^{2+}	Das et al. 2007
	Human lung	G2/M arrest, apoptosis	Cdk1, cyclin B1, Cdc25C, Bax:Bcl2, Ca^{2+}	Sakamoto et al. 1997; Xiao et al. 2009
SAMC	Human colon	G2/M or M arrest, apoptosis	Microtubule depolymerization, JNK, acetylo-H3, H4	Shirin et al. 2001; Lea et al. 2002; Xiao et al. 2003, 2005b
	Human leukemia	G2/M arrest, apoptosis	Acetylo-H3, H4	Sigounas et al. 1997; Lea et al. 2002
Ajoene	Human leukemia	G2/M arrest, apoptosis	ROS, Bcl2	Dirsch et al. 1998, 2002; Li et al. 2002a, b
	Human gastric	Apoptosis	Bax	Lee 2008
	Human skin	Apoptosis	Bcl2	Tilli et al. 2003
	Murine melanoma	Apoptosis	Caspase-3	Ledezma et al. 2004

DAS: diallyl sulfide, DADS: diallyl disulfide, DATS: diallyl trisulfide, ROS: reactive oxygen species, SAMC: S-allylmercaptocysteine

The mechanism of OSC-induced G2/M arrest has been well elucidated for DATS in a prostate cancer cell line model (Herman-Antosiewicz and Singh 2005; Antosiewicz et al. 2006; Herman-Antosiewicz et al. 2007, 2010; Xiao et al. 2005a, 2006b). The DATS-induced Tyr-15 phosphorylation of Cdk1 and inhibition of Cdk1/cyclin B1 activity in PC-3 cells was accompanied by down-regulation of total Cdc25C protein level an increased inhibitory phosphorylation of Cdc25C (Ser216), which was mediated by checkpoint kinase-1 (Chk1) (Herman-Antosiewicz and Singh 2005; Xiao et al. 2005a). Activating phosphorylations of Chk1 (Ser317) and other kinases engaged in DNA damage checkpoint such as Chk2 (Thr68) and ATM (Ser1981) as well as histone H2A.X, whose phosphorylation at Ser139 serves as an indicator of double strand break, points toward the genotoxic activity of DATS (Herman-Antosiewicz and Singh 2005). Whether DATS directly reacts with DNA or its effect is indirect through generation of reactive oxygen species needs to be elucidated. Antioxidants such as N-acetylcysteine (NAC) or EUK134 protected against the decrease in Cdc25C level and cell cycle arrest induced by DATS (Antosiewicz et al. 2006; Xiao et al. 2005a). Genotoxic activity of DADS was also reported by Aquilano et al. (2007) in SH-SY5Y human neuroblastoma cells and it was counteracted by over-expression of neuronal nitric oxide synthase.

It was also documented that DATS induced prometaphase arrest with characteristic changes in the tubulin network, chromatin condensation and increased Ser10 phosphorylation of histone H3 (Herman-Antosiewicz and Singh 2005) as well as accumulated anaphase promoting complex (APC) substrates (securin and cyclin B1) and hyperphosphorylated APC components (Herman-Antosiewicz et al. 2007). Interestingly, Chk1 and its upstream kinase, ATR, appeared to be mediators of DATS-induced mitotic arrest (Herman-Antosiewicz and Singh 2005; Herman-Antosiewicz et al. 2007).

Mitotic block or apoptosis induced by garlic OSCs might be the result of a disruption of the microtubule network in cancer cells by these compounds. For example, treatment of SW480 human colon cancer cells or NIH 3T3 mouse fibroblasts with 150 μmol/L SAMC caused rapid microtubule depolymerization and cytoskeleton disruption in interphase cells and perturbed spindle formation in mitotic cells. SAMC (300 and 1,000 μmol/L) induced microtubule depolymerization and inhibited *de novo* tubulin polymerization *in vitro* presumably because of direct interaction with sulfhydryl groups on tubulin (Xiao et al. 2003, 2005b). DADS (112 or 280 μmol/L) arrested SW480 in both G2 and M phase of the cell cycle. It was also shown that DADS at high concentrations (560 and 1,120 μmol/L) caused a 30% inhibition of *de novo* tubulin polymerization but did not induce depolymerization of polymerized tubulin *in vitro* (Xiao et al. 2005b). Interphase NIH 3T3 cells treated with 56 μmol/L DADS revealed disorganized microtubule cytoskeleton network. Interestingly, although mitotic spindles looked normal, "lagging" chromosomes were observed in DADS-treated cells, which were suggested to cause mitotic block and appearance of multinucleated cells (Xiao et al. 2005b). DATS, but not DADS, has been shown to induce mitotic arrest in HCT-15 and DLD-1 human colon cancer cells and it was associated with the disruption of the microtubule network in interphase cells and inhibition of spindle formation in mitotic cells (Hosono et al. 2005).

This study revealed that DATS-treated β tubulins were oxidatively modified at residues Cys12 and Cys354 to form S-allylmercaptocysteine (Hosono et al. 2005). Another oil soluble garlic compound, Z-ajoene, caused G2/M cell cycle arrest and dose-dependent reduction in microtubule network of PtK2 normal marsupial kidney cells and inhibited tubulin polymerization *in vitro* (Li et al. 2002a).

OSCs may affect cancer cell proliferation through modification of the histone acetylation level, which has an impact on gene expression. It has been reported that DADS and more potently its metabolite, allyl mercaptan, increased acetylation of H4 and H3 histones in DS19 cells and K562 human leukemic, rat hepatoma and human breast cancer cells and it was suggested that histone acetylation mediates the differentiation process of erythroleukemia cells (Lea et al. 1999). Growth inhibitory effects of allicin, SAMC and SAC on DS19 cells and SAMC on Caco-2 human colon and T47D human breast cancer cells have been correlated with increased histone acetylation, although its mechanism remains to be characterized (Lea et al. 2002). DADS-induced accumulation of Caco-2 and HT-29 colon tumor cells in G2/M phase of the cell cycle was also associated with the inhibition of histone deacetylase, hyperacetylation of H3 and H4 histones and up-regulation of mRNA and protein levels of p21, an inhibitor of cyclin-dependent kinases engaged in the progression of G1 and G2 phases of the cell cycle (Druesne et al. 2004a, b). The question as of the crucial role of p21 in DADS-induced cell cycle arrest awaits further investigation. The increase in the p21 protein level has been observed in DATS-treated PC-3 cells, however, silencing of *p21* expression through specific antisense RNA had no effect on DATS-induced G2/M cell cycle arrest (Xiao et al. 2005a).

Garlic-derived OSCs have been reported to induce programmed cell death of cancer cells and a majority of them activated the so called intrinsic (mitochondrial) apoptotic pathway, which is characterized by a loss of mitochondrial membrane potential, release of pro-apoptotic molecules, activation of caspase-9 and -3 and is regulated by the Bcl-2 family members. Numerous studies showed that treatment with garlic compounds decreases the ratio of anti-apoptotic (e.g. Bcl-2, Bcl-x_L) to pro-apoptotic (e.g. Bax, Bak, Bid) Bcl-2 family members (Table 10.4). For example, in DAS or DADS-treated H460 and H1299 lung cancer cells as well as SH-SY5Y neuroblastoma cells the ratio of Bax/Bcl-2 was increased in comparison to control cells (Hong et al. 2000; Karmakar et al. 2007). A time-dependent up-regulation of Bax protein levels with concomitant down-regulation of Bcl-x_L protein levels has been reported for DADS-treated MDA-MB-231 breast cancer cells (Nakagawa et al. 2001). Z-ajoene-induced apoptosis of HL-60 cells was associated with caspase-mediated cleavage of Bcl-2, which was inhibited by the antioxidant, NAC (Li et al. 2002b). Another study revealed an increase in production of peroxide in ajoene-treated HL-60 cells and NAC partially prevented ajoene-induced ROS generation as well as apoptosis (Dirsch et al. 1998). DATS-induced cell death of PC-3 and DU-145 prostate cancer cells, which was caused by a decrease in Bcl-2 level and its JNK-mediated hyperphosphorylation. This resulted in reduced Bcl-2:Bax interaction and activation of mitochondrial pathway of apoptosis (Xiao et al. 2004). Moreover, DATS-induced JNK activation and apoptosis were inhibited by over-expression of catalase which implies the involvement of hydrogen peroxide in apoptosis induction (Xiao et al. 2004). Other members of the Bcl-2 family have been shown to regulate OSC-induced apoptosis as well. DATS, reducing Akt activity in PC-3 and DU145 cells and consequently the phosphorylation of its substrate Bad, caused translocation of Bad to mitochondria which contributed to the activation of the intrinsic cell death pathway (Xiao and Singh 2006). DADS or garlic extract induced JNK-dependent phosphorylation of pro-apoptotic BimEL that resulted in its translocation to the mitochondria and apoptosis induction in MDA-MB-435 breast cancer cells (Lund et al. 2005).

Oxidative stress is a common phenomenon contributing to apoptosis induction by OSCs in cancer cells. For instance, DADS induced hydrogen peroxide formation and apoptosis in HL-60 leukemia (Kwon et al. 2002) and T-24 bladder cancer cells (Lu et al. 2004). DADS-induced ROS formation in SH-SY5Y neuroblastoma cells was evident as early as 15 minutes after its administration and it was followed by oxidation of cellular lipids and proteins (Filomeni et al. 2003). It was associated with JNK pathway activation. Over-expression of Cu, Zn superoxide dismutase or pretreatment with the spin trapping molecule 5,5'-dimethyl-1-pyrroline *N*-oxide protected against DADS-induced ROS generation, oxidative damage of cellular macromolecules and apoptosis. Moreover, cell permeable JNK inhibitor I protected against DADS-induced apoptosis but not G2/M cell cycle arrest (Filomeni et al. 2003). JNK1 pathway has also been activated and played a role in apoptosis induction by the water-soluble garlic compound, SAMC in SW480 and HT-29 colon cancer cells (Shirin et al. 2001; Xiao et al. 2003).

An increase in intracellular free calcium concentration is another response of cancer cells to garlic OSCs (Table 10.4). Park et al. reported biphasic DADS-induced elevation in Ca^{2+}: rapid with peak value at 3 minutes and slow and sustained elevation till 3 hours after the addition of DADS (Park et al. 2002). It was followed by an increase in hydrogen peroxide level and caspase-3 activation. Interestingly, although NAC or catalase treatment prevented the accumulation of H_2O_2 and apoptosis, they had no effect on Ca^{2+} elevation. On the other hand, cellular calcium chelator abolished the DADS-induced elevation of intracellular calcium, H_2O_2 levels and caspase-3 activation which strongly suggested that changes in calcium homeostasis might be the earliest events in DADS-induced cytotoxicity (Park et al. 2002). It has been shown that both DAS and DADS cause an increase in Ca^{2+} in SH-SY5Y cells which led to activation of calpains, the non-caspase cysteine proteases, which contribute to cell death by the induction of the mitochondrial pathway of apoptosis (Karmakar et al. 2007). DAS, DADS, and DATS induced apoptosis of human glioblastoma T98G and U87MG cells which was accompanied by an increase in cytosolic free Ca^{2+}, expression of calreticulin and activation of caspase-4 which indicates involvement of endoplasmic reticulum stress (Das et al. 2007).

Summarizing, garlic OSCs are able to increase ROS formation and/or modify cellular proteins (S-thiolation). They induce oxidative and genotoxic stress and abnormalities in cytoskeleton function in cancer cells which lead to G2 and M checkpoint activation, disruption of calcium homeostasis and induction of the mitochondrial pathway of apoptosis.

In vitro studies demonstrate anti-angiogenic and anti-metastasis properties of garlic preparations. For instance, AGE reduced invasive potential and motility of human and rat endothelial cells by 30% in matrigel chemo-invasion assays as well as capillary-like tube formation in 3-dimensional collagen matrix assays (Matsuura et al.

2006). DATS decreased capillary-like tube formation and migration of human umbilical vein endothelial cells and at the molecular level it was correlated with Akt inactivation and reduced level of VEGF receptor 2 and secretion of VEGF (Xiao et al. 2006b). Also DADS and DAS inhibited endothelial cell proliferation and migration reducing matrix metalloproteinases MMP-2 and MMP-9 (Thejass and Kuttan 2007a, b). Water soluble SAC inhibited MDA-MB-231 breast cancer cell motility and invasion, which was accompanied by increased expression of E-cadherin and reduced MMP-2 level and activity (Gapter et al. 2008).

10.6 Antioxidant and Pro-oxidant Activity of Garlic and OSCs

There are several reports demonstrating that garlic exerts direct or indirect antioxidant activity (Imai et al. 1994). Indirect activity is related to the ability of garlic to stimulate the activity of glutathione peroxidase (GPx), glutathione reductase and superoxide dismutase in different experimental models. Moreover, AGE, SAC and SAMC exhibited direct radical scavenging activity. In addition, it was found that AGE had the capacity to chelate copper ions and inhibit copper dependent lipid peroxidation (Dillon et al. 2003). Dietary supplementation with AGE for 14 days reduced plasma and urine concentrations of 8-iso-prostaglandin F (2 alpha) by 29% and 37% in non-smokers and by 35% and 48% in smokers (Dillon et al. 2002). There are several other published data demonstrating antioxidant activity of garlic and garlic preparations, however, as mentioned below, garlic can exert some adverse effects, which could be related to oxidative stress. There is an increasing amount of evidence that depending on the experimental model, the concentration of the compounds and other possible factors, garlic preparations act either as antioxidants or pro-oxidants. Filomeni et al. (2003) demonstrated that DADS increased ROS formation in cancer cells. In addition, DATS-induced prostate cancer cell death has been related to increased ROS formation (Xiao et al. 2004). Protective effects of over-expressed catalase or low molecular weight antioxidants against DATS-induced cell death support the role of ROS in cancer cell death. The dual role of antioxidant compounds exhibiting pro-oxidant activity is quite puzzling. In the case of the pro-oxidant activity of garlic our knowledge still is not complete. Interestingly, it was observed that in genetically modified prostate cancer cells, in a way that they could not activate c-jun kinases (JNK), DATS acted as an antioxidant. These data suggested that DATS-induced ROS formation is JNK-dependent. In fact, it was observed that JNK activation led to ferritin degradation (Antosiewicz et al. 2006). Ferritin is a multisubunit protein, whose main function is sequestering iron. Ferritin degradation leads to liberation of iron which in reactions with hydrogen peroxide, lipid hydroperoxides and other compounds can generate several radical species including hydroxyl radical. DATS-induced ferritin degradation was accompanied by an increase in labile iron pool (LIP). An iron chelator completely abolished DATS-induced ROS formation which suggested that it is an iron-dependent process (Antosiewicz et al. 2006).

The exact mechanism of JNK-stimulated ferritin degradation is not known. Interestingly, overproduction of ROS leads to the activation of JNK and there is some evidence that DATS-induced JNK activation is ROS-dependent. These data suggest that DATS, and possible other OSCs, in addition to iron-dependent ROS, stimulate formation of other free radical species. On the other hand it was shown that DADS treatment increased the ferritin H protein levels in the liver (Thomas et al. 2002). Altogether, these data indicate that OSCs might act as pro-oxidants by increasing ferritin degradation and LIP or as antioxidants when ferritin levels are increased.

10.7 Adverse Effects of Garlic and Its Interactions with Drugs

Considering the fact that garlic has been consumed by humans for centuries, it is generally accepted to be a safe food. However, few studies reported on some toxic effects of garlic. Some of these effects are possibly related to its pro-oxidant activity. Gastrointestinal irritation may occur after the consumption of raw garlic, fresh garlic juice, dehydrated raw garlic powder or even after the intake of garlic tablets (Hoshino et al. 2001; Borrelli et al. 2007). Clinical studies have shown that garlic at therapeutic doses may cause mild gastric mucosa damage and around 6% of patients complained of nausea and 0.8% of bloating (Borrelli et al. 2007). Experiments performed on animals also showed several adverse effects of garlic. Rats that were fed with high amounts of raw garlic, DATS or DADS developed anaemia and demonstrated lysis of red blood cells (Munday et al. 2003). Garlic extract on one hand has been shown to reduce the clastogenic effect of some toxins in mice, on the other hand at higher concentrations it induced a clastogenic effect (Das et al. 1996). Studies performed on dogs showed that raw garlic powder caused severe damage, including erosion of gastric mucosa. Dehydrated, boiled garlic powder also caused reddening of the mucosa, whereas AGE did not cause any undesirable effects (Hoshino et al. 2001). In rats raw garlic juice at a dose of 5 ml/kg resulted in the death of some animals due to stomach injury after 21 days (Nakagawa et al. 1980). The surviving rats exhibited swelling of the liver, hypertrophy of the spleen and adrenal glands, and the decrease of erythrocytes and various morphological changes were clearly observed after 3 and 8 days in the group administered with high doses of raw garlic juice. Such changes were not observed at any time with the use of extracted-aged garlic juice (Nakagawa et al. 1980). Similar observations were done on animals treated with aqueous garlic extract (200 g/L drinking water) for 10 days, where liver injury and local inflammatory response was observed (Joseph et al. 1989). AGE has been reported to protect against heat-induced inhibition of spermatogenesis in rats whereas the administration of garlic powder (50 mg/day) led to the inhibition of spermatogenesis in rats and decreased Leydig cell function (Dixit and Joshi 1982; Hammami et al. 2008).

Despite several studies demonstrating adverse effects of garlic it can be concluded that it is a relatively safe food component. In most animal studies a very high dose of garlic was applied, which is not normally consumed. In addition, it can be assumed that consumption of garlic with other food might reduce eventual toxic effect of the former one.

The biological effects of garlic are also related to the xenobiotic metabolism. The process of detoxification of pro-cancerogens consists of two phases. Phase I, catalyzed by the isoforms of P450 cytochromes, relies on reactions of hydroxylation, oxidation, hydrolysis, demethylation and some others yielding modified xenobiotics. Unfortunately, some products of Phase I reactions could be more toxic than parent compounds. Reactions catalyzed by Phase II detoxification enzymes promote the elimination of drugs and carcinogens from the body. Therefore, the activity of Phase II enzymes such as glutation S-transferase (GST), epoxide hydrolase, quinone oxidoreductase (NQO1) and glucuronate transferase may prevent cancerogenesis by increasing the clearance rate of toxic chemicals. The main goal of both phases of xenobiotic metabolism is to increase polarity, solubility and their subsequent excretion.

In the case of Phase II enzymes the data are consistent and clearly show that garlic as well as garlic derived OSCs increase their activity. For example, Chen et al. (2004) reported that the gene expression of NQO1 and heme oxygenase increased upon treatment with DADS or DATS. Many reports showed that OSCs increased GST activity and that allyl derivatives are more potent inducers of GST than propyl-containing OSCs (Hu et al. 1997).

The effect of OSCs on Phase I enzymes is controversial. For example, Davenport and Wargovich demonstrated that in rats, gastric intubation with a single dose of 200 mg/kg DAS, DADS or allyl methyl sulfide (AMS) decreased hepatic P450–2E1 protein by 45, 25, and 47%, respectively. On the other hand, DAS and AMS increased hepatic P450–1A2 protein levels by 282 and 70%, and DAS increased P450–1A1 protein levels by 684% (Davenport and Wargovich 2005). 1,2-dimethylhydrazine is a carcinogen activated by P450–2E1, therefore inhibition of its activity by OSCs seems to be the main mechanism of chemo-prevention (Wargovich 1987). Interestingly, it was observed that mice lacking expression of *CYP1A1, CYP1B1, CYP1A2,* and *CYP2E1* do not differ from wild-type mice. However, the *CYP* null mice have altered responses to the toxic and carcinogenic effects of chemicals as compared with wild-type mice (Gonzalez and Kimura 2003). Therefore, it can be speculated that increased activity of some P450 isoenzymes induced by OSCs could elevate toxicity of some xenobiotics but on the other hand can be also protective. Induction of Phase I and II enzymes by garlic may increase clearness of some chemotherapeutics. For example, garlic lowered the serum level of saquinavir, which is an anti-HIV drug, by as much as 50% (Piscitelli et al. 2002).

10.8 Other Therapeutic Applications

The therapeutic use of antitumor drugs has been limited by their toxicity. Treated patients often experience side effects such as nausea, vomiting, diarrhea, stomatitis, gastrointestinal ulceration and mucositis. For example, methotrexate (MTX), is an anticancer drug that has been shown to be highly effective against various kinds of cancers, however, it is highly toxic to the rapidly dividing cells of intestinal crypts. Interestingly, weight loss in rats treated with MTX and severity of jejunal damage was reduced by provision of AGE. Moreover, experiments performed on intestinal epithelial, IEC-6 cells showed that AGE almost completely suppressed MTX-induced cell death (Horie et al. 2001). These data suggest that AGE could be a supplement of choice for patients experiencing adverse effects of chemotherapy.

The mechanism of anticancer activity of garlic is far from being completely understood, however, potentiating the immune system could be a very important one. Decline in immune function is observed in advanced cancer patients and AGE has been shown to prevent the decline of NK cell number and activity in these patients (Ishikawa et al. 2006). Furthermore, AGE stimulated immune functions, such as proliferation of lymphocyte, cytokine release, NK activity and phagocytosis. Moreover, a protein fraction from garlic has been demonstrated to enhance the cytotoxicity of human peripheral blood lymphocytes against M14 melanoma cells (Morioka et al. 1993).

10.9 Summary and Conclusions

Inhibition of cancer development is a remarkable property of garlic and its constituents documented in a large body of research showing that OSCs suppress each stage of the carcinogenesis process (Fig. 10.2). The most convincing data come from

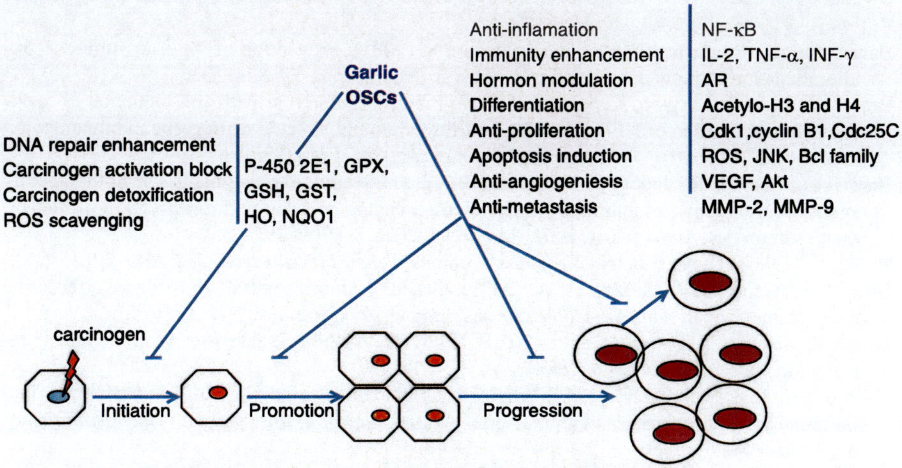

Fig. 10.2 Processes modulated by garlic and garlic-derived OSCs and their molecular targets in relation to multistage cancer development. AR: androgen receptor, Cdk1: cyclin dependent kinase 1, Cdc25C: cell division cycle 25C phosphatase, GPX: glutathione peroxidase, GSH: glutathione, GST: glutathione transferase, HO: heme oxygenase, IL-2: interleukin 2, INF-γ: interferon γ, JNK: c-jun kinase, MMP: metalloproteinase, NF-κB: nuclear factor κ B, NQO1: quinone oxidoreductase 1, ROS: reactive oxygen species, TNF-α: tumor necrosis factor α, VEGF: vascular endothelial growth factor

pre-clinical studies which demonstrate preventive activity of OSCs in chemically induced cancers in animal models as well as the ability of garlic constituents to inhibit cancer cell division, survival and metastasis both *in vitro* (cell cultures) and *in vivo* (xenograft models). However, epidemiological data and intervention trials in humans are still scarce or lacking. Thus, in order to use garlic constituents/preparations for cancer patients, additional efforts are required to: (1) assess clinical potential of OSCs in carefully designed clinical trials; (2) determine plasma and tissue concentration of OSCs in humans to relate it to effective anticancer concentrations used *in vitro*; (3) evaluate toxic amount of OSCs and their side effects on normal tissue.

References

Antosiewicz, J., Herman-Antosiewicz, A., Marynowski, S. W., et al. (2006). c-Jun NH(2)-terminal kinase signaling axis regulates diallyl trisulfide-induced generation of reactive oxygen species and cell cycle arrest in human prostate cancer cells. *Cancer Research, 66,* 5379–5386.

Aquilano, K., Filomeni, G., Baldelli, S., et al. (2007). Neuronal nitric oxide synthase protects neuroblastoma cells from oxidative stress mediated by garlic derivatives. *Journal of Neurochemistry, 101,* 1327–1337.

Arunkumar, A., Vijayababu, M. R., Gunadharini, N., et al. (2007). Induction of apoptosis and histone hyperacetylation by diallyl disulfide in prostate cancer cell line PC-3. *Cancer Letters, 251,* 59–67.

Arunkumar, A., Vijayababu, M. R., Srinivasan, N., et al. (2006a). Garlic compound, diallyl disulfide induces cell cycle arrest in prostate cancer cell line PC-3. *Molecular and Cellular Biochemistry, 288,* 107–113.

Arunkumar, A., Vijayababu, M. R., Venkataraman P, et al. (2006b). Chemoprevention of rat prostate carcinogenesis by diallyl disulfide, an organosulfur compound of garlic. *Biological & Pharmaceutical Bulletin, 29,* 375–379.

Balasenthil, S., Ramachandran, C. R., & Nagini, S. (2001). Prevention of 4-nitroquinoline 1-oxide-induced rat tongue carcinogenesis by garlic. *Fitoterapia, 72,* 524–531.

Ban, J. O., Yuk, D. Y., Woo, K. S., et al. (2007). Inhibition of cell growth and induction of apoptosis via inactivation of NF-kappaB by a sulfurcompound isolated from garlic in human colon cancer cells. *Journal of Pharmacological Sciences, 104,* 374–383.

Ban, J. O., Lee, H. S., Jeong, H. S., et al. (2009). Thiacremonone augments chemotherapeutic agent-induced growth inhibition in human colon cancer cells through inactivation of nuclear factor-{kappa}B. *Molecular Cancer Research: MCR, 7,* 870–879.

Block, E. (1985). The chemistry of garlic and onions. *Scientific American, 252,* 114–119.

Borrelli, F., Capasso, R., & Izzo, A. A. (2007). Garlic (Allium sativum L.): Adverse effects and drug interactions in humans. *Molecular Nutrition & Food Research, 51,* 1386–1397.

Busch, C., Jacob, C., Anwar, A., et al. (2010). Diallylpolysulfides induce growth arrest and apoptosis. *International Journal of Oncology, 36,* 743–749.

Cavallito, C. J., & Bailey, J. H. (1944). Alicin, the antibacterial principle of Allium sativum. I. Isolation, physical properties and antibacterial action. *Journal of the American Chemical Society, 66,* 1950–1951.

Challier, B., Perarnau, J. M., & Viel, J. F. (1998). Garlic, onion and cereal fibre as protective factors for breast cancer: A French case-control study. *European Journal of Epidemiology, 14,* 737–747.

Chen, C., Pung, D., Leong, V., et al. (2004). Induction of detoxifying enzymes by garlic organosulfur compounds through transcription factor Nrf2: Effect of chemical structure and stress signals. *Free Radical Biology & Medicine, 37,* 1578–1590.

Cheng, J. Y., Meng, C. L., Tzeng, C. C., et al. (1995). Optimal dose of garlic to inhibit dimethylhydrazine-induced colon cancer. *World Journal of Surgery, 19,* 621–655 (discussion 25–6).

Chu, Q., Lee, D. T., Tsao, S. W., et al. (2007). S-allylcysteine, a water-soluble garlic derivative, suppresses the growth of a human androgen-independent prostate cancer xenograft, CWR22R, under in vivo conditions. *BJU International, 99,* 925–932.

Cohen, L. A., Zhao, Z., Pittman, B., et al. (1999). S-allylcysteine, a garlic constituent, fails to inhibit N-methylnitrosourea-induced rat mammary tumorigenesis. *Nutrition and Cancer, 35,* 58–63.

Das, A., Banik, N. L., & Ray, S. K. (2007). Garlic compounds generate reactive oxygen species leading to activation of stress kinases and cysteine proteases for apoptosis in human glioblastoma T98G and U87MG cells. *Cancer, 110,* 1083–1095.

Das, T., Choudhury, A. R., Sharma, A., et al. (1996). Effects of crude garlic extract on mouse chromosomes in vivo. *Food and Chemical Toxicology: An International Journal Published for the British Industrial Biological Research Association, 34,* 43–47.

Davenport, D. M., & Wargovich, M. J. (2005). Modulation of cytochrome P450 enzymes by organosulfur compounds from garlic. *Food and Chemical Toxicology: An International Journal Published for the British Industrial Biological Research Association, 43,* 1753–1762.

Dillon, S. A., Lowe, G. M., Billington, D., et al. (2002). Dietary supplementation with aged garlic extract reduces plasma and urine concentrations of 8-iso-prostaglandin F(2 alpha) in smoking and nonsmoking men and women. *The Journal of Nutrition, 132,* 168–171.

Dillon, S. A., Burmi, R. S., Lowe, G. M., et al. (2003). Antioxidant properties of aged garlic extract: An in vitro study incorporating human low density lipoprotein. *Life Sciences, 72,* 1583–1594.

Dirsch, V. M., Gerbes, A. L., & Vollmar, A. M. (1998). Ajoene, a compound of garlic, induces apoptosis in human promyeloleukemic cells, accompanied by generation of reactive oxygen species and activation of nuclear factor kappaB. *Molecular Pharmacology, 53,* 402–407.

Dirsch, V. M., Antlsperger, D. S., Hentze, H., et al. (2002). Ajoene, an experimental anti-leukemic drug: Mechanism of cell death. *Leukemia, 16,* 74–83.

Dixit, V. P., & Joshi, S. (1982). Effects of chronic administration of garlic (Allium sativum Linn) on testicular function. *Indian Journal of Experimental Biology, 20,* 534–536.

Dorant, E., van den Brandt, P. A., & Goldbohm, R. A. (1994) A prospective cohort study on Allium vegetable consumption, garlic supplement use, and the risk of lung carcinoma in The Netherlands. *Cancer Research, 54,* 6148–6153.

Dorant, E., van den Brandt, P. A., & Goldbohm, R. A. (1995). Allium vegetable consumption, garlic supplement intake, and female breast carcinoma incidence. *Breast Cancer Research and Treatment, 33,* 163–170.

Dorant, E., van den Brandt, P. A., & Goldbohm, R. A. (1996a). A prospective cohort study on the relationship between onion and leek consumption, garlic supplement use and the risk of colorectal carcinoma in The Netherlands. *Carcinogenesis, 17,* 477–484.

Dorant, E., van den Brandt, P. A., & Goldbohm, R. A., et al. (1996b). Consumption of onions and a reduced risk of stomach carcinoma. *Gastroenterol, 110,* 12–20.

Druesne, N., Pagniez, A., Mayeur, C., et al. (2004a). Diallyl disulfide (DADS) increases histone acetylation and p21(waf1/cip1) expression in human colon tumor cell lines. *Carcinogenesis, 25,* 1227–1236.

Druesne, N., Pagniez, A., Mayeur, C., et al. (2004b). Repetitive treatments of colon HT-29 cells with diallyl disulfide induce a prolonged hyperacetylation of histone H3 K14. *Annals of the New York Academy of Sciences, 1030,* 612–621.

Dwivedi, C., Rohlfs, S., Jarvis, D., et al. (1992). Chemoprevention of chemically induced skin tumor development by diallyl sulfide and diallyl disulfide. *Pharmaceutical Research, 9,* 1668–1670.

Filomeni, G., Aquilano, K., Rotilio, G., et al. (2003). Reactive oxygen species-dependent c-Jun NH2-terminal kinase/c-Jun signaling cascade mediates neuroblastoma cell death induced by diallyl disulfide. *Cancer Research, 63,* 5940–5949.

Galeone, C., Pelucchi, C., Levi, F., et al. (2006). Onion and garlic use and human cancer. *The American Journal of Clinical Nutrition, 84,* 1027–1032.

Gao, C. M., Takezaki, T., Ding, J. H., et al. (1999). Protective effect of allium vegetables against both esophageal and stomach cancer: A simultaneous case-referent study of a high-epidemic area in Jiangsu Province, China. *Japanese Journal of Cancer Research: Gann, 90,* 614–621.

Gapter, L. A., Yuin, O. Z., & Ng, K. Y. (2008). S-Allylcysteine reduces breast tumor cell adhesion and invasion. *Biochemical and Biophysical Research Communications, 367,* 446–451.

Giovannucci, E., Rimm, E. B., Stampfer, M. J., et al. (1994). Intake of fat, meat, and fiber in relation to risk of colon cancer in men. *Cancer Research, 54,* 2390–2397.

Gonzalez, F. J., & Kimura, S. (2003). Study of P450 function using gene knockout and transgenic mice. *Archives of Biochemistry and Biophysics, 409,* 153–158.

Ha, M. W., Ma, R., Shun, L. P., et al. (2005). Effects of allitridi on cell cycle arrest of human gastric cancer cells. *World Journal of Gastroenterology: WJG, 11,* 5433–5437.

Hammami, I., Nahdi, A., Mauduit, C., et al. (2008). The inhibitory effects on adult male reproductive functions of crude garlic (Allium sativum) feeding. *Asian Journal of Andrology, 10,* 593–601.

Hansson, L. E., Nyren, O., Bergstrom, R., et al. (1993). Diet and risk of gastric cancer. A population-based case-control study in Sweden. *International Journal of Cancer. Journal International du Cancer, 55,* 181–189.

Hayes, M. A, Rushmore, T. H., & Goldberg, M. T. (1987). Inhibition of hepatocarcinogenic responses to 1,2-dimethylhydrazine by diallyl sulfide, a component of garlic oil. *Carcinogenesis, 8,* 1155–1157.

Herman-Antosiewicz, A., & Singh, S. V. (2005). Checkpoint kinase 1 regulates diallyl trisulfide-induced mitotic arrest in human prostate cancer cells. *The Journal of Biological Chemistry, 280,* 28519–28528.

Herman-Antosiewicz, A., Stan, S. D., Hahm. E. R., et al. (2007). Activation of a novel ataxia-telangiectasia mutated and Rad3 related/checkpoint kinase 1-dependent prometaphase checkpoint in cancer cells by diallyl trisulfide, a promising cancer chemopreventive constituent of processed garlic. *Molecular Cancer Therapeutics, 6,* 1249–1261.

Herman-Antosiewicz, A., Kim, Y. A., Kim, S. H., et al. (2010). Diallyl trisulfide-induced G2/M phase cell cycle arrest in DU145 cells is associated with delayed nuclear translocation of cyclin-dependent kinase 1. *Pharmaceutical Research, 27,* 1072–1079.

Hong, Y. S., Ham, Y. A., Choi, J. H., et al. (2000). Effects of allyl sulfur compounds and garlic extract on the expression of Bcl-2, Bax, and p53 in non small cell lung cancer cell lines. *Experimental & Molecular Medicine, 32,* 127–134.

Horie, T., Awazu, S., Itakura, Y., et al. (2001). Alleviation by garlic of antitumor drug-induced damage to the intestine. *The Journal of Nutrition, 131,* 1071S–1074S.

Hoshino, T., Kashimoto, N., & Kasuga, S. (2001). Effects of garlic preparations on the gastrointestinal mucosa. *The Journal of Nutrition, 131,* 1109S–11013S.

Hosono, T., Fukao, T., Ogihara, J., et al. (2005). Diallyl trisulfide suppresses the proliferation and induces apoptosis of human colon cancer cells through oxidative modification of beta-tubulin. *The Journal of Biological Chemistry, 280,* 41487–41493.

Hosono, T., Hosono-Fukao, T., Inada, K., et al. (2008). Alkenyl group is responsible for the disruption of microtubule network formation in human colon cancer cell line HT 29 cells. *Carcinogenesis, 29,* 1400–1406.

Howard, E. W., Ling ,M. T., Chua, C. W., et al. (2007). Garlic-derived S-allylmercaptocysteine is a novel in vivo antimetastatic agent for androgen-independent prostate cancer. *Clinical Cancer Research: An Official Journal of the American Association for Cancer Research, 13,* 1847–1856.

Howard, E. W., Lee, D. T., Chiu, Y. T., et al. (2008). Evidence of a novel docetaxel sensitizer, garlic-derived S-allylmercaptocysteine, as a treatment option for hormone refractory prostate cancer. *International Journal of Cancer. Journal International du Cancer, 122,* 1941–1948.

Hsing, A. W., Chokkalingam, A. P., Gao, Y. T., et al. (2002). Allium vegetables and risk of prostate cancer: A population-based study. *Journal of the National Cancer Institute, 94,* 1648–1651.

Hu, J., Nyren, O., Wolk, A., et al. (1994). Risk factors for oesophageal cancer in northeast China. *International Journal of Cancer. Journal International du Cancer, 57,* 38–46.

Hu, X., Benson, P. J., Srivastava, S. K., et al. (1997). Induction of glutathione S-transferase pi as a bioassay for the evaluation of potency of inhibitors of benzo(a)pyrene-induced cancer in a murine model. *International Journal of Cancer. Journal International du Cancer, 73,* 897–902.

Imai, J., Ide, N., Nagae, S., et al. (1994). Antioxidant and radical scavenging effects of aged garlic extract and its constituents. *Planta Medica, 60,* 417–420.

Ip, C., Lisk, D. J., & Stoewsand, G. S. (1992). Mammary cancer prevention by regular garlic and selenium-enriched garlic. *Nutrition and Cancer, 17,* 279–286.

Ishikawa, H,. Saeki, T., Otani, T., et al. (2006). Aged garlic extract prevents a decline of NK cell number and activity in patients with advanced cancer. *The Journal of Nutrition, 136,* 816S–820S.

Joseph, P. K., Rao, K. R., & Sundaresh, C. S. (1989). Toxic effects of garlic extract and garlic oil in rats. *Indian Journal of Experimental Biology, 27,* 977–979.

Karmakar, S., Banik, N. L., Patel, S. J., et al. (2007). Garlic compounds induced calpain and intrinsic caspase cascade for apoptosis in human malignant neuroblastoma SH-SY5Y cells. *Apoptosis, 12,* 671–684.

Key, T. J., Silcocks, P. B., Davey, G. K., et al. (1997). A case-control study of diet and prostate cancer. *British Journal of Cancer, 76,* 678–687.

Kim, J. Y., & Kwon, O. (2009). Garlic intake and cancer risk: An analysis using the Food and Drug Administration's evidence-based review system for the scientific evaluation of health claims. *The American Journal of Clinical Nutrition, 89,* 257–264.

Kim, H. J., Chang, W. K., Kim, M. K., et al. (2002) Dietary factors and gastric cancer in Korea: A case-control study. *International Journal of Cancer. Journal International du Cancer, 97,* 531–535.

Kirsh, V. A., Peters, U., Mayne, S. T., et al. (2007). Prospective study of fruit and vegetable intake and risk of prostate cancer. *Journal of the National Cancer Institute, 99,* 1200–1209.

Knowles, L. M., & Milner, J. A. (1998). Depressed p34cdc2 kinase activity and G2/M phase arrest induced by diallyl disulfide in HCT-15 cells. *Nutrition and Cancer, 30,* 169–174.

Knowles, L. M., & Milner, J. A. (2000). Diallyl disulfide inhibits p34(cdc2) kinase activity through changes in complex formation and phosphorylation. *Carcinogenesis, 21,* 1129–1134.

Kwon, K. B., Yoo, S. J., Ryu, D. G., et al. (2002). Induction of apoptosis by diallyl disulfide through activation of caspase-3 in human leukemia HL-60 cells. *Biochemical Pharmacology, 63,* 41–47.

Lan, H., & Lu, Y. Y. (2004). Allitridi induces apoptosis by affecting Bcl-2 expression and caspase-3 activity in human gastric cancer cells. *Acta Pharmacologica Sinica, 25,* 219–225.

Lawson, L. D., & Wang, Z. J. (2005). Allicin and allicin-derived garlic compounds increase breath acetone through allyl methyl sulfide: Use in measuring allicin bioavailability. *Journal of Agricultural and Food Chemistry, 53,* 1974–1983.

Lawson, L. D., Wang, Z. J., & Hughes, B. G. (1991). Identification and HPLC quantitation of the sulfides and dialk(en)yl thiosulfinates in commercial garlic products. *Planta Medica, 57,* 363–370.

Le Marchand, L., Hankin, J. H., Wilkens, L. R., et al. (1997). Dietary fiber and colorectal cancer risk. *Epidemiology (Cambridge, Mass.), 8,* 658–665.

Lea, M. A, Randolph, V. M., & Patel, M. (1999). Increased acetylation of histones induced by diallyl disulfide and structurally related molecules. *International Journal of Oncology, 15,* 347–352.

Lea, M. A., Rasheed, M., Randolph, V. M., et al. (2002). Induction of histone acetylation and inhibition of growth of mouse erythroleukemia cells by S-allylmercaptocysteine. *Nutrition and Cancer, 43,* 90–102.

Ledezma, E., Apitz-Castro, R., Cardier, J. (2004). Apoptotic and anti-adhesion effect of ajoene, a garlic derived compound, on the murine melanoma B16F10 cells: Possible role of caspase-3 and the alpha(4)beta(1) integrin. *Cancer Letters, 206,* 35–41.

Lee, Y. (2008). Induction of apoptosis by S-allylmercapto-L-cysteine, a biotransformed garlic derivative, on a human gastric cancer cell line. *International Journal of Molecular Medicine, 21,* 765–770.

Levi, F., Pasche, C., La Vecchia, C., et al. (1999). Food groups and colorectal cancer risk. *British Journal of Cancer, 79,* 1283–1287.

Li, H., Li, H. Q., Wang, Y., et al. (2004). An intervention study to prevent gastric cancer by micro-selenium and large dose of allitridum. *Chinese Medical Journal, 117,* 1155–1160.

Li, M., Ciu, J. R., Ye, Y., et al. (2002a). Antitumor activity of Z-ajoene, a natural compound purified from garlic: Antimitotic and microtubule-interaction properties. *Carcinogenesis, 23,* 573–579.

Li, M., Min, J. M., Cui, J. R., et al. (2002b). Z-ajoene induces apoptosis of HL-60 cells: Involvement of Bcl-2 cleavage. *Nutrition and Cancer, 42,* 241–247.

Lin, Y. T., Yang, J. S., Lin, S. Y., et al. (2008). Diallyl disulfide (DADS) induces apoptosis in human cervical cancer Ca Ski cells via reactive oxygen species and Ca2+-dependent mitochondria-dependent pathway. *Anticancer Research, 28,* 2791–2799.

Liu, J., Lin, R. I., & Milner, J. A. (1992). Inhibition of 7,12-dimethylbenz[a]anthracene-induced mammary tumors and DNA adducts by garlic powder. *Carcinogenesis, 13,* 1847–1851.

Lu, H. F., Sue, C. C., Yu, C. S., et al. (2004). Diallyl disulfide (DADS) induced apoptosis undergo caspase-3 activity in human bladder cancer T24 cells. *Food and Chemical Toxicology: An International Journal Published for the British Industrial Biological Research Association, 42,* 1543–1552.

Lund, T., Stokke, T., Olsen, O. E., et al. (2005). Garlic arrests MDA-MB-435 cancer cells in mitosis, phosphorylates the proapoptotic BH3-only protein BimEL and induces apoptosis. *British Journal of Cancer, 92,* 1773–1781.

Malki, A., El-Saadani, M., & Sultan, A. S. (2009). Garlic constituent diallyl trisulfide induced apoptosis in MCF7 human breast cancer cells. *Cancer Biology & Therapy, 8,* 2175–2185.

Matsuura, N., Miyamae, Y., Yamane, K., et al. (2006). Aged garlic extract inhibits angiogenesis and proliferation of colorectal carcinoma cells. *The Journal of Nutrition, 136,* 842S–846S.

Morioka, N., Sze, L. L., Morton, D. L., et al. (1993). A protein fraction from aged garlic extract enhances cytotoxicity and proliferation of human lymphocytes mediated by interleukin-2 and concanavalin A. *Cancer Immunology, Immunotherapy: CII, 37,* 316–322.

Munday, R., Munday, J. S., & Munday, C. M. (2003). Comparative effects of mono-, di-, tri-, and tetrasulfides derived from plants of the Allium family: Redox cycling in vitro and hemolytic activity and Phase 2 enzyme induction in vivo. *Free Radical Biology & Medicine, 34,* 1200–1211.

Nakagawa, H., Tsuta, K., Kiuchi, K., et al. (2001). Growth inhibitory effects of diallyl disulfide on human breast cancer cell lines. *Carcinogenesis, 22,* 891–897.

Nakagawa, S., Masamoto, K., Sumiyoshi, H., et al. (1980). Effect of raw and extracted-aged garlic juice on growth of young rats and their organs after peroral administration (author's transl). *The Journal of Toxicological Sciences, 5,* 91–112.

Nishino, H., Iwashima, A., Itakura, Y., et al. (1989). Antitumor-promoting activity of garlic extracts. *Oncology, 46,* 277–280.

Park, E. K., Kwon, K. B., Park, K. I., et al. (2002). Role of Ca(2+) in diallyl disulfide-induced apoptotic cell death of HCT-15 cells. *Experimental & Molecular Medicine, 34,* 250–257.

Peng, C. Y., Graves, P. R., Thoma, R. S., et al. (1997). Mitotic and G2 checkpoint control. Regulation of 14-3-3 protein binding by phosphorylation of Cdc25C on serine-216. *Science, 277,* 1501–1505.

Piscitelli, S. C., Burstein, A. H., Welden, N., et al. (2002). The effect of garlic supplements on the pharmacokinetics of saquinavir. *Clinical Infectious Diseases: An Official Publication of the Infectious Diseases Society of America 2002, 34,* 234–238.

Reddy, B. S., Rao, C. V., Rivenson, A., et al. (1993). Chemoprevention of colon carcinogenesis by organosulfur compounds. *Cancer Research, 53,* 3493–3498.

Rivlin, R. S. (2001). Historical perspective on the use of garlic. *The Journal of Nutrition, 131,* 951S–954S.

Sadhana, A. S., Rao, A. R., Kucheria, K., et al. (1988). Inhibitory action of garlic oil on the initiation of benzo[a]pyrene-induced skin carcinogenesis in mice. *Cancer Letters, 40,* 193–197.

Sakamoto, K., Lawson, L. D., Milner, J. A. (1997). Allyl sulfides from garlic suppress the in vitro proliferation of human A549 lung tumor cells. *Nutrition and Cancer, 29,* 152–156.

Sanchez, Y., Wong, C., Thoma, R. S., et al. (1997). Conservation of the Chk1 checkpoint pathway in mammals: Linkage of DNA damage to Cdk regulation through Cdc25. *Science, 277,* 1497–1501.

Schaffer, E. M., Liu, J. Z., Green, J., et al. (1996). Garlic and associated allyl sulfur components inhibit N-methyl-N-nitrosourea induced rat mammary carcinogenesis. *Cancer Letters, 102,* 199–204.

Shirin, H., Pinto, J. T., Kawabata, Y., et al. (2001). Antiproliferative effects of S-allylmercaptocysteine on colon cancer cells when tested alone or in combination with sulindac sulfide. *Cancer research, 61,* 725–731.

Sigounas, G., Hooker, J. L., Li, W., et al. (1997). S-allylmercaptocysteine, a stable thioallyl compound, induces apoptosis in erythroleukemia cell lines. *Nutrition and Cancer, 28,* 153–159.

Singh, A., Arora, A., & Shukla, Y. (2004). Modulation of altered hepatic foci induction by diallyl sulphide in Wistar rats. *European Journal Of Cancer Prevention: The Official Journal of the European Cancer Prevention Organisation (ECP), 13,* 263–269.

Singh, A., & Shukla, Y. (1998). Antitumor activity of diallyl sulfide in two-stage mouse skin model of carcinogenesis. *Biomedical and Environmental Sciences: BES, 11,* 258–263.

Singh, S. V., Mohan, R. R., Agarwal, R., et al. (1996). Novel anti-carcinogenic activity of an organosulfide from garlic: Inhibition of H-RAS oncogene transformed tumor growth in vivo by diallyl disulfide is associated with inhibition of p21H-ras processing. *Biochemical and Biophysical Research Communications, 225,* 660–665.

Song, J. D., Lee, S. K., Kim, K. M., et al. (2009). Molecular mechanism of diallyl disulfide in cell cycle arrest and apoptosis in HCT-116 colon cancer cells. *Journal of Biochemical and Molecular Toxicology, 23,* 71–79.

Sparnins, V. L., Mott, A. W., Barany, G., et al. (1986). Effects of allyl methyl trisulfide on glutathione S-transferase activity and BP-induced neoplasia in the mouse. *Nutrition and Cancer, 8,* 211–215.

Sparnins, V. L., Barany, G., & Wattenberg, L. W. (1988). Effects of organosulfur compounds from garlic and onions on benzo[a]pyrene-induced neoplasia and glutathione S-transferase activity in the mouse. *Carcinogenesis, 9,* 131–134.

Steinmetz, K. A., Kushi, L. H., Bostick, R. M., et al. (1994). Vegetables, fruit, and colon cancer in the Iowa Women's Health Study. *American Journal of Epidemiology, 139,* 1–15.

Sumioka, I., Matsura, T., Kasuga, S., et al. (1998). Mechanisms of protection by S-allylmercaptocysteine against acetaminophen-induced liver injury in mice. *Japanese Journal of Pharmacology, 78,* 199–207.

Sumiyoshi, H., & Wargovich, M. J. (1990). Chemoprevention of 1,2-dimethylhydrazine-induced colon cancer in mice by naturally occurring organosulfur compounds. *Cancer Research, 50,* 5084–5087.

Sundaram, S. G., & Milner, J. A. (1996a). Diallyl disulfide induces apoptosis of human colon tumor cells. *Carcinogenesis, 17,* 669–673.

Sundaram, S. G., & Milner, J. A. (1996b). Diallyl disulfide inhibits the proliferation of human tumor cells in culture. *Biochimica et Biophysica Acta, 1315,* 15–20.

Sundaram, S. G., & Milner, J. A. (1996c). Diallyl disulfide suppresses the growth of human colon tumor cell xenografts in athymic nude mice. *The Journal of Nutrition, 126,* 1355–1361.

Suzui, N., Sugie, S., Rahman, K. M., et al. (1997). Inhibitory effects of diallyl disulfide or aspirin on 2-amino-1-methyl-6-phenylimidazo[4,5-b]pyridine-induced mammary carcinogenesis in rats. *Japanese Journal of Pharmacology, 88,* 705–711.

Takahashi, S., Hakoi, K., Yada, H., et al. (1992). Enhancing effects of diallyl sulfide on hepatocarcinogenesis and inhibitory actions of the related diallyl disulfide on colon and renal carcinogenesis in rats. *Carcinogenesis, 13,* 1513–1518.

Takezaki, T., Gao, C. M., Ding, J. H., et al. (1999). Comparative study of lifestyles of residents in high and low risk areas for gastric cancer in Jiangsu Province, China; with special reference to allium vegetables. *Journal of Epidemiology/Japan Epidemiological Association, 9,* 297–305.

Tan, L. M., Zhang, M. X., Luo, H. M., et al. (2004). The initiation of G2/M checkpoint by diallyl disulfide requires the activation of p38 MAP kinase in HL-60 cells. *Zhonghua Xue Ye Xue Za Zhi, 25,* 273–276.

Tanaka, S., Haruma, K., Kunihiro, M., et al. (2004). Effects of aged garlic extract (AGE) on colorectal adenomas: A double-blinded study. *Hiroshima Journal of Medical Sciences, 53,* 39–45.

Taylor, P., Noriega, R., Farah, C., et al. (2006). Ajoene inhibits both primary tumor growth and metastasis of B16/BL6 melanoma cells in C57BL/6 mice. *Cancer Letters, 239,* 298–304.

Thejass, P., & Kuttan, G. (2007a). Antiangiogenic activity of Diallyl Sulfide (DAS). *International Immunopharmacology, 7,* 295–305.

Thejass, P., & Kuttan, G. (2007b). Inhibition of angiogenic differentiation of human umbilical vein endothelial cells by diallyl disulfide (DADS). *Life Sciences, 80,* 515–521.

Thomas, M., Zhang, P., Noordine, M. L., et al. (2002). Diallyl disulfide increases rat h-ferritin, L-ferritin and transferrin receptor genes in vitro in hepatic cells and in vivo in liver. *The Journal of Nutrition, 132,* 3638–3641.

Tilli, C. M., Stavast-Kooy, A. J., Vuerstaek, J. D., et al. (2003). The garlic-derived organosulfur component ajoene decreases basal cell carcinoma tumor size by inducing apoptosis. *Archives of Dermatological Research, 295,* 117–123.

Uda, N., Kashimoto, N., Sumioka, I., et al. (2006). Aged garlic extract inhibits development of putative preneoplastic lesions in rat hepatocarcinogenesis. *The Journal of Nutrition, 136,* 855S-860S.

Wang, H. C., Yang, J. H., Hsieh, S. C., et al. (2010a). Allyl sulfides inhibit cell growth of skin cancer cells through induction of DNA damage mediated G2/M arrest and apoptosis. *Journal of Agricultural and Food Chemistry, 58,* 7096–7103.

Wang, Y. B., Qin, J., Zheng, X. Y., et al. (2010b). Diallyl trisulfide induces Bcl-2 and caspase-3-dependent apoptosis via downregulation of Akt phosphorylation in human T24 bladder cancer cells. *Phytomedicine: International Journal of Phytotherapy and Phytopharmacology, 17,* 363–368.

Wargovich, M. J. (1987). Diallyl sulfide, a flavor component of garlic (Allium sativum), inhibits dimethylhydrazine-induced colon cancer. *Carcinogenesis, 8,* 487–489.

Wargovich, M. J., Woods, C., Eng, V. W., et al. (1988). Chemoprevention of N-nitrosomethylbenzylamine-induced esophageal cancer in rats by the naturally occurring thioether, diallyl sulfide. *Cancer Research, 48,* 6872–6875.

Wattenberg, L. W., Sparnins, V. L., & Barany, G. (1989). Inhibition of N-nitrosodiethylamine carcinogenesis in mice by naturally occurring organosulfur compounds and monoterpenes. *Cancer Research, 49,* 2689–2692.

Weinberg, D. S., Manier, M. L., Richardson, M. D., et al. (1993). Identification and quantification of organosulfur compliance markers in a garlic extract. *Journal of Agriculture and Food Chemistry, 41,* 37–41.

Witte, J. S., Longnecker, M. P., Bird, C. L., et al. (1996). Relation of vegetable, fruit, and grain consumption to colorectal adenomatous polyps. *American Journal of Epidemiology, 144,* 1015–1025.

Wu, C. C., Chung, J. G., Tsai, S. J., et al. (2004). Differential effects of allyl sulfides from garlic essential oil on cell cycle regulation in human liver tumor cells. *Food and Chemical Toxicology: An International Journal Published for the British Industrial Biological Research Association, 42,* 1937–1947.

Wu, X. J., Kassie, F., & Mersch-Sundermann, V. (2005). The role of reactive oxygen species (ROS) production on diallyl disulfide (DADS) induced apoptosis and cell cycle arrest in human A549 lung carcinoma cells. *Mutation Research, 579,* 115–124.

Xiao, D., & Singh, S. V. (2006). Diallyl trisulfide, a constituent of processed garlic, inactivates Akt to trigger mitochondrial translocation of BAD and caspase-mediated apoptosis in human prostate cancer cells. *Carcinogenesis, 27,* 533–540.

Xiao, D., Pinto, J. T., Soh, J. W., et al. (2003). Induction of apoptosis by the garlic-derived compound S-allylmercaptocysteine (SAMC) is associated with microtubule depolymerization and c-Jun NH(2)-terminal kinase 1 activation. *Cancer Research, 63,* 6825–6837.

Xiao, D., Choi, S., Johnson, D. E., et al. (2004). Diallyl trisulfide-induced apoptosis in human prostate cancer cells involves c-Jun N-terminal kinase and extracellular-signal regulated kinase-mediated phosphorylation of Bcl-2. *Oncogene, 23,* 5594–5606.

Xiao, D., Herman-Antosiewicz, A., Antosiewicz, J., et al. (2005a). Diallyl trisulfide-induced G(2)-M phase cell cycle arrest in human prostate cancer cells is caused by reactive oxygen species-dependent destruction and hyperphosphorylation of Cdc 25 C. *Oncogene, 24,* 6256–6268.

Xiao, D., Pinto, J. T., Gundersen, G. G., et al. (2005b). Effects of a series of organosulfur compounds on mitotic arrest and induction of apoptosis in colon cancer cells. *Molecular Cancer Therapeutics, 4,* 1388–1398.

Xiao, D., Lew, K. L., Kim, Y. A., et al. (2006a). Diallyl trisulfide suppresses growth of PC-3 human prostate cancer xenograft in vivo in association with Bax and Bak induction. *Clinical Cancer Research: An Official Journal of the American Association for Cancer Research, 12,* 6836–6843.

Xiao, D., Li, M., Herman-Antosiewicz, A., et al. (2006b). Diallyl trisulfide inhibits angiogenic features of human umbilical vein endothelial cells by causing Akt inactivation and down-regulation of VEGF and VEGF-R2. *Nutrition and Cancer, 55,* 94–107.

Xiao, D., Zeng, Y., Hahm, E. R., et al. (2009). Diallyl trisulfide selectively causes Bax- and Bak-mediated apoptosis in human lung cancer cells. *Environmental and Molecular Mutagenesis, 50,* 201–212.

Yang, C. S., Chhabra, S. K., Hong, J. Y., et al. (2001). Mechanisms of inhibition of chemical toxicity and carcinogenesis by diallyl sulfide (DAS) and related compounds from garlic. *The Journal of Nutrition, 131,* 1041S–1045S.

Yang, J. S., Chen, G. W., Hsia, T. C., et al. (2009). Diallyl disulfide induces apoptosis in human colon cancer cell line (COLO 205) through the induction of reactive oxygen species, endoplasmic reticulum stress, caspases casade and mitochondrial-dependent pathways. *Food and Chemical Toxicology: An International Journal Published for the British Industrial Biological Research Association, 47,* 171–179.

You, W., Chang, Y., Zhang, L., et al. (1999). An epidemiologic and preventive study on a high-risk area for gastric cancer in Linqu County, Shandong Province. *Chinese Journal of Preventive Veterinary Medicine, 33,* 52–54.

You, W. C., Blot, W. J., Chang, Y. S., et al. (1989). Allium vegetables and reduced risk of stomach cancer. *Journal of the National Cancer Institute, 81,* 162–164.

You, W. C., Zhang, L., Gail, M. H., et al. (1998). Precancerous lesions in two counties of China with contrasting gastric cancer risk. *International Journal of Epidemiology, 27,* 945–948.

You, W. C., Brown, L. M., Zhang, L., et al. (2006). Randomized double-blind factorial trial of three treatments to reduce the prevalence of precancerous gastric lesions. *Journal of the National Cancer Institute, 98,* 974–983.

Yuan, J. P., Wang, G. H., Ling, H., et al. (2004). Diallyl disulfide-induced G2/M arrest of human gastric cancer MGC803 cells involves activation of p38 MAP kinase pathways. *World Journal of Gastroenterology: WJG, 10,* 2731–2734.

Zhang, Z. M., Yang, X. Y., Deng, S. H., et al. (2007). Anti-tumor effects of polybutylcyanoacrylate nanoparticles of diallyl trisulfide on orthotopic transplantation tumor model of hepatocellular carcinoma in BALB/c nude mice. *Chinese Medical Journal, 120,* 1336–1342.

Zheng, W., Blot, W. J., Shu, X. O., et al. (1992a). Diet and other risk factors for laryngeal cancer in Shanghai, China. *American Journal of Epidemiology, 136,* 178–191.

Zheng, W., Blot, W. J., Shu, X. O., et al. (1992b). A population-based case-control study of cancers of the nasal cavity and paranasal sinuses in Shanghai. *International Journal of Cancer. Journal International du Cancer, 52,* 557–561.

Chapter 11
An Evidence-based Perspective of *Curcuma Longa* (Turmeric) for Cancer Patients

G. Bar-Sela and M. Schaffer

Abstract Curcumin [diferuloylmethane ($C_{21}H_{20}O_6$)], a polyphenol, is an active principle of the perennial herb *Curcuma longa* (commonly known as turmeric) and derived from the roots (rhizomes) of the plant. Traditionally, turmeric has been used as a foodstuff and a cosmetic and has been a component of Indian Ayurvedic medicine since 1900 BC. As a medicine, curcumin is used mainly for various allergic and inflammatory respiratory conditions, as well as for liver disorders, anorexia, rheumatism and wound healing. These folk medicinal indications are still popular and widely used as alternative agents in many parts of Southeast Asia. Extensive research over the last half-century has revealed important functions of curcumin. *In vitro* and *in vivo* research has shown curcumin's various activities, such as anti-inflammation, cytokines release, antioxidation, immunomodulation, enhancing of the apoptotic process, and anti-angiogenic properties. Curcumin has also been shown to be a mediator of chemo-resistance and radio-resistance. The anticancer effect has been seen in a few clinical trials, mainly as a chemo-prevention agent in colon and pancreatic cancer and in other high-risk premalignant conditions. Some clinical studies on healthy volunteers revealed a low bioavailability, casting doubt on the use of curcumin only as a food additive. In cancer, clinical studies have not shown significant results but the data is richer than in non-malignant conditions. Unlike chemo-therapeutic agents, including those isolated from plants, curcumin is a part of our daily food habit and its use in large quantities since ancient times has proved that it is a safe product. While the pre-clinical data is broad, clinical studies are scarce, although some are on-going. The possible clinical efficacy of this treatment as a chemo-preventive or chemo-therapeutic agent is still to be proven. This chapter ranges from a historical description to pre-clinical data and focuses on existing clinical evidence.

G. Bar-Sela (✉)
Division of Oncology, Rambam-Health Care Campus and Faculty of Medicine,
Technion-Israel Institute of Technology, POB 9602, 31096 Haifa, Israel
e-mail: g_barsela@rambam.health.gov.il

W. C. S. Cho (ed.), *Evidence-based Anticancer Materia Medica,* Evidence-based Anticancer Complementary and Alternative Medicine,
DOI 10.1007/978-94-007-0526-5_11, © Springer Science+Business Media B.V. 2011

11.1 The Plant and Its History

Curcumin [diferuloylmethane($C_{21}H_{20}O_6$)], a polyphenol, is an active principle of the perennial herb *Curcuma longa* (commonly known as turmeric) and is derived from the roots (rhizomes) of the plant. The *Curcuma* genus belongs to the division of Magnoliphyta, class of Lilipsyda, subclass of Zingiberidae, family Zingiberaceae. The Zngiber (ginger) is a member of the same family. Turmeric has roots or tubers oblong-palmate that are deep orange inside. Its leaves are about two feet long, lanceolate, long, petioled, tapering at each end, smooth, of a uniform green color. Flowers are dull yellow, three or five together, surrounded by bracteolae (Fig. 11.1). It is propagated by cuttings from the root. In its fresh state, the roots have an aromatic and spicy fragrance which, on drying, gives way to a more medicinal aroma. The bright yellow color of turmeric comes mainly from fat-soluble, polyphenolic pigments known as curcuminoids.

Turmeric grows naturally throughout the Indian subcontinent and in tropical countries, particularly Southeast Asia. It has been cultivated in China and Malaysia and has spread to Australia, Central and South America, and Africa. Traditionally, turmeric has been used as a foodstuff and a cosmetic, and has been a component of Indian Ayurvedic medicine since 1900 BC. Its use was confined to the Asian continent until the 12th–13th centuries AD. As a spice, it is used to provide curry with its distinctive yellow color and flavor. It is used as a coloring agent in cheese, butter, mustard, soft drinks, and other foods. In the international numbering system, curcumin is coded as E100, a safe coloring agent (Ammon and Wahl 1991). In Ayurvedic medicine, curcumin is used mainly for various allergic

Fig. 11.1 The various components of *Curcuma longa* (turmeric)

and inflammatory respiratory conditions (e.g. asthma, bronchitis, cough, runny nose), as well as in liver disorders, anorexia, rheumatism, and wound healing. In traditional Chinese medicine, it is used for conditions associated with abdominal pain, and in Hindu medicine it is used to treat sprains and swelling (Goel et al. 2008). These folk medicinal indications are still popular and are widely used as alternative agents in many parts of Southeast Asia. Turmeric and curcumin have at least 76 synonyms listed in a WHO monograph of 1999, the most common of which are *haldi* (Hindi), *haridra* or *gauri* (Sanskrit), *chiang huang* (Chinese), *ukon* (Japanese), *kurkum* (Arabic), Indian saffron, and yellow root (Strimpakos and Sharma 2008). Curcumin was first isolated in 1815, but identified as diferuloylmethane only at the end of the century. In 1910, Lampe was the first to synthesize the feruloylmethane skeleton of curcumin (Lampe and Milobedzka 1913). The first clinical study was published in Lancet in 1937. Curcumin in escalated doses was given to patients with biliary diseases. Symptomatic improvement was reported in all the patients with radiological improvement by cholecystogram in a third (Oppenheimer 1937).

It took a long time until curcumin entered scientific clinical trials; Phase I and II clinical studies were conducted in the last two decades. Nowadays, curcumin is one of the most popular "nutritional supplements". Beside its use as an anti-cancerous agent, it has been reported to have beneficial effects in arthritis, allergy, asthma, atherosclerosis, heart disease, Alzheimer's disease, and diabetes. While basic and laboratory research in the field of phytochemicals and especially curcumin is growing extensively, the gap between research and widespread use by the public is growing as well. Those regions of uncertainties of curcumin effectiveness are gaps that should be filled by clinical studies. Nevertheless, the high safety profile of curcumin has allowed those gaps to exist without hindering its use.

11.2 Activities and Mechanism of Curcumin as an Anticancer Agent

Curcumin, chemically known as bis-α,β-unsaturated β-diketone [diferuloylmethane ($C_{21}H_{20}O_6$)] that exhibits ket-enol tautomerism, has a predominant keto form in acidic and neutral solutions and a stable enol form in alkaline media (Fig. 11.2). Diferuloylmethane is the main curcuminoid found in turmeric and is considered to be the most active one, but other curcuminoids are found in turmeric as well, including demethoxycurcumin and bisdemethoxycurcumin (Bar-Sela et al. 2010).

The mechanisms implicated in the inhibition of tumorigenesis by curcumin are diverse and appear to involve a combination of anti-inflammatory, antioxidant, immunomodulatory, proapoptotic, and anti-angiogenic properties *via* pleiotrophic effects on genes and cell-signaling pathways at multiple levels (Hatcher et al. 2008; Strimpakos and Sharma 2008; Bar-Sela et al. 2010).

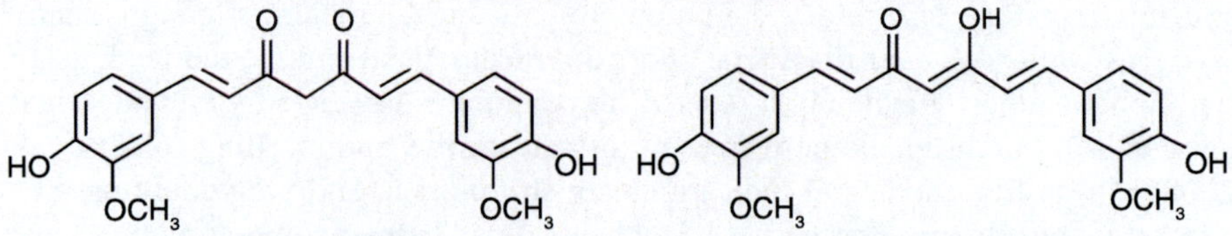

Fig. 11.2 Chemical structure. *Curcuma longa* (turmeric) contains up to 5% essential oils and up to 3% curcumin, a polyphenol. It is the active substance of turmeric and it is also known as C.I. 75,300, or Natural Yellow 3. The systematic chemical name is (1*E*,6*E*)-1,7-bis(4-hydroxy-3-methoxyphenyl)-1,6-heptadiene-3,5-dione. It can exist at least in two tautomeric forms, keto and enol. The keto form is preferred in solid phase and the enol form in solution

The paradigm of inflammatory cells infiltrating into the stromal microenvironment of tumors, where pro-inflammatory cytokines play important roles in promoting tumor cell proliferation, invasion, migration and metastasis, is well established in pre-clinical models. The cancer prevention effect of anti-inflammatory drugs, such as aspirin and other salicylates, is well documented, although the evidence has come from randomized controlled trials performed for purposes other than cancer reduction (Elwood et al. 2009). The anti-inflammatory activities of curcumin have been observed in *in vitro* studies that showed the inhibition of lipo-oxygenase and cyclo-oxygenase activities that can induce inflammation. COX-2 (Cyclooxygenase-2) over-expression has been implicated in the carcinogenesis of many tumors. In contrast to selective COX-2 inhibitors, such as celecoxib which inhibits the catalytic activity of the isoenzyme, curcumin inhibits the transcription of the COX-2 protein, reducing its level in the cells (Plummer et al. 1999). The gene inhibition of COX-2 is probably the main anti-inflammatory activity of curcumin. In models of colon cancer cells cutlers, the inhibition of COX-2 expression and activity leads to a reduction in prostaglandins synthesis and loss of cancer cell growth (Goel et al. 2001; Reddy et al. 2006). However, curcumin was also shown to be an effective inhibitor of cell growth in prostaglandin-synthesis deficient cancer cells HCT-15, suggesting other regular pathways (Hanif et al. 1997). Down-regulation of COX-2 by curcumin is basic and has occurred in different cancer cell lines, consistent with down-regulation of NF-κB and reduced PGE2, as shown recently in animal models (Padhye et al. 2009). Curcumin and some of its analogues also inhibit COX-1 (cyclooxygenase-1) transcription (Plummer et al. 1999), perhaps influencing the tumor environment as well. Interesting is the new evidence of synergistically augmented neoplastic cell growth inhibition of curcumin together with specific COX-2 inhibitors, such as celecoxib (Celebrex, Pfizer NY, USA) (Lev-Ari et al. 2008).

Oxidative phosphorylation is the main aerobic process in the cells, mediated by cytochrome oxidase, inside the cell mitochondria. In this process, the final product is not only water (H_2O), but also other reactive oxygen species (ROS) [e.g. superoxide anion (O_2^-), H_2O_2, hydroxyl radical (OH•)]. Those ROS play a key role in different aspects of carcinogenesis, including DNA alterations, increasing cell proliferation, apoptosis resistance, angiogenesis, and metastatic changes (López-Lázaro

2008). Paradoxically, increasing cell levels of ROS above threshold will cause cell death and, therefore, may play a role in inducing cancer cell death.

Curcumin was found to inhibit ROS formulation (Joe and Lokesh 1994), probably by influencing the level of redox homeostasis enzymes, such as glutathione peroxidase and superoxide dismutase (Kunwar et al. 2009). However, this effect is dual dependent on time and concentration, and may be attributed to changes in oxidative stress and antioxidant gene expression levels leading to inhibition or promotion of cell death (Kunwar et al. 2009). Curcumin has also been shown to scavenge superoxide anion radicals and hydroxyl radicals (Khopde et al. 1999). Interestingly, an *in vivo* study on rats showed protection against renal damage induced by acetaminophen through the antioxidative effect of curcumin. This improvement was attributed to the increased antioxidant enzyme activity and the reduction in malondialdehyde levels (Cekmen et al. 2009). On the other hand, similar to other dietary phytochemicals [(e.g. resveratrol, EGCG (epigallocatechin-3-gallate)], curcumin may possess pro-oxidant activity as well, dependent on dose and the chemical environment. This dose-dependent effect of curcumin was well established in a study on human hepatoma Hep3B cells treated with curcumin for 8 hours. In lower concentrations, a significant decrease in ROS levels was reported while, at higher doses, the ROS level was increased (Kang et al. 2005). This dose- and time-dependence was reported in other studies as well. In some basic studies, there was a difference between regular cells and tumor cells, with higher ROS production in tumor cells (Atsumi et al. 2005). This observation may explain the results of a murine model that showed how dietary curcumin ameliorates radiation-induced pulmonary fibrosis and increases mouse survival while not impairing tumor cell killing by radiation (Lee et al. 2010). Still, caution is needed when results from a study on lung cancer models suggest that curcumin may exhibit organ-specific effects to enhance ROS formation in the damaged lung epithelium of smokers and ex-smokers (Dance-Barnes et al. 2009). On-going clinical trials may be needed to exclude smokers and ex-smokers in chemo-preventive trials of curcumin.

The process of carcinogenesis is complicated, with many changes in the cells which can be divided into three steps: initiation, promotion, and progression. There are numerous studies showing the effect of curcumin on all three levels of carcinogenesis, and the stream of data is increasing with the growing knowledge of the process. In rodent models, oral curcumin administration prevented cancer development of the skin, soft palate, stomach, colon, liver, lung, and breast (Strimpakos and Sharma 2008). In a study done with a Min mouse model that mimicked familial adenomatous polyposis (FAP) in humans, the addition of curcumin in the diet for the lifetime of the animals showed a significant reduction in adenoma numbers compared with the control animals (Mahmoud et al. 2000). Chemicals models, using azoxymethane to induce colon cancer in mice or rats, have been made to study the effect of curcumin on the promotion and progression stages of carcinogenesis (Samaha et al. 1997; Kawamori et al. 1999). In both these studies, oral curcumin produced a significant increase in the apoptotic histologic index, compared to controls. Recently, curcumin was found to regulate several genes involved in cancer formation. Matrix metalloproteinases (MMPs) play an important role in the invasion and metastasis potential of cancer cells. Curcumin was found to inhibit the migration of lung cancer cells through the inhibition of MMP-2 and -9 and vascular endothelial growth factor (VEGF) (Lin et al. 2009). In a different *in vitro* model, curcumin was added to the chemotherapy for colon cancer cell resistance to 5-fluorouracil and oxaliplatin. Reduction of cell survival accompanied by a concomitant reduction in activation of HER-2, IGF-1R (insulin growth factor-1 receptor), and AKT, all carcinogenesis proteins, was seen (Patel et al. 2010). In the epigenomic level, curcumin has been found to up-regulate insulin-like growth factor binding protein-5 and CCAAT/enhancer-binding protein alpha, both suppressors of head and neck carcinogenesis (Chang et al. 2010).

Many studies have focused on the preventive aspects of curcumin against tumor formation. Different directions were taken to test its anti-inflammatory, antioxidant, anti-angiogenic, and anti-carcinogenetic qualities on tumor cells or in animal models. Another direction was to study its antitumoral effect as a single agent or in combination with other treatment approaches, such as chemotherapy or radiotherapy. Numerous studies have evaluated the efficacy of curcumin in various animal models. The first study, published in 1985, demonstrated the antitumor effects of curcumin given intra-peritoneally to mice with Dalton's lymphoma cells (Kuttan et al. 1985). More recently, others have studied the antitumoral effect of curcumin. In an orthotopic mouse model of human bladder cancer, curcumin alone significantly reduced the bladder tumor volume; maximum reduction was observed when curcumin was used in combination with gemcitabine, significantly more than gemcitabine alone. Curcumin also significantly decreased the proliferation marker Ki-67 and microvessel density CD31. Curcumin abolished the constitutive activation of NF-κB in the tumor tissue; induced apoptosis, and decreased cyclin D1, VEGF, COX-2, c-myc, and Bcl-2 expression in the bladder cancer tissue (Tharakan et al. 2010). In a similar study with a metastatic colorectal carcinoma model, curcumin inhibited the proliferation of human colorectal carcinoma cell lines, potentiated capecitabine-induced apoptosis, inhibited NF-κB activation, and suppressed NF-κB-regulated gene products *in vitro*. In nude mice, the combination of curcumin and capecitabine was found to be more effective than either agent alone in reducing tumor volume, Ki-67 proliferation index, microvessel density marker CD31, and suppressing distant metastasis formation (Kunnumakkara et al. 2009). In an experimental breast cancer murine model using MDA-MB-231 cells, combination therapy with paclitaxel and curcumin significantly reduced tumor size and decreased tumor cell proliferation, increased apoptosis, and decreased the expression of matrix metalloprotease 9 compared with either agent alone (Kang et al. 2009). The results of this group of studies clearly suggest that a curcumin-chemotherapy combination could be a novel strategy to enhance oncology treatment.

In hematological malignancies, curcumin was investigated in overcoming chemo-resistance and enhancing the activity of thalidomide and bortezomib used to treat patients with multiple myeloma *in vitro* and in a xenograft model in nude mice. The results showed that curcumin inhibited the proliferation of human multiple myeloma cells regardless of their sensitivity to dexamethasone, doxorubicin, or melphalan. Furthermore, in a nude mice model, curcumin potentiated the antitumor effects of bortezomib and this correlated with suppression of Ki-67, CD31, and vascular endothelial growth factor expression (Park et al. 2008; Sung et al. 2009).

An interesting aspect of curcumin's activity is its radioprotective effect on normal cells and radiosensitizing effects on cancer cells. This opposing mechanism is not entirely understood. It has been suggested that curcumin's ability to reduce oxidative stress and inhibit transcription of genes related to oxidative stress and inflammatory responses may afford protection against the harmful effects of radiation, while the radiosensitizing activity might be due to the up-regulation of genes responsible for cell death (Jagetia 2007; Akpolat et al. 2009). Curcumin as a protector against chemotherapy side effects is a new area of research, based on its antioxidant activity. Neurotoxicity induced by ROS can appear as an adverse effect of chemotherapy treatment with platinum compounds, such as cisplatin. Genotoxic/anti-genotoxic effects of curcumin in PC12 cells exposed to cisplatin (Mendonça et al. 2009) showed that curcumin significantly reduced the total frequency of micronuclei induced by cisplatin. Determining the cytotoxic and genotoxic/anti-genotoxic effects of curcumin in a neuronal model is important for assessing possible hazards when combined with other chemical agents, including chemotherapy drugs used in cancer therapy.

Basic research demonstrates different aspects of curcumin that are important for further clinical studies. Recently, Jiao et al. (2009) observed that, in cultured cells, curcumin exhibits properties of an iron chelator. These authors observed that curcumin repressed synthesis of hepcidin, a peptide that plays a central role in the regulation of the systemic iron balance. These results demonstrate that curcumin has the potential to affect systemic iron metabolism, particularly in a setting of subclinical iron deficiency. This may affect the use of curcumin in patients with marginal iron stores or those exhibiting the anemia of cancer and chronic disease. This phenomenon needs to be considered in any tumor treatment clinical trials.

In summary, the pleiotrophic activities of curcumin derive from its complex chemistry as well as from its ability to influence multiple signaling pathways, including survival pathways such as those regulated by NF-κB, Akt, and growth factors, cytoprotective pathways dependent on Nrf2, as well as metastatic and angiogenic pathways. Curcumin is a free radical scavenger and hydrogen donor, and exhibits both pro- and antioxidant activity. It also binds metals, particularly iron and copper, and can function as an iron chelator (Hatcher et al. 2008). Although curcumin is poorly absorbed after ingestion, multiple studies have suggested that even low levels of physiologically achievable concentrations of curcumin may be sufficient for its chemo-preventive and chemo-therapeutic activity. Thus, curcumin regulates multiple targets (multi-targeted therapy), which is needed for the treatment of most diseases. It is inexpensive and has been found to be safe in human clinical trials. In spite of its efficacy and safety, curcumin has not yet been approved as a therapeutic agent (Anand et al. 2008). This gap between the high level of many pre-clinical studies and the limitation of its right clinical use due to poor aqueous solubility, together with low bioavailability, leads to many uncertainties of its therapeutic effect in the Western population. The main activities of curcumin are illustrated in Fig. 11.3 and summarized in Table 11.1.

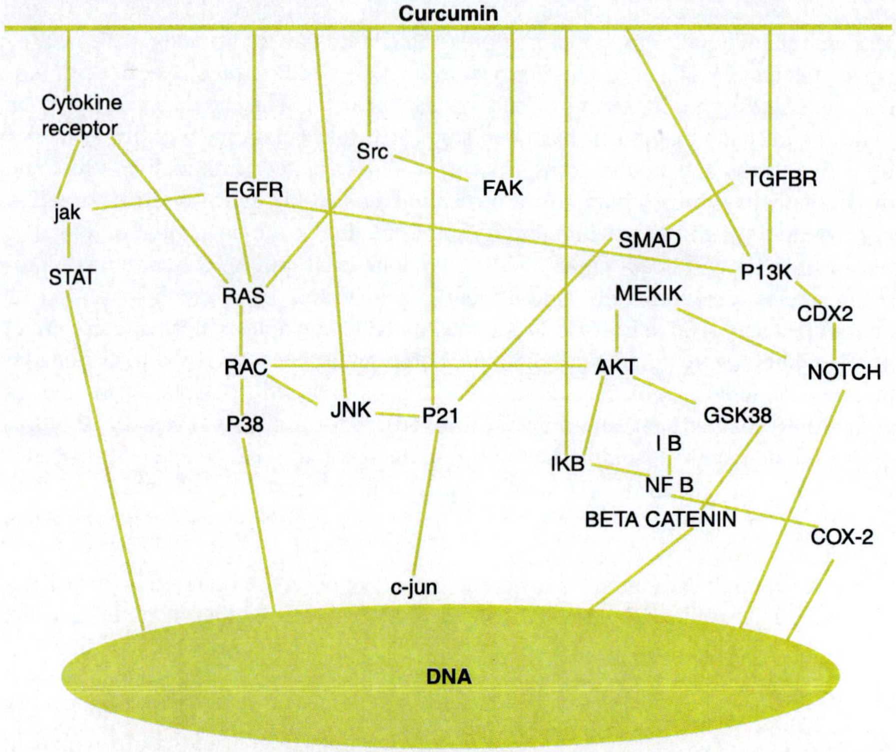

Fig. 11.3 Main activities of curcumin. Inhibits deregulation cellular proliferation, dedifferentiation, and proliferation by alerting key signaling molecules that ensured that the cell death proceeds efficiently and enhancement of apoptotic death

Table 11.1 Summary of main activities of curcumin

General activity	Mechanism
1. Immunologic modulation	1. Modulation of growth factors and their signaling pathway
2. Radio-sensitization and radio-protection	2. Induction of Phase II enzymes
3. Chemo-sensitizing activity of curcumin	3. Effects of curcumin on tumor suppressor p53
4. Chemo-preventive, chemo-therapeutic	4. Inhibition of cytokines inhibits the pro-survival kinase Akt
5. Inhibition of angiogenesis and metastasis	5. Inhibition of NF-κB
	6. Downstream of NF-κB: inhibition of COX-2
	7. Downstream of NF-κB: inhibition of cyclin D1
	8. Downstream of NF-κB: suppression of Bcl-2 and Bcl-XL
	9. Effect of curcumin on mitogen-activated protein kinases
	10. Inhibition of STAT3 activation by curcumin

11.3 Clinical Studies with Curcumin

The various properties of curcumin and its well-documented antioxidant properties were used to test its efficacy in various illnesses. The pharmacokinetic studies of curcumin indicated in general a low bioavailability of curcumin following oral application. Nevertheless, the pharmacologically active concentration of curcumin could be achieved in colorectal tissue in patients taking curcumin orally and might also be achievable in tissues such as skin and oral mucosa, which are directly exposed to the drugs applied locally or topically.

The effect of curcumin was studied in patients with rheumatoid arthritis, inflammatory eye diseases, inflammatory bowel disease, chronic pancreatitis, psoriasis and hyperlipidemia. Although preliminary results support the efficacy of curcumin in these diseases, the data to date are not conclusive (Hsu and Cheng 2007). In cancer, clinical studies have still not reached significant results but the data is richer than in non-malignant conditions. Unlike the majority of chemo-therapeutic agents, including those isolated from plants (such as taxol, vincristine, etc), curcumin is a part of our daily food habit and its use in large quantities since ancient times has proved that it is a safe product.

Curcumin treatment exhibits great promise as a therapeutic agent and is currently being investigated in human clinical trials for a variety of malignant diseases, including multiple myeloma, pancreatic cancer, myelodysplastic syndrome, and colon cancer (Table 11.2).

11.4 Cancer Prevention

A Phase I clinical study performed in Taiwan investigated curcumin's potential anticarcinogenesis activity in patients with pre-invasive malignant or high-risk premalignant conditions (Cheng et al. 2001). Twenty-five patients with recently resected superficial bladder carcinoma, Bowen's disease of the skin, uterine cervical intraepithelial neoplasia, intestinal metaplasia of the stomach, or oral leukoplakia were given doses of 1–8 g of curcumin daily for three months. Histological improvement of the premalignant lesions was noted in 1 of 2 patients with presumed bladder carcinoma *in situ*, 2 of 7 patients with oral leukoplakia, 1 of 6 patients with stomach intestinal metaplasia, 1 of 4 patients with cervical intraepithelial neoplasia, and 2 of 6 patients with Bowen's disease of the skin. Limitations for drawing definite conclusions from this study are the small number of patients with each high-risk condition, different daily doses of curcumin, and possible bias from the interpreting pathologists, as the study was not blinded.

Lichen planus is a pre-cancerous condition for oral cavity carcinoma. A randomized, double-blind, placebo-controlled trial was conducted at the University of California (San Francisco, USA). The trial was conducted on 100 patients, but interim analysis was made using data from the first 33 subjects only. Study subjects

Table 11.2 Main clinical trials

Subject	Specific cancer or pre-malignant condition	No. of patients	Amount of curcumin	Main results	References
Palliative care	Different types	62	Ointment	Reduction of necrosis smell and itching of external lesions	Kuttan et al. 1987
Cancer prevention	Different types of premalignant conditions	25	1–8 g daily	Histological improvement in some lesions	Cheng et al. 2001
Cancer prevention	Familial adenomatous polyposis	5	Curcumin 450 mg+quercetin daily	Decrease in number and size of polyps	Cruz-Correa et al. 2006
Pancreatic cancer	Pancreatic cancer	17	8 g daily, combined with gemcitabine	The combination is feasible; 45% stable disease and response	Epelbaum et al. 2010
Pancreatic cancer	Advanced pancreatic cancer (Phase II)	25	8 g/daily	Information on cytokines release and 8% clinical response	Dhillon et al. 2008
Colorectal	Advanced, refractory colon carcinoma (Phase I)	15	440–2,200 mg/day	Safe product with no side-effects; 33% had short period of stable disease	Sharma et al. 2001
Colorectal	Rectal cancer before surgery	12	450, 1,800, or 3,600 mg	Pharmacodynamic study	Garcea et al. 2005
Breast cancer	Advanced breast cancer, Phase I	14	500–8,000 mg (escalating dose)+docetaxel	8 with partial response, 3 with stable disease	Bayet-Robert et al. 2010
Oral lichen planus	Randomized, double-blind, placebo-controlled trial	33	Placebo or curcuminoids 2,000 mg/day for 7 weeks+prednisone 60 mg/day for the 1st week	The study was ended early for futility	Chainani-Wu et al. 2007

were randomized to receive either placebo or curcuminoids at 2,000 mg/day for 7 weeks. In addition, all subjects received prednisone 60 mg/day for the 1st week. The primary outcome was a change in symptoms from baseline. Secondary outcomes were changes in clinical signs and the occurrence of side effects. The interim analysis did not show a significant difference between the placebo and curcuminoid groups. Conditional power calculations suggested a less than 2% chance that the curcuminoid group would have a significantly better outcome as compared to the placebo group if the trial were continued to completion. Therefore, the study was ended early for futility (Chainani-Wu et al. 2007).

FAP is an autosomal dominant condition characterized by the development of numerous bowel adenomas that can transform to adenocarcinoma. Regression of the adenomas in this syndrome occurs with the administration of non-steroidal anti-inflammatory drugs and COX-2 inhibitors, but these compounds can have considerable side effects. In a report from the Cleveland Clinic in Florida (USA), curcumin in combination with quercetin was given to 5 patients with FAP. Quercetin is a phytochemical, acting as an antioxidant that is part of the coloring found in the skin of apples and red onions. Curcumin and quercetin were administered three times per day to these patients and all showed a decrease in the number and size of polyps compared with baseline figures (Cruz-Correa et al. 2006). The dose variability in the different studies (450 mg daily in the Cleveland study, 1–8 g daily in the Taiwan study) represents one of the main difficulties of studies done on nutritional supplements—knowing the correct effective dose.

Two studies are currently testing the chemo-prevention effect of curcumin in colon cancer. A Phase II study of patients with FAP is taking place at the NCI/Johns Hopkins University (http://clinicaltrials.gov/ct/search?term_curcumin) and another study is being done at the University of Pennsylvania where patients with previously resected adenomatous colonic polyps are being given curcumin for the prevention of colorectal cancer (http://clinicaltrials.gov/ct/search?term_curcumin).

11.5 Pancreatic Cancer

The anti-metabolite gemcitabine is the standard chemotherapy for advanced pancreatic cancer. Yet, it produces an objective response in less than 10% of patients, with a minor effect on survival. Curcumin can potentiate the antitumor effect of gemcitabine, as shown in pre-clinical models of pancreatic cancer (Fryer et al. 2009). Epelbaum et al. (2010) treated 17 patients with a combination of curcumin and gemcitabine. Gemcitabine 1,000 mg/m^2 was given on day 1 every week for 7 weeks with a 1-week break, followed by 3 weeks of treatment and a 1-week break in the next cycle. Eight grams of curcumin (500 mg capsules) were given daily, divided into 4 g twice a day (morning and evening), throughout the treatment period, and patients received a median of 2 (range, 1/3–14) cycles of gemcitabine. Five (29%) patients discontinued the curcumin after a few days to 2 weeks due to intractable abdominal fullness or pain, and 1 patient died during the 1st cycle due to an unrelated cardiac event. Curcumin and gemcitabine were delivered concomitantly for a period of 1–12 months to the remaining 11 patients. The dose of curcumin was reduced to 4 g/day in 3 because of abdominal complaints. One (9%) of the 11 evaluable patients had partial response (7 months), 4 (36%) had stable disease (2, 3, 6, and 12 months) and 6 (55%) had tumor progression. Time to tumor progression was 1–12 months (median 2), and overall survival was 1–24 months (median 6). These results suggest that a combination of gemcitabine and curcumin for patients with advanced pancreatic cancer is feasible. However, the daily oral dose of curcumin should be less than 8 g.

Dhillon et al. (2008) conducted a Phase II study of only curcumin supplement as 1st line treatment in patients with advanced pancreatic cancer. Twenty-five patients received 8 g of curcumin daily until disease progression, with restaging every 2 months. Serum cytokine levels for interleukin (IL)-6, IL-8, IL-10, and IL-1 receptor antagonists and peripheral blood mononuclear cell expression of NF-κB and COX-2 were monitored. Twenty-one patients were evaluated for response: 1 had ongoing stable disease for >18 months and, interestingly, 1 patient had a brief but marked tumor regression accompanied by significant increases in serum cytokine levels. No toxicities were observed. Curcumin down-regulated expression of NF-κB, COX-2, and phosphorylated signal transducer, and activated transcription 3 in peripheral blood mononuclear cells from the patients. In these 2 studies, curcumin was given in a daily dose of 8 g. The common side effect of intractable abdominal fullness or pain was only seen in the study that combined curcumin with chemotherapy.

Clinical knowledge of curcumin is limited. To date, Phase II studies have been performed only on pancreatic carcinoma patients. The results are interesting, raising the need for a Phase III study, combining lower doses of curcumin with chemotherapy in the treatment of pancreatic carcinoma. To the best of our knowledge, there are 2 on-going studies in pancreatic cancer. A Phase II trial of gemcitabine, curcumin, and celebrex in patients with advanced or inoperable pancreatic cancer is being carried out at the Sourasky Medical Center in Tel Aviv, Israel (http://clinicaltrials.gov/ct/search?term_curcumin), and an open label study of patients with locally advanced pancreas cancer is being conducted at MD Anderson in Houston (Texas, USA). In this trial, the daily dose of curcumin is 8 g (http://clinicaltrials.gov/ct/search?term_curcumin).

11.6 Colorectal Cancer

A Phase I clinical trial was conducted by Sharma et al. (2001) on 15 patients with advanced colorectal carcinoma, refractory to 5-fluorouracil (5-FU)-containing chemotherapy. Patients were stratified to receive various doses of curcumin once daily orally in a proprietary capsule form, at doses between 440 and 2,200 mg/day, containing 36–180 mg of curcumin, for 4 months. Oral *Curcuma* extract was well tolerated, and dose-limiting toxicity was not observed. Neither curcumin nor its metabolites were detected in blood or urine, but curcumin was recovered from feces. Curcumin sulfate was identified in the feces of 1 patient. Lymphocytic glutathione-S-transferase activity and levels of the adduct (M_1G) formed by the reaction of malondialdehyde with deoxyguanosine in DNA were assessed as biomarkers of curcumin activity. The patients receiving higher doses did not observe any change in glutathione-S-transferase activity. Correlation was not observed with levels of based M_1G on a variety of different stratifications. A decline in the cancer biomarker carcinoembryonic antigen was seen in one patient with local colon carcinoma. Radiologically stable disease was demonstrated in 5 patients for 2–4 months of treatment. The results suggest that *Curcuma* extract can be administered safely to patients at doses of up to 2.2 g daily, equivalent to 180 mg of curcumin; curcumin has low oral bioavailability in humans and may undergo intestinal metabolism.

Garcea et al. (2005) studied curcumin levels in the colorectum and the pharmacodynamics in 12 patients with confirmed colorectal cancer in various staging. Patients were assigned to 450, 1,800, or 3,600 mg of curcumin per day for 7 days prior to surgery. Biopsy samples of normal and malignant colorectal tissue were obtained at diagnosis and at 6–7 hours after the last dose of curcumin. Blood was taken 1 hour after the last dose of curcumin. The administration of curcumin 3,600 mg decreased M_1G levels from 4.8 ± 2.9 adducts per 107 nucleotides in malignant colorectal tissue to 2.0 ± 1.8 adducts per 107 nucleotides. COX-2 protein levels in malignant colorectal tissue were not affected by the curcumin. These results suggest that a daily dose of curcumin 3.6 g achieves pharmacologically efficacious levels in the colorectum with negligible distribution of curcumin outside the gut. These results also suggest that a daily dose of 3,600 mg is safe in humans. This dose has been shown to furnish agent levels in the target organ, which may be adequate to elicit antioxidative changes commensurate with long-term benefits, mainly as a chemo-preventive agent.

In a case report by Braumann et al. (2009), a treatment combination of oxaliplatin, 5-FU and leucovorin, together with 5 g of curcumin daily, was given to a 75-year old woman with colon carcinoma metastatic to the liver. Good partial remission without side effects was reported after five months of treatment.

11.7 Breast Cancer

A Phase I study on advanced breast cancer patients treated by docetaxel together with escalating doses of curcumin was published by Bayet-Robert et al. (2010). Fourteen patients were accrued. Docetaxel (100 mg/m^2) was administered every three weeks for 6 cycles. Curcumin was given orally from 500 mg/day for 7 consecutive days by cycle (from 4 days before chemotherapy to 2 days after) and escalated until dose-limiting toxicity occurred. The maximal tolerated dose of curcumin was defined as 8,000 mg/day, after 3 dose-limiting toxicities were observed and 2 of 3 patients at this dose level refused to continue treatment. Among the 8 of 14 patients with measurable lesions, 5 had partial response and 3 had stable disease (SD). Three of the 8 evaluable patients had initial inoperable breast carcinoma. Two had partial response and 1 had SD. All underwent surgery. One SD patient had only isolated tumor cells in the pathology specimen. Another SD patient's pulmonary lesions had a minor response, and the 3rd SD patient had stable multiple lesions. At the end of the study, the recommended dose of curcumin was 6,000 mg/day for 7 consecutive days every 3 weeks in combination with a standard dose of docetaxel. The authors declared that a comparative Phase II trial of this regimen plus docetaxel *versus* docetaxel alone is on-going in advanced and metastatic breast cancer patients in their center.

11.8 Other Cancers and Ongoing Studies

Kuttan et al. (1987) reported using turmeric as an ointment to treat skin cancers, breast cancer, and mucosal cancers (oral cavity, vulva). This research group found that curcumin ointment produced remarkable symptomatic relief in patients with external cancerous lesions and lead to a better quality of life. Reduction in smell (necrosis) was noted in 90% of cases, even in extensively ulcerated cases of breast cancer, and reduction of itching and dryness of weeping ulcers was observed in 70% of those cases. In a small number of patients (10%), a reduction in lesion size was reported. The effect continued for several months in many patients. An adverse reaction was noted in only 1 of the 62 patients evaluated. However, there was no control group, no assessment of anti-inflammatory activity, and no chemical analysis of the medicinal preparation. The report include only 62 of 153 treated patients and did not include any patient or tumor details, such as pathology, stage, etc.

Based on the positive results of curcumin in animal models of myeloma, a Phase I study of multiple myeloma patients is taking place at MD Anderson Cancer Center, giving curcumin with or without bioperine (http://clinicaltrials.gov/ct/search?term_curcumin).

In spite of the large number of basic research and various trials which have been conducted, the optimal dose for cancer prevention or cancer treatment is still not known.

11.9 Toxicity

Curcumin is remarkably well tolerated, but its bioavailability is poor. It does not appear to be toxic to animals (Shankar et al. 1980; NCI D 1996) or humans (NCI D 1996; Chainani-Wu 2003), even at high doses. Turmeric is generally recognized as safe by the FDA, and curcumin has been granted an acceptable daily intake level of 0.1–3 mg/kg-BW by the Joint FAO/WHO Expert Committee on Food Additives, 1996 (NCI D 1996).

In terms of dietary use in different countries, according to a study from Nepal, dietary consumption of turmeric up to 1.5 g per person per day, equivalent to 50 mg/day of curcumin, does not appear to be associated with adverse effects in humans (Eigner and Scholz 1999). In India, where the average intake of turmeric can be as high as 2–2.5 g per day (corresponding to 60–100 mg of curcumin daily), no toxicities or adverse effects have been reported at the population level (Cheng et al. 2001).

Pre-clinical models and clinical trials have documented minimal toxicity from the administration of curcumin or turmeric. In a study performed in India, the administration of 1.2–2.1 g of oral curcumin to patients with rheumatoid arthritis daily for up to 6 weeks did not cause any toxicity (Deodhar et al. 1980). Cheng et al. (2001) administrated 8 g daily of curcumin for 3 months to patients with pre-invasive malignancies, with no adverse effects. In a trial published by the NCI (Sharma et al. 2004), curcumin was well tolerated at all doses up to 3.6 g daily for up to 4 months. Adverse events were mainly nausea and diarrhea grade I-II.

The possibility of combining curcumin with chemotherapy agents provides some concern about possible drug-herb interactions. In a pivotal clinical trial with healthy volunteers, 2 weeks of 1 g of curcumin daily caused inhibition of CYP1A2 function and enhancement of CYP2A6 activity. These findings should be confirmed by other studies, but herb-drug interactions associated with curcumin should be considered in clinical trials (Chen et al. 2010).

11.10 Conclusion

Curcumin is a component of turmeric that has been used throughout the ages as a "herbal general medicine" to relieve discomfort and inflammation associated with an extraordinary spectrum of infectious and autoimmune diseases. In basic cancer research, its effects appear pertinent to all stages of carcinogenesis. Curcumin's beneficial effects have been shown in both chemical and genetic models, providing strong preliminary data for the justification of clinical studies in humans.

All these pre-clinical data lead to various, but still scarce, clinical studies (some on-going). The gap between basic research and clinical application continues to grow. The clinical studies were mainly done on and are on-going in gastrointestinal malignancies. The results published until now have shown that the treatment is safe and provide some positive clues to the value of curcumin as a chemo-preventive agent.

Various novel delivery methods are in pre-clinical development to overcome curcumin's lack of water solubility together with low systemic bioavailability after oral dosing that limits access of sufficient concentrations for pharmacologic effects in tissues outside the gastrointestinal tract. One promising *in vivo* activity encapsulated curcumin in a liposomal delivery system that allowed intravenous administration. In this approach, curcumin suppressed pancreatic carcinoma growth in murine xenograft models and inhibited tumor angiogenesis (Li et al. 2005). Another way is to incorporate curcumin into phospholipid vesicles or lipid-nanospheres, embedding this formulation to deliver curcumin into tissue macrophages through intravenous injection. This delivery method showed that curcumin was massively distributed in cells assumed as macrophages into the bone marrow and spleen of injected rates (Sou et al. 2008). The curcumin-loaded polymer-based nanoparticles approach showed promising results with serum levels of almost twice as much and a substantially longer half-life of curcumin nanoparticle compared to free curcumin in an *in vivo* model (Anand et al. 2010). A new novel approach, focusing on the chemo-prevention potential of curcumin, is to use an injectable sustained release microparticle formulation of curcumin. A biodegradable and biocompatible polymer was used to fabricate curcumin microparticles. When injected subcutaneously into mice, a single dose of microparticles sustained curcumin levels in the blood and other tissues for nearly a month. Curcumin levels in the lungs and brain were 10–30 folds higher than that in the blood. Although curcumin microparticles showed marked anticancer efficacy in nude mice xenografts compared with other controls, repeated systemic injections of curcumin were not effective in inhibiting tumor growth (Shahani et al. 2010).

References

Akpolat, M., Kanter, M., & Uzal, M. C. (2009). Protective effects of curcumin against gamma radiation-induced ileal mucosal damage. *Archives of Toxicology, 83,* 609–617.

Ammon, H. P., & Wahl, M. A. (1991). Pharmacology of curcuma longa. *Planta Medica, 57,* 1–7.

Anand, P., Nair, H. B., Sung, B., et al. (2010). Design of curcumin-loaded PLGA nanoparticles formulation with enhanced cellular uptake, and increased bioactivity in vitro and superior bioavailability in vivo. *Biochemical Pharmacology, 79,* 330–338.

Anand, P., Thomas, S. G., Kunnumakkara, A. B, et al. (2008). Biological activities of curcumin and its analogues (Congeners) made by man and Mother Nature. *Biochemical Pharmacology, 76,* 1590–1611.

Atsumi, T., Fujisawa, S., & Tonosaki K. (2005). Relationship between intracellular ROS production and membrane mobility in curcumin- and tetrahydrocurcumin-treated human gingival fibroblasts and human submandibular gland carcinoma cells. *Oral Diseases, 11,* 236–242.

Bar-Sela, G., Epelbaum, R., & Schaffer, M. (2010). Curcumin as an anticancer agent: Review of the gap between basic and clinical applications. *Current Medicinal Chemistry, 17,* 190–197.

Bayet-Robert, M., Kwiatkowski, F., Leheurteur, M., et al. (2010). Phase I dose escalation trial of docetaxel plus curcumin in patients with advanced and metastatic breast cancer. *Cancer Biology & Therapy, 9,* 8–14.

Braumann, C., Guenther, N., Loeffler, L. M, & Dubiel, W. (2009). Liver metastases after colonic carcinoma—palliative chemotherapy plus curcumin. *International Journal of Colorectal Disease, 24,* 859–860.

Cekmen, M., Ilbey, YO., Ozbek, E, et al. (2009). Curcumin prevents oxidative renal damage induced by acetaminophen in rats. *Food and Chemical Toxicology: An International Journal Published for the British Industrial Biological Research Association, 47,* 1480–1484.

Chainani-Wu, N. (2003). Safety and anti-inflammatory activity of curcumin: A component of tumeric (Curcuma longa). *Journal of Alternative and Complementary Medicine (New York, N. Y.), 9,* 161–168.

Chainani-Wu, N., Silverman, S. Jr., Reingold, A., et al. (2007). A randomized, placebo-controlled, double-blind clinical trial of curcuminoids in oral lichen planus. *Phytomedicine, 14,* 437–446.

Chang, K. W., Hung, P. S., Lin, I. Y., et al. (2010). Curcumin upregulates insulin-like growth factor binding protein-5 (IGFBP-5) and C/EBPalpha during oral cancer suppression. *International Journal of Cancer. Journal International du Cancer, 127,* 9–20.

Chen, Y., Liu, W. H., Chen, B. L., et al. (2010). Plant polyphenol curcumin significantly affects CYP1A2 and CYP2A6 activity in healthy, male Chinese volunteers. *The Annals of Pharmacotherapy, 44,* 1038–1045.

Cheng, A. L., Hsu, C. H., Lin, J. K., et al. (2001). Phase I clinical trial of curcumin, a chemopreventive agent, in patients with high-risk or pre-malignant lesions. *Anticancer Research, 21,* 2895–2900.

Cruz-Correa, M., Shoskes, D. A., Sanchez, P., et al. (2006). Combination treatment with curcumin and quercetin of adenomas in familial adenomatous polyposis. *Clinical Gastroenterology and Hepatology: The official clinical practice journal of the American Gastroenterological Association, 4,* 1035–1038.

Dance-Barnes, S. T., Kock, N. D., Moore, J. E., et al. (2009). Lung tumor promotion by curcumin. *Carcinogenesis, 30,* 1016–1023.

Deodhar, S. D, Sethi, R., & Srimal, R. C. (1980). Preliminary study on antirheumatic activity of curcumin (diferuloyl methane). *The Indian Journal of Medical Research, 71,* 632–634.

Dhillon, N., Aggarwal, B. B., Newman, R. A., et al. (2008). Phase II trial of curcumin in patients with advanced pancreatic cancer. *Clinical cancer research: An official journal of the American Association for Cancer Research, 14,* 4491–4499.

Eigner, D., & Scholz, D. (1999). Ferula asa-foetida and Curcuma longa in traditional medical treatment and diet in Nepal. *Journal of Ethnopharmacology, 67,* 1–6.

Elwood, P. C., Gallagher, A. M., Duthie, G. G., et al. (2009). Aspirin, salicylates, and cancer. *Lancet, 373,* 1301–1309.

Epelbaum, R., Schaffer, M., Vizel, B., et al. (2010). Phase II study of curcumin and gemcitabine in patients with advanced pancreatic cancer. *Nutrition and Cancer, 8,* 1137–1141.

Fryer, R. A., Galustian, C., & Dalgelish, A. G. (2009). Recent advances and developments in treatment strategies against pancreatic cancer. *Curr Clinical Pharmacology, 4,* 102–112.

Garcea, G., Berry, D. P., Jones, D. J, et al. (2005). Consumption of putative chemopreventive agent curcumin levels in the colorectum and pharmacodynamic consequences. *Cancer Epidemiology, Biomarkers & Prevention: A publication of the American Association for Cancer Research, cosponsored by the American Society of Preventive Oncology, 14,* 120–125.

Goel, A., Boland, C. R., & Chauhan, D. P. (2001). Specific inhibition of cyclooxygenase-2 (COX-2) expression by dietary curcumin in HT-29 human colon cancer cells. *Cancer Letters, 172,* 111–118.

Goel, A., Jhurani, S., & Aggarwal, B. B. (2008). Multi-targeted therapy by curcumin: How spicy is it? *Molecular Nutrition & Food Research, 52,* 1010–1030.

Hanif, R., Qiao, L., Shiff, S. J., et al. (1997). Curcumin, a natural plant phenolic food additive, inhibits cell proliferation and induces cell cycle changes in colon adenocarcinoma cell lines by a prostaglandin-independent pathway. *The Journal of Laboratory and Clinical, 130,* 576–584.

Hatcher, H., Planalp, R., Cho, J., et al. (2008). Curcumin: From ancient medicine to current clinical trials. *Cellular and Molecular Life Sciences: CMLS, 65,* 1631–1652.

Hsu, C. H., & Cheng, A. L. (2007). Clinical studies with curcumin. Advances in Experimental medicine and Biology, *595,* 471–480. http://clinicaltrials.gov/ct/search?term_curcumin.

Jagetia, G. C. (2007). Radioprotection and radiosensitization by curcumin. *Advances in experimental Medicine and Biology, 595,* 301–320.

Jiao, Y., Wilkinson, J. 4th., Di, X., et al. (2009). Curcumin, a cancer chemopreventive and chemotherapeutic agent, is a biologically active iron chelator. *Blood, 113,* 462–469.

Joe, B., & Lokesh, B. R. (1994). Role of capsaicin, curcumin and dietary n-3 fatty acids in lowering the generation of reactive oxygen species in rat peritoneal macrophages. *Biochimica et Biophysica Acta, 1224,* 255–263.

Kang, H. J., Lee, S. H., Price, J. E, et al. (2009). Curcumin suppresses the paclitaxel-induced nuclear factor-kappaB in breast cancer cells and potentiates the growth inhibitory effect of paclitaxel in a breast cancer nude mice model. *The Breast Journal, 15,* 223–229.

Kang, J., Chen, J., Shi, Y., et al. (2005). Curcumin-induced histone hypoacetylation: The role of reactive oxygen species. *Biochemical Pharmacology, 69,* 1205–1213.

Kawamori, T., Lubet, R., Steele, V. E., et al. (1999). Chemopreventive effect of curcumin, a naturally occurring anti-inflammatory agent, during the promotion/progression stages of colon cancer. *Cancer Research, 59,* 597–601.

Khopde, S., Priyadarsini, K. I., Venkatesan, P., et al. (1999). Free radical scavenging ability and antioxidant efficiency of curcumin and its substituted analogue. *Biophysical Chemistry, 80,* 85–91.

Kunnumakkara, A. B., Diagaradjane, P., Anand, P., et al. (2009). Curcumin sensitizes human colorectal cancer to capecitabine by modulation of cyclin D1, COX-2, MMP-9, VEGF and CXCR4 expression in an orthotopic mouse model. *International Journal of Cancer. Journal International du Cancer, 125,* 2187–2197.

Kunwar, A., Sandur, S. K., Krishna, M., et al. (2009). Curcumin mediates time and concentration dependent regulation of redox homeostasis leading to cytotoxicity in macrophage cells. *European Journal of Pharmacology, 611,* 8–16.

Kuttan, R., Bhanumathy, P., Nirmala, K., et al. (1985). Potential anticancer activity of turmeric (Curcuma longa). *Cancer Letters, 29,* 197–202.

Kuttan, R., Sudheeran, P. C., & Josph, C. D. (1987). Turmeric and curcumin as topical agents in cancer therapy. *Tumori, 73,* 29–31.

Lampe, V., & Milobedzka, J. (1913) Studien über Curcumin. *Berichte der deutschen chemischen Gesellschaft, 46,* 2235–2240.

Lee, J. C., Kinniry, P. A., Arguiri, E., et al. (2010). Dietary curcumin increases antioxidant defenses in lung, ameliorates radiation-induced pulmonary fibrosis, and improves survival in mice. *Radiation Research, 173,* 590–601.

Lev-Ari, S., Lichtenberg, D., & Arber, N. (2008). Compositions for treatment of cancer and inflammation. *Recent Patents on Anticancer Drug Discovery, 3,* 55–62.

Li, L., Braiteh, F. S., & Kurzrock, R. (2005). Liposome-encapsulated curcumin: In vitro and in vivo effects on proliferation, apoptosis, signaling, and angiogenesis. *Cancer, 104,* 1322–1331.

Lin, S. S., Lai, K. C., Hsu, S. C, et al. (2009). Curcumin inhibits the migration and invasion of human A549 lung cancer cells through the inhibition of matrix metalloproteinase-2 and -9 and Vascular Endothelial Growth Factor (VEGF). *Cancer Letters, 285,* 127–133.

López-Lázaro, M. (2008). Anticancer and carcinogenic properties of curcumin: Considerations for its clinical development as a cancer chemopreventive and chemotherapeutic agent. *Molecular Nutrition & Food Research, 52*(Suppl. 1), S103–S127.

Mahmoud, N. N., Carothers, A. M., Grunberger, D., et al. (2000). Plant phenolics decrease intestinal tumors in an animal model of familial adenomatous polyposis. *Carcinogenesis, 21,* 921–927.

Mendonça, L. M., Dos, Santos, G. C., Antonucci, G. A., et al. (2009). Evaluation of the cytotoxicity and genotoxicity of curcumin in PC12 cells. *Mutation Research, 675,* 29–34.

NCI, D. (1996). Clinical development plan: Curcumin. *Journal of Cellular Biochemistry. Supplement, 26,* 72–85.

Oppenheimer, A. (1937). Tumeric (curcumin) in biliary diseases. *Lancet, 229,* 619–621.

Padhye, S., Banerjee, S., Chavan, D., et al. (2009). Fluorocurcumins as cyclooxygenase-2 inhibitor: Molecular docking, pharmacokinetics and tissue distribution in mice. *Pharmaceutical Research, 26,* 2438–2445.

Park, J., Ayyappan, V., Bae, E. K., et al. (2008). Curcumin in combination with bortezomib synergistically induced apoptosis in human multiple myeloma U266 cells. *Molecular Oncology, 2,* 317–326.

Patel, B. B., Gupta, D., Elliott, A. A., et al. (2010). Curcumin targets FOLFOX-surviving colon cancer cells via inhibition of EGFRs and IGF-1R. *Anticancer Research, 30,* 319–325.

Plummer, S. M., & Holloway, K. A., (1999). Manson MM, et al. Inhibition of cyclo-oxygenase 2 expression in colon cells by the chemopreventive agent curcumin involves inhibition of NF-kappaB activation via the NIK/IKK signalling complex. *Oncogene, 18,* 6013–6020.

Reddy, S., Rishi, A. K., Xu, H., et al. (2006). Mechanisms of curcumin- and EGF-receptor related protein (ERRP)-dependent growth inhibition of colon cancer cells. *Nutrition and Cancer, 55,* 185–194.

Samaha, H. S., Kelloff, G. J., Steele, V., et al. (1997). Modulation of apoptosis by sulindac, curcumin, phenylethyl-3-methylcaffeate, and 6-phenylhexyl isothiocyanate: Apoptotic index as a biomarker in colon cancer chemoprevention and promotion. *Cancer Research, 57,* 1301–1305.

Shahani, K., Swaminathan, S. K., Freeman, D., et al. (2010). Injectable sustained release microparticles of curcumin: A new concept for cancer chemoprevention. *Cancer Research, 70,* 4443–4452.

Shankar, T., Shantha, N. V., Ramesh, H. P., et al. (1980). Toxicity studies on turmeric (Curcuma longa): Acute toxicity studies in rats, guinea pigs and monkeys. *Indian Journal of Experimental Biology, 18,* 73–75.

Sharma, R. A., Euden, S. A., Platton, S. L., et al. (2004). Phase I clinical trial of oral curcumin: Biomarkers of systemic activity and compliance. *Clinical cancer research: An official journal of the American Association for Cancer Research, 10,* 6847–6854.

Sharma, R. A., McLelland, H. R., Hill, K. A., et al. (2001). Pharmacodynamic and pharmacokinetic study of oral Curcuma extract in patients with colorectal cancer. *Clinical cancer research: An official journal of the American Association for Cancer Research, 7,* 1894–1900.

Sou, K., Inenaga, S., Takeoka, S., et al. (2008). Loading of curcumin into macrophages using lipid-based nanoparticles. *International Journal of Pharmaceutics, 352,* 287–293.

Strimpakos, A. S., & Sharma, R. A. (2008). Curcumin: Preventive and therapeutic properties in laboratory studies and clinical trials. *Antioxidants & Redox Signaling, 10,* 511–545.

Sung, B., Kunnumakkara, A. B., Sethi, G., et al. (2009). Curcumin circumvents chemoresistance in vitro and potentiates the effect of thalidomide and bortezomib against human multiple myeloma in nude mice model. *Molecular Cancer Therapeutics, 8,* 959–970.

Tharakan, S. T., Inamoto, T., Sung, B., et al. (2010). Curcumin potentiates the antitumor effects of gemcitabine in an orthotopic model of human bladder cancer through suppression of proliferative and angiogenic biomarkers. *Biochemical Pharmacology, 79,* 218–228.

Chapter 12
An Evidence-based Perspective of *Ganoderma Lucidum* (Lucid Ganoderma) for Cancer Patients

Zheng-Yuan Su and Lee-Yan Sheen

Abstract *Ganoderma lucidum* (lucid ganoderma), a traditional Chinese herb, has been used extensively in East Asian for thousand years, and the biological activities and pharmacological functions of lucid ganoderma have been successively studied. Nowadays, some researches use scientific methods and techniques to study its anticancer effect on the cancer patients in clinical trial. For example, PC-SPES extracted from a mixture of lucid ganoderma and seven herbs decreased the prostate-specific antigen in patients with chemotherapy-induced hormone-independent prostate cancer. Lucid ganoderma extracts also improved the immune-stimulating response, such as the increase of plasma interleukin (IL)-2, IL-6, and interferon-γ concentrations, the enhance of natural killer cell activity, the decrease of plasma IL-1 and tumor necrosis factor-α concentrations, etc, as well as had low adverse effects in the cancer patients. The anticancer effects of lucid ganoderma in cell and animal models could be the strong references for the clinical trials. The components with anticancer potential in lucid ganoderma include triterpenoids, steroids, polysaccharides, fatty acids, and novel proteins such as LZ-8. On the basis of *in vitro* and *in vivo* study, the anticancer mechanism of lucid ganoderma treatment against the growth of cancer cells in clinical trial might be mediated by cell cycle arrest, apoptosis, anti-invasion, anti-migration, immunomodulation, anti-angiogenesis, etc. Additionally, the combination from lucid ganoderma and other herbs or foods as an alternative treatment might exhibit synergistic anticancer efficacy. However, more studies regarding the safety and application in clinical trial need to be processed in the future for providing more evidences.

L.-Y. Sheen (✉)
Graduate Institute of Food Science and Technology, National Taiwan University, No 1, Sec 4, Roosevelt Road, 106 Taipei, Taiwan
e-mail: lysheen@ntu.edu.tw

W. C. S. Cho (ed.), *Evidence-based Anticancer Materia Medica,* Evidence-based Anticancer Complementary and Alternative Medicine,
DOI 10.1007/978-94-007-0526-5_12, © Springer Science+Business Media B.V. 2011

12.1 Introduction

Ganoderma lucidum (lucid ganoderma) is a kind of saprophytic fungus and usually grows well in a wet and ventilated environment with high temperature. Lucid ganoderma is classified as *Mycota, Basidiomycotina, Hymenomycetes, Holobasidiomycetidae, Aphyllophorales, Polyporaceae, Ganodermoideae, Ganoderma* according to the modern categorized system in Western (Shiao 1992). Lucid ganoderma is one of the Chinese precious medical fungi in the medicine since ancient times. Different cultivation methods of lucid ganoderma lead to the complicated and changeable chemical composition (Han et al. 2005; Liu and Zhang 2007).

The major constitutes contains carbohydrates such as polysaccharides (Zhu and Lin 2005), proteins such as enzymes and glycoproteins (Miura et al. 2002; Wang et al. 2002), and other bioactive compounds such as triterpenoids, steroids, and fatty acids (Akihisa et al. 2007; Fukuzawa et al. 2008). Because novel triterpenoids in numerous chemical compositions especially exists in lucid ganoderma, they have already been used to be indicators for taxonomy and quality in pharmacology (Chen et al. 1999). Many physiological functions of lucid ganoderma have been investigated, including anticancer activity (Nonaka et al. 2008), live-protective effect (Yang et al. 2006), antioxidation (Yuen and Gohel 2008), anti-blood platelet agglutination (Su et al. 2000), immune adjustment (Zhuang et al. 2009), cholesterol reduction (Hajjaj et al. 2005), anti-viral activity (Li and Wang 2006), and hypoglycemic effect (Zhang and Lin 2004).

In this chapter, we have reviewed recent researches regarding the effect of lucid ganoderma treatment on the subjects with cancer. The active components and possible mechanisms of anticancer activity *in vitro* and *in vivo* of lucid ganoderma will be discussed to expound how lucid ganoderma improved the constitutions of human being, especially cancer patients. In addition, the safety of lucid ganoderma for normal human subjects will also be described. These evidences may support the application of lucid ganoderma for the treatment of cancer patients.

12.2 Clinical Trials of Cancer Patients

According to the clinical results of de la Taille et al. (2000), it has been reported that two case reports regarding effect of PC-SPES extract, an herbal mixture contained lucid ganoderma, chrysanthemuism, isatis, licorice, ginseng, *Rabdosia rubescens*, saw palmetto, and scutellaria on hormone-refractory prostate cancer patients. A 73-year-old man with Gleason 9 (4+5) prostate cancer underwent surgery and treatment with bicalutamide for androgen blockade, but his high prostate-specific antigen (PSA) level suggested that the prostate cancer is not controlled well. After taking 3 tablets of PC-SPES per day for 6 months, his PSA level was decreased from 100 to 24 ng/ml (76% reduction) without any side effect. In the other case, an 80-year-old man with Gleason 8 T2 N0 M0 prostate cancer received radiation therapy and successively took the medicine such as leuprolide, bicalutamide, ketoconazole, and hormonal steroids, but his PSA levels did not always be lowered and maintained. However, the PSA level was decreased from 386 to 114 ng/ml (72% reduction) and remained stably after treated with PC-SPES (6 tablets/day) for 4 months, and no side effects were observed. These two cases suggested that PC-SPES might have some potential activity against hormone-independent prostate cancers. Because PC-SPES has a strong estrogenic activity *in vitro* and *in vivo*, it suggested that PC-SPES might be an alternative drug to treat hormone-independent prostate cancers (de la Taille et al. 2000; Hsieh and Wu 2002).

In the clinical trial of 47 patients (27 men and 20 women, average 48.4 ± 7.0 years old) with advanced colorectal cancer, 35 and 45 subjects received surgical resection and previous chemotherapy/radiotherapy, respectively. These patients were treated with 5.4 g/day Ganopoly polysaccharide product, a polysaccharide-enriched fraction from lucid ganoderma fruiting body, for 12 weeks, and a small increase of CD3, CD4, CD8, and CD56 lymphocytes counts, plasma interleukin (IL)-2, IL-6, interferon (IFN)-γ concentrations, and natural killer (NK) cell activity, as well as a small decrease of plasma IL-1 and tumor necrosis factor (TNF)-α concentrations in 41 assessable cancer patients compared to the baseline data. The lucid ganoderma fraction consists of 98.8% polysaccharides with glucose (61.2%), xylose (15.5%), fructose (14.4%), galactose (4.8%), and rhamnose (4.1%) linked together by β-glycosidic linkages, and the clinical results for cancer patients might be correlated with the potential immune-modulating effect of lucid ganoderma (Chen et al. 2006).

Gao et al. (2003) also reported the effect of lucid ganoderma extract containing 25% (w/w) crude polysaccharides on the immune functions of 34 patients (31–77 years old; 20 men and 14 women) with different cancer origins including lung (7 subjects), colon (6 subjects), breast (5 subjects), liver (5 subjects), prostate (4 subjects), bladder (2 subjects), brain (2 subjects), and unknown (3 subjects). Except for 2 patients without previous treatment, the patients were treated with surgery (23 subjects), chemotherapy (6 subjects), radiotherapy (10 subjects), immunotherapy (12 subjects), endocrine treatment (5 subjects), traditional Chinese medicine (19 subjects), and more than two of above treatments excluding surgery (18 subjects). The statistic results of whole subjects compared to the baseline levels showed that oral administration of lucid ganoderma extract (1,800 mg, equal to 90 g of fruiting body, three times daily before meals) for 12 weeks significantly increased the mean concentrations of plasma IL-2, IL-6, and IFN-γ ($P<0.05$) and significantly decreased the levels of IL-1 and TNF-α ($P<0.05$). In terms of the mean absolute number of lymphocyte subset, a significant increase of $CD56^+$ (NK cells) ($P<0.05$) and a small increase of $CD3^+$ (T lymphocyte), $CD4^+$ (T helper cells), and $CD8^+$ (T suppressor cells) with unchanged CD4:CD8 T cell ratios were induced by lucid ganoderma extract as compared to the baseline levels. Lucid ganoderma extract also significantly enhanced the proliferation of phytohemagglutinin-stimulated lymphocyte compared to the pretreatment baselines and increased the mean NK cells activity compared to the baselines ($P<0.05$). It indicated that lucid ganoderma extract stimulated the immune responses in patients with advanced-stage cancer in clinical trial (Gao et al. 2003).

A total of 105 cancer patients (33–84 years old) receiving chemotherapy and/or radiotherapy were subjected in the study of Zhuang et al. (2009), including 60 breast cancer patients, 24 colorectal cancer patients, 14 nasopharyngeal cancer patients, and 7 lung cancer patients with disease stages I-IV except for 5 patients with unknown stage. Chinese medicinal herb complex (CCMH) is a mixture of citronellol powder (273.6 mg) and extracts of lucid ganoderma (3 mg), *Codonopsis pilosula* (27.1 mg), and *Angelicae sinensis* (64.5 mg) in one capsule. Either 9 capsules of CCMH ($n=55$) or placebo (control group) ($n=50$) were supplied by the cancer patients every day for 6 weeks, and the mean percentages of leukocytes and neutrophils in CCMH group were significantly higher than those in control group ($P<0.05$). Supplement of CCMH also maintained the mean percentages of CD4 lymphocytes and NK cells compared to control group. Therefore, treatment with CCMH improved the immune function in the cancer patients undergoing chemotherapy and/or radiotherapy (Zhuang et al. 2009).

In an open-label study of Yoshimura et al. (2010), 17 patients (60–80 years old: 15 patients; <60 years old: 2 patients) with biochemical failure after radical treatment for non-metastasized prostate cancer with different clinical stages including B0 (3 subjects), B1 (4 subjects), B2 (7 subjects), and C1 (3 subjects). After supplement of Rokkaku Reishi, a lucid ganoderma product in Japan, for 6 months, although the results indicated that Rokkaku Reishi did not exhibit significant anticancer effects, but no patients had serious adverse effects due to Rokkaku Reishi, including blood/bone marrow, dermatology/skin, and gastrointestinal events (Yoshimura et al. 2010).

In order to understand the anticancer effect of lucid ganoderma, we prefer to discuss what active compounds are contributed to and what mechanisms are mediated. However, the evidences, such as survival prolongation, adverse effects reduction, etc, of lucid ganoderma treatment against cancer in the patients are not enough to clarify the efficacy, thus the inhibitive effect of lucid ganoderma on the cancer cells *in vitro* and *in vivo* must be explicated as well.

12.3 Anticancer Activity of Lucid Ganoderma and Its Crude Extracts

In an anticancer animal model, the growth of MM 46 mammary carcinoma inoculated in C3H/HeN mice were inhibited after taking AIN-93 M feed containing 2.5% lucid ganoderma fruiting body for 28 days (Nonaka et al. 2008). Lucid ganoderma (2.5%) containing diet also suppressed tumor growth elongated the life span of ddY mice inoculated with Sarcoma 180 cells after 100 days treatment (Nonaka et al. 2006). Hot water extracts from lucid ganoderma fruiting body (100 mg/kg bw) and spores (1,000 mg/kg bw) also possessed inhibitory activities on Sarcoma 180 cell growth in the implanted mice (Yue et al. 2008a).

In terms of cancer cell experiments, the water extract from lucid ganoderma fruiting body exhibited anti-proliferative effects on myeloid leukemia HL-60, U937, and K562 cells, lymphoblastic leukemia Blin-1 and Nalm-6 cells, and multiple myeloma RPMI8226 cells, and their ED_{50} (the effective dose which inhibited 50% growth) were 26–40 μg/ml (Muller et al. 2006). It also reported that the growth of human small cell lung cancer drug-sensitive (H69) and multi-drug resistant (VPA) cells were inhibited by the water extract from lucid ganoderma fruiting body, and the IC_{50} (the concentration of the sample to inhibit cell growth by 50%) were 60 and 80 μg/ml, respectively (Sadava et al. 2009). Against human breast cancer MCF-7 and MDA-MB-231 cell lines, the aqueous extracts of different parts of lucid ganoderma, including whole fruiting body, pileus, and stipe, have different potential inhibitive activity (Yue et al. 2006).

In addition to aqueous extract, crude methanolic extract of lucid ganoderma fruiting body caused cell death of murine cancer L1210 and 3LL cells (Tomasi et al. 2004), and the column-chromatography semipurified fraction from lucid ganoderma methanolic extract reduced cell viability of human acute promyelocytic leukemia NB4 cells and mouse IL-3 dependent lymphoma DA-1 cells (Calvino et al. 2010a, b). It has also been demonstrated that the ethanolic extract of lucid ganoderma fruiting body induced apoptosis in MCF-7 as well as human colonic carcinoma HT-29 and gastric carcinoma AGS cell lines (Hu et al. 2002; Hong et al.

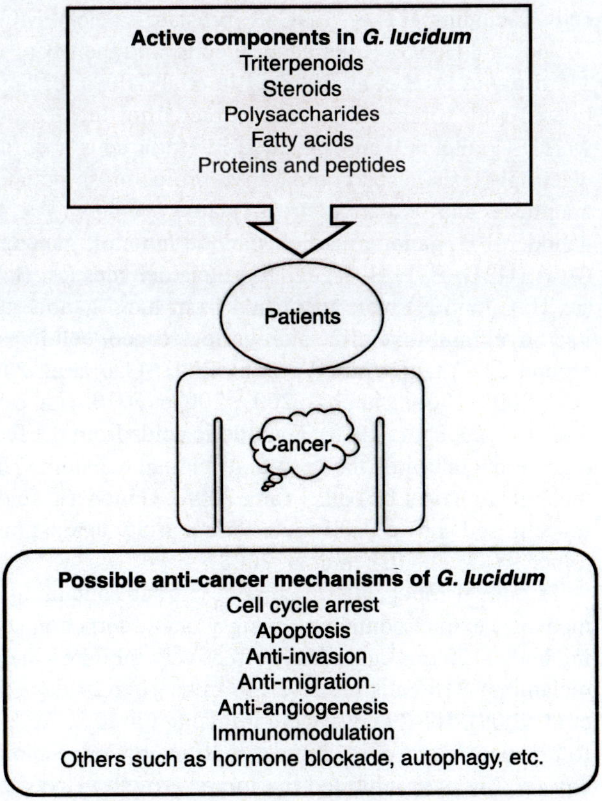

Fig. 12.1 The active components in *Ganoderma lucidum* (lucid ganoderma) and their possible anticancer mechanisms on the cancer patients

2004; Jang et al. 2010). The crude water or ethanolic extracts of lucid ganoderma spores also exhibited anticancer activity against MDA-MB-231, prostate cancer PC-3, and cervix uteri tumor HeLa cells (Zhu et al. 2000; Sliva et al. 2002, 2003; Lu et al. 2004).

Therefore, in order to support more evidences for cancer patients treated with lucid ganoderma, the active compounds with anticancer activity in lucid ganoderma and possible anticancer mechanisms of lucid ganoderma will be further discussed as follows (Fig. 12.1).

12.4 Active Components with Anticancer Activity in Lucid Ganoderma

12.4.1 Triterpenoids and Steroids

Most of the researches reported that triterpenoids have direct inhibitory activity against cancer cells *in vitro*. The triterpenoids extracts isolated from lucid ganoderma fruiting body have been demonstrated to inhibit proliferation of human cancer cells, including HT-29, LNCaP (prostate cancer), MCF-7, MDA-MB-231, HeLa, HL-60, and HepG2 (liver cancer) cells, and their IC_{50} values were among 71.6 and 219.5 μg/ml (Yue et al. 2008c; Liu et al. 2009a, b; Thyagarajan et al. 2010). After treated with triterpenoids-rich extract from lucid ganoderma fruiting body, there were 14 proteins were regulated in HeLa cells according to two-dimensional gel electrophoresis-based comparative proteomics, including proliferation, cell cycle, apoptosis, and oxidative stress-related proteins (Yue et al. 2008c). Ganoderiol F, ganoderol B, ganodermadiol, ganodermnonol, ganodermaondiol, genoderic acids (A, AM1, D, E, F, H, K, and T), lucialdehydes (A, B, and C), and lucidenic acids (A, B, C, and N) were identified from lucid ganoderma fruiting body and exhibited novel inhibitive effects on various cancer cell lines, and their IC_{50} values were around 2.8–27.9 μg/ml (Wu et al. 2001; Gao et al. 2002; Tang et al. 2006; Weng et al. 2007, 2008; Liu et al. 2007, 2009a, 2010; Hsu et al. 2008a; Jiang et al. 2008; Yue et al. 2008b, c, 2010). Ganoderic acids from the fermentation products of lucid ganoderma inhibited the growth of human hepatoma BEL7402 cells, but not a normal human liver L02 cells (Yang 2005). Ganoderic acid Me existed in fermentation mycelia and showed anti-metastatic activity against human highly metastatic lung tumor 95-D cells (Chen et al. 2008).

In animal model, the methanol extract containing total terpenoids or purified methanol extract containing mainly acidic terpenoids from lucid ganoderma fruiting body (100 mg/kg bw/day) effectively inhibited the growth of implanted murine melanoma B16 cells in C57BL/6 mice after 24 days treatment (Harhaji Trajkovic et al. 2009). HT-29 cells were implanted in male nude mice, and lucid ganoderma triterpene extract containing a mixture of lanostanoid triterpenes (6 mg/kg bw/day for 23 days) inhibited the tumor growth (Thyagarajan et al. 2010). After taking 200 mg/kg bw/day lucidenic acid-rich lucid ganoderma extract for 68 days, the growth of HepG2 cells implanted into ICR mice were suppressed (Weng et al. 2009). Ganoderic acid T has been demonstrated to suppress tumor growth in the model of male BALB/c implanted with 95-D cells and celiac injected with ganoderic acid T (12.5 mg/kg bw/day) for 12 days (Tang et al. 2006). Ganoderic acid T at 7 mg/kg bw/day (celiac injection, 10 days) also decreased tumor growth and metastasis of murine Lewis lung carcinoma (LLC) cells implanted into male C57B/6 mice (Chen et al. 2010). From lucid ganoderma fermentation mycelia, 28 mg/kg bw/day of ganoderic acid Me (16 days) were intraperitoneally administrated to inhibit both lung tumor growth and lung metastasis of Lewis lung carcinoma in C57BL/6 mice (Wang et al. 2007).

In addition to triterpenoids, ethanolic extract of lucid ganoderma fruiting body contained sterols and exhibited anti-proliferative activities against MCF-7, MDA-MB-231, HepG2, and HL-60 cell lines (Liu et al. 2009b). Two ergosterol derivatives, ergosterol peroxide, and 9,11-dehydroergosterol peroxide, form ethanolic extract of fermentation mycelia were also isolated and had cytotoxicity against human hepatocellular carcinoma Hep3B cells (Chen et al. 2009).

12.4.2 Polysaccharides

A Ganopoly polysaccharides product showed cytotoxicity against Hep3B, HepG2, SiHa (cervical carcinoma), CaSki (cervical carcinoma), HT-29, HCT116 (colorectal adenocarcinoma), and MCF-7 cells (Gao et al. 2005). Furthermore, oral administration of Ganopoly polysaccharides product (20 mg/kg bw/day) also reduced the tumor weight in C57BL/6 J mice bearing Sarcoma 180 after 21 days treatment (Gao et al. 2005). It also reported that oral supplement of lucid ganoderma polysaccharides from fruiting body at 50 mg/kg bw/day for 10 days exhibited tumor inhibitory against Sarcoma 180 in BALB/c mice (Li et al. 2008). It has been demonstrated that daily intraperitoneal injection of a glucan extracted from lucid ganoderma spores (50 mg/kg bw for 2 weeks) had tumor-suppress activity against Lewis lung cancer bearing in C57BL/6 mice (Guo et al. 2009). Additionally, sulfated and carboxymethylated lucid ganoderma polysaccharides were synthesized for stronger inhibition of the growth of sarcoma 180 tumor cells *in vitro* and *in vivo* (Wang et al. 2009).

12.4.3 Other Active Compounds

Except for triterpenoids, steroids, and polysaccharides, long chain saturated (C19:0, C18:0, C17:0, and C16:0) and unsaturated (C19:1, C18:1, C17:1, and C16:1) fatty acids in the ethanolic extract from lucid ganoderma spores inhibited the proliferation, and especially ethanolic extract and 19-carbon fatty acids induced apoptosis in HL-60 cells (Fukuzawa et al. 2008). It also reported that polysaccharides peptide (polysaccharides:peptides is 94.8:5.2) from boiling water extract of lucid ganoderma fruiting body at 50 mg/kg bw/day effectively suppressed either sarcoma 180 (10 days treatment) or human lung carcinoma PG cells (33 days treatment) implanted into BALB/c mice (Cao and Lin 2004). However, their cancer-inhibitory effects *in vivo* need to be further studied to support more adequate references for treating cancer in clinical trial.

12.5 Anticancer Mechanisms for Lucid Ganoderma

12.5.1 Cell Cycle Arrest

It has been reported that water extract of lucid ganoderma fruiting bodies induced cell cycle arrest at G2/M phase in human immune system-related cancer cell lines such as myeloid leukemia HL-60, U937, and K562 cells, lymphoblastic leukemia Blin-1 and Nalm-6 cells, and multiple myeloma RPMI8226 cells (Muller et al. 2006) and at S phase in H69 cells (Sadava et al. 2009). The ethanolic extract from lucid ganoderma fruiting body up-regulated p21/Waf1 and down-regulated cyclin D1, cdk4 (cyclin-dependent protein kinase 4) and transcription factor E2F to arrest cell cycle at G0/G1 phase in MCF-7 cells (Hu et al. 2002). A powdered mixture (containing 6% triterpenoids and 13.5% polysaccharides) from fruiting body extract and spores also induced G0/G1 phase arrest which might be correlated with the down-regulation of Akt/NF-κB signaling pathway, cyclin D1, and cdk4 in MDA-MB-231 cells (Jiang et al. 2004).

Regarding anti-proliferation potential of triterpenoids, it reported that a crude ganoderic acid-rich extract from submerged culture of lucid ganoderma blocked the cell cycle at the transition from G1 to S phase in BEL7402 cells (Yang 2005). A triterpenoids-enriched extract arrested the cell cycle of human hepatoma Huh-7 cells at G2 phase mediated by decreasing the PKC activity and JNK and p38 MAP kinases activation (Lin et al. 2003). Lucidenic acids A, C, and N from fruiting body induced cell cycle arrest at G1 phase in HL-60 cells (Hsu et al. 2008a). Both ganod eric acids A and H inhibited the activities of transcription factors AP-1 (activator protein-1) and NF-κB resulting in the decrease of Cdk4 expression and the suppression of uPA (urokinase-type plasminogen activator) secretion in MDA-MB-231 cells (Jiang et al. 2008). It has been demonstrated that ganoderic acid D induced G2/M phase arrest in HeLa cells mediated by binding six isoforms of 14-3-3 protein family, annexin A5, and aminopeptidase B (Yue et al. 2008b).

Some researches modified the structures of polysaccharides from lucid ganoderma by the method of chemical synthesis for giving better anticancer activity. For example, it also reported that chemical modified lucid ganoderma polysaccharides such as sulfated and carboxymethylated polysaccharides induced cell-cycle arrest in the G2/M phase to inhibit the proliferation of sarcoma 180 tumor cells

(Wang et al. 2009). But the safety and bioactivity of these new compounds have to be further investigated using animal models.

12.5.2 Apoptosis

Cell death of four hematopoietic cell lines such as HL-60, U937, Blin-1, and RPMI8226 cells were caused by water extract from lucid ganoderma fruiting bodies *via* apoptosis (Muller et al. 2006). A water extract from fruiting bodies increased DNA fragmentation, TUNEL staining for DNA breaks, and specific activities of caspase-3 and -9, but not caspase-8 by colorimetric assays (Sadava et al. 2009). Aqueous extract of lucid ganoderma fruiting bodies induced apoptosis accompanied with the increasing the expressions of Bax, p53, mdm2, and cleavaged caspase-3 in DA-1 cells (Calvino et al. 2010b).

A column-chromatography semipurified fraction from lucid ganoderma methanolic extract caused reduction of the Bcl2/Bax ratio, both unphosphorylated and phosphorylated Akt (protein kinase B) levels, and Erk1/2 (extracellular signal-regulated kinase) levels in NB4 cells (Calvino et al. 2010a). It has been reported that methanolic extract containing total terpenoids form fruiting body induced caspase-dependent apoptotic cell death which might be mediated by producing reactive oxygen species, up-regulated p53, and inhibited Bcl-2 expression in B16, mouse fibrosarcoma L929, and rat astrocytoma C6 cells (Harhaji Trajkovic et al. 2009). Ethanolic extract of lucid ganoderma also has pro-apoptotic function. For example, ethanolic extract of fruiting body increased the activity of caspase-3 in HT-29 cells (Hong et al. 2004). AGS cell line was treated with ethanolic extract of fruiting body, and then the phenomena of two apoptotic pathways were observed as follows: (1) mitochondria-mediated intrinsic pathway, including the activation of caspase-3 and -9, the cleavage of Bid, and the degradation of poly(ADP-ribose) polymerase (PARP); (2) death receptor-mediated extrinsic pathway, including the increase of death receptor-related proteins such as death receptor 5 and tumor necrosis factor-related apoptosis-inducing ligand, the activation of caspase-8, and the down-regulation of IAP family proteins such as XIAP and survivin (Jang et al. 2010).

In terms of pure triterpenoid compounds, lucidenic acid B induced apoptosis through mitochondria pathway in HL-60 cells, including the loss of mitochondria membrane potential, the decrease of the ratio of pro- and anti-apoptotic Bcl-2 family expressions, the release of mitochondria cytochrome c, and the activation of caspase-3 and -9, and the cleavage of PARP (Hsu et al. 2008a). Ganoderic acid T had similar pro-apoptotic mechanism against 95-D cells (Tang et al. 2006).

12.5.3 Anti-invasion and Anti-migration

Treatment of lucid ganoderma (2.5%)-containing AIN-93M (14 days) showed the anti-metastatic activity against the implanted LLC cells to the lung in C57BL/6 mice (Nonaka et al. 2008). The water extract from lucid ganoderma either spores or dried fruiting body inhibited invasion and migration of both PC-3 and MDA-MB-231 cells mediated by inhibiting the activities of transcription factors AP-1 and NF-κB, decreasing the expressions of uPA and uPAR (uPA receptor), and suppressing the secretion of uPA (Sliva et al. 2002, 2003). The lucid ganoderma fruiting body extract containing 13.5% polysaccharides and 6% triterpenoids inhibited oxidative stress-induced migration of MCF-7 cells through down-regulation of MAPK signaling pathway such as the suppression of oxidative stress stimulated Erk1/2 phosphorylation, the decrease of c-Fos expression, the inhibition of AP-1 and NF-κB activities, and the suppression of oxidative stress-mediated IL-8 secretion from MCF-7 cells (Thyagarajan et al. 2006).

In phorbol-12-myristate-13-acetate-induced invasion of HepG2 cells, lucidenic acid-rich lucid ganoderma fruiting body extract reduced the expression of MMP-9 (matrix metalloproteinase-9) as well as inhibited the phosphorylations of ERK1/2 and Akt in the cytosol and the expressions of AP-1, NF-κB, c-Jun, and c-Fos in the nucleus (Weng et al. 2008, 2009). The possible active compounds with anti-invasion in lucid ganoderma fruiting body might be lucidenic acids A, B, C, and N (Weng et al. 2007). In animal model, supplement of the lucidenic acid-rich extract (200 mg/kg bw/day for 68 days) decreased the number of tumor foci and the activities of serum MMP-2 and MMP-9 in the ICR-nu/nu mice implanted with HepG2 cells (Weng et al. 2009). Ganoderic acid Me was isolated from lucid ganoderma fermentation mycelia and showed the inhibitory effect on the cell adherence to extracellular matrix of highly metastatic 95-D cells mediated by suppressing mRNA and protein expressions of MMP-2 and MMP-9 (Chen et al. 2008). Additionally, ganoderic acid T inhibited the invasion of HCT116 cells, and the results suggested that ganoderic acid T mediated the inhibition of nuclear translocation of NF-κB and the degradation of IκB-α inhibitor resulting in the decrease of expressions of MMP-9, iNOS (inducible nitric oxide synthase), and uPA (Chen et al. 2010). In LLC cells implanted male C57B/6 mice, ganoderic acid T (7 mg/kg bw/day) *via* celiac injection for 10 days) suppressed LLC metastasis and decreased the expressions of *MMP-2* and *MMP-9* mRNA levels (Chen et al. 2010).

12.5.4 Anti-angiogenesis

Lucid ganoderma inhibits the early event in angiogenesis (Stanley et al. 2005). It reported that a mixture from fruiting body extract and spores suppressed the phosphorylation of Erk1/2 and Akt kinases, and then inhibited the secretion of vascular endothelial growth factor and TGF-β1 in highly invasive PC-3 cells (Stanley et al. 2005). The polysaccharides and peptides from lucid ganoderma fruiting body showed anti-angiogenic potential because it inhibited the proliferation of human umbilical cord vascular endothelial cells and decreased vascular endothelial growth factor in PG cells (Cao and Lin 2004, 2006).

12.5.5 Immunomodulation

AIN-93 M feed containing lucid ganoderma (2.5% for 100 days) inhibited tumor growth and elongated the life span in the C3H/He mice implanted with MM 46 mammary carcinoma mediated by decreasing splenic CD8 cell number and IFN-γ production in regional lymph nodes (Nonaka et al. 2006). A hot water extract from sporoderm-broken spores stimulated the production of cytokines such IFN-γ, IL-4, and IL-6 of spleen lymphocytes resulting in its tumor inhibitory activity in sarcoma 180 bearing mice (Yue et al. 2008a). It reported that crude triterpenoids extract suppressed the production of TNF-α and IL-6 in LPS-induced endotoxemic mice (Dudhgaonkar et al. 2009). Ganoderic acid Me enhanced the activity of NK cells and increased the expressions of IL-2 and IFN-γ in LLC cells implanted C57BL/6 mice (Wang et al. 2007).

Actually, there are more studies regarding immunomodulatiing functions of lucid ganoderma polysaccharides recently (Chen et al. 2004; Gao et al. 2005; Chan et al. 2007, 2008; Hsu et al. 2009; Lai et al. 2010). For example, the polysaccharides form fruiting body increased TNF-α and IFN-γ protein expressions and mRNA levels in splenocytes, the cytotoxicity of cytotoxic T lymphocytes, and the activity of NK cells in both healthy and sarcoma 180 bearing C57BL/6 J mice (Gao et al. 2005). A polysaccharide fraction prepared from lucid ganoderma fruiting body showed immune-stimulating activity in BALB/c mice, including increasing the number of dendritic cells and CD4, CD8, regulatory T, B, plasma, NK and NKT ($CD3^+$ NK-T/NK^+) cells in the spleen, elevating the levels of multiple cytokines and chemokines in the blood, and enhancing both Th1 and Th2 responses (Lai et al. 2010). In an *in vitro* study, it also found that the polysaccharide fraction induced the maturation of dendritic cells derived from human monocytes mediated by up-regulating CD40, CD54, CD80, CD83, CD86, and HLA-DR, enhancing mixed lymphocyte reaction, and stimulating the production of ten cytokines and six chemokines (Lai et al. 2010). A heteroglycan from lucid ganoderma fruiting body stimulated immune system to enhance the proliferation of T and B lymphocytes and the production of antibodies, but little effect on serum IgG and complement (C3) levels in inbred ICR female mice (Bao et al. 2002). The structure of the glycan was identified to be a backbone consisting of 1,4-linked α-D-glucopyranosyl residues and 1,6-linked β-D-galactopyranosyl residues with branches at *O*-6 of glucose residues and *O*-2 of galactose residues, composed of terminal glucose, 1,6-linked glucosyl residues, and terminal rhamnose (Bao et al. 2002). Additionally, supplement of lucid ganoderma extract (400 mg/kg bw/day) enhanced the recovery of immunocompetence such as the increase of the leukocytes, the relative weight of thymus, and the increase of CD4 and CD8 splenocytes in gamma-ray-irradiated ICR male mice after 28 days treatment (Chen et al. 1995a, b).

Except for polysaccharides, the protein fraction from lucid ganoderma mycelia and culture liquid also had immunomodulating activity (Jeurink et al. 2008). The immune-stimulating effect of recombinant LZ-8 (rLZ-8) expressed from the cloned lucid ganoderma *LZ-8* gene has been investigated recently, including the rLZ-8-mediated signal-transduction pathways in the regulation of *IL-2* gene expression within human T cells (Hsu et al. 2008b), the rLZ-8-modulated the production of Th1 and Th2 cytokines by peripheral blood mononuclear cells and tumor NF-α by a macrophage cell line (Yeh et al. 2008), and the rLZ-8-induced activation and maturation of immature human monocyte-derived DCs (Lin et al. 2009). Although LZ-8 or rLZ-8 may be useful in cancer treatment, there is hardly any research to confirm the antitumor activity of LZ-8 or rLZ-8 in clinical trial for human being currently.

12.5.6 Others

The cell viability of estrogen-dependent MCF-7 cell line was decreased by the mixture of lucid ganoderma fruiting body extract and spores accompanied with the down-regulation of ERα (estrogen receptor-α) and *c-myc* expressions, the inhibition of constitutive transactivation activity of ER through estrogen response element, and the decrease of TNF-α-induced NF-κB activity (Jiang et al. 2006). It also reported that ganoderiols B and F and ganodermatriol isolated from lucid ganoderma fruiting body showed a binding activity to androgen receptor which might be related to the anti-proliferative effect of lucid ganoderma on LNCaP cells (Liu et al. 2007, 2009a, 2010). Therefore, it suggested that the progress of prostate cancer in the patients affected by PC-SPES containing lucid ganoderma might be correlated to this hormone-dependent pathway (de la Taille et al. 2000).

Moreover, it is worth to mention a new report that lucid ganoderma extract containing lanostanoid triterpenoids suppressed the phosphorylation of p38 MAPK and induced the expressions of Beclin-1 and LC-3 proteins resulting in induction of autophagy, programmed cell death Type II (Thyagarajan et al. 2010). This is a new pathway of cell death induced by lucid ganoderma.

12.6 Anticancer Activity of Combination Treatment

Some researches proved the anticancer potential of lucid ganoderma combined with other herbs or foods using scientific methodology recently. For example, PC-SPES is a herbal mixture derived from eight different herbs including lucid ganoderma, *Dendranthema morifolium* Tzvel, *Panax pseudoginseng, Glycyrrhiza uralensis* Fisch, *Rabdosia rubescens* Hara, *Scutellaria baicalensis* Georgi, *Isatis indigotica* Fort, and *Serenoa repens*, and it effectively suppressed the growth of LNCaP cells (Hsieh and Wu 2002). Combined treatment with extracts of 150 μg/ml lucid ganoderma (triterpenoid-enriched fraction) and 100 μg/ml *Duchesnea chrysantha* (polysaccharide-enriched fraction) induced G1 phase arrest and apoptosis *via* mitochondria pathway in HL-60 cells, including the down-regulation of Bcl-2, the translocation of Bax, the release of mitochondrial cytochrome c, and the activation of caspase-3 (Kim et al. 2007, 2008). Lucid ganoderma combined with crocodile egg extract and ginseng inhibited proliferation and colony formation of acute myelogenous leukemia KG1a cell line (Chui et al. 2006). A mixture composed of lucid ganoderma extract containing 13.5% polysaccharides and 6% triterpenes and green tea extract containing 97% polyphenols synergistically inhibited anchorage-dependent and -independent growth, adhesion, migration, and invasion of MDA-MB-231 cells with decreases of oncogene c-myc expression and uPA secretion (Thyagarajan et al. 2007). Traditional Botanical Supplement-101 containing lucid ganoderma, *Panax ginseng*, cranberry, green tea, grape skin, grape seed, and chamomile induced apoptosis in PC-3 cells (IC_{50}=1.4 μg/ml) *in vitro* and inhibited tumor growth and invasion in nude mice implanted PC-3 tumor cells with no toxicity *in vivo* (Evans et al. 2009).

In addition, the inhibitory effects of combinative treatments of lucid ganoderma and drugs on cancer were also studied. It reported that lucid ganoderma triterpenoids extract enhanced doxorubicin-increased reactive oxygen species production and doxorubicin-decreased Ku80 protein expression resulting in apoptosis in HeLa cells (Yue et al. 2008c). Oral administration of methanolic extract of lucid ganoderma (250 and 500 mg/kg body weight) prevented mice from cisplatin (an anticancer drug)-induced nephrotoxicity, and it suggested that lucid ganoderma has a potential therapeutic effect on cancer chemotherapy (Sheena et al. 2003).

According to the theory of Chinese medicine, some combination from the herbals, drugs, or foods with different attributes might exhibit better bioactivity in bodies. Therefore, lucid ganoderma combination treatment might be a good choice for cancer patients as well if their synergistic effect is obvious without any adverse effects.

12.7 Safety of Lucid Ganoderma in Clinical Trials

The safety and bioactivities of lucid ganoderma need to be carefully evaluated in normal human clinical trials before applying lucid ganoderma for cancer patients. For example, it has been reported that no adverse effects were observed in healthy subjects after intaking lucid ganoderma extract (2 g/day) for 10 days compared to placebo group (Wicks et al. 2007).

Clinical trials in another double-blinded, placebo-controlled, and cross-over intervention study, healthy volunteers took 0.72 g/day lucid ganoderma extract (equivalent to 6.6 g/day fresh mushroom) in the form of capsules for 10 days, and their plasma lipid standardized α-tocopherol concentration and urine antioxidant capacity were significantly increased ($P<0.05$) while plasma ascorbic acid and total alpha-tocopherol concentrations and erythrocyte superoxide dismutase and glutathione peroxidase activities were slightly increased (Wachtel-Galor et al. 2004a). Supplementation with lucid ganoderma capsules (1.44 g/day; equivalent to 13.2 g/day fresh mushroom) for 4 weeks also slightly lowered plasma total cholesterol, total triglyceride, and low density lipoprotein levels while it slightly increased plasma lipid standardized α-tocopherol concentration and urine antioxidant capacity in healthy adults (Wachtel-Galor et al. 2004b).

In 40 male football players during a 28-day "living high-training low" training, lucid ganoderma capsules containing pure spores and fruiting body extract were administrated by subjects in the study of Zhang et al. (2008). The results suggested that lucid ganoderma (5 g/day) could ameliorate the variation of the $CD4^+/CD8^+$ ratio in "living high-training low" training, and the main active components contributing to immuno-modulating function might be lucid ganoderma polysaccharides (Zhang et al. 2008).

12.8 Conclusion

There are some evidences for supporting the anticancer potential of lucid ganoderma in the clinical trial of cancer patients recently. Lucid ganoderma contains active compounds contributing to inhibitory function against the growth of cancer cells in the patients, including triterpenoids, steroids, polysaccharides, fatty acids, and novel proteins. They showed direct (such as cell cycle arrest and apoptosis) and indirect (such as immunomodulation) inhibitory effects on cancer. They also exhibited anti-invasion, anti-migration, and anti-angiogenesis activities against cancer cells. Some combinations of lucid ganoderma and other herbs or foods might be applied to treat cancer patients as an alternative treatment on the basis of their synergistic efficacy. More investigations on the safety and treatment dosages for cancer patients, the immune-modulating responses for chemo-prevention in normal human being, and the different culture condition including fermentation process to produce lucid ganoderma with stronger anticancer activity are worth to be further studied in the future.

References

Akihisa, T., Nakamura, Y., Tagata, M., et al. (2007). Anti-inflammatory and anti-tumor-promoting effects of triterpene acids and sterols from the fungus Ganoderma lucidum. *Chemistry & Biodiversity, 4,* 224–231.

Bao, X. F., Wang, X. S., Dong, Q., et al. (2002). Structural features of immunologically active polysaccharides from Ganoderma lucidum. *Phytochemistry, 59,* 175–181.

Calvino, E., Manjon, J. L., Sancho, P., et al. (2010a). Ganoderma lucidum induced apoptosis in NB4 human leukemia cells: Involvement of Akt and Erk. *Journal of Ethnopharmacology, 128,* 71–78.

Calvino, E., Pajuelo, L., de Eribe Casas, J. A., et al. (2010b). Cytotoxic action of Ganoderma lucidum on interleukin-3 dependent lymphoma DA-1 cells: Involvement of apoptosis proteins. *Phytotherapy Research: PTR, 25,* 25–32.

Cao, Q. Z., & Lin, Z. B. (2004). Antitumor and anti-angiogenic activity of Ganoderma lucidum polysaccharides peptide. *Acta Pharmacologica Sinica, 25,* 833–888.

Cao, Q. Z., & Lin, Z. B. (2006). Ganoderma lucidum polysaccharides peptide inhibits the growth of vascular endothelial cell and the induction of VEGF in human lung cancer cell. *Life Sciences, 78,* 1457–1463.

Chan, W. K., Law, H. K., Lin, Z. B., et al. (2007). Response of human dendritic cells to different immunomodulatory polysaccharides derived from mushroom and barley. *International Immunology, 19,* 891–899.

Chan, W. K., Cheung, C. C., Law, H. K., et al. (2008). Ganoderma lucidum polysaccharides can induce human monocytic leukemia cells into dendritic cells with immuno-stimulatory function. *Journal of Hematology & Oncology, 1,* 9.

Chen, D. H., Shiou, W. Y., Wang, K. C., et al. (1999). Chemotaxonomy of triterpenoid pattern of HPLC of Ganoderma lucidum and Ganoderma tsugae. *Journal of the Chinese Chemical Society Taipei, 46,* 47–51.

Chen, H. S., Tsai, Y. F., Lin, S., et al. (2004). Studies on the immuno-modulating and anti-tumor activities of Ganoderma lucidum (Reishi) polysaccharides. *Bioorganic & Medicinal Chemistry, 12,* 5595–5601.

Chen, N. H., Liu, J. W., & Zhong, J. J. (2008). Ganoderic acid Me inhibits tumor invasion through down-regulating matrix metalloproteinases 2/9 gene expression. *Journal of Pharmacological Sciences, 108,* 212–216.

Chen, N. H., Liu, J. W., & Zhong, J. J. (2010). Ganoderic acid T inhibits tumor invasion in vitro and in vivo through inhibition of MMP expression. *Pharmacological Reports: PR, 62,* 150–163.

Chen, W. C., Hau, D. M., & Lee, S. S. (1995a). Effects of Ganoderma lucidum and krestin on cellular immunocompetence in gamma-ray-irradiated mice. *The American Journal of Chinese Medicine, 23,* 71–80.

Chen, W. C., Hau, D. M., Wang, C. C., et al. (1995b). Effects of Ganoderma lucidum and krestin on subset T-cell in spleen of gamma-irradiated mice. *The American Journal of Chinese Medicine, 23,* 289–298.

Chen, X., Hu, Z. P., Yang, X. X., et al. (2006). Monitoring of immune responses to a herbal immuno-modulator in patients with advanced colorectal cancer. *International Immunopharmacology, 6,* 499–508.

Chen, Y. K., Kuo, Y. H., Chiang, B. H., et al. (2009). Cytotoxic activities of 9,11-dehydroergosterol peroxide and ergosterol peroxide from the fermentation mycelia of ganoderma lucidum cultivated in the medium containing leguminous plants on Hep 3B cells. *Journal of Agricultural and Food Chemistry, 57,* 5713–5719.

Chui, C. H., Wong, R. S., Cheng, G. Y., et al. (2006). Antiproliferative ability of a combination regimen of crocodile egg extract, wild radix ginseng and natural Ganoderma lucidum on acute myelogenous leukemia. *Oncology Reports, 6,* 1313–1316.

de la Taille, A., Hayek, O. R., Burchardt, M., et al. (2000). Role of herbal compounds (PC-SPES) in hormone-refractory prostate cancer: Two case reports. *Journal of Alternative and Complementary Medicine, 6,* 449–451.

Dudhgaonkar, S., Thyagarajan, A., & Sliva, D. (2009). Suppression of the inflammatory response by triterpenes isolated from the mushroom Ganoderma lucidum. *International Immunopharmacology, 9,* 1272–1280.

Evans, S., Dizeyi, N., Abrahamsson, P. A., et al. (2009). The effect of a novel botanical agent TBS-101 on invasive prostate cancer in animal models. *Anticancer Research, 29,* 3917–3924.

Fukuzawa, M., Yamaguchi, R., Hide, I., et al. (2008). Possible involvement of long chain fatty acids in the spores of Ganoderma lucidum (Reishi Houshi) to its anti-tumor activity. *Biological & Pharmaceutical Bulletin, 31,* 1933–1937.

Gao, J. J., Min, B. S., Ahn, E. M., et al. (2002). New triterpene aldehydes, lucialdehydes A-C, from Ganoderma lucidum and their cytotoxicity against murine and human tumor cells. *Chemical & Pharmaceutical Bulletin (Tokyo), 50,* 837–840.

Gao, Y., Zhou, S., Jiang, W., et al. (2003). Effects of ganopoly (a Ganoderma lucidum polysaccharide extract) on the immune functions in advanced-stage cancer patients. *Immunological Investigations, 32,* 201–215.

Gao, Y., Gao, H., Chan, E., et al. (2005). Antitumor activity and underlying mechanisms of ganopoly, the refined polysaccharides extracted from Ganoderma lucidum, in mice. *Immunological Investigations, 34,* 171–198.

Guo, L., Xie, J., Ruan, Y., et al. (2009). Characterization and immunostimulatory activity of a polysaccharide from the spores of Ganoderma lucidum. *International Immunopharmacology, 9,* 1175–1182.

Hajjaj, H., Mace, C., Roberts, M., et al. (2005). Effect of 26-oxygenosterols from Ganoderma lucidum and their activity as cholesterol synthesis inhibitors. *Applied and Environmental Microbiology, 71,* 3653–3658.

Han, J. R., An, C. H., & Yuan, J. M. (2005). Solid-state fermentation of cornmeal with the basidiomycete Ganoderma lucidum for degrading starch and upgrading nutritional value. *Journal of Applied Microbiology, 99,* 910–915.

Harhaji Trajkovic, L. M., Mijatovic, S. A., Maksimovic-Ivanic, D. D., et al. (2009). Anticancer properties of Ganoderma lucidum methanol extracts in vitro and in vivo. *Nutrition and Cancer, 61,* 696–707.

Hong, K. J., Dunn, D. M., Shen, C. L., et al. (2004). Effects of Ganoderma lucidum on apoptotic and anti-inflammatory function in HT-29 human colonic carcinoma cells. *Phytotherapy Research: PTR, 18,* 768–770.

Hsieh, T. C., & Wu, J. M. (2002). Mechanism of action of herbal supplement PC-SPES: Elucidation of effects of individual herbs of PC-SPES on proliferation and prostate specific gene expression in androgen-dependent LNCaP cells. *International Journal of Oncology, 20,* 583–588.

Hsu, C. L., Yu, Y. S., & Yen, G. C. (2008a). Lucidenic acid B induces apoptosis in human leukemia cells via a mitochondria-mediated pathway. *Journal of Agricultural and Food Chemistry, 56,* 3973–3980.

Hsu, H. Y., Hua, K. F., Wu, W. C., et al. (2008b). Reishi immuno-modulation protein induces interleukin-2 expression via protein kinase-dependent signaling pathways within human T cells. *Journal of Cellular Physiology, 215,* 15–26.

Hsu, J. W., Huang, H. C., Chen, S. T., et al. (2009). Ganoderma lucidum polysaccharides induce macrophage-like differentiation in human leukemia THP-1 cells via caspase and p53 activation. *Evidence Based Complementary Alternative Medicine.* doi:10.1093/ecam/nep107.

Hu, H., Ahn, N. S., Yang, X., et al. (2002). Ganoderma lucidum extract induces cell cycle arrest and apoptosis in MCF-7 human breast cancer cell. *International Journal of Cancer, 102,* 250–253.

Jang, K. J., Han, M. H., Lee, B. H., et al. (2010). Induction of apoptosis by ethanol extracts of Ganoderma lucidum in human gastric carcinoma cells. *Journal of Acupuncture and Meridian Studies, 3,* 24–31.

Jeurink, P. V., Noguera, C. L., Savelkoul, H. F., et al. (2008). Immunomodulatory capacity of fungal proteins on the cytokine production of human peripheral blood mononuclear cells. *International Immunopharmacology, 8,* 1124–1133.

Jiang, J., Slivova, V., Harvey, K., et al. (2004). Ganoderma lucidum suppresses growth of breast cancer cells through the inhibition of Akt/NF-kappaB signaling. *Nutrition and Cancer, 49,* 209–216.

Jiang, J., Slivova, V., & Sliva, D. (2006). Ganoderma lucidum inhibits proliferation of human breast cancer cells by down-regulation of estrogen receptor and NF-kappaB signaling. *International Journal of Oncology, 29,* 695–703.

Jiang, J., Grieb, B., Thyagarajan, A., et al. (2008). Ganoderic acids suppress growth and invasive behavior of breast cancer cells by modulating AP-1 and NF-kappaB signaling. *International Journal of Molecular Medicine, 21,* 577–584.

Kim, K. C., Kim, J. S., Son, J. K., et al. (2007). Enhanced induction of mitochondrial damage and apoptosis in human leukemia HL-60 cells by the Ganoderma lucidum and Duchesnea chrysantha extracts. *Cancer Letters, 246,* 210–217.

Kim, K. C., Jun, H. J., Kim, J. S., et al. (2008). Enhancement of radiation response with combined Ganoderma lucidum and Duchesnea chrysantha extracts in human leukemia HL-60 cells. *International Journal of Molecular Medicine, 21,* 489–498.

Lai, C. Y., Hung, J. T., Lin, H. H., et al. (2010). Immunomodulatory and adjuvant activities of a polysaccharide extract of Ganoderma lucidum in vivo and in vitro. *Vaccine, 28,* 4945–4954.

Li, Y. B., Wang, R., Wu, H. L., et al. (2008). Serum amyloid A mediates the inhibitory effect of Ganoderma lucidum polysaccharides on tumor cell adhesion to endothelial cells. *Oncology Reports, 20,* 549–556.

Li, Y. Q., & Wang, S. F. (2006). Anti-hepatitis B activities of ganoderic acid from Ganoderma lucidum. *Biotechnology Letters, 28,* 837–841.

Lin, S. B., Li, C. H., Lee, S. S, et al. (2003). Triterpene-enriched extracts from Ganoderma lucidum inhibit growth of hepatoma cells via suppressing protein kinase C, activating mitogen-activated protein kinases and G2-phase cell cycle arrest. *Life Sciences, 72,* 2381–2390.

Lin, Y. L., Liang, Y. C., Tseng, Y. S., et al. (2009). An immunomodulatory protein, Ling Zhi-8, induced activation and maturation of human monocyte-derived dendritic cells by the NF-kappaB and MAPK pathways. *Journal of Leukocyte Biology, 86,* 877–889.

Liu, G. Q., & Zhang, K. C. (2007). Enhancement of polysaccharides production in Ganoderma lucidum by the addition of ethyl acetate extracts from Eupolyphaga sinensis and Catharsius molossus. *Applied Microbiology and Biotechnology, 74,* 572–577.

Liu, J., Shimizu, K., & Kondo, R. (2010). The effects of ganoderma alcohols isolated from Ganoderma lucidum on the androgen receptor binding and the growth of LNCaP cells. *Fitoterapia, 81,* 1067–1072.

Liu, J., Shimizu, K., Konishi, F., et al. (2007). The anti-androgen effect of ganoderol B isolated from the fruiting body of Ganoderma lucidum. *Bioorganic & Medicinal Chemistry, 15,* 4966–4972.

Liu, J., Tamura, S., Kurashiki, K., et al. (2009a). Anti-androgen effects of extracts and compounds from Ganoderma lucidum. *Chemistry & Biodiversity, 6,* 231–243.

Liu, Y. W., Gao, J. L., Guan, J., et al. (2009b). Evaluation of antiproliferative activities and action mechanisms of extracts from two species of ganoderma on tumor cell lines. *Journal of Agricultural and Food Chemistry, 57,* 3087–3093.

Lu, Q. Y., Sartippour, M. R., Brooks, M. N., et al. (2004). Ganoderma lucidum spore extract inhibits endothelial and breast cancer cells in vitro. *Oncology Reports, 12,* 659–662.

Miura, T., Yuan, L., Sun, B., et al. (2002). Isoflavone aglycon produced by culture of soybean extracts with basidiomycetes and its anti-angiogenic activity. *Bioscience, Biotechnology, and Biochemistry, 66,* 2626–2631.

Muller, C. I., Kumagai, T., O'Kelly, J. et al. (2006). Ganoderma lucidum causes apoptosis in leukemia, lymphoma and multiple myeloma cells. *Leukemia Research, 30,* 841–848.

Nonaka, Y., Shibata, H., Nakai, M., et al. (2006). Anti-tumor activities of the antlered form of Ganoderma lucidum in allogeneic and syngeneic tumor-bearing mice. *Bioscience, Biotechnology, and Biochemistry, 70,* 2028–2034.

Nonaka, Y., Ishibashi, H., Nakai, M., et al. (2008). Effects of the antlered form of Ganoderma lucidum on tumor growth and metastasis in cyclophosphamide-treated mice. *Bioscience, Biotechnology, and Biochemistry, 72,* 1399–1408.

Sadava, D., Still, D. W., Mudry, R. R., et al. (2009). Effect of Ganoderma on drug-sensitive and multidrug-resistant small-cell lung carcinoma cells. *Cancer Letters, 277,* 182–189.

Sheena, N., Ajith, T. A., & Janardhanan, K. K. (2003). Prevention of nephrotoxicity induced by the anticancer drug cisplatin, using Ganoderma lucidum, a medicinal mushroom occurring in South India. *Current Science India, 85,* 478–482.

Shiao, M. S. (1992). Triterpenoid natural products in the fungus Ganoderma Iucidum. *Journal of the Chinese Chemical Society Taipei, 39,* 669–674.

Sliva, D., Labarrere, C., & Slivova, V., et al. (2002). Ganoderma lucidum suppresses motility of highly invasive breast and prostate cancer cells. *Biochemical and Biophysical Research Communications, 298,* 603–612.

Sliva, D., Sedlak, M., Slivova, V., et al. (2003). Biologic activity of spores and dried powder from Ganoderma lucidum for the inhibition of highly invasive human breast and prostate cancer cells. *Journal of Alternative and Complementary Medicine, 9,* 491–497.

Stanley, G., Harvey, K., Slivova, V., et al. (2005). Ganoderma lucidum suppresses angiogenesis through the inhibition of secretion of VEGF and TGF-beta1 from prostate cancer cells. *Biochemical and Biophysical Research Communications, 330,* 46–52.

Su, C., Shiao, M., & Wang, C. (2000). Potentiation of ganodermic acid S on prostaglandin E(1)-induced cyclic AMP elevation in human platelets. *Thrombosis Research, 99,* 135–145.

Tang, W., Liu, J. W., Zhao, W. M., et al. (2006). Ganoderic acid T from Ganoderma lucidum mycelia induces mitochondria mediated apoptosis in lung cancer cells. *Life Sciences, 80,* 205–211.

Thyagarajan, A., Jiang, J., Hopf, A., et al. (2006). Inhibition of oxidative stress-induced invasiveness of cancer cells by Ganoderma lucidum is mediated through the suppression of interleukin-8 secretion. *International Journal of Molecular Medicine, 18,* 657–664.

Thyagarajan, A., Zhu, J., & Sliva, D. (2007). Combined effect of green tea and Ganoderma lucidum on invasive behavior of breast cancer cells. *International Journal of Oncology, 30,* 963–969.

Thyagarajan, A., Jedinak, A., Nguyen, H., et al. (2010). Triterpenes from Ganoderma Lucidum induce autophagy in colon cancer through the inhibition of p38 mitogen-activated kinase (p38 MAPK). *Nutrition and Cancer, 62,* 630–640.

Tomasi, S., Lohezic-Le Devehat, F., Sauleau, P., et al. (2004). Cytotoxic activity of methanol extracts from Basidiomycete mushrooms on murine cancer cell lines. *Pharmazie, 59,* 290–293.

Wachtel-Galor, S., Szeto, Y. T., Tomlinson, B., et al. (2004a). Ganoderma lucidum ('Lingzhi'); acute and short-term biomarker response to supplementation. *International Journal of Food Sciences and Nutrition, 55,* 75–83.

Wachtel-Galor, S., Tomlinson, B., & Benzie, I. F. (2004b). Ganoderma lucidum ("Lingzhi"), a Chinese medicinal mushroom: Biomarker responses in a controlled human supplementation study. *The British Journal of Nutrition, 91,* 263–269.

Wang, G., Zhao, J., Liu, J., et al. (2007). Enhancement of IL-2 and IFN-gamma expression and NK cells activity involved in the anti-tumor effect of ganoderic acid Me in vivo. *International Immunopharmacology, 7,* 864–870.

Wang, J., Zhang, L., Yu, Y., et al. (2009). Enhancement of antitumor activities in sulfated and carboxymethylated polysaccharides of Ganoderma lucidum. *Journal of Agricultural and Food Chemistry, 57,* 10565–10572.

Wang, Y. Y., Khoo, K. H., Chen, S. T., et al. (2002). Studies on the immuno-modulating and antitumor activities of Ganoderma lucidum (Reishi) polysaccharides: Functional and proteomic analyses of a fucose-containing glycoprotein fraction responsible for the activities. *Bioorganic & Medicinal Chemistry, 10,* 1057–1062.

Weng, C. J., Chau, C. F., Chen, K. D., et al. (2007). The anti-invasive effect of lucidenic acids isolated from a new Ganoderma lucidum strain. *Molecular Nutrition & Food Research, 51,* 1472–1477.

Weng, C. J., Chau, C. F., Hsieh, Y. S., et al. (2008). Lucidenic acid inhibits PMA-induced invasion of human hepatoma cells through inactivating MAPK/ERK signal transduction pathway and reducing binding activities of NF-kappaB and AP-1. *Carcinogenesis, 29,* 147–156.

Weng, C. J., Chau, C. F., Yen, G. C., et al. (2009). Inhibitory effects of Ganoderma lucidum on tumorigenesis and metastasis of human hepatoma cells in cells and animal models. *Journal of Agricultural and Food Chemistry, 57,* 5049–5057.

Wicks, S. M., Tong, R., Wang, C. Z., et al. (2007). Safety and tolerability of Ganoderma lucidum in healthy subjects: A double-blind randomized placebo-controlled trial. *The American Journal of Chinese Medicine, 35,* 407–414.

Wu, T. S., Shi, L. S., & Kuo, S. C. (2001). Cytotoxicity of Ganoderma lucidum triterpenes. *Journal of Natural Products, 64,* 1121–1122.

Yang, H. L. (2005). Ganoderic acid produced from submerged culture of Ganoderma lucidum induces cell cycle arrest and cytotoxicity in human hepatoma cell line BEL7402. *Biotechnology Letters, 27,* 835–838.

Yang, X. J., Liu, J., Ye, L. B., et al. (2006). In vitro and in vivo protective effects of proteoglycan isolated from mycelia of Ganoderma lucidum on carbon tetrachloride-induced liver injury. *World Journal of Gastroenterology: WJG, 12,* 1379–1385.

Yeh, C. M., Yeh, C. K., Hsu, X. Y., et al. (2008). Extracellular expression of a functional recombinant Ganoderma lucidium immunomodulatory protein by Bacillus subtilis and Lactococcus lactis. *Applied and Environmental Microbiology, 74,* 1039–1049.

Yoshimura, K., Kamoto, T., Ogawa, O., et al. (2010). Medical mushrooms used for biochemical failure after radical treatment for prostate cancer: An open-label study. *International Journal of Urology: Official Journal of the Japanese Urological Association, 17,* 548–554.

Yue, G. G., Fung, K. P., Tse, G. M., et al. (2006). Comparative studies of various ganoderma species and their different parts with regard to their antitumor and immunomodulating activities in vitro. *Journal of Alternative and Complementary Medicine, 12,* 777–789

Yue, G. G., Fung, K. P., Leung, P. C., et al. (2008a). Comparative studies on the immunomodulatory and antitumor activities of the different parts of fruiting body of Ganoderma lucidum and Ganoderma spores. *Phytotherapy Research: PTR, 22,* 1282–1291.

Yue, Q. X., Cao, Z. W, Guan, S. H., et al. (2008b). Proteomics characterization of the cytotoxicity mechanism of ganoderic acid D and computer-automated estimation of the possible drug target network. *Molecular & Cellular Proteomics: MCP, 7,* 949–961.

Yue, Q. X., Xie, F. B., Guan, S. H., et al. (2008c). Interaction of Ganoderma triterpenes with doxorubicin and proteomic characterization of the possible molecular targets of Ganoderma triterpenes. *Cancer Science, 99,* 1461–1470.

Yue, Q. X., Song, X. Y., Ma, C., et al. (2010). Effects of triterpenes from Ganoderma lucidum on protein expression profile of HeLa cells. *Phytomedicine, 17,* 606–613.

Yuen, J. W., & Gohel, M. D. (2008). The dual roles of Ganoderma antioxidants on urothelial cell DNA under carcinogenic attack. *Journal of Ethnopharmacology, 118,* 324–330.

Zhang, H. N., & Lin, Z. B. (2004). Hypoglycemic effect of Ganoderma lucidum polysaccharides. *Acta Pharmacologica Sinica, 25,* 191–195.

Zhang, Y., Lin, Z., Hu, Y., et al. (2008). Effect of Ganoderma lucidum capsules on T lymphocyte subsets in football players on "living high-training low". *British Journal of Sports Medicine, 42,* 819–822.

Zhu, H. S., Yang, X. L., Wang, L. B., et al. (2000). Effects of extracts from sporoderm-broken spores of Ganoderma lucidum on HeLa cells. *Cell Biology and Toxicology, 16,* 201–206.

Zhu, X. L., & Lin, Z. B. (2005). Effects of Ganoderma lucidum polysaccharides on proliferation and cytotoxicity of cytokine-induced killer cells. *Acta Pharmacologica Sinica, 26,* 1130–1137.

Zhuang, S. R., Chen, S. L., Tsai, J. H., et al. (2009). Effect of citronellol and the Chinese medical herb complex on cellular immunity of cancer patients receiving chemotherapy/radiotherapy. *Phytotherapy Research: PTR, 23,* 785–790.

Chapter 13
An Evidence-based Perspective of *Coriolus Versicolor* (Multicolored Polypore Mushroom) for Cancer Patients

Stanislava Stošić-Grujičić, Sanja Mijatović and Danijela Maksimović-Ivanić

Abstract In the era of molecularly targeted, rationally designed cancer therapeutics, natural products, especially medicinal mushrooms, continue to provide a rich source of anticancer agents. One potent medicinal mushroom extensively used in both traditional medicine and modern clinical practice is *Coriolus versicolor* (alternative names *Trametes versicolor, Polyporus versicolor, Polystictus versicolor,* multicolored polypore mushroom). In spite of their paucity, available data about the physical, chemical and pharmacodynamic properties of its active compounds and respectable scientific arguments support the possibility of developing at least some of the constituent molecules and metabolites into evidence-based immunotherapeutic remedies for cancer patients as part of comprehensive cancer treatment and secondary prevention strategy. In this chapter we present current knowledge about the immunological and antitumor properties of multicolored polypore mushroom constituents, and its possible applications for supportive cancer care and therapy. Modulation of innate and adaptive immunity, hematopoietic activity, and direct cytotoxic effects are reviewed with detailed analysis of molecular targets and signaling pathways affected by mushroom components and/or secondary metabolites. Data received from complementary *in vivo, ex vivo* and *in vitro* approaches are discussed. We also present some novel findings, which point to the highly different biological potency of multicolored polypore mushroom extracts and/or products from different sources and manufacturing procedures. The findings obtained from fundamental research are supplemented with information acquired from traditional usage and controlled clinical trials. Finally, we attempt to highlight the need for further evidence-based research with particular focus on potential drug interaction, contraindications and adverse effects, which should enable technological improvement of drug production.

S. Stošić-Grujičić (✉)
Department of Immunology, Institute for Biological Research "Sinisa Stankovic", Belgrade University, Bulevar Despota Stefana 142, 11060 Belgrade, Serbia
e-mail: duta@eunet.rs

W. C. S. Cho (ed.), *Evidence-based Anticancer Materia Medica,* Evidence-based Anticancer Complementary and Alternative Medicine,
DOI 10.1007/978-94-007-0526-5_13,

13.1 Traditional and Contemporary Uses

In the Orient several thousand years ago, it was recognized that many edible and certain non-edible mushrooms could have valuable health benefits. In the era of molecularly targeted, rationally designed cancer therapeutics, the natural environment continues to provide a rich source of anticancer agents, especially medicinal mushrooms. One of the most important mushrooms extensively used in both traditional herbalism and modern clinical practice is the macrofungus *Coriolus versicolor* (L. ex Fr.) Quel (alternative names *C. vesicular, C. zonztus, Trametes versicolor, Polyporus versicolor, Polystictus versicolor, Poria versicolor, Boletus versicolor, Agaricus versicolor*, multicolored polypore mushroom). Versicolor means "of several colors" and it is true that this mushroom is found in a wide variety of different hues. In China, the mushroom is called Yun Zhi ("cloud-like mushroom"), in Japan it is known as Kawaratake ("mushroom by the river bank"), and in some European countries it is named "turkey tail". Multicolored polypore mushroom belongs to the Basidiomycetes class and Polyporaceae family, with more than 120 strains recorded. This species has a long history of use in traditional Asian medicine, as recorded thousands of years ago in the first written report, the Compendium of Materia Medica and Shennong's Classic of Materia Medica (Hobbs 1995; Ng 1998). For many centuries Chinese medicine has prized the mushroom for its energizing and healing properties. Chinese folklore is filled with legends about those who discovered the 1,000 year old mushroom and became immortal. In traditional Chinese medicine, multicolored polypore mushroom was used to clear dampness, reduce phlegm, heal pulmonary disorders, strengthen the physique, increase energy and benefit people with chronic diseases. In the West, acceptance is based less on tradition and more on scientific methodology.

The experience from traditional folk medicine and the improved possibilities for genetic, pharmacological and chemical analysis let us assume that multicolored polypore mushroom has a great potential for successful bioprospecting. Over the last three decades scientific and medical studies have been carried out in Japan, China, Korea, and more recently the US, which have increasingly demonstrated the potent and unique health enhancing properties of compounds extracted from multicolored polypore mushroom. Historically, the fungus was gathered in the wild, typically in groups, growing as a mushroom body naturally on trees, but nowadays it is cultivated artificially (Monro 2003) or grown as mycelial biomass in submerged culture in bioreactors (Cui and Chisti 2003). The historical evolution of usage of this non-edible medicinal forest fungus would not have been as the whole mushroom, due to its coarse, hard texture and bitter taste, but as health tonics, tinctures, teas, soups or in herbal formulae. In Asia, multicolored polypore mushroom extracts are usually crude mixtures, available as oral proprietary products and can be purchased without a prescription. As such they should not be confused with pharmaceuticals, which are almost invariably a defined chemical preparation, the specifications for which are listed in pharmacopoeia. Regular intake of these concentrates is believed to enhance immune responses of the human body, thereby increasing resistance to disease and in some cases causing regression of the illness (Jong et al. 1991). In the context of cancer management, dietary supplements may be desirable for reduction of the hematological and non-hematological toxicities inherent with conventional chemotherapeutics. However, translating traditional Eastern practices into evidence-based Western therapies is difficult, due to different manufacturing standards, criteria of purity, and under-powered clinical trials. Nevertheless, purified bioactive compounds have increasingly been confirmed scientifically as a new source of anticancer agents with strong immunomodulating and anticancer properties (Moradali et al. 2007; Ragupathi et al. 2008). In clinical practice, purified multicolored polypore mushroom extracts are recommended by health authorities for various types of cancers, liver ailments (including hepatitis B and chronic active hepatitis), and infections of the upper respiratory, urinary, and digestive tracts (Chu et al. 2002; Monro 2003; Tsang et al. 2003). Data from multiple epidemiologic and clinical studies of immune therapy utilizing multicolored polypore mushroom constituents suggest clinical benefits that may be warranted as part of a comprehensive cancer treatment and/or secondary prevention strategy.

13.2 Active Constituents and Products

Many cultures worldwide have long recognized that hot water-soluble fractions (decoctions and essences) from medicinal mushrooms can have health-promoting benefits. Available scientific evidence does not support claims that the raw mushroom itself is an effective anticancer agent in humans. However, there is some scientific evidence that substances derived from parts of the mushroom may be useful against cancer. Among medicinal mushrooms multicolored polypore mushroom was considered to be the most suitable for further fractionation due to its high antitumor activity and stability during serial cultivation. It is believed that the anticancer properties of multicolored polypore mushroom are largely contained in its diverse chemical constituents. However, it is only within the last three decades that chemical technology has been able to isolate the relevant compounds from fruiting bodies, pure culture mycelia, and culture filtrate (culture broth) and employ them in controlled experiments. Multicolored polypore mushroom is found almost worldwide in many different climates, but its biological activity varies depending on the strain, the developmental stage of the mushroom, and the habitat in which it grows. Constituently, the gross dry matter composition for about 95% of the whole mushroom includes proteins (4.2%), carbohydrates (65.1%), fiber (23.2%), crude fat (1.1%), and salts and metals (6.4%) (Mau et al. 2001). Regarding non-volatile components, multicolored polypore mushroom contains 20.2 mg/g dry weight of total soluble sugars, 14.1 mg/g total free amino acids, and 17.0 mg/g 5'-nucleotides (Mau et al. 2001). These include mainly polysaccharides (in particular β-D-glucans), polysaccharide peptide (PSP) and protein-bound polysaccharides (PSK), but also terpenoids, phenols, lipids and a number of small molecules (Table 13.1).

Table 13.1 Bioactive products or constituents isolated from *Coriolus versicolor* (multicolored polypore mushroom)

Main compound group	Name/example	Molecular mass	Selected references
Polysaccharide	β-D-glucans	500–2,000 kDa	Bohm and BeMillar 1995; Mizuno 1999; Kidd 2000
	Verisicolor polysaccharide (VPS)	–	Coles and Toth 2005
	Heteropolysaccharides	–	Rau et al. 2009
	Dietary fiber	–	Mau et al. 2001; Smith et al. 2002
Glyco-conjugates	Polysaccharopeptide (PSP)	100 kDa	Cui and Christi 2003; Yang 1999; Kidd 2000; Mizumo 1999
	Polysaccharide protein (PSK, Krestin, Coriolan)	100 kDa	
	Proteoglycans	–	Moradali et al. 2007
Terpenes	Triterpenoids	400–600 kDa	Mizuno et al. 1995; Harhaji et al. 2008
	Coriolin	280 Da	Takeuchi et al. 1969
Phenolic compounds	Phenols	–	Mau et al. 2002; Zjawiony 2004; Harhaji et al. 2008; Kim et al. 2008
	Phenolic acids	–	
	Polyphenols	–	
Small molecules	Active hexose correlated compound (AHCC)	5 kDa	Ghoneum et al. 1995
	Lipid compounds	–	Breene 1990; Mau et al. 2001
	Vitamins	–	Mattila et al. 2001
	Mineral and trace elements	–	
	Proteins, polypeptides, amino acids, deoxycoriolic acid, nucleotides, etc	6–50 kDa	Yang et al. 1992; Mau et al. 2001; Shimokawa et al. 2008

13.2.1 Polysaccharides and Glyco-conjugates

13.2.1.1 β-D-glucans

Important bioactive components among the vast arsenal of compounds isolated from multicolored polypore mushroom are polysaccharides. In the mushroom, they occur mostly as water-soluble β-D-glucans with different types of glycosidic linkages (Bohm and BeMillar 1995; Mizuno 1999; Kidd 2000) as true heteroglucans, while others are bound to protein residues as polysaccharide-protein complexes. Unlike proteins and nucleic acids, polysaccharides consist of repetitive structural polymers of monosaccharide residues joined to each other by glycosidic linkages. The basic β-D-glucan is a fiber-form polysaccharide, with repeating D-glucose units joined together in linear chains by beta-bonds (β). These can extend from carbon 1 of one saccharide ring to carbon 3 of the next (β1→3), from carbon 1 to carbon 4 (β1→4) or from carbon 1 to carbon 6 (β1→6). Usually there is a main chain that is either β1→3, β1→4 or mixed β1→3, β1→4 with β1→6 side chains, giving a comb-like structure with various conformations. One example of an antitumor β-D-glucan isolated from cultured mycelium is compound D-II (Zjawiony 2004). The chemical structure of D-II is a (β1→3)-D-glucan in which one out of every three glucose residues is branched at C-6 with a (β1→6)-linkage. Not all β-D-glucans act against tumors. Among these preparations, greater antitumor activity is correlated with higher molecular weight, a lower level of branching, and greater water solubility of β-glucans (Mizuno 1999). Some of the water insoluble β-glucans are soluble in dilute alkali and then they can show marked antitumor activity. Versicolor polysaccharide (VPS), an extract from the mushroom, is sold as a dietary supplement in the US (Coles and Toth 2005).

13.2.1.2 Heteropolysaccharides

The polysaccharide fraction of multicolored polypore mushroom also may contain β-D-glucans with heterosaccharide residues, i.e. xylose, mannose, galactose, arabinose, ribose, glucuronic acid, and uronic acid that can combine with other components (Smith et al. 2002). In addition, multicolored polypore mushroom-derived exopolysaccharide, containing β-1,3/β-1,6-linked D-glucose, galactose, mannose, arabinose and xylose in their backbone, can be secreted in bioreactors (Rau et al. 2009).

13.2.1.3 Dietary Fibers

As many other mushrooms, multicolored polypore mushroom contains dietary fiber, high molecular weight compounds that are excreted without digestion and absorption by humans (Smith et al. 2002). Dietary fiber includes β-glucans, chitin, and heteropolysaccharides (pectinous substances, hemicellulose, polyuronides, etc). Many of these compounds have carcinostatic activity and by physicochemical interactions they may absorb possible carcinogenic substances and hasten their excretion from the intestine.

13.2.1.4 Glyco-conjugates

The best known commercial preparations of multicolored polypore mushroom are two closely related constituents, protein-bound polysaccharide (Krestin, PSK, Coriolan), and PSP. PSK and PSP are the most effective substances, produced by batch fermentation from the CM-101 and Cov-1 strains of multicolored polypore mushroom in Japan and the People's Republic of China, respectively (Ng 1998; Mizuno 1999; Yang 1999; Kidd 2000). PSK is obtained from hot water extracts by salting out with ammonium sulfate, whereas PSP is recovered by alcoholic precipitation. Both products possess molecular weights of approximately 100 kDa, and are highly water-soluble. The extracts differ mainly in the presence of fucose in PSK and rhamnose and arabinose in PSP. PSK was first isolated in Japan in the late 1960s and was commercially available from 1997, whereas PSP was isolated around 1983 in China and appeared on the market in 1987. PSP and PSK are just beginning to be available as pharmaceutical grade products in the US and Europe. Analyses performed by Ikuzawa et al. (1988) indicate that PSK contains 62% polysaccharide and 38% protein. The glucan portion of PSK consists of a $\beta 1 \rightarrow 4$ main chain and $\beta 1 \rightarrow 3$ side chains, with $\beta 1 \rightarrow 6$ side chains bound to a polysaccharide moiety through O- or N-glycosidic bonds. There is wide consensus that it is the $\beta 1 \rightarrow 3$ and $\beta 1 \rightarrow 6$ side chains that are immunologically active. The polypeptide portion is rich in aspartic, glutamic and other amino acids (Sakagami and Aoki 1991). PSP may contain at least four discrete molecules, all of which are true proteoglycans. The polysaccharide moiety of PSP is a $\beta 1 \rightarrow 3$-glucan branching at 4' and 6' positions, and consists of five different sugars including arabinose, glucose, galactose, mannose and xylose. The protein moiety of PSP is rich in aspartic and glutamic acids, but many other amino acids are also present in smaller amounts.

Proteoglycans are a special class of diverse glycoproteins that are heavily glycosylated (Moradali et al. 2007). This group of glycol-conjugates consists of a core protein with one or more covalently attached glycosaminoglycan chains.

13.2.2 Terpenes

A still increasing number of fungal metabolites have been described and hundreds of terpenes have been isolated from the fungi kingdom, the majority belonging to the group of sesquiterpenes. One multicolored polypore mushroom sesquiterpene, coriolin, a metabolite with potent antitumor and antibacterial properties, was first isolated in 1969 (Takeuchi et al. 1969). It is a fused, linear tricyclopentanoid molecule with eight stereogenic and two quaternary centers. To date, syntheses and characterization of coriolin and several derivatives of chemically modified coriolin have been reported by leading investigators (Zhou and Yang 1999; Abraham 2001). Among the terpenes, numerous triterpenoids have been exclusively found in Basidiomycetes, including multicolored polypore mushroom. At least 50 different triterpenoids have been identified, such as ganoderic, ganoderenic, ganodermic, lucidenic, trametenolic and applanoxidic acids, ganoderols, lucidumols, ganodermanontriol, etc. (Mizuno et al. 1995).

13.2.3 Phenolic Compounds

Phenolic compounds can be classified as simple phenols and phenolic acids (e.g. gallic acid, benzoic acid, syringic acid, chlorogenic acid) and polyphenols, which are divided into many groups such as flavonoids (catechin, naringin, naringenin, myricetin, quercetin, biochanin A, formononetin, hesperetin, kaempferol, rutin), tannins, stilbenes, and so on. Besides production by cultivated mushrooms, fresh fruiting bodies of medicinal mushrooms also contain varying numbers of phenolic compounds, with differences between mushroom species ranging from 15 to 30 compounds (Kim et al. 2008). The active phenolic compounds 3β,5α,9α-trihydroxyergosta-7,22-dien-6-one and 3β,5α,9α-trihydroxy-6β-methoxyergosta-7,22-diene were isolated from the crude extract of multicolored polypore mushroom by bioassay-guided fractionation using rat hepatoma cells (Zjawiony 2004).

13.2.4 Other Active Low Molecular Weight Compounds

13.2.4.1 Active Hexose Correlated Compound (AHCC)

AHCC is prepared from the co-cultivation of several Basidiomycete mushrooms including multicolored polypore mushroom (Ghoneum et al. 1995). It contains partially acetylated polysaccharides, glycoproteins, amino acids, lipids and minerals. In contrast to the other anticancer glucans, those of AHCC are low molecular weight (approximately 5 kDa) rare α-1,3 glucan structures. These details are surprising since typically low molecular weight material is normally inactive and α-glucans have minimal immuno-potentiating activity.

13.2.4.2 Lipid Compounds

While crude fat in multicolored polypore mushroom contains all the main classes of lipid compounds including free fatty acids, mono-, di- and triglycerides, sterols (including ergosterol and β-sitosterol), sterol esters and phospholipids, levels are generally low, around 1% of dry weight (Mau et al. 2001; Breene 1990).

13.2.4.3 Vitamins

Multicolored polypore mushroom contains many vitamins, especially thiamine (B_1), riboflavin (B_2), cobalamin (B_{12}), ergosterol (provitamin D_2) derivatives, niacin, folates, coumarin, biotin, and ascorbic acid (C). Ergosterol is converted into vitamin D_2 (ergocalciferol) when the mushroom is dried by light and heat (Mattila et al. 2001).

13.2.4.4 Mineral and Trace Elements and Other Low Molecular Weight Compounds

Like other mushrooms, multicolored polypore mushroom has proved to be a good source of many mineral elements, e.g. K, P, Mg, Zn, Cu, Ca, Na, Fe, and Se, but the bioavailability of the mineral and trace elements is, however, questionable. Seleno-compounds identified in the carpophore include selenocysteine, selenomethionine, Se-methylselenocysteine, selenite, and several unidentified seleno-compounds, but their proportions widely vary. Perhaps the most important question regarding minor elements concerns the amounts of the toxic trace elements, Cd, Hg, As, and Pb, whose dietary excess may be injurious to health. If mushrooms are produced in substrates contaminated with these elements, accumulation may occur.

Obviously, multicolored polypore mushroom constituently contains less amounts of other bioactive molecules, such as alkaloids, proteins (lectin, aegerolysin-like proteins), peptides, amino acids (aspartic acid, glutamic acid, alanine, leucine), nucleosides, nucleotides, furans, lignin, mycins, etc and thereby is not such a good source of these compounds as other mushrooms (Mau et al. 2001). However, some novel bioactive compounds, such as the highly methylated cyclic heptapeptide (-)-ternatin, a potent candidate for an anti-obesity drug, were recently isolated (Shimokawa et al. 2008).

13.3 Pharmacological Actions of the Active Ingredients

13.3.1 Effects on Hematopoietic Stem Cells

Cell cycle arrest and/or induction of different routes to death of cells with high dividing potential represent the main anticancer mechanisms of action of chemotherapeutics and radiation. Unfortunately, this therapy, which is designed to block uncontrolled division of transformed cells, unselectively harms normal cells that divide rapidly, such as those of the hematopoietic system. The most common side effect of chemotherapy and radiotherapy is myelosuppression. The important immunopotentiating effects of active mushroom substances, especially carbohydrates, include stimulation of hematopoietic stem cells (Moradali et al. 2007). Combined administration of PSK and growth factors such as G-CSF, GM-CSF or IL-3 was demonstrated to increase the hematological recovery of mice with 5-fluorouracil induced myelosuppression (Kohgo et al. 1994). In addition, PSP significantly increased granulocyte production in mice with cyclophosphamide-induced granulocytopenia (Mayer and Drews 1980). Enhancement of repopulation and hematopoiesis of bone marrow cells was also reported in irradiated mice. Medium conditioned by PSK-exposed leukocytes significantly stimulated formation of different types of colonies such as erythroid, granulocyte, macrophage, eosinophil, megakaryocyte, and mixed hemopoietic colonies (Tsuji et al. 1989). Together, the data suggest that PSK can restore the hematological status of an organism upon anticancer treatment.

13.3.2 Effects on Immune System

13.3.2.1 Recognition and Receptors

Multicolored polypore mushroom products are obvious immunoenhancers that potentiate the immune system in multiple ways. An extremely broad range of physiological effects has been linked with mushroom polysaccharides and polysaccharopeptides, the antitumor action of which is mediated through immunomodulatory regulation rather than by direct cytotoxicity towards tumor cells. One of the most abundant forms of fungal polysaccharides is β-glucans, recognized by the human immune system as non-self molecules, and therefore able to promote innate and adaptive responses. The immune functions of β-glucans are actually dependent on their high grade conformational complexity. It was considered that immunomodulatory properties of β-glucans, as poorly digestible carbohydrates, may be partly dependent on intestinal microbial-mediated degradation. However, β-glucans can bind directly to specific receptors on immune cells, suggesting a microbial independent immunomodulatory effect (Chan et al. 2009). The three dimensional structure of β-glucans allows lock and key interaction between the branching side chains and receptors on different immune cells. It is clarified that several receptors mediate β-glucan recognition and activities, such as dectin-1, CR3 (complement receptor 3), TLR (toll-like receptors), the lactosylceramide (LacCer), and scavenger receptors (Chen and Seviour 2007; Chan et al. 2009) (Fig. 13.1).

Dectin-1

Dectin-1 is a type II transmembrane protein, composed of an extracellular carbohydrate-recognition C-type lectin domain and a cytoplasmic domain with an immunoreceptor tyrosine-based activation motif (ITAM). It is commonly expressed on macrophages/monocytes, neutrophils, dendritic and some T-cells, but not on natural killer (NK) cells (Moradali et al. 2007). The receptor binds specifically to β-$(1 \rightarrow 3)$-glucans. This contact promotes phagocytosis, but also stimulates the

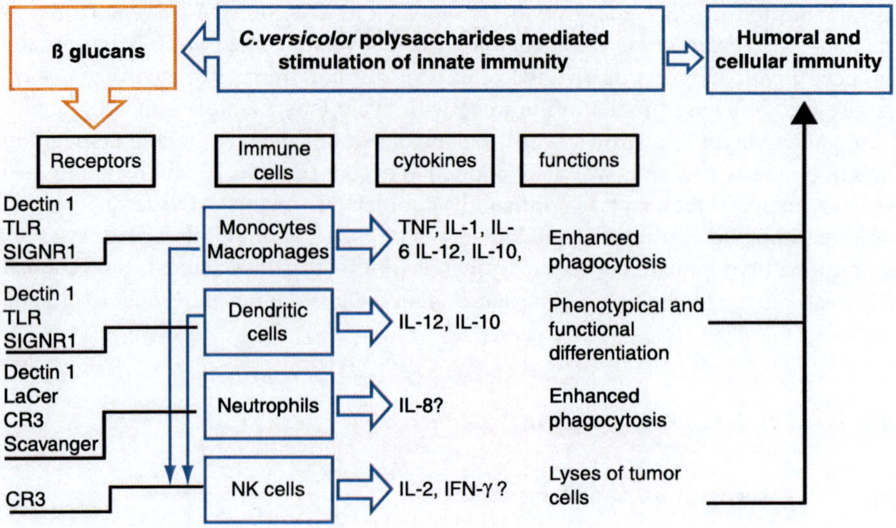

Fig. 13.1 *Coriolus versicolor* (multicolored polypore mushroom) promoted innate and adaptive immune response

production of reactive oxygen intermediates and inflammatory cytokines such as TNF-α, IL-2, and IL-12 (Chen and Seviour 2007). The biological response transduced by dectin-1 collaborated with the TLR pathway (Chan et al. 2009). Moreover, the dendritic cell-specific ICAM-3-grabbing non-integrin homolog -SIGNR1 is an additional mannose receptor expressed on macrophages that cooperates with dectin-1 in recognition of β-glucans, resulting in phagocytosis (Taylor et al. 2004). It was shown that treatment with multicolored polypore mushroom β-glucans induced remarkable dectin-1 mediated phagocytosis in RAW264.7 cells, peritoneal macrophages, and microglia, which was related to NO production and cNOS and iNOS regulation. In addition to the inflammatory response, dectin-1 binds to T lymphocytes and augments their mitogenic response by cross-linking T cell receptors (Moradali et al. 2007).

Complement Receptor 3

The CR3 receptor (CD11b/CD18) occurs on monocytes, macrophages, dendritic cells, neutrophils, and NK cells. It recognizes a large spectrum of microbial cells and also functions as an adhesion molecule. The ligand-receptor complex can be internalized, but the intracellular events that occur after glucan-receptor binding have not been fully determined till now. It was demonstrated that small β-(1 → 3)-glucan fragments, produced by digestion in bone marrow macrophages were subsequently attached by CR3 of marginated granulocytes to an epitope different from the inactivated complement 3b (iC3b) binding site (Chan et al. 2009). These granulocytes possessed enhanced capacity to kill iC3b-opsonized tumor cells after recruitment to a site of complement activation, such as tumor cells coated with monoclonal antibody (Hong et al. 2004). It was also shown that intravenous administered soluble β-glucans can be delivered directly to the CR3 on circulating granulocytes (Chan et al. 2009). The function of NK cells was also potentiated by preincubation with β-glucans. While neutrophils have both dectin-1 and CR3, NK cells have no dectin-1 receptors. Therefore, ligation of glucans from multicolored polypore mushroom extract to CR3 may be solely responsible for their activation confirmed in several studies as well as clinical trials (Ohwada et al. 2006; Sadahiro et al. 2010).

Tool-like Receptors

The family of TLRs includes a large group of transverse cell membrane proteins located on innate immune cells. TLRs are capable of recognizing diverse groups of microbes including fungi, bacteria, viruses, and protozoa (Roeder et al. 2004), and can link innate and adaptive immunity. Fungal β-glycans activate innate immunity through ligation to TLR-2, 4 and 6 present on macrophages, dendritic cells, and intestinal epithelium (Moradali et al. 2007). Recent *in vitro* data indicated that some immunologic effects of multicolored polypore mushroom are mediated through TLRs (Schmidt 2006). Enhanced TNF-α and IL-6 production in J774A.1 cell cultures, primary murine macrophages and splenocytes exposed to PSK was TLR-4 dependent (Price et al. 2010). In addition, PSP up-regulated expression of 22 genes associated with TLR signaling (such as *IFN-γ, CXCL10, TLR4, TLR5*), while 23 genes (e.g. *TLR9, TLR10, SARM1, TOLLIP*) out of 117 genes regulated by TLR ligation in peripheral blood mononuclear cells were down-regulated under the treatment (Li et al. 2010).

Lactosylceramide and Scavenger Receptors

Lactosylceramide (LacCer; CDw17) is a glycosphingolipid located on neutrophil and endothelial cells, which recognizes both microbial cells and fungal β-(1→3)-glucans (Zimmerman et al. 1998). Scavenger receptors located on myeloid and endothelial cells recognize different foreign cells (Chan et al. 2009). The role of these receptors in immunostimulation mediated by multicolored polypore mushroom components has not been fully determined yet, primarily because the chemical structure of the fungal preparation was not well defined in any experiment performed.

13.3.2.2 Effects of on the Innate Immune System

The innate immune system allows a first line of defense from infection in a non-specific manner, recognizing and responding to pathogens in a generic way. Unlike the adaptive immune system, activities performed by macrophages, polymorphonuclear leukocytes (PMN), NK cells, and dendritic cells (DC) in the context of the innate immune response do not confer stable immunity to the host (Chen and Seviour 2007). Mononuclear phagocytes, DC and NK cells are involved in recognition and destruction of abnormal cells such as cancer cells. Among these, the macrophages preferentially attack dead cells and intracellular pathogens, NK cells lyse cancer and virus-infected cells, whereas PMN play a prominent role in the overall course of infectious diseases and participate in the defense against intra- and extra-cellular pathogens such as bacteria, viruses and some protozoa (Moradali et al. 2007). However, multicolored polypore mushroom is able to affect the innate immune response through direct interactions of its constituents with cell receptors (Fig. 13.1). Multicolored polypore mushroom polysaccharides stimulate the function of macrophages and, consequently enhance overall immune function (Wong et al. 2004). β-Glucans from this mushroom induced a remarkable increase of phagocytosis in RAW264.7 cells, peritoneal macrophages, and microglia (Park et al. 2009). Importantly, in the relevant concentration PSP exerted no cytotoxicity on cultured mouse peritoneal macrophages nor on two macrophage like cell lines (PU5 and P338D1) (Smith et al. 2002). Enhanced phagocytic activity of different immune cells exposed to PSP or PSK *in vivo* was described in the earliest studies on the immunomodulatory properties of multicolored polypore mushroom. Intravenous administration of PSP resulted in significant acceleration of the phagocytic activity of cells in the blood (Cheng and Leung 2008). PSP significantly increased both clearance and phagocytic indices, as well as the production of IL-1, IL-6 and IFN-γ in mice that were intravenously challenged with charcoal particles. In addition, the anticancer effects of multicolored polypore mushroom are at least partly caused by the elevated tumoricidal potential of activated macrophages, which is mainly mediated by their products, such as TNF together with reactive nitrogen and oxygen intermediates (Nathan and Hibbs 1991). All these molecules are able to inhibit tumor progression through induction of different types of programmed or uncontrolled cell death (Simon et al. 2000). Besides the well known effects of multicolored polypore mushroom polysaccharides on macrophage activity, the potential of terpenoid and phenolic compounds to enhance the cytotoxic action of macrophages against mouse melanoma cells was described (Harhaji et al. 2008). This indicates that immunomodulation of the antitumor response is not exclusively caused by the polysaccharide component of multicolored polypore mushroom. The mechanism responsible for the observed effect could also be related to enhanced production of reactive nitrogen and oxygen intermediates but this has to be demonstrated. Finally, PSK may activate the function of PMN. However, it is not known whether PSP modulates cytokine action or affects cytokine production through modulation of intracellular signaling and subsequently, gene expression. Northern blot analysis of DNA verified that PSP triggered transcription of the *TNF-α* gene in these cells, confirming the significance of PSP in immune-defense activities. Peritoneal macrophages isolated from C57BL/6 mice and BALB/c exposed to PSP produce significantly more TNF-α and reactive nitrogen and oxygen species than non-treated controls (Liu et al. 1996, 1997).

NK cells are another important component of nonspecific immunity and are able to lyse or kill infected or tumor cells and, in parallel, alter innate and specific immune responses by production of cytokines such as IFN-γ, TNF-α, GM-CSF. They are recruited at the site of inflammation where they can be stimulated by IL-12 from activated macrophages or dendritic cells. Because of this, stimulation of NK cells by different mushroom metabolites is of crucial relevance for tumor immunity. Theoretically, multicolored polypore mushroom metabolites are capable of stimulating NK cells- directly by ligation of β-glucans from the extract to appropriate receptors. It was shown that PSK is able to stimulate NK cells independently of IFN and IL-2 (Jiménez et al. 2005). Indirectly, elevated secretion of different cytokines, such as IL-2 and IL-12, observed after PSK treatment can augment the proliferation of NK cells. Despite the finding that multicolored polypore mushroom PSP did not significantly change NK activity against K562 tumor cells *in vitro*, it markedly recovered compromised NK activity in tumors treated with cyclophosphamide. In concordance with this, clinical studies showed that PSP treatment enhanced the activity of NK cells in cancer patients by approximately 60% (Ng 1998). Importantly, local administration of PSK simultaneously affected malignant tissue and its microenvironment causing T cell infiltration and promoting tumoricidal activity of killer cells (Smith et al. 2002). Finally, PSP stimulated lymphokine-activated killer (LAK) cell proliferation and decreased the level of IL-2, which is necessary for their activation (Smith et al. 2002).

DCs are bone marrow-derived immune accessory cells, found in epithelial and lymphoid tissues. These cells are engaged by both innate and adaptive immunity systems. Macrofungi metabolites, especially β-glucans and PSK, are able to promote maturation of bone marrow-derived DC, increasing the expression of class II major histocompatibility complex molecules (MHC-II) and the costimulatory molecules—CD40 and CD80, as well as the production of IL-12 (Kanazawa et al. 2004). Moreover, PSK attenuated suppression of DC maturation provoked by tumor cell supernatant (Kanazawa et al. 2004). Maturation of DC exposed to GM-CSF and IL-4 was potentiated in the presence of PSK and manifested through enlarged expression of both MHC-II class and CD40 molecules per cell, while the presence of CD80, CD86, and CD83 was increased both per cell and in the total population (Kanazawa et al. 2004). Notably, phenotypical and functional maturation of DC are synchronized. DC are in different stages of differentiation and regulate effectors of innate immunity, such as NK cells and NK-T cells. Moreover, phenomena observed upon treatment with compounds isolated from multicolored polypore mushroom, focused attention on the involvement of DC in the development of an adaptive immune reaction. Finally, all data obtained from evaluation of the influence of multicolored polypore mushroom polysaccharide ingredients on DC behavior, anticipated its possible use in DC vaccine therapy.

13.3.2.3 Effects on Adaptive Immunity

The adaptive immune system responds to foreign antigens by the specific activity of B and T lymphocytes, allowing long-lasting protection and fast response in the next contact with the invader. Both types of specific immunity—humoral and cellular, function through the well orchestrated action of antigen presenting cells (APC) and T cells, organized in two directions. The first is MHC-I- mediated recognition of intracellular antigens by specific receptors on CD8$^+$ T cells and the second is recognition of proteolytic peptides from extracellular pathogens presented in the context of MHC-II molecules to CD4$^+$ T helper cells. Previously it was thought that fungal polysaccharides stimulate immune responses independently of T cells by activating innate immunity constituents and promoting the antigen presenting capacity of professional APC, such as dendritic cells (Fig. 13.1). Now it is known that β-glucans possess their own capacity to activate CD4$^+$ T cells through the MHC-II machinery (Cobb et al. 2004). Using a nitric oxide dependent mechanism, β-glucans are processed to low molecular weight carbohydrates capable of binding to MHC-II inside antigen presenting cells, such as DC. Together with data about the enhanced capacity of these professional antigen presenting cells to express MHC and co-stimulatory molecules in response to carbohydrates from multicolored polypore mushroom, considerable potentiation of lymphocyte proliferation and activation of naive T and B cells could be expected. Different concentrations of PSP were recently reported to promote the proliferation of T lymphocytes both in human peripheral blood and mouse splenocytes. Interestingly human lymphocytes were observed to be more sensitive than mouse lymphocytes (Smith et al. 2002). Both humoral and cellular immune functions affected by alkylating agents like cyclophosphamide, have been restored in the presence of multicolored polypore mushroom polysaccharides. It has been documented that PSP reconstitutes the ability of B lymphocytes to build up a suitable antibody response *in vivo* and *in vitro* (Smith et al. 2002). PSP was shown to enhance humoral immunity in tumor bearing mice that had been challenged with foreign antigens (i.e. sheep red blood cells) (Zhou et al. 1988). When immunity in sarcoma 180-bearing ICR mice was depressed by trinitrophenyl, antibody production could be restored by PSK administration (Oguchi et al. 1987). The same compound promoted the proliferation of phytohemagglutinin (PHA)—activated human peripheral blood lymphocytes and augmented CD4$^+$ T helper cell activation, thereby raising the CD4$^+$/CD8$^+$ ratio. Oral administration of PSK can improve the damaged antitumor CD4$^+$ T cell response in gut-associated lymphoid tissue of specific-pathogen free mice (Smith et al. 2002). However, these molecules do not exclusively polarize the immune response into either Th1, Th2, or regulatory T cells as previously mentioned. Complexity of this subject as well as cytokine pattern to which the individual strain is predisposed could not be ignored. PSK enhances the cytotoxic activity of PBL *in vivo* and *in vitro* (Ho et al. 2004; Jiménez et al. 2005). Moreover, it promoted the interaction of PBL with tumor cells when both were exposed to PSK simultaneously. Similarly to PSK, PSP stimulates T cell activation and production of IFN-γ and IL-2 with the potential to restore IL-2 production in immune-suppressed circumstances triggered by cyclophosphamide action (Quian et al. 1997; Jiménez et al. 2005). In humans, multicolored polypore mushroom polysaccharides temporarily restored cellular immune function in healthy older people but with depressed cellular immunity (Yokoe et al. 1979; Kondo and Torisu 1985). Unfortunately, the response declined to pre-treatment levels very soon after the end of the treatment.

13.3.2.4 Effects on Cytokine Production

The extreme potential of multicolored polypore mushroom ingredients to stimulate innate and adaptive immunity through specific receptor recognition, phenotypical and functional cell maturation is accompanied with enhanced cytokine production. Besides cytokines produced by particular immune cells exposed to PSP or PSK (Fig. 13.1), general expansion of IL-1, IL-6, and IFN-γ together with GCSF and GM-CSF secretion, were observed in peripheral blood mononuclear cells (Li et al. 2010). Multicolored polypore mushroom extract acted as a mitogen with Th1 oriented activity in a population of murine splenic lymphocytes (Ho et al. 2004). Prolonged hyperproduction of IL-2 and IL-12, together with time-limited expansion of IFN-γ and IL-18, were not followed by changes in production of IL-4 and IL-6. In contrast, standardized extract of PSP ("I'm-Yunity") did not affect either non-stimulated or PHA-stimulated lymphocytes, suggesting that the quality and composition of the prepared extract is crucial for this type of efficacy (Hsieh et al. 2002). Importantly, numerous tumor cell lines exposed to multicolored polypore mushroom dramatically altered their panel and quantity of cell generated cytokines. Having in mind that some of them are essential for cancer cell growth, while others are cytocidal, the observed phenomenon could be of crucial relevance for malignant cell survival. An aqueous extract of I'm-Yunity significantly increased IL-1 and IL-6, while substantially lowering IL-8 in human promyelocytic leukemic HL-60 cells, decreasing tumor cell viability (Hsieh et al. 2002). PSK suppressed the progression of QR-32 cells by reducing TGF-β and enhancing IFN-γ synthesis (Habelhah et al. 1998). Finally, besides direct effects on cytokine production, the possible influence of PSK on the effectiveness of already produced cytokines was mentioned earlier.

13.3.3 Direct Antitumor Effects, Molecular Targets and Signaling Pathways Affected by the Compounds

From the 80's until the present, numerous *in vitro* studies have confirmed efficient anticancer activity of both PSK and PSP (Ng 1998; Chu et al. 2002; Cheng and Leung 2008). In addition, a small peptide of multicolored polypore mushroom was found to possess a potent cytotoxic effect on several human tumor cell lines (HL-60, LS174-T, SMMU-7721, and SCG-7901) together with strong immunopotentiating activity (Yang et al. 1992). Moreover, we previously reported that a methanol extract rich in terpenoids and polyphenols efficiently suppressed the growth of B16 melanoma *in vitro* and *in vivo*, directly through inhibition of proliferation and induction of apoptosis and through endorsement of tumoricidal activity of macrophages (Harhaji et al. 2008).

A large spectrum of tumor cell lines, including cells derived from gastric, breast, melanoma, lung, prostate, and liver tumors, leukemia and lymphoma, are sensitive to fungal polysaccharides (Chu et al. 2002). On the other hand, a few tumor cell lines like mouse monocytes/macrophages, melanoma, sarcoma, and placental choriocarcinoma are unaffected by treatment with PSP (Ng 1998). It seems that responsiveness to multicolored polypore mushroom-derived compounds is defined by the type of tumor and is also dose dependent. However, no toxicity against normal counterparts of tumor cells such as hepatocytes, lung cells or fibroblasts was seen (Chu et al. 2002). Multicolored polypore mushroom-derived compounds display various mechanisms of antitumor action including inhibition of proliferation, as well as induction of apoptosis and differentiation. Kim et al. (1990) observed that PSK stimulated IFN-γ-induced differentiation of human myelogenous leukemic U-937 and THP-1 cells. Some authors speculated that inhibition of DNA synthesis and division rather than apoptosis is behind the antitumoral activity of multicolored polypore mushroom extract, but it seems that direct effects on tumor cells depend on their intrinsic characteristics. Cell cycle arrest with cell accumulation in the G0/G1 phase and increases in apoptosis and caspase-3 expression were proposed mechanisms of inhibition observed in various tumor cell lines derived from leukemias, melanomas, fibrosarcomas, cervix, lung, pancreas, and gastric cancers upon treatment with PSK (Jiménez-Medina et al. 2008).

Interestingly, multicolored polypore mushroom ingredients display selectivity in action between cell lines originating from the same tissue. This phenomenon is mainly defined by the cell specific mutations responsible for the malignant transformation. Thus, an ethanolic extract of multicolored polypore mushroom significantly reduced growth of hormone responsive prostate cancer LNCaP cells, down-regulated the level of secreted, but not intracellular prostate-specific antigen without affecting androgen receptors (Hsieh and Wu 2001). The effect was less pronounced on androgen independent PC-3 and DU-145 cells, where the treatment was followed with reduction of regulatory proteins, retinoblastoma and proliferating cell nuclear antigen. A standardized aqueous ethanol extract from multicolored polypore mushroom inhibited the proliferation of breast tumor cell lines T-47D, MCF-7, MDA-MB-231, while BT-20 was not affected (Ho et al. 2005). Nucleosome production was elevated in all sensitive cells, since the expression of *p53* and *Bcl-2* was differently regulated. Numerous intracellular targets for PSP and PSK have been described. Thus, PSP induced apoptosis of human promyelocytic leukemia HL-60 cells in association with a diminished *Bcl-2/Bax* ratio, mitochondrial transmembrane potential, cytochrome C release and activation of caspase-3, -8, and -9 (Yang et al. 2005). Moreover, decrease in the *p65* and *p50* forms of transcription factor NF-κB, and reduction in the expression of cyclooxygenase-2 was detected upon PSP treatment. PSP up-regulated *STAT1* (signal transducer and activator of transcription 1) and concomitantly, decreased expression of the activated form of *ERK* (extracellular signal-regulated kinase) (Hsieh et al. 2006). Zeng et al. (2005) evaluated the gene expression profile of these cells after treatment with PSP and reported that apoptosis was mediated by up-regulation of early transcription factors, such as *AP-1, EGR1, IER2*, and *IER5*, together with down-regulation of NF-κB transcription pathways. The expression of several genes involved in apoptosis and suppression of proliferation such as *GADD45A/B* and *TUSC2* was potentiated, whereas the activity of some phosphatase and kinase genes was inhibited. In addition, alteration of carcinogenesis-related gene transcripts such as *SAT, DCT, Melan-A, uPA*, and *cyclin E1* was observed.

A remarkable feature of protein-bound polysaccharides obtained from multicolored polypore mushroom is inhibition of the metastatic process. It seems that all steps during metastasis can be affected by these compounds (Kobayashi et al. 1995). Katano et al. (1987) reported suppression by PSK of tumor cell motility *in vitro* and *in vivo*. Thus, the motility of mouse leukemia EL4, Ehrlich carcinoma, human null cell leukemia, and colon 26 cells was inhibited by PSK in a dose dependent manner. PSK suppressed invasion through matrigel, chemotaxis and tumor cell adhesion to components of the basement membrane (Kobayashi et al. 1995). Moreover, it was shown to inhibit tumor cell induced platelet aggregation in *vitro*, indicating that the intravasation process is affected by the preparation (Tanaka et al. 1991). Some indirect evidence about interaction of PSK with cytoskeletal proteins, such as microtubule polymerization or myosine, suggested an influence on the extravasation process (Kobayashi et al. 1995). The antimetastatic action of PSK could be attributed to its potential to abolish matrix metalloproteases, key enzymes for degradation of the extracellular matrix, as well as other enzymes involved in metastatic formation. It was shown that decreased invasion by pancreatic and gastric tumor cell lines was associated with inhibition of TGF-β1, MMP-2, and MMP-9 at both mRNA and protein levels (Zhang et al. 2000). PSK also hampered mouse hepatoma MH134-induced angiogenesis suggesting that the anti-metastatic effect of PSK could be attributed to suppression of tumor-induced angiogenesis (Kanoh et al. 1994).

One important property of compounds derived from multicolored polypore mushroom that might determine both the antitumor and the immune stimulatory action is the ability to trap free radicals and enhance manganese superoxide dismutase activity. Regarding this, the constituents of multicolored polypore mushroom extract can influence chemotherapy or radiotherapy in the treatment of cancer (Kobayashi et al. 1993).

Besides their direct tumoricidal effect *in vitro*, protein-bound polysaccharides of multicolored polypore mushroom showed synergy with some chemotherapeutic drugs indicating their potential usefulness as an adjuvant in the treatment of different malignant diseases. Thus, it was shown that PSP treatment enhanced the cytotoxicity of doxorubicin and etoposide to human breast cancer cell line ZR-75-30 through enhancement of the apoptotic machinery of these drugs in a cell cycle-dependent manner (Wan et al. 2008). Furthermore, PSP enhanced the cytotoxic effect of cyclophosphamide on a HepG2 cancer cell line (Chan and Yeung 2006a). The effect of chemosensitization is also cell specific. While PSK prevented cisplatin induced cytotoxicity towards NRK-49F, in H4-II-E and human ovarian cancer cells the tumoricidal effect of this drug was potentiated (Kobayashi et al. 1994).

13.3.4 Other Pharmacological Effects

13.3.4.1 Antimicrobial Effects

Multicolored polypore mushroom extracts exhibit potent antimicrobial activity against different pathogens. It was demonstrated that mushroom extracts were efficient *in vitro* against *Escherichia coli, Listeria monocytogenes, Candida albicans, Staphylococcus aureus, Bacillus subtilis, Klebsiella pneumoniae*, and *Mycobacterium smegmatis* (Chu et al. 2002; Poyrazoglu Çoban et al. 2008). Moreover, its efficacy was confirmed in animals infected with *Escherichia coli, Pseudomonas aeruginosa, Listeria monocytogenes, Staphylococcus aureus, Candida albicans, Klebsiella pneumoniae*, and *Streptococcus pneumoniae* (Chu et al. 2002). It was considered that activation of polymorphonuclear cells and production of TNF-α and IL-1 mediated the described effect. Both PSP and PSK also showed antiviral activity against human immunodeficiency virus infection by modification of the viral receptor, blockade of binding to lymphocytes and stimulation of interferon production (Collins and Ng 1997). In addition, PSK efficiently overcame *in vivo* infection with ectromelia virus, cytomegalovirus or herpes simplex virus either by direct inactivation of virus or indirectly through an immunological response (Chu et al. 2002).

13.3.4.2 Hepatoprotective and Analgesic Effect

In China, multicolored polypore mushroom is believed to be functional in the treatment of hepatitis. Some data indicated that multicolored polypore mushroom extracts obstructed liver destruction induced by hepatotoxins. Thus, PSP protected liver cells from paracetamol-induced damage in rats (Yeung et al. 1994) by decreasing paracetamol-induced serum transaminases (SGPT and SGOT), hepatic enzymes considered as indicators for hepatotoxicity. In this model PSP interacted with these microsomal proteins inducing changes in their binding properties. In two other models of liver injury, CCl_4-induced lipid peroxidation damage and injury induced in mice by delayed-type hypersensitivity to picryl chloride, PSK showed beneficial effects (Yoshikawa et al. 1982; Xu et al. 1997).

Contradictory data about the influence of multicolored polypore mushroom compounds on pain have been obtained. It was previously reported that PSP possessed analgesic potential, which was related to its capacity to promote IL-2 production (Gong et al. 1998). It was also observed to elicit an analgesic effect in the hot-plate test in mice (Ng et al. 1999). However, it is described that hyperalgesia was potentiated by PSP through activation of peritoneal resident cells and increased production of inflammatory mediators (Chan and Yeung 2006b).

13.4 Clinical and Experimental Evidence for Anticancer Properties

13.4.1 Animal Studies

Numerous *in vivo* studies revealed the beneficial antitumor effect of multicolored polypore mushroom compounds. Advantages of their application include their possession of notable immune system enhancing activity, besides a direct effect on tumor cells. The panel of exerted activities is wide. First of all, it is well documented that prophylactic treatment with PSK prevents chemical carcinogen- and radiation-induced tumors, cancers developed spontaneously or by direct application of malignant cells (Kobayashi et al. 1993). Accordingly, the incidence of gastrointestinal cancers in rats provoked by subcutaneous injection of dimethylhydrazine was significantly suppressed by PSK treatment (Sakita et al. 1982). Improvement of survival upon PSK treatment was further observed in rats with 3-methyl-4-dimethylamino benzene induced hepatoma (Nakajima et al. 1990). In addition, suppression of cancerogenesis by PSK was observed in cancers of the stomach, intestine, esophagus, liver, lung, mammary gland, bladder, and brain as well as sarcoma induced by various chemical carcinogens (Kobayashi et al. 1993). Matsunaga et al. (2000) showed that neonatal inoculation of PSK increased the resistance of adult animals to challenges with syngeneic cells (colon adenocarcinoma, Lewis lung carcinoma, B16 melanoma) and reduced colonic precancerous lesions induced in F344 rats by azoxymethane. Pre-treatment with PSK prolonged survival and inhibited metastasis formation in mice injected with sarcoma cells (Algarra et al. 1999). Similarly, PSP administration for two weeks before sarcoma 180 inoculation in nude mice decreased the incidence of tumor growth and tumor size (Dong et al. 1996). Two weeks pretreatment with a small polypeptide from multicolored polypore mushroom diminished the incidence of tumors in nude mice inoculated with different human tumor cell lines (Yang et al. 1992). A beneficial effect of PSK was also observed in the model of radiation-induced lymphoma in mice (Kobayashi et al. 1993). In C3H/OuJ mice in which breast cancer spontaneously develops, PSK significantly suppressed the incidence and improved survival (Kobayashi et al. 1993). As previously mentioned, multicolored polypore mushroom methanol extract treatment inhibited tumor growth in C57BL/6 mice inoculated with syngeneic B16 tumor cells (Harhaji et al. 2008).

Besides the benefit of prophylactic application, numerous studies in experimental animals have confirmed the therapeutic efficacy of multicolored polypore mushroom extract. The list of tumors for which it is known to be efficacious in animals is long and includes adenosarcoma, fibrosarcoma, mastocytoma, prostate adenocarcinoma, plasmacytoma, melanoma, sarcoma, mammary cancer, colon cancer, and lung cancer. For example, in tumor-bearing nude mice, administration of PSP five days after sarcoma 180 inoculations, significantly suppressed the growth of tumor mass (Dong et al. 1996). PSK remarkably extended the life span of C3H/He mice bearing syngeneic plasmacytoma X5563 in a schedule- and dose-dependent manner (Matsunaga et al. 1992). Therapeutic treatment with PSP slowed progression of H238 tumors in BALB/c mice through direct cytotoxic effects, as well as immunomodulation (Mao et al. 1996). Interestingly, in a double grafted tumor system where BALB/c mice received simultaneous inoculations of Meth-A fibrosarcoma on both sides of the body it was revealed that PSP treatment of tumor on one side led to regression of the distant tumor. These data suggested that activation of the immune response is responsible for the protective effect of the treatment (Ebina and Kohya 1988).

Beneficial effects of both PSP and PSK on solid tumors were further shown in several metastatic models. Thus, it was reported that PSK suppressed pulmonary metastasis of sarcoma induced in mice by methylcholanthrene, inoculation of B16 melanoma, or prostate cancer DU-145M and PC-3 cells (Kobayashi et al. 1995). In addition, it inhibited lymphatic metastasis of mouse leukemia P388 and hepatoma MH134, as well as liver metastasis in animals injected with lung cancer AOI, hepatoma AH60C, colon26, and leukemia RL male 1 cells (Kobayashi et al. 1995).

Finally, as mentioned earlier, synergistic action between multicolored polypore mushroom compounds and different chemo-therapeutic drugs, such as tamoxifen, docetaxel, uracil, carboquinone, tegafur, etc. was reported in different animal models of tumors (Mickey 1985; Iino et al.1992; Katoh and Ooshiro 2007; Yamasaki et al. 2009).

13.4.2 Clinical Studies

Multicolored polypore mushroom was first recorded during the Ming Dynasty of China, and subsequently in 1965 in a Japanese report about a patient with stomach cancer. Since then numerous clinical studies in Japan and China identified PSK and PSP as the most valuable constituents of multicolored polypore mushroom for the treatment of digestive system, lung, breast and nasopharyngeal cancers. The majority of studies were done with PSK, while the few investigations on the beneficial effect of PSP in concomitant treatment with chemo- and radiotherapy were reviewed by Ng (1998) and Kidd (2000). In addition, detailed information about currently active projects or those at the stage of recruiting patients are available at US NIH. Most research was dedicated to evaluation of polysaccharopeptide effectiveness as a supplement to chemotherapy and/or radiation. Clinical trials were focused on assessing the ability of the extracts to prolong remission and upgrade the prognosis, as well as on examining the effect on restoration of the immune system, since its destruction is a common side effect of conventional cancer therapies.

13.4.2.1 PSP Trials

Clinical trials with PSP started soon after its isolation in 1986, at first in patients with stomach, esophageal and non-small lung cancer. PSP utilization was not associated with serious adverse effects and it offered protection against some unpleasant effects of cancer and the toxic therapies used for its treatment. In 1992 Liu and Zhou (1993) studied at few hospitals in Shanghai, 274 patients with stomach, esophageal and lung cancer. Patients received conventional (radio- and/or chemo) anticancer therapy after surgery. PSP was given at 3.1 g/day for two months orally. They observed marked improvement of clinical symptoms, enhanced NK activity, increased IL-2 production, CD4/CD8 ratio and white blood cell count. The success of this trial led to its extension and later findings for 189 patients confirmed the beneficial effects of PSP (Liu et al. 1999). Wu et al. (1993) obtained similar results using the same protocol for patients with esophageal cancer undergoing radiotherapy. Several studies published in 1993 showed significant immunological improvement in gastric patients who received PSP after surgery and concomitantly with chemotherapy (Ng 1998; Kidd 2000). In addition, Shiu et al. (1993) reported increased appetite and normalization of white blood cell and platelet counts in patients with breast cancer that received chemotherapy in parallel. PSP also improved the quality of life of patients with ovarian and endometrial cancers (Kidd 2000). Tsang et al. (2003) described a randomized controlled double blinded trial in patients with advanced and inoperable non-small cell lung cancer with life expectancy more than 12 weeks and TNM stage III or IV. Radiotherapy and chemotherapy was completed at least four weeks prior to the PSP treatment. Four weeks of treatment in group of 34 patients led to a significant increase in the level of IgG and IgM ($P=0.02$ and 0.01), total leukocyte and neutrophil counts ($P=0.003$ and 0.005) and body fat content ($P=0.02$). Finally, Wong et al. (2005) reported that regular oral consumption of capsules containing multicolored polypore mushroom (50 mg/kg body weight, 100% PSP) and *Salvia miltiorrhiza* (20 mg/kg body weight) for 6 months could be beneficial for promoting immunological function during post-treatment of breast cancer patients. These authors showed that counts of $CD4^+$ T cells, CD4/CD8 ratio, as well as counts of B lymphocytes were significantly elevated.

13.4.2.2 PSK Trials

Breast Cancer

A five-year randomized PSK trial was carried out on 914 women with stage IIA, IIB and IIIA primary breast cancer who received an extended, standard or modified radical mastectomy (Toi et al. 1992). Tamoxifen was applied to standard chemotherapy for estrogen receptor-positive patients, while estrogen receptor-negative patients received PSK (3 g/day for 24 months) as adjuvant therapy besides chemotherapy. The results showed that PSK significantly extended the survival of patients with node-negative, ER-negative and Stage IIA T2N1 cancer ($P<0.0017$). Morimoto et al. (1996) conducted a 5 year-post-operative adjuvant randomized trial comparing chemo-endocrine therapy, chemotherapy, and immunotherapy for patients with stage II breast cancer. The authors evaluated 364 ER-patients who received ftorafur chemotherapy with or without PSK, but did not obtain a statistically significant difference between the groups.

Another large randomized control trial involved 227 patients with operable breast cancer with vascular invasion in the tumor and/or metastatic lymph nodes (Iino et al. 1995). Patients were randomized into three groups receiving: (1) 5-fluorouracil + cyclophosphamide + mitomycin C + predonisolone (FEMP), (2) FEMP + levamisole and (3) FEMP + PSK (3 g/day). Each treatment was applied at six-month intervals for five years. Concerning disease-free and overall survival, there were no significant differences between the tested groups, while the risk ratio (RR) was lower in the FEMP + PSK group compared to the FEMP group. Side effects of leucopenia and nausea were mild and tolerable. The same authors reanalyzed the findings for 134 patients with vascular invasion, who were separated into FEMP and FEMP + PSK groups (Yokoe et al. 1997). Patients received two 28-day courses of these treatments a year for five years. Each group was stratified according to the presence of HLA B40. Since the disease free survival rates for the FEMP + PSK group of B40-positive patients at 5- and 10-years were 100%, while for B40-negative patients rates were 76% and 55%, it was concluded that HLA typing may be useful for predicting effects of PSK application in patients with operable breast cancer.

Lung Cancer

The first study was conducted in Japan in 1990, but was published in Japanese, so only the abstract is available in English. In this randomized controlled study chemotherapy (CDDP and VDS) was combined with PSK (3 g/day) for stage III and IV adenocarcinoma of the lung. The survival curve indicated that the immuno-chemotherapy group was superior to the chemotherapy group (log-rank test $P=0.075$) (Nishiwaki et al. 1990). Three years later, Hayakawa et al. (1993) reported an evaluation of the effect of PSK adjuvant treatment after radical radiotherapy in 185 patients with I, II and III stage of non-small cell lung cancer. They found significantly longer disease-free intervals and survival for lung cancer patients treated with PSK. PSK alone or in combination with G-CSF restored leukocyte and platelet counts in patients with lung cancers during chemotherapy (Katoh et al. 1997).

Leukemia

In the early 80's, 28 patients with acute leukemia were randomly divided into chemotherapy and chemo-immunotherapy groups (Nagao et al. 1981). Remission was induced with combined chemotherapy (neocarzinostatin, cytosine arabinoside, prednisolone or vinicristine, daunorubicin, prednisolone) after two to three courses of mercaptopurine chemotherapy with or without PSK until relapse. The median duration of complete remission and survival were longer and the complete remission rate of the second induction was higher in the chemo-immunotherapy group than in the chemotherapy group. Another prospective randomized cooperative trial included 31 patients with acute non-lymphocytic leukemia who received PSK in combination with chemotherapy (Ohno et al. 1984). PSK treatment for six months extended remission and survival time, but prolonged analysis did not show statistical significance. Patients who achieved a remission of more than 270 days survived longer if they received PSK ($P < 0.06$). Kawa et al. (1991) studied 125 children with acute lymphoblastic leukemia, who were in continuous remission for three years on chemotherapy. After treatment suspension, the groups of patients were treated with different biological modifiers including PSK. However, the relapse rate in the PSK group was found to be similar to that observed for children who were off therapy.

Colorectal Cancer

In the last 20 years numerous clinical trials have been performed to evaluate multicolored polypore mushroom extract in patients with colorectal cancers. A selection is presented in Table 13.2. In order to assess the benefit of immune-chemotherapy with PSK for patients with curatively resected colorectal cancer, Sakamoto et al. (2006) reported meta-analysis of data with center randomization for 1,094 patients in three clinical trials. Adjuvant immune-chemotherapy with PSK was shown to improve both survival and disease-free survival of the patients. The overall survival RR for suitable patients was 0.71 and the disease-free survival RR was 0.72. Although the effect of PSK on disease survival has been investigated many times, there are few studies concerning its influence on the immune system. After corroboration that PSK increased the 5-year disease-free survival rate and reduced the risk of recurrence in patients with stage II and III colorectal cancer, Ohwada et al. (2006) attempted to elucidate the disease-free survival benefits of PSK and to determine which immunological markers could indicate PSK responders. They found that PSK decreased the mean serum immunosuppressive acidic protein level, and increased the mean population of NK cells compared with the controls. Serum concentrations of sIL-2R and the production of IL-10, which are elevated in patients with advanced colorectal cancer, were reduced after 2 months of treatment with PSK concomitantly with chemotherapy (Shibata et al. 2002). Recent data from patients with adenocarcinoma of the rectum revealed significant increases of NK cell count in the peripheral blood and cytotoxic T cell count in the peri-tumoral and normal mucosa, and a significant decrease of serum immunosuppressive acidic protein level in the PSK-treated group combined with preoperative chemo-radiotherapy (Sadahiro et al. 2010).

Gastric Cancer

A major cause of mortality in China and Japan is stomach cancer and since 1970 numerous trials have been undertaken to estimate the beneficial effect of immunotherapy. Some are described here or summarized in Table 13.3. One prospective controlled study involving 739 patients with gastric cancer, who were subjected to gastrectomy followed by adjuvant chemotherapy and PSK, revealed that PSK induced significantly longer survival in HLA-A2 positive patients (Ogoshi et al. 1997). The same group performed a retrospective study on 872 gastric cancer patients with resections and chemotherapy with or without PSK. Treatment groups were analyzed according to preoperative serum levels of carcinoembryonic antigen (CEA) and acute phase reactants (APR). Patients with abnormal levels of CEA and one or more abnormal APR levels with PSK had significantly longer survival (log-rank test, $P=0.0015$; Breslow test, $P=0.0042$) (Ogoshi et al. 1998). In a meta-analysis including 8,009 patients from eight randomized controlled trials, Oba et al. (2007) performed central randomization and analyzed the survival benefit of PSK after curative gastric resections. The overall hazard ratio for appropriate patients

Table 13.2 Selected clinical trials for colorectal cancer

Study design	Sample size	Treatment	Results	References
Randomized controlled trial	111 patients 56 cases 55 controls	1. Surgery + placebo 2. Surgery + PSK in decreasing doses over 3 years	Remission and surviving at 10- years were significantly higher ($P<0.05$) in the PSK group. Polymorphonuclear leukocytes from PSK-treated patients showed enhancement in chemotactic locomotion and phagocytic activity.	Torisu et al. 1990
Randomized controlled trial	448 patients in 35 institutions in the Kanagawa prefecture 221 cases 227 controls	Mitomycin C after surgery 1. 5-FU for over 6 months 2. 5-FU + PSK 3 years	Disease-free survival and survival ($P=0.013$) were statistically significant.	Mitomi et al. 1992
Prospective cohort with controls	58 patients with stages II or III colorectal cancer 48 cases 10 controls	Postoperative 1. Fluoropyrimidine therapy and PSK (3 g/day) for at least 12 months 2. Fluoropyrimidine	3-year disease-free survival rate in the PSK + chemotherapy group was significant in comparison to control ($P=0.0467$). Serum type IV collagen levels as indicator of destruction in basement membrane were significantly higher ($P=0.0072$) in the chemotherapy group.	Kudo et al. 2002
Randomized controlled study	558 patients enrolled at 38 centers with stage II or III primary colorectal cancer 277 patients in group 1 281 patients in group 2	Mitomycin C after surgery 1. 5-DFUR + PSK (3 g/day) 2. 5-FU + PSK (3 g/day)	5'-DFUR + PSK regimen was more useful for patients with stage II and III (RR, 1.451; $P=0.048$) with tumor depth of pT3 or pT4 (RR, 1.568; $P=0.02$).	Koda et al. 2003

Table 13.2 (continued)

Study design	Sample size	Treatment	Results	References
Randomized controlled trial	441 patients from 93 cooperating institutions in Japan with primary colon cancer and lymph node metastasis 220 cases 221 controls	5-FU 3–4 weeks postoperative 1. PSK 4 weeks followed by 4 weeks of 5-FU as one course 2. 4 weeks of rest followed by 4 weeks of 5-FU as one course Total of 10 courses	7-year overall survival and disease free rates showed non-significant differences between groups. 7-year survival rate until cancer death for the PSK group, was significantly higher than the control group ($P=0.019$). No toxic effects observed.	Ito et al. 2004
Randomized controlled trial	205 patients from 19 hospitals in Japan with stage II and III 137 cases 68 controls	Mitomycin C postoperative 1. PSK + UFT 2. UFT 2 years	5-year disease free survival was significantly higher for the PSK group ($P=0.016$). 5-year disease free survival and overall survival from PSK group with stage III was significantly greater ($P=0.002$, $P=0.003$).	Ohwada et al. 2004
Randomized controlled trial	205 patients with stage II and III 137 cases 68 controls	Mitomycin C postoperative 1. PSK + UFT 2. UFT 2 years	PSK decreased the mean serum immunosuppressive acidic protein (IAP) level, and increased the mean population of natural killer cells.	Ohwada et al. 2006
Randomized controlled trial	202 patients with primary colon cancer 99 cases 103 controls	1. 5-FU 2. 5-FU + PSK	The presence of diffuse nuclear accumulation of activated β-catenin, observed in patients received immunotherapy was related with improved recurrence-free survival, cancer death survival and overall survival rates. Therefore, this type of accumulation of beta catenin in nucleus identified patients who better respond to PSK treatment.	Yamashita et al. 2007

Table 13.3 Selected clinical trials for gastric and esophageal cancers

Study design	Sample size	Treatment	Results	Reference
Randomized controlled study	579 patients from 97 hospitals in the Kyushu and Chugoku district of Japan with gastric cancer	Gastrectomy 1. Mitomycin C, futraful 2. Mitomycin C, futraful and PSK (3 g/day) 1 year	PSK group showed a significant increase in 5-year survival ($P<0.05$); PSK group had best survival in cases with positive lymph node metastases, positive serosal invasion and undifferentiated carcinoma by histological type.	Niimoto et al. 1988
Randomized controlled study	47 patients 29 gastric 18 colorectal	1. Surgery 2. Surgery + PSK (3 g/day before surgery, daily or every other day in short or prolonged regimen)	The results showed that the effect of PSK was significantly influenced by duration but not frequency of the treatment. In short term PSK group, enhanced activity of peripheral blood cells as well as decreased proportion of $CD9^+$ $11b^+$ suppressor T cells in regional lymph node were observed. However, in long term PSK group, increased proportion and cytotoxicity of peripheral blood $CD16^+$ cells together with enlarged number of $CD4^+$ Leu8 helper T cells in regional lymph node were determined.	Nio et al. 1992
Randomized controlled study	253 patients with gastric cancer	Post gastrectomy 1. Mitomycin C + fluorouracil 2. Chemotherapy + PSK	PSK treatment increased 5-year disease-free period and 5-year survival ($P<0.05$ and $P=0.04$, respectively). No characteristic toxic effect for PSK treated group was observed.	Nakazato et al. 1994

Table 13.3 (continued)

Study design	Sample size	Treatment	Results	Reference
Randomized controlled study	174 patients with esophageal cancer	Esophagectomy Radiotherapy ± PSK (3 months) Radiotherapy + chemotherapy ± PSK (3 months)	Tendency for longer survival on PSK was observed, but statistical significance was not reached.	Ogoshi et al. 1995a
Randomized controlled study, stratified reanalysis	158 patients with esophageal cancer	Esophagectomy Radiotherapy ± PSK Radiotherapy + chemotherapy ± PSK	The survival of the PSK group was significantly better. Preoperative serum levels of alpha 1-antichymotrypsin and sialic acid may possibly predict the effectiveness of immunotherapy using PSK.	Ogoshi et al. 1995b
Reanalysis of randomized controlled study	253 patients I-IV stage of gastric cancer	Post gastrectomy 1. Mitomycin C + fluorouracil 2. Chemotherapy + PSK	In the group with low immunosuppressive acidic protein levels, survival was better with PSK with and without splenectomy, but only the low level group without splenectomy was significantly better ($P=0.024$).	Saji et al. 1999
Reanalysis of randomized controlled study	751 patients with gastric cancer	1. Surgery + chemotherapy 2. Surgery + chemotherapy + PSK	Only patients with granulocyte/lymphocyte ≥2.0 treated with PSK had significant benefit in 5-year survival ($P=0.007$).	Toge et al. 2000
Randomized controlled study	21 patients with stage III gastric cancer 10 cases 11 controls	1. Surgery + UFT 2. Surgery + UFT + PSK (3 g/day, 1 year)	PSK improves overall survival of stage III gastric cancer patients ($P=0.038$). Proportion of $CD57^{+}T$ (poor prognostic marker) cells was significantly reduced in PSK treated patients ($P=0.0486$). There was no difference between the survival rate of CD57-high and CD57-low patients treated with PSK.	Akagi et al. 2010

was 0.88 (95% confidence interval, 0.79–0.98; $P=0.018$). The results confirmed that co-treatment with PSK improved the survival of these patients.

13.5 Safety Studies

Multicolored polypore mushroom constituents and metabolites are considered as a safe and useful approach for disease treatment, particularly as they can be added to the diet and used orally, without the need to go through Phase I/II/III trials as an ordinary medicine. Nevertheless, most of the major multicolored polypore mushroom polysaccharides have already been subjected to extensive preclinical toxicology tests. A number of modern toxicological experiments conducted on monkeys, dogs, mice, rats, and guinea pigs, have given negative results for acute, chronic, genetic, reproductive, and teratogenic toxicity (Cheng and Leung 2008). The LD_{50} of crude multicolored polypore mushroom extract administered orally in mice was greater than 18 g/kg (Cheng and Leung 2008). Similarly, the LD_{50} value of PSP administered by intraperitoneal injection was 26–300 mg/kg, and considered to be non-toxic according to the Procedures for Toxicological Assessment on Food Safety Acute Toxicity Test (GB15193.3-94). The results of subchronic and chronic toxicity studies showed no toxic symptoms, death, or obvious developmental, hematological, and pathohistological changes after a 2–3-month dosing period with either crude multicolored polypore mushroom extract (Chu et al. 2002), or PSP (Cheng and Leung 2008) at 50–200 times the oral clinical dose. Evaluation of cytogenotoxicity, including chromosomal aberration, Ames and micronucleus tests on mammalian somatic cells, indicated that PSP showed no evidence of mutagenic potential (Cheng and Leung 2008). The possible effects of PSP on reproductive physiology and embryonic development were also examined, but in mice no adverse effects have been observed (Cheng and Leung 2008). Researchers demonstrated that PSP did not cause sperm aberration even at a dosage 100 times higher than the usual human dose. Furthermore, no deleterious effects of PSP have been observed on ovarian follicular development, steroidogenesis, quality of oocytes, pregnancy, and embryonic development (Ng and Chan 1997).

There are few documented examples of adverse effects of multicolored polypore mushroom constituents on man, and as such they can be generally considered as safe, irrespective of age and gender. Low-grade toxicities have been reported when administered in conjunction with chemotherapeutic agents. However, such effects may be caused by the chemo agents themselves. Rare side effects include nausea, vomiting, loss of appetite, and diarrhea. Even less common are darkening of the fingernails and low blood cell counts. No metabolic demand on the liver or stress on the kidneys has been found. However, in some cases the use of multicolored polypore mushroom preparations may be contraindicated. This includes patients receiving bone marrow transplants or suffering from autoimmune disorders (Chu et al. 2002).

Compared to conventional chemotherapy, multicolored polypore mushroom derivatives appear to be benign. Thus, no clearly defined drug interaction has been reported for multicolored polypore mushroom to date. Investigators have begun to understand the putative immune-enhancing effects of some polysaccharide constituents, which can potentially counteract the action of any co-administered immunosuppressant. However, these compounds were shown to promote the therapeutic efficacy of mAbs (e.g. trastuzumab, rituximab, and cetuximab) (Cheung et al. 2002; Modak et al. 2005; Standish et al. 2008). Besides evident medical benefits of multicolored polypore mushroom preparations, patients currently taking other drugs should exert caution since complications can occur. This underscores the urgent need to collect clinical safety data from large scale clinical trials systematically and to evaluate potential drug-drug interactions of the mushroom-derived anticancer pharmaceuticals in relation to safety and efficacy.

13.6 Translating the Pharmacologic Potential of Multicolored Polypore Mushroom into an Anticancer Agent

Multicolored polypore mushroom constituents and metabolites represent an interesting case of crossover from traditional herbal medicine to pharmaceutical-grade products. Major research on the isolation of pharmacologically active compounds comes from those Eastern countries with the historically best established tradition of employing medicinal mushrooms. Among the biologically active compounds, several polysaccharide preparations from multicolored polypore mushroom have found their way to the market and have been introduced into cancer therapy in combination with standard radiotherapy and chemotherapy. Unfortunately, North America and Europe are poorly represented in this field. For example, PSK is approved for clinical use in Japan, but was rejected by the Ministry of Health in Spain and by the FDA in the US. As the example of multicolored polypore mushroom demonstrates, translating traditional Eastern practices into acceptable evidence-based Western therapies is difficult, due to different manufacturing standards, criteria of purity, and under-powered clinical trials. Considering, however, the leading role of the US in the study of natural products worldwide, this gap could soon be filled. The National Cancer Institute has studied *Coriolus sp* and has deemed them effective in stimulating the function of the immune system in every animal species tested. Right now, human studies are taking place in Italy, England, Spain, and California.

Academic researchers are not the only ones testing the efficacy of traditional remedies. In the US, pharmaceutical companies are examining their ability to compete in the future FDA-approved herbal pharmaceutical market. The WHO has encouraged the rational use of traditional, plant based medicines by member states of the World Health Assembly and, to this purpose, has developed technical guidelines for the assessment of herbal medicines (WHO 2000). The application of standard pharmaceutical methods for quality assurance, safety assessment, and efficacy testing of crude extracts, active fractions, or pure multicolored polypore mushroom-derived compounds constitutes the first step in the process of bringing them from the field into the clinic, dispensary, and hospital. It is essential to ensure good manufacturing practice standards with batch-to-batch consistency, consistent analytical profiles, stability of end products, and relevant toxicity data. Furthermore, randomized, double-blind, placebo-controlled clinical trials of traditional mushroom remedies are essential to meet the standards of the US FDA and to produce data that the clinical community will accept.

A variety of multicolored polypore mushroom extracts, including PSP and VPS, or raw mushroom extract, are available as dietary supplements. Unlike approved drugs (which must be tested before being allowed to be sold), the companies making these preparations are not required to prove to the FDA that their supplements are safe or effective, as long as there is no claim that the supplements can prevent, treat, or cure any specific disease. However, some such products may not contain the amount of compounds written on the label, and some may include other substances (contaminants). Actual amounts per dose may vary between brands or even between different batches of the same brand. Therefore, the key is making sure that the new, mushroom-based formulations meet the same standards of safety and efficacy as conventional pharmaceuticals.

Since many of the associated compounds may be essential cofactors for mushroom component function, a central feature of the purported chemopreventive or medicinal role of the mushroom extracts may lie in the synergistic interaction of multiple active principles (Sulivan et al. 2006). Nonetheless, in order to ensure high quality standards, medicinal mushrooms need to have botanical and analytical reference materials with strict quality assurance and quality control measures.

13.7 Concluding Remarks

In spite of many hundred years of traditional practice in China, Japan, Korea, and Slav regions, medicinal mushrooms were recognized as a potentially important new source of anticancer agents only within the past two decades. Scientific evaluation of mushroom water and alcohol extracts have verified the remedial capacity of polysaccharides, glycoproteins, proteoglycans, terpenoids, fatty acids, proteins, lectins, and other physiologically active compounds. Most of them are cell wall and other organ constituents or secondary metabolites. Experience from the traditional and conventional treatment of patients with cancer has indicated impressive immunostimulatory potential of multicolored polypore mushroom, certifying this mushroom as one of the most potent natural therapeutics for immunosuppression. However, besides the confirmed efficacy of polysaccharide fractions in restoring the immune system in cancer patients, potent anticancer activity of multicolored polypore mushroom ingredients is based on their ability to inhibit tumor cell proliferation, induce cell death, and suppress metastasis. Hundreds of *in vitro* and *in vivo* animal and human studies have undoubtedly confirmed that mushrooms are a potentially important new source of anticancer agents and therefore their assimilation into Western drug discovery programs and clinical trials is desirable. Although traditional medicines in cancer treatment, including multicolored polypore mushroom, are not accepted in many countries of the world, we believe that they will play an important role in human health care in the near future, since intercommunion and cooperative activities are enhanced in areas of science and technology.

Acknowledgment This work was supported by the Serbian Ministry of Science (Grant 173013).

References

Abraham, W. R. (2001). Bioactive sesquiterpenes produced by fungi: Are they useful for humans as well? *Current Medicinal Chemistry, 8,* 583–606.

Akagi, J., & Baba, H. (2010). PSK may suppress CD57(+) T cells to improve survival of advanced gastric cancer patients. *International Journal of Clinical Oncology/Japan Society of Clinical Oncology, 15,* 145–152.

Algarra, I., Collado, A., Garcia Lora, A., et al. (1999). Differential effect of protein-bound polysaccharide (PSK) on survival of experimental murine tumors. *Journal of Experimental & Clinical Cancer Research: CR, 18,* 39–46.

Bohm, J. A., & BeMillar, J. N. (1995). (1-3)-βD-glucans as biological response modifiers: A review of structure-functional activity relationships. *Carbohydrate Polymers, 28,* 3–14.

Breene, W. (1990). Nutritional and medicinal value of speciality mushrooms. *Journal of Food Production, 53,* 883–894.

Chan, S. L., & Yeung, J. H. (2006a). Effects of polysaccharide peptide (PSP) from Coriolus versicolor on the pharmacokinetics of cyclophosphamide in the rat and cytotoxicity in HepG2 cells. *Food and Chemical Toxicology, 44,* 689–694.

Chan, S. L., & Yeung, J. H. (2006b). Polysaccharide peptides from COV-1 strain of Coriolus versicolor induce hyperalgesia via inflammatory mediator release in the mouse. *Life Sciences, 78,* 2463–2470.

Chan, G. C., Chan, W. K., & Sze, D. M. (2009). The effects of beta-glucan on human immune and cancer cells. *Journal of Hematology Oncology, 2,* 25.

Chen, J., & Seviour, R. (2007). Medicinal importance of fungal beta-(1→3), (1→6)-glucans. *Mycological Research, 111,* 635–652.

Cheng, K. F., & Leung, P. C. (2008). General review of polysaccharopeptides (PSP) from C. versicolor: Pharmacological and clinical studies. *Cancer Therapy, 6,* 117–130.

Cheung, N. K., Modak, S., Vickers, A., et al. (2002). Orally administered beta-glucans enhance anti-tumor effects of monoclonal antibodies. *Cancer Immunology Immunotherapy, 51,* 557–564.

Chu, K. K., Ho, S. S., & Chow, A. H. (2002). Coriolus versicolor: A medicinal mushroom with promising immunotherapeutic values. *Journal of Clinical pharmacology, 42,* 976–984.

Cobb, B. A., Wang, Q., Tzianabos, A. O., et al. (2004). Polysaccharide processing and presentation by the MHCII pathway. *Cell, 117,* 677–687.

Coles, M., & Toth, B. (2005). Lack of prevention of large intestinal cancer by VPS, an extract of Coriolus versicolor mushroom. *In Vivo, 19,* 867–871.

Collins, R. A., & Ng, T. B. (1997). Polysaccharopeptide from Coriolus versicolor has potential for use against human immunodeficiency virus type 1 infection. *Life Sciences, 60,* 383–387.

Cui, J., & Chisti, Y. (2003). Polysaccharopeptides of Coriolus versicolor: Physiological activity, uses, and production. *Biotechnology Advances, 21,* 109–122.

Dong, Y., Kwan, C. Y., Chen, Z. N., et al. (1996). Antitumor effects of a refined polysaccharide peptide fraction isolated from Coriolus versicolor: In vitro and in vivo studies. *Research Communications in Molecular Pathology and Pharmacology, 92,* 140–148.

Ebina, T., & Kohya, H. (1988). Antitumor effector mechanism at a distant site in the double grafted tumor system of PSK, a protein-bound polysaccharide preparation. *Japanese Journal of Cancer Research: Gann, 79,* 957–964.

Ghoneum, M., Wimbley, M., Salem, F., et al. (1995). Immunomodulatory and anticancer effects of active hemicellulose compound (AHCC). *International Journal of Immunotherapy, 11,* 23–28.

Gong, S., Zhang, H. Q., Yin, W. P., et al. (1998). Involvement of interleukin-2 in analgesia produced by Coriolus versicolor polysaccharide peptides. *Zhongguo Yao Li Xue Bao = Acta Pharmacologica Sinica, 19,* 67–70.

Habelhah, H., Okada, F., Nakai, K., et al. (1998). Polysaccharide K induces Mn superoxide dismutase (Mn-SOD) in tumor tissues and inhibits malignant progression of QR-32 tumor cells: Possible roles of interferon alpha, tumor necrosis factor alpha and transforming growth factor beta in Mn-SOD induction by polysaccharide K. *Cancer Immunology, Immunotherapy: CII, 46,* 338–344.

Harhaji, L. J., Mijatovic, S., Maksimovic-Ivanic, D., et al. (2008). Anti-tumor effect of Coriolus versicolor methanol extract against mouse B16 melanoma cells: In vitro and in vivo study. *Food and Chemical Toxicology: An International Journal Published for the British Industrial Biological Research Association, 46,* 1825–1833.

Hayakawa, K., Mitsuhashi, N., Saito, Y., et al. (1993). Effect of Krestin (PSK) as adjuvant treatment on the prognosis after radical radiotherapy in patients with non-small cell lung cancer. *Anticancer Research, 13,* 1815–1820.

Ho, C. Y., Lau, C. B., Kim, C. F., et al. (2004). Differential effect of Coriolus versicolor (Yunzhi) extract on cytokine production by murine lymphocytes in vitro. *International Immunopharmacology, 4,* 1549–1557.

Ho, C. Y., Kim, C. F., Leung, K. N., et al. (2005). Differential anti-tumor activity of coriolus versicolor (Yunzhi) extract through p53- and/or Bcl-2-dependent apoptotic pathway in human breast cancer cells. *Cancer Biology & Therapy, 4,* 638–644.

Hobbs, C. (1995). *Medicinal mushrooms. An exploration of tradition, healing and culture.* Santa Cruz: Botanica Press.

Hong, F., Yan, J., Baran, J. T., et al. (2004). Mechanism by which orally administered beta-1,3-glucans enhance the tumoricidal activity of antitumor monoclonal antibodies in murine tumor models. *Journal of Immunology (Baltimore, Md.: 1950), 173,* 797–806.

Hsieh, T. C., & Wu, J. M. (2001). Cell growth and gene modulatory activities of Yunzhi (Windsor Wunxi) from mushroom Trametes versicolor in androgen-dependent and androgen-insensitive human prostate cancer cells. *International Journal of Oncology, 18,* 81–88.

Hsieh, T. C., Kunicki, J., Darzynkiewicz, Z., et al. (2002). Effects of extracts of Coriolus versicolor (I'm-Yunity) on cell-cycle progression and expression of interleukins-1 beta,-6, and -8 in promyelocytic HL-60 leukemic cells and mitogenically stimulated and nonstimulated human lymphocytes. *Journal of Alternative and Complementary Medicine (New York), 8,* 591–602.

Hsieh, T. C., Wu, P., Park, S., et al. (2006). Induction of cell cycle changes and modulation of apoptogenic/anti-apoptotic and extracellular signaling regulatory protein expression by water extracts of I'm-Yunity (PSP). *BMC Complementary and Alternative Medicine, 6,* 30.

Iino, Y., Takai, Y., Sugamata, N., et al. (1992). PSK (krestin) potentiates chemotherapeutic effects of tamoxifen on rat mammary carcinomas. *Anticancer Research, 12,* 2101–2103.

Iino, Y., Yokoe, T., Maemura, M., et al. (1995). Immunochemotherapies versus chemotherapy as adjuvant treatment after curative resection of operable breast cancer. *Anticancer Research, 15,* 2907–2911.

Ikuzawa, M., Matsunaga, K., & Nishiyama, S. (1988). Fate and distribution of an antitumor protein-bound polysaccharide Krestin (PSK)™. *International Journal of Immunopharmacology, 10,* 415–23.

Ito, K., Nakazato, H., Koike, A., et al. (2004). Long-term effect of 5-fluorouracil enhanced by intermittent administration of polysaccharide K after curative resection of colon cancer. A randomized controlled trial for 7-year follow-up. *International Journal of Colorectal Disease, 19,* 157–164.

Jiménez, E., Garcia-Lora, A., Martinez, M., et al. (2005). Identification of the protein components of protein-bound polysaccharide (PSK) that interact with NKL cells. *Cancer Immunology Immunotherapy, 54,* 395–399.

Jiménez-Medina, E., Berruguilla, E., Romero, I., et al. (2008). The immunomodulator PSK induces in vitro cytotoxic activity in tumour cell lines via arrest of cell cycle and induction of apoptosis. *BMC Cancer, 8,* 78.

Jong, S. C., Birmingham, J. M., & Pai, S. H. (1991). Immunomodulatory substances of fungal origin. *Journal of Immunology Immunopharmacology, 3,* 115–122.

Kanazawa, M., Mori, Y., Yoshihara, K., et al. (2004). Effect of PSK on the maturation of dendritic cells derived from human peripheral blood monocytes. *Immunology Letters, 91,* 229–238.

Kanoh, T., Matsunaga, K., Saito, K., et al. (1994). Suppression of in vivo tumor-induced angiogenesis by the protein-bound polysaccharide PSK. *In Vivo, 8,* 247–250.

Katano, M., Yamamoto, Y., & Torisu, M. (1987). A suppressive effect of PSK, protein-bound polysaccharide preparation, on tumor growth: A new effect of PSK on cell motility. *Japanese Journal of Cancer & Chemotherapy, 14,* 2321–2326.

Katoh, R., & Ooshiro, M. (2007). Enhancement of antitumor effect of tegafur/uracil (UFT) plus leucovorin by combined treatment with protein-bound polysaccharide, PSK, in mouse models. *Cellular & Molecular Immunology, 4,* 295–299.

Katoh, R., Takenoshita, S., Shimizu, Y., et al. (1997). Changes in serum soluble IL-2 receptors (sIL-2R) and immunosuppressive acidic protein (IAP) associated with chemotherapy for lung cancer. *Anticancer Research, 17,* 3787–3792.

Kawa, K., Konishi, S., Tsujiino, G., et al. (1991). Effects of biological response modifiers on childhood ALL being in remission after chemotherapy. *Biomedecine & Pharmacotherapy, 45,* 113–116.

Kidd, P. M. (2000). The use of mushroom glucans and proteoglycans in cancer treatment. *Alternative Medicine Review: A Journal of Clinical Therapeutic, 5,* 4–27.

Kim, F., Sakagami, H., Tanuma, S., et al. (1990). Stimulation of interferon-gamma-induced human myelogenous leukemic cell differentiation by high molecular weight PSK subfraction. *Anticancer Research, 10,* 55–58.

Kim, M. Y., Seguin, P., Ahn, J. K., et al. (2008). Phenolic compound concentration and antioxidant activities of edible and medicinal mushrooms from Korea. *Journal of Agricultural and Food Chemistry, 56,* 7265–7270.

Kobayashi, H., Matsunaga, K., & Fujii, M. (1993). PSK as a chemopreventive agent. *Cancer Epidemiology, Biomarkers & Prevention: A Publication of the American Association for Cancer Research, Cosponsored by the American Society of Preventive Oncology, 2,* 271–276.

Kobayashi, Y., Kariya, K., Saigenji, K., et al. (1994). Enhancement of anti-cancer activity of cis-diaminedichloroplatinum by the protein-bound polysaccharide of Coriolus versicolor QUEL (PS-K) in vitro. *Cancer Biother, 9,* 351–358.

Kobayashi, H., Matsunaga, K., & Oguchi, Y. (1995). Antimetastatic effects of PSK (Krestin), a protein-bound polysaccharide obtained from basidiomycetes: An overview. *Cancer Epidemiology, Biomarkers & Prevention: A Publication of the American Association for Cancer Research, Cosponsored by the American Society of Preventive Oncology, 4,* 275–281.

Koda, K., Miyazaki, M., Sarashina, H., et al. (2003). A randomized controlled trial of postoperative adjuvant immunochemotherapy for colorectal cancer with oral medicines. *International Journal of Oncology, 23,* 165–172.

Kohgo, Y., Hirayama, Y., Sakamaki, S., et al. (1994). Improved recovery of myelosuppression following chemotherapy in mice by combined administration of PSK and various cytokines. *Acta Haematologica, 92,* 130–135.

Kondo, M., & Torisu, M. (1985). Evaluation of an anticancer activity of a protein-bound polysaccharide PS-K (Krestin). In M. Torisu & T. Yoshida (Eds.), *Basic mechanisms and clinical treatment of tumor metastasis* (p. 623). New York: Academic Press.

Kudo, S., Tanaka, J., Kashida, H., et al. (2002). Effectiveness of immunochemotherapy with PSK, a protein bound polysaccharide, in colorectal cancer and changes of tumor marker. *Oncology Reports, 9,* 635–638.

Li, W., Liu, M., Lai, S., et al. (2010). Immunomodulatory effects of polysaccharopeptide (PSP) in human PBMC through regulation of TRAF6/TLR immunosignal-transduction pathways. *Immunopharmacology and Immunotoxicology, 32,* 576–584.

Liu, J. X., & Zhou, J. Y. (1993). Phase II clinical trial for PSP capsules. In Q. Y. Yang & C. Y. Kwok (Eds.), *Proceedings of PSP International Symposium* (pp. 183–208). Shanghai: Fudan University Press.

Liu, F., Fung, M. C., Ooi, V. E. C., et al. (1996). Induction in the mouse of gene expression of immunomodulating cytokines by mushroom polysaccharide-protein complexes. *Life Sciences, 58,* 1795–1803.

Liu, F., Ooi, V. E., & Chang, S. T. (1997). Free radical scavenging activities of mushroom polysaccharide extracts. *Life Sciences, 60,* 763–771.

Liu, J., Zhou, J., & Liu, T. (1999). Phase III clinical trial for Yun Zhi polysaccharopeptide (PSP) capsules. In Q. Yang (Ed.), *Advanced research in PSP* (pp. 295–303). Hong Kong: Hong Kong Association for Health Care Ltd.

M. D. Anderson Cancer Center. (2010). http://www.coriolusversicolor.us/2010/07/25/md-anderson-cancer-center-coriolus-versicolor/. Accessed 31 July 2010.

Mao, X. W., Archambeau, J. O., & Gridley, D. S. (1996). Immunotherapy with low-dose interleukin-2 and a polysaccharopeptide derived from Coriolus versicolor. *Cancer Biotherapy & Radiopharmaceuticals, 11,* 393–403.

Matsunaga, K., Iijima, H., Aota, M., et al. (1992). Enhancement of effector cell activities in mice bearing syngeneic plasmacytoma X5563 by a biological response modifier, PSK. *Journal of Clinical & Laboratory Immunology, 37,* 21–37.

Matsunaga, K., Iijima, H., & Kobayashi, H. (2000). Neonatal inoculation with the protein-bound polysaccharide PSK increases resistance of adult animals to challenge with syngeneic tumor cells and reduces azoxymethane-induced precancerous lesions in the colon. *Cancer Epidemiology, Biomarkers & Prevention: A Publication of the American Association for Cancer Research, Cosponsored by the American Society of Preventive Oncology, 9,* 1313–1322.

Mattila, P., Könkö, K., Eurola, M., et al. (2001). Contents of vitamins, mineral elements, and some phenolic compounds in cultivated mushrooms. *Journal of Agricultural and Food Chemistry, 49,* 2343–2348.

Mau, J. L., Lin, H. C., & Chen, C. C. (2001). Non-volatile components of several medicinal mushrooms. *Food Research International, 34,* 521–526.

Mau, J. L., Lin, H. C., & Chen, C. C. (2002). Antioxidant properties of several medicinal mushrooms. *Journal of Agricultural and Food Chemistry, 50,* 6072–6077.

Mayer, P., & Drews, J. (1980). The effect of a protein-bound polysaccharide from Coriolus versicolor on immunological parameters and experimental infections in mice. *Infection, 8,* 13–21.

Mickey, D. D. (1985). Combined therapeutic effects of an immunomodulator, PSK, and chemotherapy with carboquone on rat bladder carcinoma. *Cancer Chemotherapy and Pharmacology, 15,* 54–58.

Mitomi, T., Tsuchiya, S., Iijima, N., et al. (1992). Randomized, controlled study on adjuvant immunochemotherapy with PSK in curatively resected colorectal cancer. *Diseases of the Colon and Rectum, 35,* 123–130.

Mizuno, T. (1999). The extraction and development of antitumour-active polysaccharides from medicinal mushrooms in Japan. *International Journal of Medicinal Mushrooms, 1,* 9–29.

Mizuno, T., Saito, H., Nishitoba, T., et al. (1995). Antitumor active substances from mushrooms. *Food Reviews International, 11,* 23–61.

Modak, S., Koehne, G., Vickers, A., et al. (2005). Rituximab therapy of lymphoma is enhanced by orally administered (1–>3),(1–>4)-D-beta-glucan. *Leukocyte Research, 29,* 679–683.

Monro, J. A. (2003). Treatment of cancer with mushroom products. *Archives of Environmental Health, 58,* 533–537.

Moradali, M. F., Mostafavi, H., Ghods, S., et al. (2007). Immunomodulating and anticancer agents in the realm of macromycetes fungi (macrofungi). *International Immunopharmacology, 7,* 701–724.

Morimoto, T., Ogawa, M., Orita, K., et al. (1996). Postoperative adjuvant randomised trial comparing chemoendocrine therapy, chemotherapy and immunotherapy for patients with stage II breast cancer: 5-year results from the Nishinihon Cooperative Study Group of Adjuvant Chemoendocrine Therapy for Breast Cancer (ACETBC) of Japan. *European Journal of Cancer, 32,* 235–242.

Nagao, T., Komatsuda, M., Yamauchi, K., et al. (1981). Chemoimmunotherapy with Krestin in acute leukemia. *The Tokai Journal of Experimental and Clinical Medicine, 6,* 141–146.

Nakajima, T., Ichikawa, S., Uchida, S., et al. (1990). Effects of a protein-bound polysaccharide from a basidiomycetes against hepatocarcinogenesis induced by 3'-methyl-4-dimethylaminobenzene in rats. *Clinical Therapeutics, 12,* 385–392.

Nakazato, H., Koike, A., Saji, S., et al. (1994). Efficacy of immunochemotherapy as adjuvant treatment after curative resection of gastric cancer. *Lancet, 343,* 1122–1126.

Nathan, C. F., & Hibbs, J. B. (1991). Role of nitric oxide synthesis in macrophage antimicrobial activity. *Current Opinion in Immunology, 3,* 65–70.

Ng, T. B. (1998). A review of research on the protein-bound polysaccharide (polysaccharopeptide, PSP) from the mushroom Coriolus versicolor (Basidiomycetes: Polyporaceae). *General Pharmacology, 30,* 1–4.

Ng, T. B., & Chan, W. Y. (1997). Polysaccharopeptide from the mushroom Coriolus versicolor possesses analgesic activity but does not produce adverse effects on female reproductive or embryonic development in mice. *General Pharmacology, 29,* 269–273.

Ng, T. B., Wang, H. X., Liu, F., et al. (1999). Polysaccharopeptide (PSP) from mycelia of the mushroom *Coriolus versicolor*: an updated review. In Q. Y. Yang (Ed.), *Advanced Research in PSP* (pp. 80–87). Hong Kong: Hong Kong Association for Health Care.

Niimoto, M., Hattori, T., Tamada, R., et al. (1988). Postoperative adjuvant immunochemotherapy with mitomycin C, futraful and PSK for gastric cancer. An analysis of data on 579 patients followed for five years. *The Japanese Journal Surgery, 18,* 681–686.

Nio, Y., Tsubono, M., Tseng, C. C., et al. (1992). Immunomodulation by orally administered protein-bound polysaccharide PSK in patients with gastrointestinal cancer. *Biotherapy, 4,* 117–128.

Nishiwaki, Y., Furuse, K., Fukuoka, M., et al. (1990). A randomized controlled study of PSK combined immuno-chemotherapy for adenocarcinoma of the lung. The advanced lung cancer immuno-chemotherapy study group. Gan to kagaku ryoho. *Cancer & Chemotherapy, 17,* 131–136.

Oba, K., Teramukai, S., Kobayashi, M., et al. (2007). Efficacy of adjuvant immunochemotherapy with polysaccharide K for patients with curative resections of gastric cancer. *Cancer Immunology, Immunotherapy: CII, 56,* 905–911.

Ogoshi, K., Satou, H., Isono, K., et al. (1995a). Immunotherapy for esophageal cancer: A randomized trial in combination with radiotherapy and radiochemotherapy. *American Journal of Clinical Oncology, 18,* 216–222.

Ogoshi, K., Satou, H., Isono, K., et al. (1995b). Possible predictive markers of immunotherapy in esophageal cancer: Retrospective analysis of a randomized study. *Cancer Investigation, 13,* 363–369.

Ogoshi, K., Tajima, T., Mitomi, T., et al. (1997). HLA-A2 antigen status predicts metastasis and response to immunotherapy in gastric cancer. *Cancer Immunology, Immunotherapy: CII, 45,* 53–59.

Ogoshi, K., Miyaji, M., Nakamura, K., et al. (1998). Immunotherapy and combined assay of serum levels of carcinoembryonic antigen and acute-phase reactants. *Cancer Immunology, Immunotherapy: CII, 46,* 14–20.

Oguchi, Y., Ando, T., Matsunaga, K., et al. (1987). Effect of PSK on cytotoxicity against sarcoma-180 in tumor-bearing mice. *Anticancer Research, 7,* 681–684.

Ohno, R., Yamada, K., Masaoka, T., et al. (1984). A randomized trial of chemoimmunotherapy of acute nonlymphocytic leukemia in adults using a protein-bound polysaccharide preparation. *Cancer Immunology, Immunotherapy: CII, 18,* 149–154.

Ohwada, S., Ikeya, T., Yokomori, T., et al. (2004). Adjuvant immunochemotherapy with oral Tegafur/Uracil plus PSK in patients with stage II or III colorectal cancer: A randomised controlled study. *British Journal of Cancer, 90,* 1003–1010.

Ohwada, S., Ogawa, T., Makita, F., et al. (2006). Beneficial effects of protein-bound polysaccharide K plus tegafur/uracil in patients with stage II or III colorectal cancer: Analysis of immunological parameters. *Oncology Reports, 15,* 861–868.

Park, S., Jang, S. A., Lim, J. D., et al. (2009). Phagocytic effects of β-glucan from Coriolus versicoloar are related with Dectin-1 signaling and NOS and Ikaros regulation in macrophages. *Meeting Abstract The FASEB Journal: Official Publication of the Federation of American Societies for Experimental Biology, 23,* 1003–1009.

Poyrazoglu Çoban, E., Isman, B., & Biyik, H. H. (2008). Effects of protein-bound polysaccharide against some microorganisms. *Türk Mikrobiyoloji Cemiyeti/Turkish Microbiological Society, 38,* 150–155.

Price, L. A., Wenner, C. A., Sloper, D. T., et al. (2010). Role for toll-like receptor 4 in TNF-alpha secretion by murine macrophages in response to polysaccharide Krestin, a Trametes versicolor mushroom extract. *Fitoterapia, 81,* 914–919.

Quian, Z. M., Xu, M. F., & Tang, P. L. (1997). Polysaccharide peptide (PSP) restores immunosuppression induced by cyclophosphamide in rats. *The American Journal of Chinese Medicine, 25,* 27–35.

Ragupathi, G., Yeung, K. S., Leung, P. C., et al. (2008). Evaluation of widely consumed botanicals as immunological adjuvants. *Vaccine, 26,* 4860–4865.

Rau, U., Kuenz, A., Wray, V., et al. (2009). Production and structural analysis of the polysaccharide secreted by Trametes (Coriolus) versicolor ATCC 200801. *Applied Microbiology and Biotechnology, 81,* 827–837.

Roeder, A., Kirschning, C. J., Rupec, R. A., et al. (2004). Toll-like receptors as key mediators in innate antifungal immunity. *Medical Mycology: Official Publication of the International Society for Human and Animal Mycology, 42,* 485–498.

Sadahiro, S., Suzuki, T., Maeda, Y., et al. (2010). Effects of preoperative immunochemoradiotherapy and chemoradiotherapy on immune responses in patients with rectal adenocarcinoma. *Anticancer Research, 30,* 993–999.

Saji, S., Sakamoto, J., Teramukai, S., et al. (1999). Impact of splenectomy and immunochemotherapy on survival following gastrectomy for carcinoma: Covariate interaction with immunosuppressive acidic protein, a serum marker for the host immune system. Tumor Marker Committee for the Study Group of Immunochemotherapy with PSK for Gastric Cancer. *Surgery Today, 29,* 504–510.

Sakagami, H., & Aoki, T. (1991). Induction of immunopotentiation activity by a protein-bound polysaccharide. PSK (review). *Anticancer Research, 11,* 993–1000.

Sakamoto, J., Morita, S., Oba, K., et al. (2006). Meta-Analysis Group of the Japanese Society for Cancer of the Colon Rectum. Efficacy of adjuvant immunochemotherapy with polysaccharide K for patients with curatively resected colorectal cancer: A meta-analysis of centrally randomized controlled clinical trials. *Cancer Immunology, Immunotherapy: CII, 55,* 404–411.

Sakita, M., Imai, H., Yamane, T., et al. (1982). Effect of PSK on dimethyl hydnazine (DMH) induction of intestinal tumors in rats. *Japanese Journal of Cancer Chemotherapy, 9,* 1216–1225.

Schmidt, C. (2006). Immune system's Toll-like receptors have good opportunity for cancer treatment. *Journal of the National Cancer Institute, 98,* 574–575.

Shibata, M., Nezu, T., Kanou, H., et al. (2002). Immunomodulatory effects of low dose cis-Diaminedichloroplatinum (cisplatin) combined with UFT and PSK in patients with advanced colorectal cancer. *Cancer Investigation, 20,* 166–173.

Shimokawa, K., Mashima, I., Asai, A., et al. (2008). Biological activity, structural features, and synthetic studies of (-)-ternatin, a potent fat-accumulation inhibitor of 3T3-L1 adipocytes. *Chemistry, an Asian Journal, 3,* 438–446.

Shiu, W. C. T., Leung, T. W., & Tao, M. (1993). A clinical study of PSP on peripheral blood cell counts during chemotherapy. In Q. Y. Yang & C. Y. Kwok (Eds.), *Proceedings of PSP International Symposium* (pp. 216–220). Shanghai: Fudan University Press.

Simon, H. U., Haj-Yehia, A., & Levi-Schaffer, F. (2000). Role of reactive oxygen species (ROS) in apoptosis induction. *Apoptosis, 5,* 415–418.

Smith, J. E., Rowan, N., & Sullivan, R. (2002). Medicinal mushrooms: Their therapeutic properties and current medicinal usage with special emphasis on cancer treatment. http://sci.cancerresearchuk.org/labs/med_mush/med_mush.html.

Standish, L. J., Wenner, C. A., Sweet, E. S., et al. (2008). Trametes versicolor mushroom immune therapy in breast cancer. *Journal of the Society for Integrative Oncology, 6*, 122–128.

Sullivan, R., Smith, J. E., & Rowan, N. J. (2006). Medicinal mushrooms and cancer therapy; translating a traditional practice into Western medicine. *Perspectives in Biology and Medicine, 49*, 159–170.

Takeuchi, T., Iinuma, H., Iwanaga, J., et al. (1969). Coriolin, a new Basidiomycetes antibiotic. *The Journal of Antibiotics, 22*, 215–217.

Tanaka, K., Wakasugi, J., Seki, F., et al. (1991). Effect of PSK on cell growth, chemotaxis and platelet aggregating activity of colon 26 cell line. Proceedings of the 50th annual Japanese cancer association meeting, p. 263.

Taylor, P. R., Brown, G. D., Herre, J., et al. (2004). The role of SIGNR1 and the beta-glucan receptor (dectin-1) in the nonopsonic recognition of yeast by specific macro-phages. *Journal of Immunology (Baltimore, Md.: 1950), 172*, 1157–1162.

Toge, T., & Yamaguchi, Y. (2000). Protein-bound polysaccharide increases survival in resected gastric cancer cases stratified with a preoperative granulocyte and lymphocyte count. *Oncology Reports, 7*, 1157–1161.

Toi, M., Hattori, T., Akagi, M., et al. (1992). Randomized adjuvant trial to evaluate the addition of Tamoxifen and PSK to chemotherapy in patients with primary breast cancer. 5-year results from the Nishi-Nippon group of the adjuvant chemoendocrine therapy for breast cancer organization. *Cancer, 70*, 2475–2483.

Torisu, M., Hayashi, Y., Ishimitsu, T., et al. (1990). Significant prolongation of disease-free period gained by oral polysaccharide K (PSK) administration after curative surgical operation of colorectal cancer. *Cancer Immunology, Immunotherapy: CII, 31*, 261–268.

Tsang, K. W., Lam, C. L., Yan, C., et al. (2003). Coriolus versicolor polysaccharide peptide slows progression of advanced non-small cell lung cancer. *Respiratory Medicine, 97*, 618–624.

Tsuji, K., Takagi, M., Kobayashi, T., et al. (1989). Effect of a protein-bound polysaccharide, PSK, on human hemopoietic progenitors. *Nippon Ketsueki Gakkai Zasshi: Journal of Japan Haematological Society, 52*, 594–600.

Wan, J. M., Sit, W. H., & Louie, J. C. (2008). Polysaccharopeptide enhances the anticancer activity of doxorubicin and etoposide on human breast cancer cells ZR-75-30. *International Journal of Oncology, 32*, 689–699.

WHO (2000). General Guidelines for Methodologies on Research and Evaluation of Traditional Medicines. *World Health Organization*, Geneva.

Wong, C. K., Tse, P. S., Wong, E. L., et al. (2004). Immunomodulatory effects of yun zhi and danshen capsules in health subjects—a randomized, double-blind, placebo-controlled, crossover study. *International Immunopharmacology, 4*, 201–211.

Wong, C. K., Bao, Y. X., Wong, E. L., et al. (2005). Immunomodulatory activities of Yunzhi and Danshen in post-treatment breast cancer patients. *The American Journal of Chinese Medicine, 33*, 381–395.

Wu, Z., Chen, C., & Chen, Y. M. (1993). Clinical observation of esophageal cancer treatment by radiotherapy together with the oral administration of PSP. In Q. Y. Yang & C. Y. Kwok (Eds.), *Proceedings of PSP International Symposium* (pp. 265–268). Shanghai: Fudan University Press.

Xu, Q., Lu, J., Wang, R., et al. (1997). Liver injury model induced in mice by a cellular immunologic mechanism—study for use in immunopharmacological evaluations. *Pharmacological Research: The Official Journal of the Italian Pharmacological Society, 35*, 273–278.

Yamasaki, A., Shoda, M., Iijima, H., et al. (2009). A protein-bound polysaccharide, PSK, enhances tumor suppression induced by docetaxel in a gastric cancer xenograft model. *Anticancer Research, 29*, 843–850.

Yamashita, K., Ougolkov, A. V., Nakazato, H., et al. (2007). Adjuvant immunochemotherapy with protein-bound polysaccharide K for colon cancer in relation to oncogenic beta-catenin activation. *Diseases of the Colon and Rectum, 50*, 1169–1181.

Yang, Q. Y. (1999). Yun Zhi polysaccharopeptide (PSP) and the general aspects of its research. In Q. U. Yang (Ed.), *Advanced Research in PSP* (pp. 29–38). Hong Kong: The Hong Kong Assocation for Health Care Ltd.

Yang, M. M., Chen, Z., & Kwok, J. S. (1992). The anti-tumor effect of a small polypeptide from Coriolus versicolor (SPCV). *The American Journal of Chinese Medicine, 20,* 221–232.

Yang, X., Sit, W. H., Chan, D. K., et al. (2005). The cell death process of the anticancer agent polysaccharide-peptide (PSP) in human promyelocytic leukemic HL-60 cells. *Oncology Reports, 13,* 1201–1210.

Yeung, J. H., Chiu, L. C., & Ooi, V. E. (1994). Effect of polysaccharide peptide (PSP) on glutathione and protection against paracetamol-induced hepatotoxicity in the rat. *Methods and Findings in Experimental and Clinical Pharmacology, 16,* 723–729.

Yokoe, Y., Kato, H., Takemura, S., et al. (1979). Effect of long-term administration of OK-432 (picibanil), PSK (Krestin) and levamisole on immune responses in aged subjects without malignancy. *Japanese Journal of Clinical Oncology, 9,* 209–214.

Yokoe, T., Iino, Y., Takei, H., et al. (1997). HLA antigen as predictive index for the outcome of breast cancer patients with adjuvant immunochemotherapy with PSK. *Anticancer Research, 17,* 2815–2818.

Yoshikawa, T., Furukawa, Y., Wakamatsu, Y., et al. (1982). Protection by BCG, levamisole, PS-K, OK-432, and vitamin E against carbon tetrachloride hepatotoxicity. *Hepatogastroenterology, 29,* 240–242.

Zeng, F., Hon, C. C., Sit, W. H., et al. (2005). Molecular characterization of Coriolus versicolor PSP-induced apoptosis in human promyelotic leukemic HL-60 cells using cDNA microarray. *International Journal of Oncology, 27,* 513–523.

Zhang, H., Morisaki, T., Matsunaga, H., et al. (2000). Protein-bound polysaccharide PSK inhibits tumor invasiveness by down-regulation of TGF-beta1 and MMPs. *Clinical & Experimental Metastasis, 18,* 343–352.

Zhou, J. X., Li, X. Y., & Shen, X. L. (1988). The anti-tumor and immunomodulating activity of PSP in mice. *Journal of Shanghai Teachers University (Natural Sciences Education), 3,* 72.

Zhou, Y. L., & Yang, Q. Y. (1999). Active principles from Coriolus sp. In Q. U. Yang (Ed.), *Advances research in PSP* (pp. 111–124). Hong Kong: Hong Kong Association for Health Care.

Zimmerman, J. W., Lindermuth, J., Fish, P. A., et al. (1998). A novel carbohydrate-glycosphingolipid interaction between a beta -(1-3)-glucan immunomodulator, PGG-glucan, and lactosylceramide of human leukocytes. *The Journal of Biological Chemistry, 273,* 22014–22020.

Zjawiony, J. (2004). Biologically active compounds from Aphyllophorales (Polypore) fungi. *Journal of Natural Products, 67,* 300–310.

Chapter 14
An Evidence-based Perspective of *Lentinus Edodes* (Shiitake Mushroom) for Cancer Patients

Jiao Shen, Jia-fei Yao, Mamoru Tanida, Yuko Horii and Katsuya Nagai

Abstract *Lentinus edodes*, known as shiitake mushroom in Japan and Xiang Gu or fragrant mushroom in China, is an edible mushroom that is commonly cultivated worldwide. This mushroom has been consumed for centuries as a delicacy and for its beneficial effects on human health. Early in the fourteenth century, the Chinese physician Wu Rui recorded that shiitake was beneficial for the treatment of various forms of malignancy. Since 1969, groundbreaking investigations have performed in this subject, and a large number of high quality scientific studies have demonstrated the immunomodulatory, antitumor, antiviral, and cholesterol-regulating effects of this mushroom. In addition, the use of shiitake in combination with chemotherapy and radiotherapy has reported in Japan. Since cancer is a leading cause of death worldwide and it is exceptionally difficult to cure malignant tumors, cancer prevention may be a more effective strategy to control and ultimately overcome cancer. Here, we will summarize the antitumor activity and immunomodulating action of biologically active polysaccharides, mostly β-glucans, from shiitake mushroom fruit bodies and/or cultured mycelium based on experimental and clinical findings. The following aspects of shiitake mushroom in relation to cancer would be discussed: mechanisms of action, optimal dosage, and timing of administration for clinical use of its antitumor and immunostimulating properties, prospective use in clinical therapy, clinical safety, and possible adverse effects.

14.1 Introduction

Cancer is a major public health concern and a leading cause of death worldwide and it is exceptionally difficult to cure. Therefore, cancer prevention is an important strategy to reduce and control the incidence of human cancers. Many

J. Shen (✉)
ANBAS Corporation, 4th Floor, Nakajima Building, 4-12-17, Toyosaki, Kita-Ku, Osaka 531-0072, Japan
e-mail: jshen@anbas.co.jp

W. C. S. Cho (ed.), *Evidence-based Anticancer Materia Medica,* Evidence-based Anticancer Complementary and Alternative Medicine,
DOI 10.1007/978-94-007-0526-5_14,

dietary constituents markedly influence carcinogenic agents (Dennis et al. 2009). For over 40 years, mushrooms have been investigated for their medical effects in both *in vivo* and *in vitro* model systems (Table 14.1), and many new antitumor and immunomodulating polysaccharides have been identified and put into practical use (Wasser and Weis 1999a; Ikekawa 2001). Epidemiological evidence has shown a correlation between daily mushroom consumption and a low rate of cancer mortality in Japan (Borchers et al. 1999). Among mushrooms, shiitake mushroom (*Lentinus edodes*, family *Tricholomataceae*, order *Agaricales*, and class *Basidomycetes*) has been studied most extensively. Shiitake mushroom, also known as Xiang Gu or fragrant mushroom in China, is the second most popular edible mushroom in the global market after *Agaricus bisporus*. A large number of high quality scientific studies have demonstrated the immunomodulatory (Tabata et al. 1992; Kobayashi et al. 1998; Wasser and Weis 1999b) and antitumor effects (Sugano et al. 1985; Sorimachi et al. 1990; Minato et al. 1999) of shiitake mushroom.

Polysaccharide-containing mushrooms can boost immunological responses and produce antitumor effects *via* a mechanism that is similar to the activation of macrophages in response to the polysaccharides on the surface of bacteria (Wasser 2002). Several polysaccharides have been identified from the carbohydrate portion of shiitake mushroom, namely, water-soluble polysaccharides (1–5% of dried mushroom), lentinan, β-1,3;1,6-D-glucans, glycogen-like polysaccharides, α-1,4;1,6-D-glucans, 1,6-β-D-glucans with 1,3- and 1,4-β-bonded heteroglucans, heterogalactan, heteromannan, and xyloglucan (Jong et al.1991).

The first antitumor compound isolated from the shiitake mushroom was lentinan, which is a purified β-1,3-D-glucan with β-1,6 branches in a triple helical structure (Chihara et al. 1970a). This structural feature is reportedly necessary for antitumor action (Ng and Yap 2002; Wasser 2002). The uses of the shiitake mushroom mycelial extract (SMM), which also reportedly inhibits tumor growth (Shen et al. 2009a), are extremely similar to the uses of the fruit body. The KS-2 antitumor active polysaccharide has also been isolated from SMM (Fujii et al. 1978). In addition to polysaccharides, shiitake contains amino acids such as lysine and arginine, which inhibit tumor growth when administered orally (Shen et al. 2009b).

The anti-carcinogenic properties of mushroom polysaccharides may be attributable to the modulation of both innate and adaptive immunity by enhancement of the numbers and/or functions of macrophages (Kim et al. 2004), natural killer (NK) cells (Akramiene et al. 2007), and subsets of T cells (Miyakoshi et al. 1984).

To improve water solubility, purity, and clinical qualities, some extractions and/or derivates of lentinan such as active hexose correlated compound (AHCC) (Spierings et al. 2007), superfine dispersed lentinan (Isoda et al. 2009; Hazama et al. 2009; Shimizu et al. 2009; Yoshino et al. 2010), polysaccharide L-II (Ruan et al. 2005), β-glu6 (Wang et al. 2010), and sulfonated derived glucans (Unursaikhan et al. 2006; Guo et al. 2009) have been produced and well studied for their antitumor activities.

Table 14.1 Summarization of some important findings

No	Bioactive compounds from shiitake mushroom	Important findings	References
1	Lentinan	Inhibition of mouse sarcoma 180 by lentinan	Chihara et al. 1969, 1970a; Aoki 1984; Yap and Ng 2002
		Effects of lentinan on immune system	Fruehauf et al. 1982; Aoki 1984; Izawa et al. 1984; Miyakoshi et al. 1984; Suzuki et al. 1994; Adachi et al. 1999; Kim et al. 2004; Akramiene et al. 2007; Zhou et al. 2009; McCormack et al. 2010
2	SMM	SMM affected sympathetic activity and cancer cell proliferation	Shen et al. 2009a
3	KS-2	A new antitumor polysaccharide, KS-2 is isolated and characterized	Fujii et al. 1978
4	Polysaccharide isolated from Phellinus linteus (PL)	PL act as an effective immunomodulator and enhances the antitumoral activity of peritoneal macrophages	Kim et al. 2004
5	Superfine dispersed lentinan (SDL)	SDL improved the quality of life, prognosis and prolonged the survival times of patients with advanced gastric cancer	Yoshino et al. 2010
		SDL improved the quality of life, prognosis and prolonged the survival times of patients with hepatocellular carcinoma	Isoda et al. 2009
		SDL improve the quality of life, prognosis and prolong the survival times of patients with pancreatic cancer	Shimizu et al. 2009
6	Hexose correlated compound (AHCC)	The safety of AHCC	Spierings et al. 2007
7	O-sulfonated derivated (OSDs) glucans	OSDs exhibited dose dependently and higher antitumor activates than the native glucan	Unursaikhan et al. 2006; Guo et al. 2009
8	β-1,3;1,6-glucan	Finding of CR3 receptor of β-1,3;1,6-glucan	Ross et al. 1987; Thornton et al. 1996; Xia et al. 1999
		Finding of Dectin-1 receptor of β-1,3;1,6-glucan	Brown and Gordon 2001; Taylor et al. 2002

14.2 Testing Antitumor Activity in Animals

Chihara and co-workers (1969, 1970a, b) were the first to isolate a water-soluble antitumor polysaccharide from the fruiting bodies of shiitake mushroom, which was named "lentinan" after the generic name of this mushroom. The physical and chemical properties of lentinan have been widely characterized (Sasaki and Takasuka 1976; Saito et al. 1977; Chihara et al. 1987). This structure, a purified β-1,3-D-glucan with β-1,6 branches in a triple helical structure, illustrates the importance of the cell-mediated immunoresponse in tumor development (Ng and Yap 2002; Wasser 2002). The effect of lentinan was originally tested using sarcoma 180-transplanted CD-1/ICD mice, and resulted in complete tumor regression when given intraperitoneally (Chihara et al. 1969, 1970a). Lentinan later showed prominent antitumor activity against allogenic tumors and various synergic and autochthonous tumors (Zakany et al. 1980; Suga et al. 1984).

SMM, a water-soluble extract from shiitake mushroom, offers a source of cancer preventive food. SMM-treated rats simultaneously fed a carcinogen as a part of their food supply showed as much as 50% less proliferation of rat-ascites hepatoma (Sugano et al. 1982). BALB/c athymic nude mice implanted with human breast cancer (MCF-7) cells or human colon cancer (HCT116) cells given 100 mg/ml SMM solution showed significantly diminished tumor growth relative to tumor volume (Shen et al. 2009a).

KS-2, a water-soluble, peptide-containing polysaccharide that was prepared by ethanol precipitation of the hot water extract of SMM, has potent antitumor activity and is active by the oral route. A small oral dose of KS-2 (1 mg/kg) had a very strong inhibitory effect against the sarcoma 180 tumor cells, and all the mice survived. Mice with a malignant tumor that is much more difficult to survive (Ehrlich ascite tumors) required a higher oral dose of KS-2 (140 mg/kg), but 70% of mice survived after 50 days, which is 20 days longer than most antitumor studies are conducted (Fujii et al. 1978).

McCormack et al. (2010) reported that a combination of lentinan with idarubicin and cytarabine, the standard care for acute myeloid leukemia, increased average survival compared to monotherapy, and reduced cachexia in Brown-Norway rats with Brown-Norway acute myelocytic leukemia.

O-sulfonated derivates (OSDs) from water-insoluble (1,3)-α-D-glucans that were isolated from fruiting bodies of shiitake mushroom showed antitumor activity in BALB/c mice and inhibited S-180 tumor cell proliferation *in vitro*. 5-Fluorouracil (5-FU), native α-glucans, and OSDs from shiitake were injected intraperitoneally once daily for 10 days at 24 hours after subcutaneous tumor inoculation in BALB/c mice. OSDs exhibited dose-dependent antitumor activities that were higher than native α-glucans but slightly lower than 5-FU. However, body weight enhancement ratios were much higher for OSD than for 5-FU, suggesting that the polysaccharide derivates are less toxic than 5-FU, which kills cancer cells as well as normal cells (Unursaikhan et al. 2006).

14.3 Epidemiological Data from Clinical Studies

Clinical studies showed that the addition of lentinan to standard chemotherapy offers a significant advantage over chemotherapy alone in terms of survival for patients with advanced, unresected/recurrent gastric cancer. A meta-analysis study using individual data from 650 patients in five trials evaluated the effect of immunochemotherapy with lentinan compared to chemotherapy alone in patients with advanced gastric cancer. Lentinan significantly prolonged the overall survival (stratified log-rank $P=0.011$). The overall hazard ratio was 0.8 (95% confidence interval = 0.68–0.95) and there was no heterogeneity between trials (Oba et al. 2009). This study provides evidence for the acceptance of immunochemotherapy to treat advanced gastric cancer.

Lentinan also exerts dramatic antitumor effects in hepatocellular carcinoma (Yang et al. 2008) and squamous cell carcinoma patients (Harada et al. 2010). In hepatocellular carcinoma patients, a combined therapy of lentinan and transcatheter arterial chemoembolization (TACE) and radiofrequency ablation (RFA) resulted in the following observations based on computed tomography scan data: (1) significantly increased tumor necrosis (88.6%, 37.5%, 47.8%, and 60.3% in the combination group, TACE-only group, RFA-only group, and TACE/RFA group, respectively), (2) low tumor recurrence rate (17.8%, 45.8%, 34.7%, and 29.0% in the combination group, TACE-only group, RFA-only group, and TACE/RFA group, respectively), and (3) prolonged mean survival duration (28.2, 14.9, 18.8, and 21.9 months in the combination group, TACE-only group, RFA-only group, and TACE/RFA group, respectively) (Yang et al. 2008).

Lentinan is also useful for the treatment of lymph node metastases, which is a frequent cause of shortened survival in patients with cancer. Adoptive immunotherapy with lentinan containing only β-glucan maintained the condition of a 67-year-old woman with ovarian cancer for 5 months. She had undergone post-operative adjuvant chemotherapy and experienced recurrent cancer in the lymph nodes (Fujimoto et al. 2006). This result is consistent with the meta-analysis study of Oba et al. (2009), which suggested that lentinan was more effective in patients with lymph-node metastasis than in non-node metastasis.

Lentinan is a large protein (400,000–1,000,000 Da), resulting in limited oral bioavailability (Kidd 2000). To overcome this issue, micellary mushroom extracts containing superfine particles of β-1,3-glucan were prepared. The crude-sized β-glucan ranged from 10 to 1,000 μm in diameter, whereas micellary β-glucan ranged from 0.05 to 0.2 μm in diameter. Smaller amounts of micellary β-1,3-glucan more effectively potentiated immunity than crude-sized β-glucan (Shen et al. 2007). Earlier reports showed that crude sized β-glucan was required at the level of 10–30 mg/(day/mouse) for immunopotentiation (Tsukada et al. 2003). However, less than 1 mg/(day/mouse) of micellary β-glucan was effective for the same level of potentiation, suggesting that a small amount of micellary β-glucan is sufficient to potentiate intestinal immunity, and micellary β-glucan might be useful in patients who have difficulty in consuming a large amount of β-glucan for digestive tract malignancy. Three-year follow-up survey results showed that oral treatments of superfine dispersed lentinan (SDL) appeared to improve the quality of life, prognosis, and prolong the survival times of patients with advanced gastric cancer (Yoshino et al. 2010), hepatocellular carcinoma (Isoda et al. 2009), and pancreatic cancer (Shimizu et al. 2009) when combined with chemotherapy or taken as a supplementary food. Similar effects of SDL have been reported in patients with advanced colorectal cancer (Hazama et al. 2009).

14.4 Determination of Optimal Dosage and Administration Timing

For any drug or supplement, the treatment dosage should be optimized to achieve the desired effects. Chihara and co-workers (1969) and Aoki (1984) showed that 0.5 mg/kg of lentinan led to total regression of sarcoma 180 tumors in 80% of mice (Chihara et al. 1970a). Similarly, 10 days of intraperitoneal injection of 1 mg/kg lentinan caused complete regression of sarcoma 180 transplanted into mice, whereas a larger dose of 80 mg/kg for 5 days yielded no antitumor activity in comparison with untreated control mice (Bohn and BeMiller 1995). Yap and Ng (Yap and Ng 2001) demonstrated that 150 mg/kg lentinan is the most effective dose when administrated orally and achieved a tumor inhibition rate of 94.44% in murine lymphoma. Higher or lower does resulted in a lower tumor inhibition rate (76–85%).

Studies (Mori et al.1987; Nanba et al. 1987; Suzuki et al. 1990) using oral administration of dried shiitake mushroom fruiting bodies powder showed that tumor-growth-inhibitory activity increased when the concentration of the powder was increased. When feed containing 10% powdered shiitake fruit bodies (L-feed) was used, the rate of tumor inhibition was 39.6%, but inhibition rates of 53.2% and 58.9% were achieved with 20% L-feed and 30% L-feed, respectively. Thus, the degree of inhibition was proportional to the concentration of the powder in the experimental diet.

The administration schedule of lentinan also influenced the rate of tumor inhibition. Male ARK mice (5–6 weeks old) with tumors induced by the K36 murine lymphoma cell line were force-fed 3 mg of lentinan resuspended in buffer daily, resulting in prevention of tumor development in 83–94% of mice (Yap and Ng 2001, 2002). Seven days of lentinan pre-feeding was the most effective regimen compared to simultaneous and 7 days post-feeding of lentinan.

However, in another study, patients with prostate cancer were given a shiitake mushroom extract consisting of oligosaccharide (α-1,4-glucan), polysaccharide (β-1,3-glucan), and proteins (deVere White et al. 2002). Although several patients had stable prostate-specific antigen levels, none of the patients exhibited a complete or partial response of at least 50% reduction of prostate-specific antigen.

These observations suggest a possible relationship between shiitake mushroom or its components, the type of cancer under study, the route of administration, and the host's immune system, whereby a saturation threshold of lymphocyte activation exists. Therefore, determining the optimal dosage is a crucial consideration in treatment with shiitake mushroom or its components.

14.4.1 Mechanisms of Antitumor and Immunomodulating Action by Lentinan (Fig. 14.1)

The immunopotentiating ability of lentinan was postulated to play a key role in the antitumor process. Daily lentinan supplementation in BN rats, which are well-known for their high capacity for IgE production and hyper-responsiveness to exposure to allergens or other chemicals, resulted in weight gain, increased white blood cells, monocytes, and circulating cytotoxic T cells, and reduction of the anti-inflammatory cytokines interleukin (IL)-4, IL-10, and IL-6 (McCormack et al. 2010). Polysaccharide L-II, which was isolated and purified from the fruiting body of shiitake mushroom, increased the levels of tumor necrosis factor-alpha (TNF-α) and interferon-gamma (IFN-γ) in sarcoma 180-bearing mice (Ruan et al. 2005).

Lentinan β-glucan binding to the surface layer of lymphocytes or specific serum proteins presumably induces lentinan phagocytosis by cells of macrophage lineage present in organs such as liver, spleen, lung, and gut-associated lymphoid tissue (GALT), which subsequently activates macrophages (Kim et al. 2004; Mushiake et al. 2005), T cells (Miyakoshi et al. 1984), NK cells (Akramiene et al. 2007), and other effector cells, and increases production of antibodies (Adachi et al. 1999), ILs (Vetvicka et al. 2008; McCormack et al. 2010), TNF-α (Vetvicka et al. 2008), and INFs (Zhou et al. 2009). Macrophages are the first cells to recognize foreign bodies and they convey this information to lymphocytes to activate the immune system. Suzuki et al. (1994) showed that the antitumor effect of lentinan was dependent on CD-positive T cells. Lentinan can restore the suppressed activity of helper T cells in the tumor-bearing host to their normal state, leading to complete restoration of humoral immune responses (Ooi and Liu 1999). Lentinan can also activate the nor-

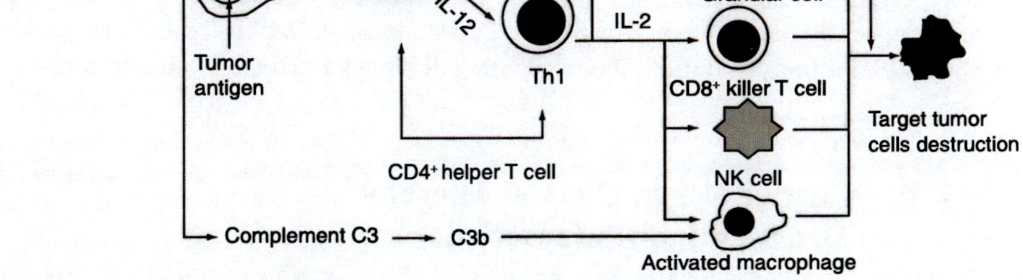

Fig. 14.1 Possible immune mechanism of lentinan and/or β-glucan action. MHC: major histocompatibility complex, IL: interleukin, IFN-γ: interferon-gamma, produced by Th1 CD4+ cells, activates NK cells, macrophages, and killer cells, TNF-α: tumor necrosis factor-alpha, produced mainly by monocytes, activates endothelial cells and other cells of immune and non-immune system

mal and alternative pathways of the complement system and can induce the split of C3 into C3a and C3b, thereby increasing peritoneal macrophage cytotoxicity against metastatic tumors (Aoki 1984; Hobbs 2000). The augmentation of immune mediators such as ILs, colony-stimulating factor(s) (Izawa et al. 1984), migration inhibitory factor (Zakany et al.1980; Fruehauf et al. 1982; Izawa et al. 1984), and the ability to increase the capacity of peripheral blood mononuclear cells (Arinaga et al. 1992a, b; Hazama et al. 2009) contributed additively to antitumor efficacy.

Lentinan's immune-activating ability may link to its modulation of hormonal factors, which are known to play a role in tumor growth. The antitumor activity of lentinan was strongly reduced by administration of thyroxin or hydrocortisone (Aoki 1984). Anti-angiogenesis may be one of the pathways through which lentinan reduces tumor proliferation and prevents tumor metastasis (Akramiene et al. 2007).

14.4.2 Molecular Mechanism by Which Shiitake Mushroom Affects Immunity Response

It is known that lentinan is a purified β-1,3;1,6-glucan and 2 membrane β-1,3-glucan receptors that mediate biological responses to β-1,3-glucan have been characterized at the molecular level. The first to be reported was the inactivated C3b (iC3b)[6] receptor, which is known as CR3 (Mac-1, CD11b/CD18, or $\alpha_M\beta_2$ integrin) (Ross et al.1987; Thornton et al. 1996; Xia et al. 1999). CR3 is highly expressed on neutrophils, monocytes, and NK cells, but is less expressed in macrophages (Ross 2000). Dectin-1 was the second β-1,3-glucan receptor to be described at the molecular level (Brown and Gordon 2001; Taylor et al. 2002). Dectin-1 is preferentially expressed on macrophages rather than granulocytes, and is absent from NK cells (Taylor et al. 2002). β-1,3-glucan can be ingested by gastrointestinal macrophages and shuttled to the bone marrow where soluble degradation fragments are released and taken up by granulocytes *via* membrane CR3. These granulocytes with β-1,3-glucan-primed CR3 could potentially kill iC3b-coated tumor cells. Oral β-1,3-glucan-mediated tumor regression required the presence of iC3b on tumors and CR3 on granulocytes. The requirement for CR3 was confirmed by the significantly reduced tumor-killing activity of granulocytes from CR3-deficient mice that had been given oral β-1,3-glucan (Hong et al. 2004), and cytotoxicity was blocked by an anti-CR3 antibody and did not occur with leukocytes from CR3- deficient mice.

14.5 Prospective Use in Clinical Therapy and Drug-lentinan Interaction

Lentinan was reported to balance Th1 and Th2 responses, thus canceling Th2-dominant condition in patients with digestive cancer (Yoshino et al. 2000). β-glu6 is derived from lentinan but has several advantages over lentinan such as possessing a defined structure and high purity, increased ability to recruit $CD4^+$ and $CD8^+$ T cells to the spleen (Wang et al. 2010), and an enhanced specific immune response to the hepatitis B surface antigen (Dong et al. 2007). This makes lentinan and its derivatives suitable candidate adjuvants for different types of vaccines, providing a potential clinical application.

Clinical tests showed that combined treatment of lentinan/micellary β-1,3-glucan and chemotherapy prolonged the lifespan of cancer patients (Nakano et al. 1999; Oba et al. 2009; Yoshino et al. 2010). Surgery often successfully reduces tumor mass and chemotherapy or radiation therapy may further reduce detectable tumors and minimize invasiveness and metastasis. However, these toxic therapies invariably damage host immunity and small invasive masses or malignant cell clumps predictably survive the best efforts to eradicate them. Mushroom immunoceuticals have the capability to mobilize the immune system against formed tumors and tumor metastases and to decrease adverse side effects of conventional therapies, and should therefore offer a clinically attractive option to the thinking oncologist (Kidd 2000). The major side effect of most chemo-therapeutic drugs is neutropenia because the administration of anticancer drugs impairs blood cell formation. These functions are important for maintenance of the defense system of the patient. As a result, chemotherapy may accelerate the risk of infections that decrease the quality of life for cancer patients. Oral administration of a water-soluble glucan every day for 24 days to mice irradiated with 6 Gy increased serum levels of radioprotective cytokines such as IL-1β, IL-6, and granulocyte macrophage colony-stimulating factor, while the level of TNF-α, which is increased because of tissue injury and anemia due to radiation, was decreased over time (Jin et al. 2003). In addition, the number of colony forming cells was similar to the level observed in non-irradiated mice at day 8, and this similarity was maintained during the 24-day period. These data suggest that glucan may be used to modulate the dysregulation of cytokine production by radiation damage.

Helping the body to mount an immune cell attack against disease is now a practical reality called immunotherapy. Immunotherapy using monoclonal antibodies (mAbs) is a novel cancer treatment strategy. Immunotherapy holds great promise because of its ability to target cancer cells specifically and to minimize damage to normal tissues, which is an important advantage over chemotherapy and radiotherapy. Some mAbs are already used in clinical oncology for treatment of malignant diseases. Lentinan or β-1,3-glucan may function as a potent adjuvant for antitumor mAbs, leading to enhanced tumor regression and patient survival. Antitumor mAbs bind to tumors and activate complement, coating the tumor with iC3b. β-1,3-glucan primes the inactivated CR3 of circulating granulocytes, which enables CR3 to trigger cytotoxicity of the iC3b-coated tumor. The use of antitumor mAbs and β-1,3-glucan together should not interfere with the function of CTL-based immunotherapy, and targeting tumors simultaneously with both CTL and granulocytes should provide a more effective means of eliminating tumors and developing a long-term tumor-specific immunity that prevents tumor recurrence (Hong et al. 2004). Various tumor models were described in this and previous reports in which specific mAbs given alone had little or no effect on tumor regression and yet mediated complete remission when given together with either oral or iv β-1,3-glucan (Yan et al. 1999; Cheung et al. 2002a, b; Hong et al. 2003).

Lentinan is insoluble in cold water (Chihara et al. 1970b; Saito et al. 1977) and different approaches must be used to improve the antitumor activity of mushroom polysaccharides by chemical modification. These modifications are also necessary to improve water solubility, clinical qualities, and the ability of mushroom polysaccharides to permeate the stomach walls after oral ingestion. The main procedures used for chemical improvement are as follows: Smith degradation (oxydo-reducto-hydrolysis), formolysis, and carboxymethylation (Wasser 2002). In addition to SDL, some studies suggested that sulfation increases the biological activities and even potentiates new effects (Wang et al. 2004). Sulfated modification of lentinan enhanced the adjuvanticity of lentinan, enhanced serum antibody titer, and promoted lymphocyte proliferation, thus improving the immune effect of the Newcastle disease vaccine compared to non-modified lentinan (Unursaikhan et al. 2006; Guo et al. 2009). Since the triple helical structure plays an important role in life science and is necessary for antitumor action, controlling and further producing this structure by denaturation-renaturation using dimethylsulfoxide, a powerful hydrogen bond acceptor that works by breaking associative hydrogen bonds in the polysaccharide, may be useful in the future for cancer therapy (Xu et al. 2010).

Therefore, we expect that modified-lentinan or other polysaccharides possess better efficacy than standard agents and will be new types of immune adjuvants.

14.6 Toxicity and Adverse Effects

Millions of people enjoy the mushroom every day without any complaint. AHCC, which is an extract of shiitake mushroom of the basidiomycete family of fungi rich in alpha glucans, has been used for many years as a dietary supplement to enhance the immune system and is in clinical trials as an adjunctive treatment for hepatocellular cancer. A Phase I trial investigated the clinical safety and tolerability of AHCC. Nine grams of AHCC (150 ml of the currently available liquid AHCC) was administered orally per day for 14 days to 26 healthy male or female subjects aged 18–61 years. Adverse effects, including nausea, diarrhea, bloating, headache, fatigue, and foot cramps, occurred in 6 subjects (20%) but they were mild and transient. There were no abnormalities in the laboratory parameters, and the dose was tolerated by 85% of the subjects (Spierings et al. 2007).

Mushroom immunoceuticals are safe to take for long periods, and they improve energy levels, speed regeneration of damaged bone marrow, support the liver (Yeung and Chiu 1994), reduce the side effects of toxic anticancer therapies, and generally increase the well-being in cancer patients. Proteoglycan mushroom immunoceuticals offer hope for cancer patients. These substances are pro-homeostatic and uniquely effective immune boosters that pose no threat of autoimmune backlash. As dietary supplements, they are safe, clinically proven, and exhibit near-perfect benefit-risk profiles. Therefore, mushroom immunoceuticals are a potential boon for individuals with cancer who also have impaired immunity (Kidd 2000).

However, as is true of almost any food, some people will experience allergic reactions. Some components of the fruit-body extract may cause temporary diarrhea. Animal studies revealed that the LD_{50} is over 2,500 mg/kg by intraperitoneal and 250–500 mg/kg by intravenous administration in mice and rats (Chihara et al. 1987). However, when undesirable side effects are experienced, one has to temporarily stop supplementing or reduce the dose.

Collectively, the literature corroborate that the shiitake mushroom and its extractions have potent antitumor actions. However, the time of harvest and the storage conditions may influence the biological activity and polysaccharide content of shiitake mushrooms (Mizuno 2000). Detailed information regarding growing, harvesting, processing, and extraction procedures should be well established. Human cancer is a disease with great diversity compared to infectious diseases, and the tumor-host relationship may differ, depending on the tumor growth stage and the course of treatment. Therefore, the use of lentinan for human cancer should be based on strict administration timing and dosage schedules in compliance with immunological and biological changes.

References

Adachi, Y., Miura, N. N., Ohno, N., et al. (1999). Enzyme immunoassay system for estimating the ultrastructure of 1,6-branched-1,3-β-glucans. *Carbohydrate Polymers, 39,* 225–229.

Akramiene, D., Kondrotas, A., Didziapetriene, J., et al. (2007). Effects of beta-glucans on the immune system. *Medicina (Kaunas), 43,* 597–606.

Aoki, T. (1984). Lentinan. In R. L Fenichel & M. A. Chirgis (Eds.), *Immune modulation agents and their mechanisms* (Immunological Studies, 25, 62–77). New York: Madison Avenue.

Arinaga, S., Karimine, N., TakamuKu, K., et al. (1992a). Enhanced production of interleukin 1 and tumor necrosis factor by peripheral monocytes after lentinan administration in patients with gactric carcinoma. *International Journal of Immunopharmacology, 14,* 43–47.

Arinaga, S, Karimine, N., TakamuKu, K., et al. (1992b). Enhanced induction of lympholine-activated killer activity after lentinan administration in patients with gastric carcinoma. *International Journal of Immunopharmacology, 14,* 535–539.

Bohn, J. A., & BeMiller, J. N. (1995). (1→3)-β-D-glucans as biological response modifiers: A review of structure-functional activity relationships. *Carbohydrate Polymers, 28,* 3–14.

Borchers, A. T., Stern, J. S., Hackman, R. M., et al. (1999). Mushrooms, tumors, and immunity. *Proceedings Society Experimental Biology and Medicine, 221,* 281–293.

Brown, G. D., & Gordon, S. (2001). Immune recognition: A new receptor for β-glucans. *Nature, 413,* 36.

Cheung, N. K., & Modak, S. (2002a). Oral (1,3;1,4)-β-D-glucan synergizes with antiganglioside GD2 monoclonal antibody 3F8 in the therapy of neuroblastoma. *Clinical Cancer Research, 8,* 1217.

Cheung, N. K., Modak, S., Vickers, A., et al. (2002b). Orally administered β-glucans enhance anti-tumor effects of monoclonal antibodies. *Cancer Immunology, Immunotherapy, 51,* 557.

Chihara, G., Maeda, Y., Hamuro, J., et al. (1969). Inhibition of mouse sarcoma 180 by polysaccharides from lentinus edodes (berk) sing. *Nature, 222,* 687–688.

Chihara, G., Hamuro, J., Maeda, Y. Y., et al. (1970a). Fraction and purification of the polysaccharides with marked antitumor activity, especially lentinan, from Lentinan edodes (berk.) sing. (an edible mushroom). *Cancer Research, 30,* 2776–2781.

Chihara, G., Hamuro, J., Maeda, Y. Y., et al. (1970b). Antitumor polysaccharide derived chemically from natural glucan (pachyman). *Nature, 225,* 943–944.

Chihara, G., Hamura, J., Maeda, Y. Y., et al. (1987). Antitumor and metastasis-inhibitory activities of lentinan as an immunomodulator: An overview. *Cancer Detection and Prevention, 1,* 423–443.

Dennis, T., Fanous, M., & Mousa, S. (2009). Natural products for chemopreventive and adjunctive therapy in oncologic disease. *Nutrition and Cancer, 61,* 587–597.

deVere White, R. W., Hackman, R. M., et al. (2002). Effects of a mushroom mycelium extract on the treatment of prostate cancer. *Urology, 60,* 640–644.

Dong, S. F., Chen, J. M., Zhang, W., et al. (2007). Specific immune response to HBsAg is enhanced by beta-glucan oligosaccharide containing an alpha-(1→3)-linked bond and biased towards M2/Th2. *International Immunopharmacology, 7,* 725–733.

Fujii, T., Maeda, H., Suzuki, F., et al. (1978). Isolation and characterization of a new antitumor polysaccharide, KS-2, extracted from culture mycelia of Lentinus edodes. *The Journal of Antibiotics, 31,* 1079–1090.

Fujimoto, K., Tomonaga, M., & Goto, S. (2006). A case of recurrent ovarian cancer successfully treated with adoptive immunotherapy and lentinan. *Anticancer Research, 26,* 4015–4018.

Fruehauf, J. P., Bonnard, G. D., & Herberman, R. B. (1982). The effect of lentinan on production of interleukin-1 by human monocytes. *Immunopharmacology, 5,* 65–74.

Guo, Z., Hu, Y., Wang, D., et al. (2009). Sulfated modification can enhance the adjuvanticity of lentinan and improve the immune effect of ND vaccine. *Vaccine, 27,* 660–665.

Harada, K., Itashiki, Y., Takenawa, T., et al. (2010). Effects of lentinan alone and in combination with fluoropyrimidine anticancer agent on growth of human oral squamous cell carcinoma in vitro and in vivo. *International Journal of Oncology, 37,* 623–631.

Hazama, S., Watanabe, S., Ohashi, M., et al. (2009). Efficacy of orally administered superfine dispersed lentinan (beta-1,3-glucan) for the treatment of advanced colorectal cancer. *Anticancer Research, 29,* 2611–2617.

Hobbs, C. (2000). Medicinal value of Lentinus edodes (Berk.) Sing.(Agaricomycetideae). A literature review. *International Journal for Medicinal Mushrooms, 2,* 287–302.

Hong, F, Hansen, R. D., Yan, J., et al. (2003). β-glucan functions as an adjuvant for monoclonal antibody immunotherapy by recruiting tumoricidal granulocytes as killer cells. *Cancer Research, 63,* 9023.

Hong, F., Yan, J., Baran, J. T, et al. (2004). Mechanism by which orally administered beta-1,3-glucans enhance the tumoricidal activity of antitumor monoclonal antibodies in murine tumor models. *Journal of Immunology, 173,* 797–806.

Ikekawa, T. (2001). Beneficial effects of edible and medicinal mushrooms in health care. *International Journal for Medicinal Mushrooms, 3,* 291–298.

Isoda, N., Eguchi, Y., Nukaya, H., et al. (2009). Clinical efficacy of superfine dispersed lentinan (beta-1,3-glucan) in patients with hepatocellular carcinoma. *Hepatogastroenterology, 56,* 437–441.

Izawa, M., Ohno, K., Amikura, K., et al. (1984). Lentinan auguments the production of IL-3 and CSF (s) by T-cells. In T. Aoki, E. Tsubura, & I. Urushizaki (Eds.), *Manipulation of Host Defence Mechanisms* (pp. 59–69). Amsterdam: Excerpta Medica.

Jin, M., Jeon, H., Jung, H. J., et al. (2003). Enhancement of repopulation and hematopoiesis of bone marrow cells in irradiated mice by oral administration of PG101, a water-soluble extract from Lentinus lepideus. *Experimental Biology and Medicine, 228,* 759–766.

Jong, S. C., Birmingham, J. M., & Pai, S. H. (1991). Immuno-modulatory substances of fungal origin. *Journal of Immunology Immunopharmacology, 9,* 115–122.

Kidd, P. M. (2000). The use of mushroom glucans and proteoglycans in cancer treatment. *Alternative Medicine Review, 5,* 4–27.

Kim, G. Y., Choi, G. S., Lee, S. H., et al. (2004). Acidic polysaccharide isolated from Phellinus linteus enhances through the upregulation of nitric oxide and tumor necrosis factor-alpha from peritoneal macrophages. *Journal of Ethnopharmacology, 95,* 69–76.

Kobayashi, H., Imanaka, M., Inokuchi, N., et al. (1998). Characterization of ribonucleases from culture medium of lentinus edodes. , *Biotechnology, and Biochemistry, 62,* 1604–1608.

McCormack, E., Skavland, J., Mujic, M., et al. (2010). Lentinan: Hematopoietic, immunological, and efficacy studies in a syngeneic model of acute myeloid leukemia. *Nutrition and Cancer, 62,* 574–583.

Minato, K., Mizuno, M., Terai, H., et al. (1999). Autolysis of lentinan, an antitumor polysaccharide, during storage of Lentinus edodes, shiitake mushroom. *Journal of Agricultural and Food Chemistry, 47,* 1530–1532.

Miyakoshi, H., Aoki, T., & Mizukoshi, M. (1984). Acting mechanisms of lentinan in human: Enhancement of non-specific cell-mediated cytoxicity as an interferon inducer. *International Journal of Immunopharmacology, 6,* 373–379.

Mizuno, M. (2000). Anti-tumor polysaccharides from mushrooms during storage. *BioFactors, 12,* 275–281.

Mori, K., Toyomasu, T., Nanba, H., et al. (1987). Anti-tumour action of fruit bodies of edible mushrooms orally administerated to mice. *Mushroom Journal for the Tropics, 7,* 121–126.

Mushiake, H., Tsunoda, T., Nukatsuka, M., et al. (2005). Dendritic cells might be one of key factors for eliciting antitumor effect by chemoimmunotherapy in vivo. *Cancer Immunology, Immunotherapy, 54,* 120–128.

Nakano, H., Namatame, K., Nemoto, H., et al. (1999). A multi-institutional prospective study of lentinan in advanced gastric cancer patients with unresectable and recurrent diseases: Effect on prolongation of survival and improvement of quality of life. *Hepatogastroenterology, 46,* 2662–2668.

Nanba, H., Mori, K., Toyomasu, T., et al. (1987). Antitumor action of shiitake (lentinus edodes) fruit bodies orally administered to mice. *Chemical & Pharmaceutical Bulletin, 35,* 2453–2458.

Ng, M. L., & Yap, A. T. (2002). Inhibition of human colon carcinoma development by lentinan from shiitake mushrooms (Lentinus edodes). *Journal of Alternative and Complementary Medicine, 8,* 581–589.

Oba, K., Kobayashi, M., Matsui, T., et al. (2009). Individual patient based meta-analysis of lentinan for unresectable/recurrent gastric cancer. *Anticancer Research, 29,* 2739–2745.

Ooi, V. E, & Liu, F. (1999). A review of pharmacological activities of mushroom polysaccharides. *International Journal for Medicinal Mushrooms, 1,* 195–206.

Ross, G. D. (2000). Regulation of the adhesion versus cytotoxic functions of the Mac-1/CR3/$\alpha_M\beta_2$ -integrin glycoprotein. *Critical Reviews in Immunology, 20,* 197.

Ross, G. D, Cain, J. A., Myones, B. L., et al. (1987). Specificity of membrane complement receptor type three (CR3) for β-glucans. *Complement and Inflammation, 4,* 61.

Ruan, Z., Su, J., Dai, H. C., et al. (2005). Characterization and immunomodulating activities of polysaccharide from Lentinus edodes. *International Immunopharmacology, 5,* 811–820.

Sasaki, T., & Takasuka, N. (1976). Futher study of the structure of lentinan, an antitumor polysaccharide from lentinus edodes. *Carbohydrate Research, 47,* 99–104.

Saito, H., Ohki, T., Takaoka, N., et al. (1977). A ^{13}C-NMR-spectral study of a gelforming, branched (1→3)-β-D-glucan (lentinan) from Lentinus edodes, and its acid degraded fractions: Structure and dependence of confirmation on the molecular weight. *Carbohydrate Research, 58,* 293–305.

Shen, J. W., Ren, H. W., Tomiyama-Miyaji, C., et al. (2007). Potentiation of intestinal immunity by micellary mushroom extracts. *Biomedical Research, 28,* 71–77.

Shen, J., Mamoru, T., Fujisaki, Y., et al. (2009a). Effect of the culture extract of Lentinus edodes mycelia on splenic sympathetic activity and cancer cell proliferation. *Autonomic Neuroscience, 145,* 50–54.

Shen, J., Horii, Y., Fujisaki, Y., et al. (2009b). Effects of L-arginine and L-lysine mixtures on splenic sympathetic nerve activity and tumor proliferation. *Autonomic Neuroscience, 147,* 86–90.

Shimizu, K., Watanabe, S., Watanabe, S., et al. (2009). Efficacy of oral administered superfine dispersed lentinan for advanced pancreatic cancer. *Hepatogastroenterology, 56,* 240–244.

Sorimachi, K., Yamazaki, S., Toda, S., et al. (1990). Morphological changes of tumor cells caused by macrophages treated with lignin derivatives. *Agricultural and Biological Chemistry, 54,* 1611–1613.

Spierings, E. L, Fujii, H., Sun, B., et al. (2007). A Phase I study of the Safety of the Nutritional Supplement, Active Hexose Correlated Compound, AHCC, in Healthy Volunteers. *Journal of Nutritional Science and Vitaminology, 53,* 536–539.

Sugano, N., Hibino, Y., Choji, Y., et al. (1982). Anticarcinogenic actions of water-soluble and alcohol-insoluble fractions from culture medium of Lentinus edodes mycelia. *Cancer Letters, 17,* 109–114.

Sugano, N., Choji, Y., Hibino, Y., et al. (1985). Anticarcinogenic action of an alcohol-insoluble fraction (LAP1) from culture medium of Lentinus edodes mycelia. *Cancer Letters, 27,* 1–6.

Suga, T., Shiio, T., Maeda, Y. Y, et al. (1984). Antitumor activity of lentinan in murine syngeneic and autochthonous hosts and its suppressive effect on 3-methylcholanthrene-induced carcinohenesis. *Cancer Research, 44,* 5132–5137.

Suzuki, I., Tanaka, H., Kinoshita, A., et al. (1990). Effect of orally administered β-glucan on macrophage function in mice. *International Journal of Immunopharmacology, 12,* 675–678.

Suzuki, M., Kikuchi, T., Takatsuki, F., et al. (1994). Curative effects of combination therapy with lentinan and interleukin-2 against estabolished murine tumors, and the role of CD8-positive T cells. *Cancer Immunology, Immunotherapy, 38,* 1–8.

Tabata, T., Watanabe, W., Horita, K., et al. (1992). Mitogenic activities of heteroglycan and heteroglycan-protein fractions culture medium of Lentinus edodes mycelia. *Immunopharmacology, 24,* 57–63.

Taylor, P. R., Brown, G. D., Reid, D. M., et al. (2002). The β-glucan receptor, dectin-1, is predominantly expressed on the surface of cells of the monocyte/macrophage and neutrophil lineages. *Journal of Immunology, 169,* 3876.

Thornton, B. P., Vetvicka, V., Pitman, M., et al. (1996). Analysis of the sugar specificity and molecular location of the β-glucan-binding lectin site of complement receptor type 3 (CD11b/CD18). *Journal of Immunology, 156,* 1235.

Tsukada, C., Yokoyama, H., Miyaji, C., et al. (2003). Immunopotentiation of intraepithelial lymphocytes in the intestine by oral administrations of beta-glucan. *Cellular Immunology, 221,* 1–5.

Unursaikhan, S., Xu, X., Zeng, F., et al. (2006). Antitumor activities of O-sulfonated derivatives of (1–3)-alpha-D-glucan from different Lentinus edodes. *Bioscience, Biotechnology, and Biochemistry, 70,* 38–46.

Vetvicka, V., Vashishta, A., Saraswat-Ohri, S., et al. (2008). Immunological effects of yeast- and mushroom-derived beta-glucans. *Journal of Medicinal Food, 11,* 615–622.

Wang, J., Dong, S. F., Liu, C. H., et al. (2010). β-glucan oligosaccharide enhances CD8+ T cells immune response induced by a DNA vaccine encoding hepatitis B virus core antigen. *Journal of Biomedicine & Biotechnology, 2010,* 645213.

Wang, Y., Zhang, L., Li, Y, et al. (2004). Correlation of structure to antitumor activities of five derivates of a β-glucan from Poria cocos sclerotium. *Carbohydrate Research, 339,* 2567.

Wasser, S. P. (2002). Medicinal mushrooms as a source of antitumor and immunomodulating polysaccharides. *Applied Microbiology and Biotechnology, 60,* 258–274.

Wasser, S. P., & Weis, A. L. (1999a). Medicinal properties of substances occurring in Higher Basidiomycetes mushrooms: Current perspectives. *International Journal for Medicinal Mushrooms, 1,* 31–62.

Wasser, S. P., & Weis, A. L. (1999b). Therapeutic effects of substances occurring in higher basidiomycestes mushrooms; a modern perspective. *Critical Reviews in Immunology, 19,* 65–96.

Xia, Y., Vetvicka, V., Yan, J., et al. (1999). The β-glucan-binding lectin site of mouse CR3 (CD11b/CD18) and its function in generating a primed state of the receptor that mediates cytotoxic activation in response to iC3b-opsonized target cells. *Journal of Immunology, 162,* 2281.

Xu, X. J., Wang, X. H., Cai, F., et al. (2010). Renaturation of triple helical polysaccharide lentinan in water-diluted dimethylsulfoxide solution. *Carbohydrate Research, 345,* 419–424.

Yan, J., Vetvicka, V., Xia, Y., et al. (1999). β-glucan, a "specific" biologic response modifier that uses antibodies to target tumors for recognition by complement receptor type 3 (CD11b/CD18). *Journal of Immunology, 163,* 3045.

Yang, P., Liang, M., Zhang, Y., et al. (2008). Clinical application of a combination therapy of lentinan, multi-electrode RFA and TACE in HCC. *Advances in Therapy, 25,* 787–794.

Yap, A. T., & Ng, M. L. (2001). An improved method for the isolation of lentinan from the edible and medicinal shiitake mushroom, lentinan edoded (Berk.) Sing. (Agaricomycetideae). *International Journal for Medicinal Mushrooms, 3,* 9–19.

Yap, A. T., & Ng, M. L. (2002). Inhibition of human colon sarcoma by lentinan from shiitake mushroom (lentinus edodes). *Journal of Alternative and Complementary Medicine, 8,* 581–589.

Yeung, J. H., & Chiu, L. C. (1994). Effect of polysaccharide peptide (PSP) on glutathione and protection against paracetamol-induced hepatotoxicity in the rat. *Methods and Findings in Experimental and Clinical Pharmacology, 16,* 723–729.

Yoshino, S., Tabata, T., Hazama, S., et al. (2000). Immunoregulatory effects of the antitumor polysaccharide lentinan on Thl/Th2 balance in patients with digestive cancers. *Anticancer Research, 20,* 4707–4711.

Yoshino, S., Watanabe, S., Imano, M., et al. (2010). Improvement of QOL and prognosis by treatment of superfine dispersed lentinan in twenty-seven patients with advanced gastric cancer. *Hepatogastroenterology, 57,* 172–177.

Zakany, J., Chihara, G., & Fachet, J. (1980). Effects of lentinan on tumor growth in murine allogenic and syngenic hosts. *International Journal of Cancer, 25,* 371–376.

Zhou, L. D., Zhang, Q. H., Zhang, Y., et al. (2009). The shiitake mushroom-derived immunostimulant lentinan protects against murine malaria blood-stage infection by evoking adaptive immune-responses. *International Immunopharmacology, 9,* 455–462.

Chapter 15
Molecular Targets of Resveratrol in Carcinogenesis

Seung J. Baek and Nichelle C. Whitlock

Abstract Resveratrol, a dietary phytoalexin readily available in the diet, is reported to possess antitumorigenic and anti-inflammation activity in several cancers. However, the molecular mechanisms and/or targets underlying the antitumorigenic activity of this compound remain largely unknown. As a potential signaling pathway modulator, resveratrol is shown to affect a multitude of signaling transduction pathways, and it is likely that this collective activity may play an important role in resveratrol-induced antitumorigenesis. Therefore, the elucidation of molecular targets of resveratrol is necessary to understand how this compound is beneficial in antitumorigenic processes. Studies to identify molecular markers have recently spawned increasing examination of the possible targets of resveratrol as related to anticancer proteins. These include, but are not limited to: (1) inhibition of carcinogenic activation and induction of detoxification by Phase I and Phase II xenobiotic metabolizing enzymes, respectively, (2) modulation of cell survival and apoptosis, (3) suppression of pro-inflammatory signaling pathways, (4) inhibition of angiogenesis, invasion, and metastasis, and (5) modulation of hormonal receptors. This chapter summarizes the diverse molecular targets of resveratrol in prevention and treatment of carcinogenesis.

15.1 Introduction

Cancer is an important public health problem in the Western world and is the second leading cause of death in the United States. It has become increasingly obvious that the problem of cancer cannot be solved by traditional treatment alone and an alternative approach is needed. The traditional approaches to cancer prevention have included attempts to eliminate carcinogenic agents and to detect and remove precancerous lesions; many of these efforts are focused on interrupting,

---

S. J. Baek (✉)
Department of Pathobiology, College of Veterinary Medicine, University of Tennessee, 2407 River Dr, Knoxville, TN 37996, USA
e-mail: sbaek2@utk.edu

W. C. S. Cho (ed.), *Evidence-based Anticancer Materia Medica,* Evidence-based Anticancer Complementary and Alternative Medicine,
DOI 10.1007/978-94-007-0526-5_15,

reversing, or delaying the neoplastic process. However, using dietary compounds with minimal or no side effects may prevent cancer from reaching its invasive and metastatic stages. The importance of dietary compounds in cancer prevention was first pointed out during the 1960s and 1970s by the experiments of Wattenberg and colleagues (Wattenberg 1966), in which they demonstrated that various compounds inhibit chemically-induced tumors in laboratory animals. Current research also suggests various dietary phytochemicals function as chemo-preventive and/or adjuvant chemo-therapeutic agents, adding to the paradigm that a diet high in fruit and vegetable content confers protection against chronic diseases, including cancer. Epidemiological studies further support this fact and illustrate the important role nutrition plays in the development of carcinogenesis (Surh 2003). Thus, further molecular mechanism-based studies are required to evaluate the use of dietary components in cancer chemo-prevention and chemo-therapeutic research.

One such phytochemical is resveratrol (3, 4′, 5-trihydroxystilbene), a naturally occurring phytoalexin readily available in the diet and to which a plethora of health-promoting effects have been ascribed. Resveratrol, first identified as a bioactive compound in 1992, is found in several plants, particularly in the skin of red grapes (Baur and Sinclair 2006). This compound has elicited much attention as a potential cancer chemo-preventive and/or chemo-therapeutic compound since its inhibitory effect on carcinogenic processes was first reported in 1997 (Jang et al. 1997). Subsequently, numerous studies have illustrated the anti-proliferative effect of resveratrol on cancer cells, believed attributable to induction of cell cycle arrest in the G1/S and G2/M phase and induction of apoptosis and its related proteins (Fan et al. 2008; Harikumar and Aggarwal 2008). Treatment with resveratrol has also been shown to result in induction of autophagocytosis and senescence in ovarian and colorectal cancer cells, respectively (Opipari et al. 2004; Heiss et al. 2007). Furthermore, resveratrol treatment resulted in the inhibition of angiogenic, invasive, and metastatic factor release (Kundu and Surh 2008), as well as inhibition of tumorigenesis *in vivo* (Athar et al. 2007; Wang et al. 2008; Roy et al. 2009). Thus, resveratrol modulates multiple signaling pathways that interrupt the carcinogenic process and are also capable of extending one or more stages of this process (Fig. 15.1). Tables 15.1 and 15.2 contain recently reported *in vitro* and *in vivo* evidence for the anticancer activities of resveratrol.

Despite substantial progress in understanding the molecular basis of resveratrol's anticancer activities, few clinical trials have been undertaken to confirm its use as a chemo-preventive and/or adjuvant chemo-therapeutic agent. Pre-clinical studies have demonstrated the inhibitory effects of resveratrol in different cancers (Table 15.2) (Bishayee 2009) and its ability to act as an adjuvant to traditional chemo-therapeutics (Bhardwaj et al. 2007; Shankar et al. 2007a, b; Gatouillat et al. 2010; Harikumar et al. 2010). As an adjuvant, resveratrol potentiates the cytotoxic effects of anticancer drugs given at subtoxic levels as well as alleviate associated side effects. A Phase I clinical study is underway to determine potential adverse resveratrol-drug interaction(s) with emphasis on CYP3A4 activity; CYP3A4 is involved in metabolism of transplant medications/chemotherapies and is inhibited by resveratrol. Other clinical trials seek to determine the effects of resveratrol on Wnt

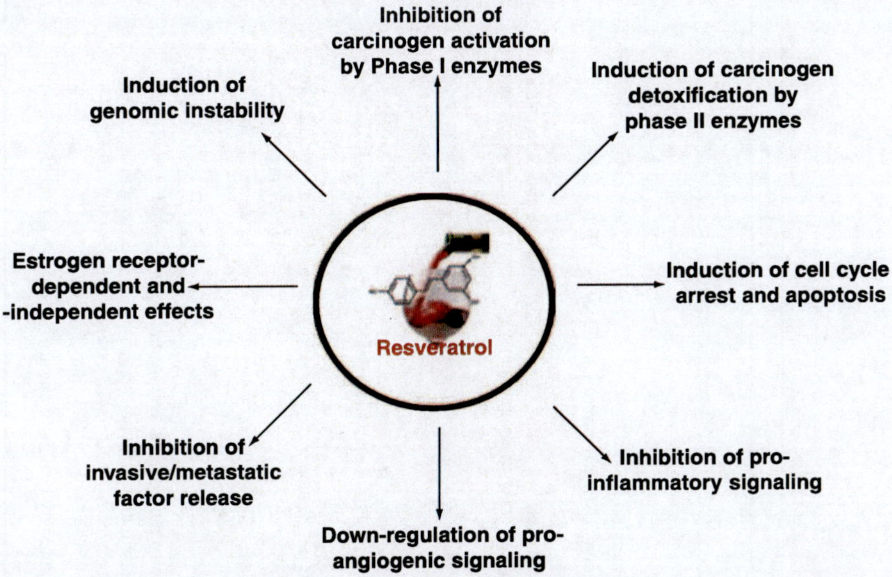

Fig. 15.1 Biochemical mechanisms attributed to the chemo-preventive and chemo-therapeutic effects of resveratrol

signaling in normal mucosa and colon cancers (www.clinicaltrials.gov). Although shown to be well tolerated in humans, the bioavailability of resveratrol remains poor (Boocock et al. 2007). Thus, in order to recommend use of resveratrol as a chemo-preventive or adjuvant chemo-therapeutic, further studies are necessary to enhance bioavailability of the phytochemical as well as to validate identified *in vitro/vivo* targets of resveratrol. Yet, resveratrol remains a promising candidate for the prevention and treatment of cancers.

15.2 Regulation of Phase I and Phase II Xenobiotic Metabolizing Enzymes

Oxidative metabolism by Phase I enzymes converts pro-carcinogens to reactive electrophilic intermediates, which are further metabolized by Phase II enzymes *via* conjugation of hydrophilic moieties. Resultant metabolites are detoxified and eliminated. However, inadequate detoxification by Phase II enzymes potentiates genotoxicity (e.g. DNA and protein adduct formation) of Phase I products, thus initiating the carcinogenic process (Nebert and Dalton 2006; Androutsopoulos et al. 2009). Modification of Phase I and Phase II xenobiotic metabolizing enzyme expression is one mechanism by which the dietary anticancer agent resveratrol exerts its chemo-preventive effects. Among those Phase I and II enzymes, the expression

Table 15.1 *In vitro* and *in vivo* molecular targets of resveratrol: anticancer effects

Molecular targets	Experimental model	Mechanisms	References
Phase I enzymes			
CYP1A1, CYP1B1	Dioxin-treated MCF-7, HepG2 cells	↓CYP1A1/1B1 expression; ↓AhR/ARNT recruitment to promoter	Beedanagari et al. 2009
	TCDD-treated keratinocytes	↓CYP1A1 expression	Du et al. 2006
	DMBA-treated keratinocytes	↓CYP1A1 and 1B1 expression	Kowalczyk et al. 2009
	B[a]P-treated BEAS, BEP2D cells	↓CYP1A1/1B1 expression; ↓B[a]P metabolism, DNA adduct formation	Berge et al. 2004
	TCDD-treated T47D, MCF-7 cells	↑CYP1A1 mRNA degradation	Lee and Safe 2001
	DMBA-treated MCF10A cells	↓CYP1A1/1B1 expression; ↓DMBA-induced DNA adduct formation	Leung et al. 2009
	TCDD or estradiol-treated MCF10F cells	↓CYP1B1 expression; ↓estrogen metabolism, DNA adduct formation	Lu et al. 2008
	TCDD-treated AGS cells	↓CYP1A1 expression	Peng et al. 2009
	B[a]P-treated Balb/c mice (lung cancer)	↓CYP1A1 expression; ↓B[a]P diol epoxidation-DNA adduct formation	Revel et al. 2003
CYP1A2, CYP2B	B[a]P-treated Balb/c mice (skin cancer)	↓CYP1A1/1A2/1B1/2B expression	Szaefer et al. 2008
CYP2E1	Pyrogallol-treated Swiss albino mice	↓CYP2E1 catalytic activity	Upadhyay et al. 2008
CYP3A4	Sf9 insect microsomes	↓CYP3A4 activation	Chan and Delucchi 2000
Phase II enzymes			
Nrf2	DENA-induced liver carcinogenesis	↑Nrf2 expression	Bishayee et al. 2010a
NQO1	4-OHE2 or E2-3,4-Q-treated MCF10F cells	↑NQO1 expression, ↓estrogen-DNA adduct formation	Zahid et al. 2008
	Cigarette smoke extract-treated HBE1 cells	↑NQO1 expression; mediated by Nrf2 activation	Zhang et al. 2010
	Oxidative stress exposed primary hepatocytes	↑NQO1 expression; mediated by Nrf2 nuclear translocation and activation	Rubiolo 2008
GCL	Cigarette smoke extract-treated A549 cells	↑GCL expression; mediated by Nrf2 activation; ↑glutathione	Kode et al. 2008

Table 15.1 (continued)

Molecular targets	Experimental model	Mechanisms	References
Cell cycle/apoptosis machinery			
Cell cycle proteins	T24 cells	↓cyclin D1, CDK4, Rb hyper-phosphorylation; ↑p21 expression	Bai et al. 2010
	MDA-MB-231 cells	↓cyclin D1, cylin E, CDK4, ribonucleotide reductase expression	Pozo-Guisado et al. 2002
	HT29, SW480 cells	↓cyclin D1, ↑p27, ↑p53-mediated apoptosis; mediated by ↓IGF1R, Akt, WNT signaling	Vanamala et al. 2010
	HepG2cells	↓cylin D1 expression; mediated by ↓p38 M APK, Akt, PAK1 activation	Parekh et al. 2010
	LNCaP, PC3 cells	↓cyclin D1, cylin E, CDK4 expression; ↑p21, p27, p53 expression	Benitez et al. 2007
	B16 cells	↓cyclin D1, CDK4 expression; ↑p53-mediated apoptosis	Gatouillat et al. 2010)
	A431 cells	↓cyclin D1, CDK6 expression, Rb hyper-phosphorylation; ↑p21, p27 expression; mediated by MAPK inactivation	Kim et al. 2006
Apoptosis-related proteins	Pancreatic cancer cells	↓Bcl2, BclXL, cyclin D1 expression	Harikumar et al. 2010
	A431, MCF-7 cells	↑Bax, mitochondrial membrane depolarization, caspase-dependent apoptosis; ↓Bcl2	Madan et al. 2008; Singh et al. 2009
	LNCaP/PC3, U251, NHL/MM cells	Activation of intrinsic pathway	Jazirehi et al. 2004; Jiang et al. 2005; Shankar et al. 2007a, b
	Colon cancer, multiple myeloma cells	Redistribution of death receptors to membrane lipid rafts; sensitization to death receptor-mediated apoptosis	Delmas et al. 2004; Reis-Sobreiro et al. 2009

Table 15.1 (continued)

Molecular targets	Experimental model	Mechanisms	References
Inflammatory mediators			
COX-2	ELISA-based kinetic assay	↓COX-2 expression/activity	Calamini et al. 2010
	HT29 cells	↓COX-2 expression/activity; associated with ↓colony formation, anchorage independent growth	Zykova et al. 2008
	DENA-induced liver carcinogenesis	↓COX-2 expression; mediated by ↓NFκB activation	Bishayee et al. 2010b
	TPA-induced skin carcinogenesis	↓COX-2 expression; mediated by ↓MAPK signaling, NFκB activation	Kundu et al. 2006
TNF-α, ILs	DSS-induced colitis (mouse model)	↓TNF-α, IL-6, IL-1β, COX-2 expression	Sánchez-Fidalgo et al. 2010; Singh et al. 2010
	Acute myeloid leukemia cells	↓IL-1β expression, NFκB activation	Estrov et al. 2003
	Streptozotocin-nicotinamide-induced diabetic rats	↓TNF-α, IL-1β, IL-6 expression in pancreatic β cells	Palsamy and Subramanian 2010
	Turbot (Psetta maxima)	↓TNF-α, IL-1β mRNA processing	Leiro et al. 2010
Angiogenesis, invasion, metastasis			
HIF-1α and VEGF	Ovarian cancer cells	↓HIF-1α, VEGF expression; mediated by PI3K/Akt & protein translation regulator inactivation; ↑HIF-1α protein degradation	Cao et al. 2004; Park et al. 2007
	Tongue squamous cell carcinoma, hepatoma cells	↓HIF-1α, VEGF expression; mediated by MAPK/Akt; ↑HIF-1α protein degradation	Zhang et al. 2005
MMP2, 9	MCF-7 cells	↓heregulin-β1-induced MMP9 expression; mediated by ERK1/2 inactivation	Tang et al. 2008a
	HepG2 cells	↓TNF-α-mediated MMP9 expression; mediated by NFκB inactivation	Yu et al. 2008
	A549 cells	↓HO-1 induced MMP expression; mediated by NFκB inactivation	Liu et al. 2010b
	MDA-MB-435 cells	↓IGF-1-mediated MMP2 expression; mediated by PI3K/Akt inactivation	Tang et al. 2008b

Table 15.1 (continued)

Molecular targets	Experimental model	Mechanisms	References
Estrogen receptors			
ERα, β	CHO-K1 cells	↑ERE-driven reporter gene activity ER-expressing cells	Bowers et al. 2000
	Ishikawa cells	↓ERα expression and activity in the presence of 17β-estradiol	Bhat and Pezzuto 2001
	MCF-7, MDA-MB-231 cells	↑ERα reporter assays	Gehm et al. 1997, 2004
	MCF-7 cells	Biphasic effect on ERα-dependent PI3 K activity; ↑low concentrations, ↓high concentrations	Pozo-Guisado et al. 2004
Other targets			
Topoisomerase II	Glioblastoma cells	↓topoisomerase activity	Leone et al. 2010
Telomerase	U2OS, A549 cells	↓telomere stability; ↑H2AX, p53 phosphorylation; ↑DNA damage	Rusin et al. 2009
	MCF-7	↓telomerase activity, hTERT nuclear expression	Lanzilli et al. 2006
	HT29, WiDr cells	↓telomerase activity, hTERT nuclear expression	Fuggetta et al. 2006
ATF3	HCT-116, HT29 cells	↑ATF3 expression, apoptosis	Whitlock et al. 2011
NAG-1	HCT-116, U2OS, A549 cells	↑NAG-1 expression, apoptosis	Baek et al. 2004
	CD18, S2–013 cells	↑NAG-1 expression; ↓cell proliferation	Golkar et al. 2007
DDIT3	HT29, K562 cells	↑DDIT3-mediated cell cycle arrest/apoptosis	Woo et al. 2007; Liu et al. 2010a; Um et al. 2010

Table 15.2 Chemo-preventive/therapeutic effects of resveratrol: *in vivo* evidence

Experimental model	Dose/route	Effects	Mechanisms	References
Breast cancer				
SD rats (♀), DMBA-induced	10 ppm, 127 day; diet	↑latency period, ↓incidence/multiplicity	↓NF-κB activation, ↓COX-2, ↓MMP-9	Banerjee et al. 2002
SD rats (♀), N-methyl-N- nitrosourea-induced	100 mg/kg, 127 days; ig	↑latency period, ↓incidence/multiplicity	↑apoptosis	Bhat et al. 2001
Nude mice (♀), BRCA1 mutant cell line	5–10 μg/30 g BW, 3 weeks; ip	↓tumor initiation, ↓tumor growth	↓proliferation, ↑apoptosis (↓survivin mediated by SirT1)	Wang et al. 2008
Colon cancer				
C57B/6 mice, AOM+DSS-induced	300 ppm, 70 days; diet	↓incidence/multiplicity, ↓tumor volume	↓COX-2, ↓TNF-α, ↓iNOS	Cui et al. 2010
Liver cancer				
SD rats (♀), DENA+phenobarbital-induced	50–300 mg/(kg/days), 20 weeks; diet	↓incidence/multiplicity	↓proliferation, ↑Bax, ↓Bcl2	Bishayee and Dhir 2009
Pancreatic cancer				
Nude mice (♂), MIA PaCa2 cells	40 mg/kg, 4 weeks; po	↓tumor formation, ↓volume/growth	↓proliferation, ↓microvessel density, ↓NF-κB, ↓cyclin D1, ↓COX-2, ↓MMP9, ↓survivin	Harikumar et al. 2010
Skin cancer				
B6D2F1 mice (♀), doxorubicin-resistant B16 melanoma cells	12.5 mg/kg, 30 days; sc	↓tumor growth, ↑survival of tumor bearing mice		Gatouillat et al. 2010
SENCAR mice (♀), DMBA-induced	1–10 μM, 2/week for 4 weeks; topically	↓epidermal hyperplasia		Kowalczyk et al. 2009
SKH-1 hairless mice (♀), UVB-induced	10 μM, 7 treatments on alternate days; topically	↓epidermal hyperplasia	↓leukocyte infiltration, ↓proliferation, ↓cdk2/6/4, ↓cyclin D1/D2, ↑p53/p21, ↓MAPK pathway	Reagan-Shaw et al. 2004
Swiss albino mice (♀), DMBA-induced	25–50 μM, 3/week for 28 weeks; topically	↓incidence/multiplicity, ↓tumor volume	↑apoptosis: ↑p53, ↑Bax, ↑cytochrome *c* release, ↑caspase activation, ↓Bcl2, ↓survivin, ↓PI3K/Akt pathway	Roy et al. 2009

of cytochrome P450 enzymes and the transcription factor Nrf2 (nuclear factor E2-related factor 2) have been reported to be altered by resveratrol.

15.2.1 Cytochrome P450 (CYP) Enzymes

Metabolic activation of carcinogens, including polycyclic aromatic hydrocarbons and poly-halogenated hydrocarbons (PAH/PHAH), by CYP1A1 and CYP1B1 Phase I enzymes is considered a hallmark of the initiation of carcinogenesis. Several studies demonstrate that resveratrol suppresses PAH/PHAH-induced CYP1A1 and CYP1B1 expression and DNA damage *in vitro* and *in vivo*. Two primary mechanisms of this suppression were identified: (1) inhibition of aryl hydrocarbon receptor (AhR)-mediated activation, and (2) direct inhibition of enzyme expression and activity. Resveratrol was shown to impair recruitment of AhR and the AhR nuclear translocator to the *CYP1A1/1B1* and *CYP1A1* promoter in MCF-7 breast and HepG2 liver cancer cells, respectively, resulting in decreased expression (Beedanagari et al. 2009). Resveratrol also effectively blocked CYP1A1 and/or CYP1B1 expression and activity induced by B[a]P (benz[*a*]pyrene), DMBA (7, 12-dimethylbenz[*a*]pyrene), dioxin, or TCDD (2,3,7,8-tetrachlorodibenzo-*p*-dioxin) in differentiating keritinocytes (Du et al. 2006; Kowalczyk et al. 2009), immortalized bronchial epithelial (Berge et al. 2004), breast cells (Lee and Safe 2001; Lu et al. 2008; Leung et al. 2009), and gastric cells (Peng et al. 2009). Resveratrol also suppressed CYP1A1/1B1 expression *in vivo* utilizing the Balb/c animal model (Revel et al. 2003; Szaefer et al. 2008). Moreover, Mikstacka et al. (2008) suggest addition of thiol- and methoxy-groups to the resveratrol backbone increased selectivity for and inhibition of CYP1A1 and CYP1B1. In addition to modulation of CYP1A1/CYP1B1 expression and activity, resveratrol is demonstrated to negatively regulate other CYP family members: inhibition of B[a]P-induced CYP1A2 and CYP2B expression (Szaefer et al. 2008) and abrogating CYP2E1 catalytic activity (Upadhyay et al. 2008) *in vivo*. Moreover, CYP3A4, which is predominantly over-expressed in colon and liver cancers, has been found to be inhibited by resveratrol (Chan and Delucchi 2000). Since resveratrol affects CYPs, it is noted that resveratrol might decrease how quickly the liver breaks down some medications including lovastatin and fexofenadine.

15.2.2 Nrf2

Induction of Nrf2 signaling by resveratrol is thought to confer protection against Phase I-activated carcinogens and associated carcinogenicity *via* the trans-activation of Phase II (detoxifying) enzymes. Under basal conditions, Keap1 sequesters Nrf2 in the cytoplasm, targeting the protein for proteasomal degradation. However, when induced by dietary agents such as resveratrol, Nrf2 dissociates from Keap1 and translocates to the nucleus, where it dimerizes with Maf proteins and activates Phase II enzymes containing antioxidant response elements (Jana and Mandlekar 2009; Kensler and Wakabayashi 2010). Bishayee et al. (2010a) demonstrated that attenuation of diethylnitrosamine (DENA)-induced liver carcinogenesis by resveratrol was mediated by increased Nrf2 expression. Resveratrol induction of NAD(P) H:quinone oxidoreductase (NQO1) expression in TCDD-treated MCF10F immortalized breast cells involved Nrf2, leading to the suppression of DNA adduct formation (Lu et al. 2008); NQO1 was also increased by resveratrol after 4-hydroxyestradiol (4-OHE2) and estradiol-3,4-quinone (E2-3,4-Q) treatment (Zahid et al. 2008). Additionally, induction of Nrf2 signaling by resveratrol resulted in increased expression of NQO1 and glutamate cysteine ligase (GCL) in bronchial epithelial cells (Zhang et al. 2010); NQO1, glutathione peroxidase and glutathione-S-transferase in primary hepatocytes (Rubiolo et al. 2008); glutathione-S-transferase, glutathione peroxidase, and glutathione reductase in pyrogallol-treated mice (Upadhyay et al. 2008); and GCL in A549 lung cancer cells treated with cigarette smoke extract (Kode et al. 2008). Moreover, resveratrol synergized with epigallocatechin gallate and γ-tocotrienol to increase NQO1 expression in ER-positive MCF-7 breast cancer cells (Hsieh and Wu 2008). These results indicate that Nrf2 is a key protein that controls resveratrol-induced antitumorigenesis in several cancers.

15.3 Modulation of Cell Cycle Progression and Apoptosis

The deregulation of cell cycle machinery and evasion of apoptosis, which contribute to aberrant cell proliferation and survival, are considered a fundamental hallmark of cancer development (Brown and Attardi 2005; Meeran and Katiyar 2008). Therefore, identification of agents that target this abnormal signaling offers an important strategy for both chemo-prevention and chemotherapy. Resveratrol is one such compound; several studies have demonstrated the anti-proliferative effect of resveratrol on cancer cells attributable to inhibition of multiple signaling pathways (Fan et al. 2008; Harikumar and Aggarwal 2008). Here, we focus on the effect of resveratrol on the cell cycle proteins and its effect on the apoptotic machinery.

15.3.1 Cell Cycle Proteins

Eukaryotic cell replication involves a series of processes collectively known as the cell cycle. The cell cycle is divided into two major events: (1) interphase (divided into 4 phases: G0, G1, S, and G2), in which the cell grows and accumulates nutrients, and (2) mitosis, wherein the cells divide into two daughter cells. Proper functioning of the cell cycle is important; the presence of numerous internal checkpoints allows the cell to detect genetic damage and stalls cell cycle progression until these aberrations are repaired or apoptosis is triggered. Failure of DNA quality control by checkpoints (and loss of regulatory balance) facilitates uncontrolled cell proliferation and the development of cancer (Vermeulen et al. 2003).

Cyclins and cylin-dependent kinases (CDKs) are key regulatory molecules governing the ordered processes of the cell cycle. Activation of CDKs by cyclins results in the phosphorylation of target proteins, leading to the activation or inactivation of these target proteins, and facilitates progression into the next phase of the cell cycle. Of the many cyclins, cyclin D1 is an important cell cycle regulator involved in the G1/S transition and in the regulation of proliferation and differentiation. Cyclin D1, produced in response to growth factors, is often increased in various cancers due to deregulated degradation (Meeran and Katiyar 2008).

Resveratrol is reported to induce G1/S or G2/M phase cell cycle arrest in many cancer cell lines, attributable to the modulation of cyclin-associated CDK proteins (Harikumar and Aggarwal 2008). Experiments in cell culture revealed that down-regulation of cyclin D1 and/or CDK4/6 by resveratrol resulted in the inhibition of cell proliferation, growth arrest, and subsequent induction of apoptosis. Such an effect was observed in studies of the bladder (Bai et al. 2010), breast (Pozo-Guisado et al. 2002), colon (Joe et al. 2002), liver (Parekh et al. 2010), multiple myeloma (Bhardwaj et al. 2007), pancreas (Harikumar et al. 2010), prostate (Benitez et al. 2007; Hudson et al. 2007), and skin cancer cells (Kim et al. 2006; Gatouillat et al. 2010). These results were also observed *in vivo* utilizing an UVB-induced model of skin tumorigenesis (Reagan-Shaw et al. 2004). Resveratrol treatment of colorectal cancer cells also inhibited expression of cyclin D1 and its complex formation with CDK4 (Wolter et al. 2001); it is likely that resveratrol-induced cyclin D1 suppression may result from β-catenin inactivation in colorectal cancer cells and should be considered a molecular target in these cancers (Joe et al. 2002). Moreover, treatment with resveratrol potentiated the cytotoxic effects of the chemo-therapeutic drugs doxorubicin and gemcitabine in chemo-resistant B16 melanoma and pancreatic cells, respectively, *in vitro* and inhibited tumor growth *in vivo* (Gatouillat et al. 2010; Harikumar et al. 2010).

The CDK inhibitor p21 interacts with cyclin-CDK complexes to negatively regulate their expression and is responsible for the G1 phase arrest phenotype (Meeran and Katiyar 2008). In contrast to cyclin D1, resveratrol-mediated cell cycle arrest is associated with p53-mediated p21 expression. Activation of p21 or p27 by resveratrol was observed in the bladder (Bai et al. 2010), breast (Pozo-Guisado et al. 2002), liver (Kuo et al. 2002), lung (Kubota et al. 2003), prostate (Benitez et al. 2007), and skin cancer cells (Reagan-Shaw et al. 2004; Kim et al. 2006). Again, cell cycle arrest was associated with apoptosis. Interestingly, Castello and Tessitore (2005) demonstrated resveratrol inhibition of U937 leukemic monocyte lymphoma cell proliferation was associated with increased cyclin A and E expression, concomitant p21 and ribonucleotide reductase inhibition, and the absence of apoptosis.

15.3.2 *Apoptosis*

Apoptosis, or programmed cell death, involves a series of biochemical events that lead to a variety of morphological changes such as blebbing, cell membrane asymmetry and attachment, cell shrinkage, nuclear fragmentation, chromatin condensation, and chromosomal DNA fragmentation. Apoptosis removes damaged and unwanted cells and maintains tissue integrity and function (Ghobrial et al. 2005). The tumor suppressor protein p53 is a critical transcription factor involved in the regulation of cell proliferation and apoptosis; as such, it is a key mediator in the prevention of carcinogenesis (Farnebo et al. 2010). It has been shown that resveratrol can cause apoptosis through p53-dependent and p53-independent pathways (Fan et al. 2008). Recently, it has been suggested that resveratrol also acetylates p53 protein, resulting in the subsequent activation of pro-apoptotic genes (Kai et al. 2010).

Utilizing both *in vitro* and *in vivo* models of tumorigenesis and carcinogenesis, resveratrol has been demonstrated to induce apoptosis *via* activation of both the intrinsic (mitochondria mediated) and extrinsic (death receptor mediated) pathways of cell death (Fan et al. 2008). Resveratrol altered the Bax:Bcl2 ratio in the A431 epidermoid carcinoma cell line, leading to mitochondrial membrane depolarization and subsequent induction of caspase-dependent apoptosis (Madan et al. 2008). Similar results were observed in breast (Singh et al. 2009) and in a DENA-initiated, phenobarbital-promoted *in vivo* model of hepato-carcinogenesis (Bishayee and Dhir 2009). Resveratrol increased expression of pro-apoptotic proteins (e.g. PUMA, Noxa, Bax, Bak, etc) and the release of mitochondria proteins (e.g. cytochrome *c*, Smac/DIABLO, etc) to the cytosol thus triggering suppression of inhibitors of apoptosis proteins and caspase-9 and caspase-3 activation (Jazirehi and Bonavida 2004; Jiang et al. 2005; Shankar et al. 2007a, b). In addition, resveratrol induced the redistribution of death receptors to membrane lipid rafts in colorectal and multiple myeloma/T cell leukemia cancer cells, thus sensitizing these cells to death receptor-mediated apoptosis (Delmas et al. 2004; Reis-Sobreiro et al. 2009). Furthermore, resveratrol sensitized TNF-related apoptosis-inducing ligand-dependent cell death in androgen-sensitive and androgen-insensitive prostate cancer cells (Shankar et al. 2007a, b).

15.4 Suppression of the Inflammatory Signaling Pathway

Chronic inflammation contributes to the development of many diseases, including cancer, arthritis, heart diseases, and Alzheimer's disease. For example, chronic colitis is associated with an increased risk of developing colorectal cancer, and the susceptibility to cancer increases when the tissue is chronically inflamed. Many tumors contain non-cancer cells, including immune and vascular cells that are important in inflammation. The critical molecular interactions between neoplastic cells and inflammatory cells remain unknown, but both pro- and anti-inflammatory mediators are likely involved. Thus, it is clear that the link between inflammation and cancer is strong and that the interplay between the immune system and the development and remission of cancer is critical. These inflammatory processes contribute to the development and progression of carcinogenesis, including tumor growth, angiogenesis, invasion, and metastasis (Wu and Zhou 2009). Nuclear factor-κB (NF-κB) signaling is a key mediator of cellular transformation induced by inflammation. Pro-inflammatory cytokines are also implicated in these processes and serve as targets of anticancer agents (Balkwill and Mantovani 2010). Resveratrol is demonstrated to inhibit the expression of various components of inflammation. In this section, we discuss the inhibitory effects of resveratrol on COX-2 (cyclooxygenase 2) and on the pro-inflammatory cytokines TNF-α (tumor necrosis factor-alpha) and the interleukins (ILs).

15.4.1 COX-2

Over-expression of COX-2 is suggested to contribute to tumor development *via* modulation of the multiple stages of tumorigenesis: cell survival signaling pathways, angiogenesis, invasion, and metastasis (Danese and Mantovani 2010). Resveratrol inhibited the expression and activity of COX-2 through direct interaction with the enzyme's catalytic domain (Calamini et al. 2010). Zykova et al. (2008) demonstrated inhibition of COX-2 activity (prostaglandin E2 synthesis), inhibition of colony formation, and suppression of anchorage-independent growth in HT29 colon cancer cells by this compound; COX-2 inhibition was absolutely required for resveratrol-mediated activity. Resveratrol also reduced COX-2 expression in DENA-initiated liver (Bishayee et al. 2010b), DMBA-initiated mammary (Banerjee et al. 2002), and TPA (12-*o*-tetradeconoylphrobol-13-acetate)-initiated skin (Kundu et al. 2006) carcinogenesis. Furthermore, using a dextran sodium sulfate (DSS)-induced mouse model of colitis, resveratrol treatment was shown to reduce COX-2 expression (Sánchez-Fidalgo et al. 2010; Singh et al. 2010); similar results were observed in a DSS-induced, azoxymethane-promoted model of colitis-driven colon cancer (Cui et al. 2010).

15.4.2 Pro-inflammatory Cytokines: TNF-α and ILs

Pro-inflammatory cytokines, such as TNF-α and ILs, are important mediators of both inflammation-related cancer and cancer-related inflammation. Expression of these cytokines are associated with carcinogenesis mediated *via* activation of the NF-κB signaling pathway (Wu and Zhou 2009, 2010). Resveratrol reversed colitis-associated expression of TNF-α, IL-6, and IL-1β induced by DSS (Cui et al. 2010; Sánchez-Fidalgo et al. 2010; Singh et al. 2010). Resveratrol suppressed IL-1β-stimulated cell proliferation, reduced IL-1β production, and IL-1β-induced NF-κB activation in acute myeloid leukemia cells (Estrov et al. 2003). Using streptozotocin-nicotinamide-induced diabetic rats, Palsamy and Subramanian (2010) reported that resveratrol ameliorates induction of pro-inflammatory cytokines (e.g. TNF-α, IL-1β, and IL-6) in pancreatic β-cells. Leiro et al. (2010) proposed the anti-inflammatory activity ascribed to resveratrol is related, in part, to the inhibition of TNF-α and IL-1β mRNA processing, which resulted in reduced migration of head and kidney leukocytes in response to inflammatory stimuli.

15.5 Inhibition of Angiogenesis, Invasion, and Metastasis

Angiogenesis, the formation of new blood vessels, is critical to tumor development and growth, invasion, and metastasis (Zetter and Bruce 1998). Neovascularization facilitates cell dissociation from the primary tumor site, entry into circulation, and establishment of the tumor(s) at a secondary site (a process termed metastasis) (John and Tuszynski 2001; Duffy et al. 2008; Kessenbrock et al. 2010). Disruption of the angiogenic and metastatic signaling cascades is another mechanism by which resveratrol exerts its chemo-therapeutic effect. In this section, we focus on HIF-1α (hypoxia inducible factor-1α), VEGF (vascular endothelial growth factor), and MMP2/MMP9 (matrix metalloproteinase 2/9), each of which is involved in the processes of angiogenesis, invasion, and metastasis.

15.5.1 HIF-1α and VEGF

Adaptation to hypoxia by tumor cells is mediated by HIF-1α, which regulates the expression of proteins, such as VEGF, associated with angiogenic and invasive signaling cascades (Koh et al. 2010). Treatment with resveratrol reduced expression of HIF-1α and VEGF in ovarian cancer cells; resveratrol-mediated activity was attributed to inactivation of P13K/Akt and protein translational regulators, enhancement of HIF-1α proteasomal degradation, and subsequent suppression of VEGF (Cao et al. 2004). These results were also observed in tongue squamous cell carcinoma and lysophophatic acid-stimulated ovarian cancer cells (Zhang et al. 2005; Park et al. 2007). VEGF expression was also reduced in multiple myeloma (Sun et al. 2006) and endometrial cancer cells (Dann et al. 2009).

15.5.2 MMP2 and MMP9

Cell invasion and metastasis are facilitated *via* the proteolytic degradation of basement membrane and connective tissue; this degradation of extracellular matrices creates a channel by which tumors escape into circulation. MMPs are essential factors in this process (John and Tuszynski 2001; Wu and Zhou 2009). Numerous reports illustrate the anti-invasive and anti-metastatic properties of resveratrol. Treatment with resveratrol inhibited heregulin-β1-mediated MMP9 expression in breast cancer cells through the inactivation of ERK1/2 signaling resulting in suppression of invasion (Tang et al. 2008a). TNF-α-mediated MMP9 expression and invasion was also reduced in resveratrol-treated hepatocellular carcinoma cells; this suppression was associated with down-regulation of NF-κB signaling (Yu et al. 2008). Inactivation of NF-κB signaling in lung adenocarcinoma cells also contributed to the anti-invasive effect of resveratrol and led to inhibition of heme oxygenase-1-induced MMP9 expression, migration, and invasion (Liu et al. 2010b). Resveratrol blocked IGF-1 (insulin growth factor-1)-mediated cell migration *in vitro* through inactivation of the PI3K/Akt signaling pathway and down-regulation of MMP2 induced by IGF-1 (Tang et al. 2008b). Moreover, resveratrol blunted TNF-α-induced monocyte adhesion and migration through modulation of adhesion molecules and MMP2 enzymatic activity (Kim et al. 2007).

15.6 Effect as a Selective Estrogen Receptor Modulator

Resveratrol, due to its structural similarity to diethylstilbestrol, is characterized as a phytoestrogen, although its role as an estrogen agonist and/or antagonist remains controversial (Corre et al. 2005). Thus, resveratrol is often cited to function as an estrogen mixed agonist/antagonist, and its effects in breast cancer appear to involve both estrogen receptor (ER)-dependent and ER-independent mechanisms. Bowers et al. (2000) observed comparable binding affinity by resveratrol with ERα and ERβ, but this binding was 7,000 fold lower than E_2 (17β-estradiol). These authors also showed specificity of resveratrol for ERβ. Resveratrol was demonstrated to behave as a super-agonist of estrogen in MCF-7 (ERα positive) and MDA-MB-231 (ERα negative) breast cancer cells stably transfected with wild-type *ERα* reporter construct and subsequent activation of hormone receptor-mediated gene transcription (Gehm et al. 1997, 2004). However, Bhat et al. (2001) observed a weak estrogenic response after resveratrol treatment in MCF-7 cells and antagonism when combined with E_2; in T47D (ER positive), resveratrol functioned solely as an antagonist. Similar results were seen in resveratrol-treated Ishikawa (ER positive) endometrial cancer cells; resveratrol treatment also led to the down-regulation of ERα expression and subsequent inhibition of cell proliferation, which may occur *via* ER-dependent and ER-independent mechanisms (Bhat and Pezzuto 2001). Furthermore, Pozo-Guisado et al. (2004) suggested that the pro-apoptotic effect of resveratrol in MCF-7 cells was dependent on ERα, its effect was not seen in MDA-MB-231 cells. Such contrasts may be explained by considering alteration of non-genotropic activities of steroid receptor complexes by resveratrol. For example, the compound is suggested to function as an agonist for the cAMP/PKA pathway, a well documented pro-apoptotic mechanism in breast cancer cells (El-Mowafy and Alkhalaf 2003). Resveratrol also displayed a biphasic effect on ERα-dependent PI3K activity associated with modulation of cell cycle regulator proteins; a low concentration of resveratrol increased ERα-dependent signaling whereas higher concentrations led to the inhibition of this signaling (Pozo-Guisado et al. 2004). Thus, resveratrol may also exert anti-estrogenic action by triggering parallel pathways that inhibit estrogen-induced cellular outcomes, such as cell proliferation, tumor transformation, and progression.

Resveratrol may therefore be considered a natural selective estrogen receptor modulator, although the balance between pro-survival genotropic and opposing non-genotropic activities is not clearly predictable due to the role of a broad array of intervening factors. In general, at the low doses provided by dietary intake, resveratrol may act as a weak estrogen competitor, dependent on receptor expression and hormonal status of tissues; it counteracts the proliferative effects of hormones and provides a balancing antitumor activity. Tissue-specific expression of α and β ERs, cofactors regulating DNA binding and different gene promoters, are important factors in deciding resveratrol's effect. In the absence of endogenous hormones and according to cellular specificity, the agonistic activity of resveratrol may act in an opposite manner and prevent tissue senescence and apoptosis. When stress signals overcome proliferative signals, or when the latter are missing (absence of hormones), the polyphenol-induced pathway may switch to apoptosis.

15.7 Other Molecular Targets

15.7.1 Topoisomerase II

Topoisomerases are ubiquitously expressed enzymes that regulate the topological state of DNA; these enzymes remove knots and tangles from the genome through the winding and unwinding of DNA, generating transient double-stranded breaks in the DNA double helix to control cell division and proliferation. In malignant tumor cells, this enzyme is over-expressed when compared to normal, quiescent cells; thus topoisomerase II has been suggested as a potential chemo-preventive target (Degrassi et al. 2004). Resveratrol is demonstrated to inhibit topoisomerase activity. Treatment of glioblastoma cells with the compound interfered with decatenation (i.e. unwinding of DNA) by topoisomerase; these authors suggest resveratrol acted as a "topoisomerase poison" due to the prolongation of the topoisomerase-DNA complex (Leone et al. 2010).

15.7.2 Human Telomerase Reverse Transcriptase (hTERT)

Addition of hexametric repetitive sequences (referred to as telomeres) to chromosomal ends by telomerase, ensures and maintains genomic stability. With each cell division, telomeres progressively shorten until a critical length is reached and the cell undergoes senescence or apoptosis. In most cancers, telomerase is highly activated resulting in tumor cell immortalization due to the maintenance of telomeres. This immortalization is associated with increased expression of the telomerase catalytic subunit, hTERT (Artandi and DePinho 2010). Therefore, targeting telomerase and its catalytic subunit hTERT provides an additional mechanism by which anticancer agents exert chemo-preventive/chemo-therapeutic effects. Resveratrol has been shown to suppress telomerase activity in breast (Lanzilli et al. 2006) and colon (Fuggetta et al. 2006) cancer cells associated with decreased nuclear levels of hTERT. Furthermore, resveratrol treatment of osteosarcoma and lung cancer cells induced telomeric instability, phosphorylation of histone H2AX and p53, and activation of DNA damage signaling (Rusin et al. 2009).

15.8 Identification of Novel Molecular Targets

To identify novel molecular targets that mediate the anticancer activities of resveratrol, our laboratory performed microarray analysis using HCT-116 human colorectal cancer cells treated with resveratrol (50 μM) for 24 h (Table 15.3). Among

Table 15.3 Alteration of gene expression profiles in HCT-116 human colorectal cancer cells by resveratrol

Gene symbol	Gene name	GenBank no	Up/down
DDIT3	DNA-damage-inducible transcript 3	NM_004083	Up
ATF3	Activating transcription factor 3	NM_004024	Up
COBRA1	Cofactor of BRCA1	NM_015456	Up
OR8B8	Olfactory receptor, family 8, subfamily B, member 8	NM_012378	Up
AADACL1	Arylacetamide deacetylase-like 1	NM_020792	Up
PLK3	Polo-like kinase 3 (Drosophila)	NM_004073	Up
PCK2	Phophoenolpyruvate carboxykinase 2 (mitochondrial)	NM_004563	Up
NAG-1	NSAID-activated gene-1	NM_004864	Up
WNT16	Wingless-type MMTV integration site family, member 16	NM_016087	Down
C4orf18	Chromosome 4 open reading frame 18	NM_016613	Down
FANCC	Fanconi anemia, complementation group C	NM_000136	Down
PDE4B	Phosphodiesterase 4B	NM_001037	Down
IGFBP7	Insulin-like growth factor binding protein 7	NM_001553	Down

HCT-116 colorectal cancer cells were treated with DMSO or resveratrol (50 μM) for 24 h. Total RNA was amplified using Agilent Low RNA Input Fluorescent Linear Amplification Kit protocol (Santa Clara, CA). Starting with 500 ng of total RNA, Cy3-, or Cy5-labeled cRNA was produced according to manufacturer's protocol. For each two-color comparison, 750 ng of each Cy3- and Cy5-labeled cRNA was mixed and fragmented using Agilent *in situ* Hybridization Kit protocol. Hybridizations were performed for 16 h in a rotating hybridization oven using the Agilent 60-mer oligo microarray processing protocol. Data was obtained using Agilent Feature Extraction Software (v7.5) using defaults for all parameters. Intensity plots were rated for each ratio experiment, and genes were considered "signature genes" if the *P* value was less than 0.001. Functional annotation of genes was performed according to the Gene Ontology Consortium (http://www.geneontology.org/index/html) by Gene Spring (v7.3) the 519 genes altered by resveratrol, three genes, *ATF3* (activating transcription factor 3), *NAG-1* (non-steroidal anti-inflammatory drug-activated gene 1), and *DDIT3* (DNA-damage-inducible transcript 3) were of particular interest due to their suggested involvement as negative regulators of tumorigenesis and carcinogenesis.

15.8.1 ATF3 (Activating Transcription Factor 3)

ATF3, a member of the ATF/CREB (cAMP response element binding) family of bZIP transcription factors, is characterized as a stress inducible and/or adaptive response gene (Lu et al. 2007). Much controversy exists as to the physiological role ATF3 has in tumorigenesis and ATF3 is demonstrated to be a positive or negative effecter in tumor progression. Recently, a dichotomous role was reported for ATF3 in cancer development; the authors concluded its role as a tumor suppressor or oncogene is largely dependent on cellular context and extent of malignancy (Yin et al. 2007). However, several lines of evidence suggest ATF3 may function as a tumor suppressor. First, ATF3 expression is markedly reduced in cancer tissues when compared to normal adjacent tissue (Yan and Boyd 2006). Secondly, ATF3 over-expression is demonstrated to elicit a number of cellular responses, including induction of cell cycle arrest and inhibition of proliferation (Fan et al. 2002), induction of apoptosis *in vitro* and *in vivo* (Lu et al. 2006; Yamaguchi et al. 2006; Huang et al. 2008; Turchi et al. 2008), inhibition of invasion (Yan et al. 2002; Stearns et al. 2004; Bottone et al. 2005), and retardation of tumor formation *in vivo* (Bottone et al. 2005; Lu et al. 2006). Finally, ATF3 is reported to mediate or enhance induction of apoptosis by compounds demonstrated to possess antitumor properties (Mashima et al. 2001; Lee et al. 2005; Yan et al. 2005; Yamaguchi et al. 2006; Wang et al. 2009).

Recently, we performed microarray analysis to identify novel molecular targets of resveratrol and identified *ATF3* as a gene up-regulated by the treatment (Table 15.3). Indeed, resveratrol increased ATF3 at both the mRNA and protein level. We found that Egr-1 (early growth response-1) and KLF4 (Krüppel-like factor 4) mediate *ATF3* trans-activation by resveratrol. We also found Egr-1 and KLF4 interaction by resveratrol, which may facilitate *ATF3* transcriptional regulation. Moreover, resveratrol-induced apoptosis is mediated, in part, by ATF3 (Whitlock et al. 2011).

15.8.2 NAG-1 (Non-steroidal Anti-inflammatory Drug-activated Gene-1)

NAG-1, a divergent member of the transforming growth factor β superfamily of cytokines, is suggested to play a role in a number of cellular processes, including inflammation and cell survival. NAG-1 was first identified by Baek et al. (2001) through PCR-based subtractive hybridization from an non-steroidal anti-inflammatory drug (NSAID)-induced library in human colorectal cancer cells, NAG-1 was shown to be highly expressed in mature intestinal epithelial cells and significantly reduced in human colorectal carcinomas and neoplastic intestinal polyps of *Min* mice (Kim et al. 2002). NAG-1 over-expression reduced cell viability of breast cancer cells by up to 80% (Li et al. 2000) and treatment of prostate cancer cells with purified NAG-1 induced apoptosis (Liu et al. 2003). In addition, NAG-1 transgenic mice are resistant to the development of intestinal tumors following treatment with azoxymethane or by introduction of a mutant *APC* gene (Baek et al. 2006). Furthermore, subsequent studies have identified NAG-1 induction by a number of dietary and naturally occurring compounds demonstrated to possess anticancer activity. These include genistein (Wilson et al. 2003), epicatechin gallate (Baek et al. 2004), indole-3-carbinol (Lee et al. 2005), conjugated linoleic acid (Lee et al. 2006), berberine (Piyanuch et al. 2007; Auyeung and Ko 2009), 6-gingerol (Lee et al. 2008), platycodon D (Shin et al. 2009), capsaicin (Lee et al. 2010), *Astragalus saponins* (Auyeung et al. 2009), and the isoflavonoid formononetin (Auyeung and Ko 2010). Increased NAG-1 expression contributed to the inhibition of cell proliferation and induction of apoptosis by these compounds. Additionally, we and others have shown NAG-1 up-regulation by resveratrol resulted in increased cell death in colorectal, lung, and ostersarcoma cancer cells (Baek et al. 2002) and inhibition of cell proliferation in pancreatic cancer cells (Golkar et al. 2007).

15.8.3 DDIT3 (DNA-Damage-Inducible Transcript 3)

DDIT3 is a critical mediator in the induction of cell cycle arrest and/or apoptosis in response to ER stress. Transcriptional activation by ER stress results in downstream regulation of genes involved in the modulation of cell survival. For example, DDIT3 expression inhibits Bcl2 and leads to Bax translocation to the mitochondria resulting in activation of the intrinsic pathway of apoptosis (Oyadomari and Mori 2003). In several studies, resveratrol is demonstrated to increase DDIT3 expression in colon (Woo et al. 2007; Um et al. 2010) and leukemia (Liu et al. 2010a) cancer cells. Increased DDIT3 expression contributed to induction of cell cycle arrest and/or induction of apoptosis by resveratrol.

15.8.4 Other Potential Targets

Although gene alteration by resveratrol has yet to be validated, each of the aforementioned genes (refer to Table 15.3) is suggested to act as either a negative or positive regulator of tumorigenesis. COBRA1, an integral component of negative elongation factor complex, is demonstrated to inhibit estrogen-dependent transcription and growth of breast cancer cells *in vitro*. Furthermore, loss of COBRA1 is associated with breast cancer progression and metastasis (Sun et al. 2008). Plk3, whose expression is rapidly increased in response to DNA damage and cellular stress, is believed to behave as a tumor suppressor gene and serve as a "functional link" in p53-induced cell cycle arrest and apoptosis (van de Weerdt and Medema 2006). Thus, induction of COBRA1 or Plk3 may play a role in resveratrol-mediated inhibition of proliferation or induction of apoptosis in colorectal and other cancers. On the other hand, down-regulation of WNT16, FANCC, or PDE4B by resveratrol may lead to enhanced inhibition of cell growth or sensitization to chemotherapy. For example, over-expression of WNT16 is associated with oncogene-induced senescence *via* regulation of p53 activity (Binet et al. 2009) and enhanced cell proliferation and clonogenicity of basal cell carcinomas (Teh et al. 2007). Disruption of FANCC, which results in deactivation of the Fanconi anemia pathway, sensitizes cancer cells to DNA damage and growth inhibition induced by DNA interstrand-cross-linking agents thereby abrogating chemo-resistance (Gallmeier et al. 2006; Kachnic et al. 2010; Palagyi et al. 2010). Moreover, increased PDE4B expression is associated with inactivation of cAMP and abrogation of growth inhibitory effects of cAMP (Smith et al. 2005; Narita et al. 2007).

15.9 Conclusion

Identification of dietary phytochemicals and their related derivatives with chemopreventive and/or chemo-therapeutic activities offers an alternate and complementary approach to the prevention and treatment of cancers. Studies investigating the use of these compounds for cancer prevention or as adjuvants to traditional treatment have revealed several potential benefits: (1) suppression of tumorigenesis and carcinogenesis *in vitro* and *in vivo*, (2) sensitization of tumor/cancer cells to drug-induced growth inhibition, and (3) minimize adverse side effects associated with conventional therapies (Garg et al. 2005). Usage of nutritional supplements by cancer patients is on the rise as the result of dietary phytochemicals' ability to alleviate side effects and toxicities of traditional drugs (Dennis et al. 2009). Moreover, dietary phytochemicals are demonstrated to target multiple signal transduction pathways involved in tumorigenesis and carcinogenesis, an important advantage due to the inherent heterogeneity of cancers.

It is not surprising that the number of studies involving these dietary compounds has dramatically increased in the past decade; this is especially true for the dietary phytoalexin resveratrol. As discussed in previous sections, resveratrol is demonstrated to modulate the expression and/or activity of proteins involved in critical pathways of carcinogenesis including carcinogen activation and detoxification, inflammation, angiogenesis, invasion, and metastasis. Resveratrol is demonstrated to participate in both pro-survival and pro-death cellular mechanisms, either by favoring the preservation of the functional status of cells and possibly elongating cellular life span or inducing death of those cells whose physiological conditions have become deranged. Resveratrol also affects the transcriptional machinery resulting in trans-activation of key regulatory proteins. Although numerous molecular targets have been identified, the underlying mechanisms involved in the antitumor/anticancer activities of resveratrol are not completely understood. As a result, researchers continue to investigate the molecular and cellular effects of resveratrol in cancer in hopes of unraveling the mysteries of this fascinating and promising dietary molecule.

References

Androutsopoulos, V., Tsatsakis, A., & Spandidos, D. (2009). Cytochrome P450 CYP1A1: Wider roles in cancer progression and prevention. *BMC Cancer, 9,* 187.

Artandi, S. E., & DePinho, R. A. (2010). Telomeres and telomerase in cancer. *Carcinogenesis, 31,* 9–18.

Athar, M., Back, J. H., Tang, X., et al. (2007). Resveratrol: A review of preclinical studies for human cancer prevention. *Toxicology and applied pharmacology, 224*(3), 274–283.

Auyeung, K., & Ko, J. (2009). Coptis chinensis inhibits hepatocellular carcinoma cell growth through nonsteroidal anti-inflammatory drug-activated gene activation. *International Journal of Molecular Medicine, 24,* 571–577.

Auyeung, K., & Ko, J. (2010). Novel herbal flavonoids promote apoptosis but differentially induce cell cycle arrest in human colon cancer cell. *Investigational New Drugs, 28,* 1–13.

Auyeung, K. K., Cho, C. H., & Ko, J. K. (2009). A novel anticancer effect of Astragalus saponins: Transcriptional activation of NSAID-activated gene. *International Journal of Cancer, 125,* 1082–1091.

Baek, S. J., Kim, K. S., Nixon, J. B., et al. (2001). Cyclooxygenase inhibitors regulate the expression of a TGF-β superfamily member that has proapoptotic and antitumorigenic activities. *Molecular Pharmacology, 59,* 901–908.

Baek, S. J., & Wilson, L. C., & Eling, T. E. (2002). Resveratrol enhances the expression of nonsteroidal anti-inflammatory drug-activated gene (NAG-1) by increasing the expression of p53. *Carcinogenesis, 23,* 425–432.

Baek, S. J., Kim, J. S., Jackson, F. R., et al. (2004). Epicatechin gallate-induced expression of NAG-1 is associated with growth inhibition and apoptosis in colon cancer cells. *Carcinogenesis, 25,* 2425–2432.

Baek, S. J., Okazaki, R., Lee, S. -H., et al. (2006). Nonsteroidal anti-inflammatory drug-activated gene-1 over expression in transgenic mice suppresses intestinal neoplasia. *Gastroenterology, 131,* 1553–1560.

Bai, Y., Mao, Q. Q., Qin, J., et al. (2010). Resveratrol induces apoptosis and cell cycle arrest of human T24 bladder cancer cells in vitro and inhibits tumor growth in vivo. *Cancer Science, 101,* 488–493.

Balkwill, F., & Mantovani, A. (2010). Cancer and inflammation: Implications for pharmacology and therapeutics. *Clinical Pharmacology and Therapeutics, 87,* 401–406.

Banerjee, S., Bueso-Ramos, C., & Aggarwal, B. B. (2002). Suppression of 7,12-dimethylbenz(a) anthracene-induced mammary carcinogenesis in rats by resveratrol. *Cancer Research, 62,* 4945–4954.

Baur, J. A., & Sinclair, D. A. (2006). Therapeutic potential of resveratrol: The in vivo evidence. *Nature Reviews Drug Discovery, 5,* 493–506.

Beedanagari, S. R., Bebenek, I., Bui, P., et al. (2009). Resveratrol inhibits dioxin-induced expression of human CYP1A1 and CYP1B1 by inhibiting recruitment of the aryl hydrocarbon receptor complex and RNA polymerase II to the regulatory regions of the corresponding genes. *Toxicological Sciences, 110,* 61–67.

Benitez, D. A., Pozo-Guisado, E., Alvarez-Barrientos, A., et al. (2007). Mechanisms involved in resveratrol-induced apoptosis and cell cycle arrest in prostate cancer-derived cell lines. *Journal of Andrology, 28,* 282–293.

Berge, G., Ovrebo, S., Botnen, I. V., et al. (2004). Resveratrol inhibits benzo[a]pyrene-DNA adduct formation in human bronchial epithelial cells. *British Journal of Cancer, 91,* 333–338.

Bhardwaj, A., Sethi, G., Vadhan-Raj, S., et al. (2007). Resveratrol inhibits proliferation, induces apoptosis, overcomes chemoresistance through down-regulation of STAT3 and nuclear factor-κB-regulated antiapoptotic and cell survival gene products in human multiple myeloma cells. *Blood, 109,* 2293–2302.

Bhat, K. P. L., & Pezzuto, J. M. (2001). Resveratrol exhibits cytostatic and antiestrogenic properties with human endometrial adenocarcinoma (Ishikawa) cells. *Cancer Research, 61,* 6137–6144.

Bhat, K. P. L., Lantvit, D., Christov, K., et al. (2001). Estrogenic and antiestrogenic properties of resveratrol in mammary tumor models. *Cancer Research, 61,* 7456–7463.

Binet, R., Ythier, D., Robles, A. I., et al. (2009). WNT16B is a new marker of cellular senescence that regulates p53 activity and the phosphoinositide 3-kinase/Akt pathway. *Cancer Research, 69,* 9183–9191.

Bishayee, A. (2009). Cancer prevention and treatment with resveratrol: From rodent studies to clinical trials. *Cancer Prevention Research, 2,* 409–418.

Bishayee, A., & Dhir, N. (2009). Resveratrol-mediated chemoprevention of diethylnitrosamine-initiated hepatocarcinogenesis: Inhibition of cell proliferation and induction of apoptosis. *Chemico Biological Interactions, 179,* 131–144.

Bishayee, A., Barnes, K. F., Bhatia, D., et al. (2010a). Resveratrol suppresses oxidative stress and inflammatory response in diethylnitrosamine-initiated rat hepatocarcinogenesis. *Cancer Prevention Research, 3,* 753–763.

Bishayee, A., Waghray, A., Barnes, K., et al. (2010b). Suppression of the inflammatory cascade is implicated in resveratrol chemoprevention of experimental hepatocarcinogenesis. *Pharmaceutical Research, 27,* 1080–1091.

Boocock, D. J., Faust, G. E. S., Patel, K. R., et al. (2007). Phase I dose escalation pharmacokinetic study in healthy volunteers of resveratrol, a potential cancer chemopreventive agent. *Cancer Epidemiology, Biomarkers & Prevention, 16,* 1246–1252.

Bottone, F. G., Moon, Y., Kim, J. S., et al. (2005). The anti-invasive activity of cyclooxygenase inhibitors is regulated by the transcription factor ATF3 (activating transcription factor 3). *Molecular Cancer Therapeutics, 4,* 693–703.

Bowers, J. L., Tyulmenkov, V. V., Jernigan, S. C., et al. (2000). Resveratrol Acts as a Mixed Agonist/Antagonist for Estrogen Receptors α and β. *Endocrinology, 141,* 3657–3667.

Brown, J. M., & Attardi, L. D. (2005). The role of apoptosis in cancer development and treatment response. *Nature Reviews Cancer, 5,* 231–237.

Calamini, B., Ratia, K., Malkowski, M. G., et al. (2010). Pleiotropic mechanisms facilitated by resveratrol and its metabolites. *The Biochemical Journal, 429,* 273–282.

Cao, Z., Fang, J., Xia, C., et al. (2004). Trans-3,4,5′-trihydroxystibene inhibits hypoxia-inducible factor 1α and vascular endothelial growth factor expression in human ovarian cancer cells. *Clinical Cancer Research, 10,* 5253–5263.

Castello, L., & Tessitore, L. (2005). Resveratrol inhibits cell cycle progression in U937 cells. *Oncology Reports, 13,* 133–137.

Chan, W. K., & Delucchi, A. B. (2000). Resveratrol, a red wine constituent, is a mechanism-based inactivator of cytochrome P450 3A4. *Life Sciences, 67,* 3103–3112.

Corre, L. L., Chalabi, N., Delort, L., et al. (2005). Resveratrol and breast cancer chemoprevention: Molecular mechanisms. *Molecular Nutrition & Food Research, 49,* 462–471.

Cui, X., Jin, Y., & Hofseth, A. B., et al. (2010). Resveratrol suppresses colitis and colon cancer associated with colitis. *Cancer Prevention Research, 3,* 549–559.

Danese, S., & Mantovani, A. (2010). Inflammatory bowel disease and intestinal cancer: A paradigm of the Yin-Yang interplay between inflammation and cancer. *Oncogene, 29,* 3313–3323.

Dann, J. M., Sykes, P. H., Mason, D. R., et al. (2009). Regulation of vascular endothelial growth factor in endometrial tumour cells by resveratrol and EGCG. *Gynecologic Oncology, 113,* 374–378.

Degrassi, F., Fiore, M., & Palitti, F. (2004). Chromosomal aberrations and genomic instability induced by topoisomerase- targeted antitumour drugs. *Current Medicinal Chemistry Anti Cancer, 4,* 317–325.

Delmas, D., Rébé, C., Micheau, O., et al. (2004). Redistribution of CD95, DR4 and DR5 in rafts accounts for the synergistic toxicity of resveratrol and death receptor ligands in colon carcinoma cells. *Oncogene, 23,* 8979–8986.

Dennis, T., Fanous, M., & Mousa, S. (2009). Natural products for chemopreventive and adjunctive therapy in oncologic disease. *Nutrition and Cancer, 61,* 587–597.

Du, L., Neis, M. M., Ladd, P. A., et al. (2006). Differentiation-specific factors modulate epidermal CYP1–4 gene expression in human skin in response to retinoic acid and classic aryl hydrocarbon receptor ligands. *The Journal of Pharmacology and Experimental Therapeutics, 319,* 1162–1171.

Duffy, M., McGowan, P., & Gallagher, W. (2008). Cancer invasion and metastasis: Changing views. *The Journal of Pathology, 214,* 283–293.

El-Mowafy, A. M., & Alkhalaf, M. (2003). Resveratrol activates adenylyl-cyclase in human breast cancer cells: A novel, estrogen receptor-independent cytostatic mechanism. *Carcinogenesis, 24,* 869–873.

Estrov, Z., Shishodia, S., Faderl, S., et al. (2003). Resveratrol blocks interleukin-1β-induced activation of the nuclear transcription factor NF-κB, inhibits proliferation, causes S-phase arrest, induces apoptosis of acute myeloid leukemia cells. *Blood, 102,* 987–995.

Fan, E., Jiang, S., Zhang, L., et al. (2008). Molecular mechanism of apoptosis induction by resveratrol, a natural cancer chemopreventive agent. *International Journal for Vitamin and Nutrition Research, 78,* 3–8.

Fan, F., Jin, S., Amundson, S. A., et al. (2002). ATF3 induction following DNA damage is regulated by distinct signaling pathways and over-expression of ATF3 protein suppresses cells growth. *Oncogene, 21,* 7488–7496.

Farnebo, M., Bykov, V. J. N., & Wiman, K. G. (2010). The p53 tumor suppressor: A master regulator of diverse cellular processes and therapeutic target in cancer. *Biochemical and Biophysical Research Communications, 396,* 85–89.

Fuggetta, M. P., Lanzilli, G., Tricarico, M., et al. (2006). Effects of resveratrol on proliferation and telomerase activity of human colon cancer cells in vitro. *Journal of Experimental & Clinical Cancer Research, 25,* 189–193.

Gallmeier, E., Calhoun, E. S., Rago, C., et al. (2006). Targeted disruption of FANCC and FANCG in human cancer provides a preclinical model for specific therapeutic options. *Gastroenterology, 130,* 2145–2154.

Garg, A. K., Buchholz, T. A., & Aggarwal, B. B. (2005). Chemosensitization and radiosensitization of tumors by plant polyphenols. *Antioxidants & Redox Signaling, 7,* 1630–1647.

Gatouillat, G., Balasse, E., Joseph-Pietras, D., et al. (2010). Resveratrol induces cell-cycle disruption and apoptosis in chemoresistant B16 melanoma. *Journal of Cellular Biochemistry, 110,* 893–902.

Gehm, B. D., McAndrews, J. M., Chien, P. Y., et al. (1997). Resveratrol, a polyphenolic compound found in grapes and wine, is an agonist for the estrogen receptor. *Proceedings of the National Academy of Sciences of the United States of America, 94,* 14138–14143.

Gehm, B. D., Levenson, A. S., Liu, H., et al. (2004). Estrogenic effects of resveratrol in breast cancer cells expressing mutant and wild-type estrogen receptors: Role of AF-1 and AF-2. *The Journal of Steroid Biochemistry and Molecular Biology, 88,* 223–234.

Ghobrial, I. M., Witzig, T. E., & Adjei, A. A. (2005). Targeting apoptosis pathways in cancer therapy. *CA: A Cancer Journal for Clinicians, 55,* 178–194.

Golkar, L., Ding, X. Z., Ujiki, M. B., et al. (2007). Resveratrol inhibits pancreatic cancer cell proliferation through transcriptional induction of macrophage inhibitory cytokine-1. *The Journal of Surgical Research, 138,* 163–169.

Harikumar, K. B., & Aggarwal, B. (2008). Resveratrol: A multitargeted agent for age-associated chronic diseases. *Cell Cycle, 7,* 1020–1035.

Harikumar, K. B., Kunnumakkara, A. B., Sethi, G., et al. (2010). Resveratrol, a multitargeted agent, can enhance antitumor activity of gemcitabine in vitro and in orthotopic mouse model of human pancreatic cancer. *International Journal of Cancer, 127,* 257–268.

Heiss, E. H., Schilder, Y. D. C., & Dirsch, V. M. (2007). Chronic treatment with resveratrol induces redox stress- and ataxia telangiectasia-mutated (ATM)-dependent senescence in p53-positive cancer cells. *The Journal of Biological Chemistry, 282,* 26759–26766.

Hsieh, T., & Wu, J. (2008). Suppression of cell proliferation and gene expression by combinatorial synergy of EGCG, resveratrol and gamma-tocotrienol in estrogen receptor-positive MCF-7 breast cancer cells. *International Journal of Cancer, 33,* 851–859.

Huang, X., Li, X., & Guo, B. (2008). KLF6 Induces Apoptosis in Prostate Cancer Cells through Up-regulation of ATF3. *The Journal of Biological Chemistry, 283,* 29795–29801.

Hudson, T. S., Hartle, D. K., Hursting, S. D., et al. (2007). Inhibition of prostate cancer growth by muscadine grape skin extract and resveratrol through distinct mechanisms. *Cancer Research, 67,* 8396–8405.

Jana, S., & Mandlekar, S. (2009). Role of Phase II drug metabolizing enzymes in cancer chemoprevention. *Current Drug Metabolism, 10,* 595–616.

Jang, M., Cai, L., Udeani, G. O., et al. (1997). Cancer chemopreventive activity of resveratrol, a natural product derived from grapes. *Science, 275,* 218–220.

Jazirehi, A. R., & Bonavida, B. (2004). Resveratrol modifies the expression of apoptotic regulatory proteins and sensitizes non-Hodgkin's lymphoma and multiple myeloma cell lines to paclitaxel-induced apoptosis. *Molecular Cancer Therapeutics, 3,* 71–84.

Jiang, H., Zhang, L., Kuo, J., et al. (2005). Resveratrol-induced apoptotic death in human U251 glioma cells. *Molecular Cancer Therapeutics, 4,* 554–561.

Joe, A. K., Liu, H., Suzui, M., et al. (2002). Resveratrol induces growth inhibition, s-phase arrest, apoptosis, changes in biomarker expression in several human cancer cell lines. *Clinical Cancer Research, 8,* 893–903.

John, A., & Tuszynski, G. (2001). The role of matrix metalloproteinases in tumor angiogenesis and tumor metastasis. *Pathology Oncology Research, 7,* 14–23.

Kachnic, L. A., Li, L., Fournier, L., et al. (2010). Fanconi Anemia pathway heterogeneity revealed by cisplatin and oxaliplatin treatments. *Cancer Letters, 292,* 73–79.

Kai, L., Samuel, S. K., & Levenson, A. S. (2010). Resveratrol enhances p53 acetylation and apoptosis in prostate cancer by inhibiting MTA1/NuRD complex. *International Journal of Cancer, 126,* 1538–1548.

Kensler, T. W., & Wakabayashi, N. (2010). Nrf2: Friend or foe for chemoprevention? *Carcinogenesis, 31,* 90–99.

Kessenbrock, K., Plaks, V., & Werb, Z. (2010). Matrix metalloproteinases: Regulators of the tumor microenvironment. *Cell, 141,* 52–67.

Kim, A. L., Zhu, Y., Zhu, H., et al. (2006). Resveratrol inhibits proliferation of human epidermoid carcinoma A431 cells by modulating MEK1 and AP-1 signalling pathways. *Experimental Dermatology, 15,* 538–546.

Kim, D. S., Kwon, H. M., Choi, J. S., et al. (2007). Resveratrol blunts tumor necrosis factor-α-induced monocyte adhesion and transmigration. *Nutrition Research and Practice, 1,* 285–290.

Kim, K. -S., Baek, S. J., & Flake, G. P., et al. (2002). Expression and regulation of nonsteroidal anti-inflammatory drug-activated gene (NAG-1) in human and mouse tissue. *Gastroenterology, 122,* 1388–1398.

Kode, A., Rajendrasozhan, S., Caito, S., et al. (2008). Resveratrol induces glutathione synthesis by activation of Nrf2 and protects against cigarette smoke-mediated oxidative stress in human lung epithelial cells. *American Journal of Physiology Lung Cellular and Molecular Physiology, 294,* L478–L488.

Koh, M., Spivak-Kroizman, T., & Powis, G. (2010). HIF-1alpha and cancer therapy. *Recent Results in Cancer Research, 180,* 15–34.

Kowalczyk, M. C., Walaszek, Z., Kowalczyk, P., et al. (2009). Differential effects of several phytochemicals and their derivatives on murine keratinocytes in vitro and in vivo: Implications for skin cancer prevention. *Carcinogenesis, 30,* 1008–1015.

Kubota, T., Uemura, Y., Kobayashi, M., et al. (2003). Combined effects of resveratrol and paclitaxel on lung cancer cells. *Anticancer Research, 23,* 4039–4046.

Kundu, J. K., & Surh, Y. J. (2008). Cancer chemopreventive and therapeutic potential of resveratrol: Mechanistic perspectives. *Cancer Letters, 269,* 243–261.

Kundu, J. K., Shin, Y. K., Kim, S. H., et al. (2006). Resveratrol inhibits phorbol ester-induced expression of COX-2 and activation of NF-κB in mouse skin by blocking IκB kinase activity. *Carcinogenesis, 27,* 1465–1474.

Kuo, P. L., Chiang, L. C., & Lin, C. C. (2002). Resveratrol-induced apoptosis is mediated by p53-dependent pathway in Hep G2 cells. *Life Sciences, 72,* 23–34.

Lanzilli, G., Fuggetta, M. P, Tricarico, M., et al. (2006). Resveratrol down-regulates the growth and telomerase activity of breast cancer cells in vitro. *International Journal of Oncology, 28,* 641–648.

Lee, J. E., & Safe, S. (2001). Involvement of a post-transcriptional mechanism in the inhibition of CYP1A1 expression by resveratrol in breast cancer cells. *Biochemical Pharmacology, 62,* 1113–1124.

Lee, S. H., Kim, J. S., Yamaguchi, K., et al. (2005). Indole-3-carbinol and 3,3′-diindolylmethane induce expression of NAG-1 in a p53-independent manner. *Biochemical and Biophysical Research Communications, 328,* 63–69.

Lee, S. H., Yamaguchi, K., Kim, J. S., et al. (2006). Conjugated linoleic acid stimulates an anti-tumorigenic protein NAG-1 in an isomer specific manner. *Carcinogenesis, 27,* 972–981.

Lee, S. H., Cekanova, M., & Baek, S. J. (2008). Multiple mechanisms are involved in 6-gingerol-induced cell growth arrest and apoptosis in human colorectal cancer cells. *Molecular Carcinogenesis, 47,* 197–208.

Lee, S. H., Krisanapun, C., & Baek, S. J. (2010). NSAID-activated gene-1 as a molecular target for capsaicin-induced apoptosis through a novel molecular mechanism involving GSK3β, C/EBPβ and ATF3. *Carcinogenesis, 31,* 719–728.

Leiro, J. M., Varela, M., Piazzon, M. C., et al. (2010). The anti-inflammatory activity of the polyphenol resveratrol may be partially related to inhibition of tumour necrosis factor-α (TNF-α) pre-mRNA splicing. *Molecular Immunology, 47,* 1114–1120.

Leone, S., Cornetta, T., Basso, E., et al. (2010). Resveratrol induces DNA double-strand breaks through human topoisomerase II interaction. *Cancer Letters, 295,* 167–172.

Leung, H. Y., Yung, L. H., Shi, G., et al. (2009). The red wine polyphenol resveratrol reduces polycyclic aromatic hydrocarbon-induced DNA damage in MCF-10A cells. *The British Journal of Nutrition, 102,* 1462–1468.

Li, P., Wong, J., Ayed, A., et al. (2000). Placental transforming growth factor-beta is a downstream mediator of the growth arrest and apoptotic response of tumor cells to DNA damage and p53 overexpression. *The Journal of Biological Chemistry, 275,* 20127–20135.

Liu, B. Q., Gao, Y. Y., Niu, X. F., et al. (2010a). Implication of unfolded protein response in resveratrol-induced inhibition of K562 cell proliferation. *Biochemical and Biophysical Research Communications, 391,* 778–782.

Liu, P. L., Tsai, J. R., Charles, A. L., et al. (2010b). Resveratrol inhibits human lung adenocarcinoma cell metastasis by suppressing heme oxygenase 1-mediated nuclear factor-kappaB pathway and subsequently downregulating expression of matrix metalloproteinases. *Molecular Nutrition & Food Research, 54,* S196–S204.

Liu, T., Bauskin, A. R., Zaunders, J., et al. (2003). Macrophage inhibitory cytokine 1 reduces cell adhesion and induces apoptosis in prostate cancer cells. *Cancer Research, 63,* 5034–5040.

Lu, D., Wolfgang, C., & Hai, T. (2006). Activating transcription factor 3, a stress-inducible gene, suppresses Ras-stimulated tumorigenesis. *The Journal of Biological Chemistry, 281,* 10473–10481.

Lu, D., Chen, J., & Hai, T. (2007). The regulation of ATF3 gene expression by mitogen-activated protein kinases. *The Biochemical Journal, 401,* 559–567.

Lu, F., Zahid, M., Wang, C., et al. (2008). Resveratrol prevents estrogen-DNA adduct formation and neoplastic transformation in MCF-10F cells. *Cancer Prevention Research, 1,* 135–145.

Madan, E., Prasad, S., Roy, P., et al. (2008). Regulation of apoptosis by resveratrol through JAK/STAT and mitochondria mediated pathway in human epidermoid carcinoma A431 cells. *Biochemical and Biophysical Research Communications, 377,* 1232–1237.

Mashima, T., Udagawa, S., & Tsuruo, T. (2001). Involvement of transcriptional repressor ATF3 in acceleration of caspase protease activation during DNA damaging agent-induced apoptosis. *Journal of Cellular Physiology, 188,* 352–358.

Meeran, S. M., & Katiyar, S. K. (2008). Cell cycle control as a basis for cancer chemoprevention through dietary agents. *Frontiers in Bioscience, 13,* 2192–2202.

Mikstacka, R., Baer-Dubowska, W., Wieczorek, M., et al. (2008). Thiomethylstilbenes as inhibitors of CYP1A1, CYP1A2 and CYP1B1 activities. *Molecular Nutrition & Food Research, 52,* S77–S83.

Narita, M., Murata, T., Shimizu, K., et al. (2007). A role for cyclic nucleotide phosphodiesterase 4 in regulation of the growth of human malignant melanoma cells. *Oncology Reports, 17,* 1133–1139.

Nebert, D. W., & Dalton, T. P. (2006). The role of cytochrome P450 enzymes in endogenous signalling pathways and environmental carcinogenesis. *Nature Reviews Cancer, 6,* 947–960.

Opipari, A. W. Jr., Tan, L., Boitano, A. E., et al. (2004). Resveratrol-induced autophagocytosis in ovarian cancer cells. *Cancer Research, 64,* 696–703.

Oyadomari, S., & Mori, M. (2003). Roles of CHOP//GADD153 in endoplasmic reticulum stress. *Cell Death and Differentiation, 11,* 381–389.

Palagyi, A., Neveling, K., Plinninger, U., et al. (2010). Genetic inactivation of the Fanconi anemia gene FANCC identified in the hepatocellular carcinoma cell line HuH-7 confers sensitivity towards DNA-interstrand crosslinking agents. *Molecular Cancer, 9,* 127.

Palsamy, P., & Subramanian, S. (2010). Ameliorative potential of resveratrol on proinflammatory cytokines, hyperglycemia mediated oxidative stress, pancreatic beta-cell dysfunction in streptozotocin-nicotinamide-induced diabetic rats. *Journal of Cellular Physiology, 224,* 423–432.

Parekh, P., Motiwale, L., Naik, N., et al. (2010). Downregulation of cyclin D1 is associated with decreased levels of p38 MAP kinases, Akt/PKB and Pak1 during chemopreventive effects of resveratrol in liver cancer cells. *Experimental and Toxicologic Pathology, 63,* 167–173.

Park, S. Y., Jeong, K. J., Lee, J., et al. (2007). Hypoxia enhances LPA-induced HIF-1α and VEGF expression: Their inhibition by resveratrol. *Cancer Letters, 258,* 63–69.

Peng, T. -L., Chen, J., Mao, W., et al. (2009). Aryl hydrocarbon receptor pathway activation enhances gastric cancer cell invasiveness likely through a c-Jun-dependent induction of matrix metalloproteinase-9. *BMC Cell Biology, 10,* 27.

Piyanuch, R., Sukhthankar, M., Wandee, G., et al. (2007). Berberine, a natural isoquinoline alkaloid, induces NAG-1 and ATF3 expression in human colorectal cancer cells. *Cancer Letters, 258,* 230–240.

Pozo-Guisado, E., Alvarez-Barrientos, A., Mulero-Navarro, S., et al. (2002). The antiproliferative activity of resveratrol results in apoptosis in MCF-7 but not in MDA-MB-231 human breast cancer cells: Cell-specific alteration of the cell cycle. *Biochemical Pharmacology, 64,* 1375–1386.

Pozo-Guisado, E., Lorenzo-Benayas, M. J., & Fernández-Salguero, P. M. (2004). Resveratrol modulates the phosphoinositide 3-kinase pathway through an estrogen receptor alpha-dependent mechanism: Relevance in cell proliferation. *International Journal of Cancer, 109,* 167–173.

Reagan-Shaw, S., Afaq, F., Aziz, M. H., et al. (2004). Modulations of critical cell cycle regulatory events during chemoprevention of ultraviolet B-mediated responses by resveratrol in SKH-1 hairless mouse skin. *Oncogene, 23,* 5151–5160.

Reis-Sobreiro, M., Gajate, C., & Mollinedo, F. (2009). Involvement of mitochondria and recruitment of Fas/CD95 signaling in lipid rafts in resveratrol-mediated antimyeloma and antileukemia actions. *Oncogene, 28,* 3221–3234.

Revel, A., Raanani, H., Younglai, E., et al. (2003). Resveratrol, a natural aryl hydrocarbon receptor antagonist, protects lung from DNA damage and apoptosis caused by benzo[a]pyrene. *Journal of Applied Toxicology, 23,* 255–261.

Roy, P., Kalra, N., Prasad, S., et al. (2009). Chemopreventive potential of resveratrol in mouse skin tumors through regulation of mitochondrial and PI3K/Akt signaling pathways. *Pharmaceutical Research, 26,* 211–217.

Rubiolo, J. A., Mithieux, G., & Vega. F. V. (2008). Resveratrol protects primary rat hepatocytes against oxidative stress damage: Activation of the Nrf2 transcription factor and augmented activities of antioxidant enzymes. *European Journal of Pharmacology, 591,* 66–72.

Rusin, M., Zajkowicz, A., & Butkiewicz, D. (2009). Resveratrol induces senescence-like growth inhibition of U-2 OS cells associated with the instability of telomeric DNA and upregulation of BRCA1. *Mechanisms of Ageing and Development, 130,* 528–537.

Sánchez-Fidalgo, S., Cárdeno, A., Villegas, I., et al. (2010). Dietary supplementation of resveratrol attenuates chronic colonic inflammation in mice. *European Journal of Pharmacology, 633,* 78–84.

Shankar, S., Chen, Q., Siddiqui, I., et al. (2007a). Sensitization of TRAIL-resistant LNCaP cells by resveratrol (3, 4′, 5 tri-hydroxystilbene): Molecular mechanisms and therapeutic potential. *Journal of Molecular Signaling, 2,* 7.

Shankar, S., Siddiqui, I., & Srivastava, R. (2007b). Molecular mechanisms of resveratrol (3,4,5-tri-hydroxy-trans-stilbene) and its interaction with TNF-related apoptosis inducing ligand (TRAIL) in androgen-insensitive prostate cancer cells. *Molecular and Cellular Biochemistry, 304,* 273–285.

Shin, D. Y., Kim, G. Y., Li, W., et al. (2009). Implication of intracellular ROS formation, caspase-3 activation and Egr-1 induction in platycodon D-induced apoptosis of U937 human leukemia cells. *Biomedicine & Pharmacotherapy, 63,* 86–94.

Singh, N., Nigam, M., Ranjan, V., et al. (2009). Caspase mediated enhanced apoptotic action of cyclophosphamide- and resveratrol-treated MCF-7 cells. *Journal of Pharmacological Sciences, 109,* 473–485.

Singh, U. P., Singh, N. P., Singh, B., et al. (2010). Resveratrol (trans-3,5,4′-trihydroxystilbene) induces silent mating type information regulation-1 and down-regulates nuclear transcription factor-κb activation to abrogate dextran sulfate sodium-induced colitis. *The Journal of Pharmacology and Experimental Therapeutics, 332,* 829–839.

Smith, P. G., Wang, F., Wilkinson, K. N., et al. (2005). The phosphodiesterase PDE4B limits cAMP-associated PI3K/AKT-dependent apoptosis in diffuse large B-cell lymphoma. *Blood, 105,* 308–316.

Stearns, M. E., Kim, G., Garcia, F., et al. (2004). Interleukin-10 induced activating transcription factor 3 transcriptional suppression of matrix metalloproteinase-2 gene expression in human prostate CPTX-1532 cells. *Molecular Cancer Research, 2,* 403–416.

Sun, C., Hu, Y., Liu, X., et al. (2006). Resveratrol downregulates the constitutional activation of nuclear factor-κB in multiple myeloma cells, leading to suppression of proliferation and invasion, arrest of cell cycle, induction of apoptosis. *Cancer Genetics and Cytogenetics, 165,* 9–19.

Sun, J., Watkins, G., Blair, A. L., et al. (2008). Deregulation of cofactor of BRCA1 expression in breast cancer cells. *Journal of Cellular Biochemistry, 103,* 1798–1807.

Surh, Y. J. (2003). Cancer chemoprevention with dietary phytochemicals. *Nature Reviews Cancer, 3,* 768–780.

Szaefer, H., Krajka-Kuzniak, V., & Baer-Dubowska, W. (2008). The effect of initiating doses of benzo[a]pyrene and 7,12-dimethylbenz[a]anthracene on the expression of PAH activating enzymes and its modulation by plant phenols. *Toxicology, 251,* 28–34.

Tang, F. Y., Chiang, E. P., & Sun, Y. C. (2008a). Resveratrol inhibits heregulin-β1-mediated matrix metalloproteinase-9 expression and cell invasion in human breast cancer cells. *The Journal of Nutritional Biochemistry, 19,* 287–294.

Tang, F. Y., Su, Y. C., Chen, N. C., et al. (2008b). Resveratrol inhibits migration and invasion of human breast-cancer cells. *Molecular Nutrition & Food Research, 52,* 683–691.

Teh, M. T., Blaydon, D., Ghali, L. R., et al. (2007). Role for WNT16B in human epidermal keratinocyte proliferation and differentiation. *Journal of Cell Science, 120,* 330–339.

Turchi, L., Aberdam, E., Mazure, N., et al. (2008). Hif-2alpha mediates UV-induced apoptosis through a novel ATF3-dependent death pathway. *Cell Death and Differentiation, 15,* 1472–1480.

Um, H., Bae, J., Park, J., et al. (2010). Differential effects of resveratrol and novel resveratrol derivative, HS-1793, on endoplasmic reticulum stress-mediated apoptosis and Akt inactivation. *International Journal of Oncology, 36,* 1007–1013.

Upadhyay, G., Singh, A. K., Kumar, A., et al. (2008). Resveratrol modulates pyrogallol-induced changes in hepatic toxicity markers, xenobiotic metabolizing enzymes and oxidative stress. *European Journal of Pharmacology, 596,* 146–152.

Vanamala, J., Reddivari, L., Radhakrishnan, S., et al. (2010). Resveratrol suppresses IGF-1 induced human colon cancer cell proliferation and elevates apoptosis via suppression of IGF-1R/Wnt and activation of p53 signaling pathways. *BMC Cancer, 10,* 238.

van de Weerdt, B. C., & Medema, R. H. (2006). Polo-like kinases: A team in control of the division. *Cell Cycle, 5,* 853–864.

Vermeulen, K., Van Bockstaele, D. R., & Berneman, Z. N. (2003). The cell cycle: A review of regulation, deregulation and therapeutic targets in cancer. *Cell Proliferation, 36,* 131–149.

Wang, Q., Mora-Jensen, H., Weniger, M. A., et al. (2009). ERAD inhibitors integrate ER stress with an epigenetic mechanism to activate BH3-only protein NOXA in cancer cells. *Proceedings of the National Academy of Sciences of the United States of America, 106,* 2200–2205.

Wang, R. H., Zheng, Y., Kim, H. S., et al. (2008). Interplay among BRCA1, SIRT1, survivin during BRCA1-associated tumorigenesis. *Molecular Cell, 32,* 11–20.

Wattenberg, L. W. (1996). Chemoprophylaxis of carcinogenesis: A review. *Cancer Research, 26*(7 Part 1), 1520–1526.

Whitlock, N. C., Bahn, J. H., Lee, S. H., et al. (2011). Resveratrol-induced apoptosis is mediated by early growth response-1, Kruppel-like factor 4, and activating transcription factor 3. *Cancer Prevention Research, 4,* 116–127.

Wilson, L. C., Baek, S. J., Call, A., et al. (2003). Nonsteroidal anti-inflammatory drug-activated gene (NAG-1) is induced by genistein through the expression of p53 in colorectal cancer cells. *International Journal of Cancer, 105,* 747–753.

Wolter, F., Akoglu, B., Clausnitzer, A., et al. (2001). Downregulation of the cyclin D1/cdk4 complex occurs during resveratrol-induced cell cycle arrest in colon cancer cell lines. *The Journal of Nutrition, 131,* 2197–2203.

Woo, K. J., Lee, T. J., Lee, S. H., et al. (2007). Elevated gadd153/chop expression during resveratrol-induced apoptosis in human colon cancer cells. *Biochemical Pharmacology, 73,* 68–76.

Wu, Y., & Zhou, B. (2009). Inflammation: A driving force speed cancer metastasis. *Cell Cycle, 8,* 3267–3273.

Wu, Y., & Zhou, B. P. (2010). TNF-α/NF-κB/Snail pathway in cancer cell migration and invasion. *British Journal of Cancer, 102,* 639–644.

Yamaguchi, K., Lee, S. H., Kim, J. S., et al. (2006). Activating transcription factor 3 and early growth response 1 are the novel targets of LY294002 in a phosphatidylinositol 3-kinase--independent pathway. *Cancer Research, 66,* 2376–2384.

Yan, C., & Boyd, D. D. (2006). ATF3 regulates the stability of p53: A link to cancer. *Cell Cycle, 5,* 926–929.

Yan, C., Wang, H., & Boyd, D. D. (2002). ATF3 represses 72-kDa type IV collagenase (MMP-2) expression by antagonizing p53-dependent trans-activation of the collagenase promoter. *The Journal of Biological Chemistry, 277,* 10804–10812.

Yan, C., Jamaluddin, M. S., Aggarwal, B., et al. (2005). Gene expression profiling identifies activating transcription factor 3 as a novel contributor to the proapoptotic effect of curcumin. *Molecular Cancer Therapeutics, 4,* 233–241.

Yin, X., DeWille, J. W., & Hai, T.(2007). A potential dichotomous role of ATF3, an adaptive-response gene, in cancer development. *Oncogene, 27,* 2118–2127.

Yu, H., Pan, C., Zhao, S., et al. (2008). Resveratrol inhibits tumor necrosis factor-[alpha]-mediated matrix metalloproteinase-9 expression and invasion of human hepatocellular carcinoma cells. *Biomedicine & Pharmacotherapy, 62,* 366–372.

Zahid, M., Gaikwad, N. W., Ali, M. F., et al. (2008). Prevention of estrogen-DNA adduct formation in MCF-10F cells by resveratrol. *Cancer Prevention Research, 45,* 136–145.

Zetter, P., & Bruce, R. (1998). Angiogenesis and tumor metastasis. *Annual Review of Medicine, 49,* 407–424.

Zhang, H., Shih, A, Rinna, A., et al. (2010). Exacerbation of tobacco smoke mediated apoptosis by resveratrol: An unexpected consequence of its antioxidant action. *The International Journal of Biochemistry & Cell Biology.* doi:10.1016/j.biocel.2009.12.022.

Zhang, Q., Tang, X., Lu, Q. Y., et al. (2005). Resveratrol inhibits hypoxia-induced accumulation of hypoxia-inducible factor-1α and VEGF expression in human tongue squamous cell carcinoma and hepatoma cells. *Molecular Cancer Therapeutics, 4,* 1465–1474.

Zykova, T. A., Zhu, F., Zhai, X., et al. (2008). Resveratrol directly targets COX-2 to inhibit carcinogenesis. *Molecular Carcinogenesis, 47,* 797–805.

Chapter 16
An Evidence-based Perspective of *Camellia Sinensis* (Green Tea) for Cancer Patients

Joshua D. Lambert

Abstract *Camellia sinensis*, Theaceae (green tea) and its major polyphenolic constituent, (-)-epigallocatechin-3-gallate (EGCG) have been extensively studied for their preventive activity against cancer. Studies in animal models of carcinogenesis have shown that green tea and EGCG can inhibit the development of cancer in various organ systems at the initiation, promotion and progression stages. This inhibition of tumorigenesis is associated with decreased cell proliferation, increased apoptosis, and inhibition of angiogenesis. Various mechanisms of action have been proposed based on studies with human cell lines and cell-free systems. These mechanisms include induction of oxidative stress, inhibition of key enzyme systems, inhibition of growth factor signaling and others. Few of these mechanisms have been clearly demonstrated to play a role in the prevention of cancer by tea *in vivo*. Although a large number of epidemiological studies have been conducted on the relationship between tea consumption and cancer risk, the results remain mixed. Very few human intervention studies have been conducted, yet at least one demonstrated that tea has promise for the prevention of prostate cancer. Further controlled intervention studies are needed to clearly test the cancer preventive effects of green tea in high risk human subjects. Additionally, carefully designed, mechanism-based animal model studies in conjunction with biochemical and immunohistochemical studies of human samples will be critical for unraveling the key underlying mechanisms of action. Although there is promise for green tea as a cancer chemopreventive agent, considerable work is needed to fully realize its potential impact on public health.

J. D. Lambert (✉)
The Department of Food Science, 332 Food Science Building, The Pennsylvania State University, University Park, PA 16802, USA
e-mail: jdl134@psu.edu

W. C. S. Cho (ed.), *Evidence-based Anticancer Materia Medica,* Evidence-based Anticancer Complementary and Alternative Medicine,
DOI 10.1007/978-94-007-0526-5_16, © Springer Science+Business Media B.V. 2011

16.1 Introduction

Camellia sinensis, Theaceae (green tea) is a beverage with worldwide popularity second only to water. Currently per capita consumption of tea worldwide exceeds 105 L (Wolf et al. 2008). Although all tea is derived from the leaves of *C. sinensis*, there is significant variation in the processing of the leaves which results in the three major types of commercially-available tea beverage: green tea, oolong tea, and black tea. These differences in processing affect the color, flavor, and phytochemistry (Balentine et al. 1997).

Green tea accounts for approximately 20% of world consumption and is prepared by pan-frying or steaming fresh tea leaves in order to inactivate polyphenol oxidase and other enzymes. Chemically, green tea is characterized by the presence of large amounts of flavan-3-ols or catechins (Fig. 16.1). The four major catechins are (-)-epigallocatechin-3-gallate (EGCG), (-)-epigallocatechin (EGC), (-)-epicatechin-3-gallate (ECG), and (-)-epicatechin (EC). A typical cup of green tea (2.5 g of leaves in 250 ml water) contains 240–320 mg of catechins and 20–50 mg of caffeine (Balentine et al. 1997). EGCG is the most abundant catechin and may represent 30–50% of the extracted catechins.

By contrast, in the preparation of black tea, leaves are crushed and allowed to undergo a polyphenol oxidase-mediated oxidation known as "fermentation". This process results in the polymerization of the catechins to form higher molecular weight compounds including theaflavins and thearubigins. These higher molecular weight compounds represent 3–6% and >20% of the water extractable material in black tea, respectively, and are responsible for the characteristic orange-brown color of black tea (Yang et al. 2002). Approximately 80% of tea consumed worldwide is consumed as black tea.

Oolong tea, which represents approximately 2% of world consumption, is intermediate to green tea and black tea in terms of processing. The leaves are allowed to undergo a partial oxidation which results in the formation of interesting higher molecular weight compounds, but allows greater preservation of the monomeric catechins (Yang et al. 2002).

Tea and tea polyphenols have been extensively studied for their potential beneficial biological effects. Numerous studies have suggested that tea and tea preparations may be useful for the prevention of chronic disease including: type 2 diabetes, cardiovascular disease, neurodegenerative disease, and cancer (Wolfram 2007; Yang et al. 2007; Grove and Lambert 2010). To a greater or lesser extent, each of these potential indications is supported by data generated by *in vitro* or *in vivo* laboratory studies, epidemiological studies, or human intervention studies. In most cases, unequivocal data supporting efficacy for disease prevention by tea in humans is lacking.

Numerous *in vitro* studies have been conducted and they have been extensively reviewed (Siddiqui et al. 2007; Halder et al. 2008; Tachibana 2009; Yang et al. 2009). In this chapter, I will discuss the evidence for the prevention of cancer by green tea and green tea polyphenol (GTP). I will focus on studies in laboratory animal models, human epidemiological studies, and human intervention studies. The potential mechanisms suggested by studies in animal models and human subjects will be dis-

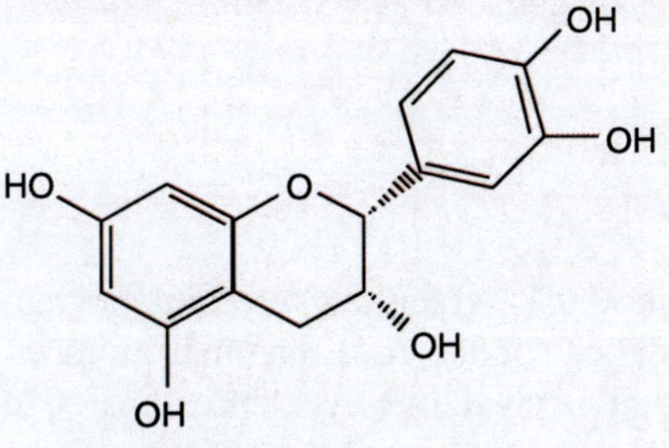

(-)-Epicatechin (EC)

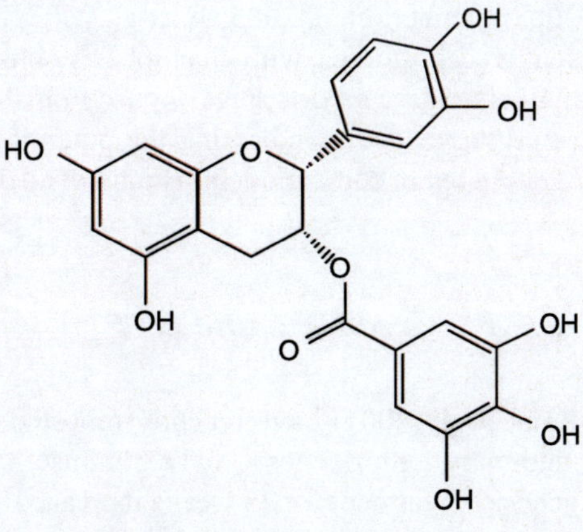

(-)-Epicatechin-3-gallate (ECG)

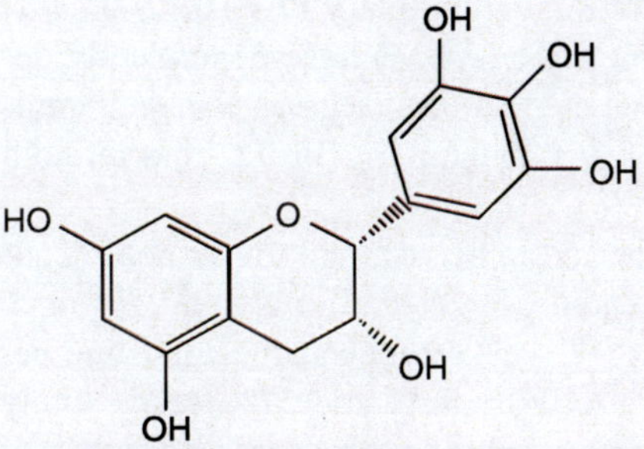

(-)-Epigallocatechin (EGC)

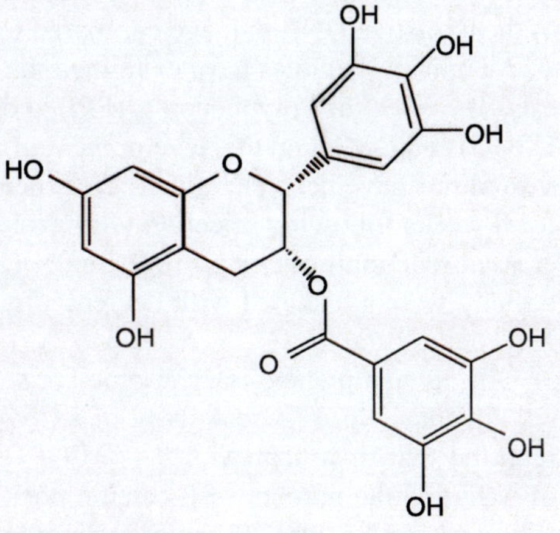

(-)-Epigallocatechin-3-gallate (EGCG)

Fig. 16.1 The structure of the major tea catechins

cussed, and supporting *in vitro* data will be presented. I will also discuss the emerging evidence for toxicity resulting from high oral doses of GTP and the implications of this toxicity for the potential usefulness of green tea as a cancer preventive agent in humans. This chapter is meant to inform readers new to the field, to point out areas in need of further study, and to stimulate new efforts to more fully determine the efficacy and mechanism of action of green tea as a cancer preventive agent.

16.2 Evidence of Cancer Prevention from Animal Studies

Green tea and GTP have been shown to inhibit carcinogenesis in a number of animal models of cancer of the oral cavity, gastrointestinal tract, liver, lung, prostate, mammary gland, and skin (reviewed in Yang et al. 2002; Chung et al. 2003; Saleem et al.

2003; Lambert et al. 2005; Ju et al. 2007). In the present chapter, I will summarize some recent studies, with an emphasis on those reporting potential mechanistic biomarkers. Most reports have focused on the polyphenols in tea, although there is emerging evidence indicating the potential role of caffeine as a cancer preventive agent in tea in some models (Huang et al. 1997; Chung 1999; Lu et al. 2006).

16.2.1 Green Tea and GTP

Kaur et al. (2007) have recently reported that 0.05% green tea catechins prevent mammary tumorigenesis in TAg mouse model of spontaneous mammary tumorigenesis. Treatment for 25 weeks increased mean survival time by 4.8% compared to water-treated controls and reduced tumor burden by 25%. Inhibition of tumorigenesis was associated with increased tumor cell apoptosis as well as decreased levels of the oxidative DNA damage compared to water-treated controls.

A number of studies have examined the effect of GTP against colon carcinogenesis. The effect of Polyphenon E (PPE, a defined catechin mixture containing 65% EGCG) and EGCG on the development of aberrant crypt foci (ACF) in carcinogen-treated rats have been examined. Treatment of rats with dietary PPE (0.12–0.24%) for 8 weeks following injection with azoxymethane (AOM) dose-dependently decreased ACF multiplicity by 16.3–36.9% (Xiao et al. 2008). Decreases in ACF multiplicity were associated with decreased nuclear β-catenin, cyclin D_1, and retinoid X receptor α staining.

Obesity and metabolic syndrome are suggested to be risk factors for colon cancer development. A recent study of AOM-treated C57BL/KsJ-*db/db* (*db/db*) mice reported that treatment with 0.1% EGCG for 7 weeks reduced the total number of ACF and the number of β-catenin positive ACF by 33% and 57%, respectively (Shimizu et al. 2008). These reductions were associated with decreased levels of β-catenin, cyclooxygenase (COX)-2, and cyclin D_1 protein in the colonic mucosa. EGCG treatment also decreased colonic levels of total and phosphorylated IGF-1 (insulin-like growth factor-1) receptor. Conversely, EGCG treatment increased plasma levels of IGF-1 binding protein (IGFBP)-3, a negative regulator of IGF-1 signaling. The authors suggest that this modulation of IGF-1 signaling may represent the underlying mechanism for the cancer preventive activity of EGCG in this model.

IGF-1 signaling has also been suggested as a potential target of green tea in a mouse model of prostate cancer. Oral GTP have been shown to inhibit prostate cancer in the transgenic adenocarcinoma of the mouse prostate (TRAMP) mouse model. Treatment with 0.1% GTP for 24 weeks delayed tumor development, reduced prostate weight, and decreased tumor cell proliferation (Gupta et al. 2001). Western blot and immunohistochemical analysis showed that GTP treatment decreased the IGF-1 to IGFBP-3 ratio in tumor tissue by 70–83%. PI3 K (phosphotidylinositol-3-kinase) and phosphorylation of ERK (extracellular responsive kinase) 1/2 and AKT were also reduced (Adhami et al. 2004). IGF-1 signaling has been shown to play an important role in metastasis and has been shown to increase MMP (matrix metalloproteinase) expression. Adhami et al. (2004) have shown that GTP treatment reduces the expression of MMP-2 and MMP-9 by 68% and 60%, respectively.

GTP has not been shown to have universally beneficial effects in animal models of carcinogenesis. A recent study by Kim et al. (2007) suggests that, under certain conditions, these compounds may have a deleterious effect. The authors report that in the dextran sulfate sodium (DSS) mouse model of acute colitis, treatment with 0.5% and 1% GTP enhanced the inflammatory response. Mice treated with these doses of GTP had significant body weight loss, enhanced colon shortening and increased levels of interleukin (IL)-1β compared to mice treated with 2% DSS only. By contrast, treatment with 0.01% GTP had no negative effect on body weight and decreased IL-1β levels compared to DSS-treated mice. These results may be specific to this model, or they may indicate that under conditions of high oxidative stress (e.g. colitis) higher doses of GTP can exacerbate inflammation and oxidative stress. This would be support previous *in vitro* studies demonstrating the pro-oxidant effect of EGCG under cell culture conditions (Hou et al. 2005). Further studies are needed to more thoroughly assess the potential utility of GTP for the prevention of colitis and colitis-associated carcinogenesis.

16.2.2 Caffeine

Several studies have shown the potential role for caffeine in the cancer preventive activity of tea in several animal models. A study in 4-(methylnitrosamino)-1-(3-pyridyl)-1-butanone (NNK)-induced A/J mice showed that caffeine may play a key role in the cancer preventive effects of green tea (Lu et al. 2006). Treatment of adenoma-bearing A/J mice with caffeine (0.044%) or PPE (0.5%) as the sole source of drinking fluid for 12 weeks reduced progression of adenoma to adenocarcinoma. Incidence of adenocarcinoma was reduced by 48% and 52% in caffeine and PPE-treated mice, respectively. Adenocarcinoma multiplicity was also reduced by 49% and 63% in caffeine and PPE-treated mice, respectively. Immunohistochemistry revealed that inhibition of progression was associated with decreased phosphorylation of c-jun and ERK 1/2, and increased apoptotic index in treated tumor cells.

Topical application or oral administration of caffeine has been shown to inhibit skin tumor formation and induced tumor cell apoptosis in ultraviolet light B-treated SKH-1 mice (Huang et al. 1997; Lou et al. 1999). Irradiation of these mice for 20 weeks increases risk of skin tumor formation and results in the formation of cellular patches in the epidermis that are immunoreactive to antibodies that recognize mutated *p53* (Lu et al. 2005). These patches represent a possible precancerous lesion. Discontinuation of UVB-treatment results in a disappearance of these patches over time, and treatment with topical caffeine has been shown to enhance this disappearance (Kramata et al. 2005; Lu et al. 2005).

A recent study by Castro et al. (2008) has reported that green tea and caffeine, but not decaffeinated green tea or EGCG, reduces dibenzo[*a,l*]pyrene (DBP)-induced transplancental carcinogenesis in mice. Treatment of pregnant B6129SF1 mice with

DBP induces aggressive T-cell lymphoma in the off-spring that results in premature death post-partum. Pre-treatment with 0.5% green tea or the equivalent amount of caffeine or EGCG for 17 days prior to injection with DBP significantly increased the survival of the pups post-partum. Caffeine increased the surviving fraction from 7.7% to 23% compared to water-treated controls. The authors found that caffeine induced cytochrome P450 (CYP)1B1 activity in the liver of the mother, and hypothesized that this may reduce the bioavailability DBP and therefore reduce lymphoma formation in the offspring.

16.2.3 *Effects on In Vivo Antioxidants and Carcinogen Metabolism*

Studies have suggested that GTP can inhibit the expression and activity of carcinogen activating enzymes such as CYP and increase the expression of detoxification enzymes. In some cases, these studies may suggest generalized mechanisms for the prevention of cancer. In others, the effects may be specific for chemical carcinogen-induced cancers.

Krishnan et al. (2005) have shown that pretreatment of Swiss mice with 1% tea polyphenols blunted benzo[a]pyrene (B[a]P)-induced increase in expression of CYP1A1 and 1A2. This is likely due to interference at the aromatic hydrocarbon receptor level. Western blot analysis showed that B[a]P-induced expression of these two isoforms were decreased by 55–58% and 79–86% by tea polyphenols in the liver and lung, respectively. Conversely, long-term treatment with green tea resulted in increased expression of hepatic CYP1A1 and 1A2 in rats in the absence of carcinogen treatment (Sohn et al. 1994; Xu et al. 1996). It has been suggested that these effects on CYP1A1 and 1A2 activity is likely due to the caffeine (Chen et al. 1996).

Treatment of piglets with 0.2% green tea extract (45% EGCG) for 3 weeks increased glutathione conjugation of aflatoxin (AF)B_1 by small intestinal microsomes in *ex vivo* studies (Tulayakul et al. 2007). Maliakal et al. (2001) have reported that treatment of female Wistar rats with 2% green tea solution for 4 weeks was shown to increased glutathione-*S*-transferase (GST) activity in the liver. Other studies, however, found no effect (Liu et al. 2003). More recently, oral gavage treatment with EGCG (200 mg/kg) has been shown to upregulate gene expression of γ-glutamyltransferase, glutamate cysteine ligase, and hemeoxygenase 1 in C57bl/6 J mice in a nuclear factor (erythroid-derived 2)-like (Nrf2)-dependent fashion (Shen et al. 2005).

Although green tea and GTP have been historically regarded as antioxidant, and they are powerful chemical antioxidants which can chelate transition metals and directly quench free radicals, it is possible that they function by a more indirect mechanism (Lambert and Elias 2010). In the absence of pre-existing oxidative stress, the tea polyphenols may generate a low level of reactive oxygen species that stimulate upregulation of the endogenous antioxidant systems by mechanisms such as the antioxidant response element/Nrf2 signaling pathway. Indeed, Yuan et al.

(2007a) have recently reported that treatment of colon cancer xenograft-bearing nude mice with dietary EGCG caused dose-dependent upregulation of Nrf2 expression in orthotopically-implanted colon tumors. Shen et al. (2005) have reported similar results in the small intestine and liver of non-tumor-bearing C57bl/6 J mice.

Recent studies have lent support to the hypothesis that EGCG exerts some of its anticancer effects *via* pro-oxidative mechanisms (Azam et al. 2004; Sang et al. 2005; Maeta et al. 2007). H_2O_2 has previously been shown to play a role in the growth inhibitory and pro-apoptotic effects of EGCG and EGC against human lung cancer cells (Yang et al. 1998, 2000). In these studies, EGCG and EGC dose-dependently generated H_2O_2 and induced apoptosis in both transformed bronchial epithelial cells and lung cancer cells. Inclusion of exogenous catalase ablated the pro-apoptotic effects of EGCG. A similar role for H_2O_2 was observed in other cancer cell lines (Weisburg et al. 2004; Chan et al. 2006).

EGCG, given either orally or by intraperitoneal injection, has recently been reported to induce oxidative stress selectively in human lung cancer xenografts in nude mice (Li et al. 2010). Tumors derived from EGCG-treated mice had higher levels of phosphorylated histone 2A.X, 8-hydroxy-2-deoxyguanosine (8-OHdG) and metallothionein I/II. These markers correlated with tumor growth inhibition, and represent both direct oxidative stress as well as the inducible cellular response to oxidative stress.

The end result of decreased carcinogen activation and increased detoxification induced by tea may be the prevention of DNA damage. Tea and tea components have been shown to inhibit carcinogen-induced DNA damage *in vitro*. For example, co-treatment of human leukocytes with EGCG (2 μM) and bleomycin (20 μg/ml) resulted in a 50% decrease in bleomycin-induced DNA damage compared to treatment with bleomycin alone (Glei and Pool-Zobel 2006). It has also been reported that green tea extract can dose-dependently protect Chang liver cells from B[a]P-induced DNA damage (Yen et al. 2004). Similar protective effects have been observed *in vivo*.

Pre-treatment of C57bl/6 Big Blue *lacI* transgenic mice with 2% green tea prior to a single dose of B[a]P was shown to reduce characteristic GC to TA transversions in the liver by 54% compared to water-treated controls (Jiang et al. 2001). In contrast, green tea administration did not inhibit DNA adduct formation in the lungs of NNK-treated A/J mice even though tea did reduce tumorigenesis (Shi et al. 1994). Lin et al. (2003) have reported that pre-treatment of rats with 3% green tea extract as the sole source of drinking fluid for 10 days reduced PhIP-DNA adduct formation. PhIP-DNA adducts were reduced by 50–63% in the colon, heart, lung, and liver, respectively by green tea treatment.

16.2.4 Green Tea in Combination with Other Compounds

The idea of using combinations of different agents, which work *via* different mechanisms, for enhanced chemopreventive effect is growing as an area of research interest. A relatively small number of studies have been conducted with GTP, but these have included combination with other dietary components as well as drugs.

Green tea in combination with the turmeric-derived polyphenol, curcumin, has yielded promising results for the prevention of cancer. Li et al. (2002) have reported that the combination of oral green tea and topically-applied curcumin reduced tumorigenesis in the 7,12-dimethylbenz[a]anthracene-induced hamster model of oral cancer. Hamsters treated with the combination had a reduction in visible tumors and tumor volume of 52.4% and 69.8%, respectively, compared to water-treated controls. The combination enhanced apoptotic index and reduced angiogenesis.

More recently, the same combination was shown to inhibit carcinogenesis in the 1,2-dimethylhydrazine (DMH)-induced rat model of colon carcinogenesis (Xu et al. 2010). Dietary treatment of rats with 0.1% tea catechins (80% EGCG, 15% EGC, and 5% ECG, EC) and 0.1% curcumin for 32 weeks reduced tumor multiplicity and tumor volume by 64.5% and 36.6%, respectively, compared to rats fed a control diet. By contrast, treatment with 0.2% catechins or 0.2% curcumin as single agents had no significant effect on either parameter. The effects appear to be the result of a reduction in tumor cell proliferation and enhanced apoptosis in the tumors of combination-treated rats. Although statistical analysis was not conducted, the results appear to be additive rather than synergistic.

Mechanistically, it is unclear how the combination of green tea and curcumin results in enhanced chemopreventive activities. If the effects are truly synergistic, the enhancement may result from the combination of the direct chemopreventive effects of the single agent, and, an enhancement in the bioavailability of the tea catechins.

Previous studies have shown that the tea catechins are substrates for multidrug resistance-associated proteins (MRP)-1 and -2 (Hong et al. 2003; Vaidyanathan and Walle 2003). MRP-2 is expressed on the apical membrane of the enterocytes and pumps compounds from the interior of the epithelium back into the intestinal lumen. Curcumin has been shown to inhibit this efflux transporter (Wortelboer et al. 2005; Chearwae et al. 2006). It is conceivable, although purely speculative at this point, that curcumin could reduce the efflux of catechins from the enterocytes of DMH-treated rats, enhance the bioavailability of these compounds, and thereby enhance the bioactivity of the combination. Such a hypothesis remains to be tested.

Green tea catechins, in combination with the cholesterol-lowering drug atorvastatin, have been shown to have synergistic inhibitory effects against lung cancer cells in culture, and to have enhanced lung cancer preventive effects *in vivo* (Lu et al. 2008). Treatment of NNK-induced A/J mice with 0.25% PPE and 200 ppm atorvastatin for 20 weeks significantly reduced both lung tumor multiplicity and lung tumor burden by 55.8% compared to treatment with either agent alone. Inhibition of tumorigenesis was associated with decreased expression of the anti-apoptotic protein myeloid leukemia cell differentiation protein (Mcl)-1 and an increase in the apoptotic index in the tumor cells. *In vitro* studies by the same group showed that co-treatment of human lung cancer cells with PPE and atorvastatin synergistically inhibited cell growth and induced apoptosis. These effects were again accompanied by decreased expression of Mcl-1 and increased apoptosis. Combination treatment also reduced the expression of cyclin D_1 and cyclin-dependent kinase-4 and -6.

Previous *in vitro* studies have shown that EGCG is able to bind directly to the BH-3 domain of members of the Bcl-2 family of proteins, of which Mcl-1 is a member (Leone et al. 2003). Using a combination of binding studies, NMR, and molecular modeling studies, Leone et al. (2003) found that EGCG can potently bind to and inhibit the activity of Bcl-X_L with an inhibitory constant value of 490 nM. Based on molecular modeling studies, the gallate ester was found to essential for inhibitory activity. It is possible that the effects observed in the NNK-treated A/J mouse and the human lung cancer cells *in vitro* are downstream effects of this direct binding, but further studies using site-directed mutagenesis and *in vivo* binding assays are needed to more fully understand the role of this potential mechanism.

The combination of GTP and the selective cyclooxygenase (COX)-2 inhibitor, celecoxib, have shown enhanced anticancer effects against human androgen-sensitive prostate cancer xenografts in nude mice (Adhami et al. 2007). Time to euthanizable tumor volume (mean = 1,300 mm^3) was increased from 28 days in control mice to 48 days in mice treated with the combination of GTP (0.1% in the drinking fluid) and celecoxib (5 mg/kg, ip, once daily). Mice treated with either celecoxib or GTP had intermediate rates of tumor growth. Biochemical analysis showed that combination treatment had significantly reduced plasma IGF-1 levels and significantly increased plasma IGFBP-3 levels compared to the control mice. Analysis also showed an increase in the ratio of BAX:BCL-2 and PARP cleavage in the tumors of combination-treated mice indicating an increase in apoptosis in the tumors. The results of this study are interesting, although a clear mechanistic hypothesis to explain the results was not presented.

Many other potential combinations of green tea and other chemopreventive agents remain to be explored. Studies by our laboratory and others have suggested that enhancement of tea polyphenol bioavailability, either by inhibition of Phase II metabolism, inhibition of protein-mediated efflux, or modulation of gastrointestinal motility may represent a means to enhance cancer preventive activity (Hong et al. 2003; Vaidyanathan and Walle 2003; Lambert et al. 2004). Additionally, EGCG has been reported to inhibit a number of key cellular pathways *in vitro*. Among these potential targets is the proteasome (Dou 2009). Proteasome inhibitors have been suggested for use in combination with other chemotherapeutic drugs (e.g. gemcitabine) for enhanced anticancer effects (Bold et al. 2001). Such results suggest several novel GTP-based combinations that could be explored.

16.3 Evidence of Cancer Prevention from Human Studies

16.3.1 Epidemiological Studies

A large number of epidemiological studies have been conducted on the potential cancer preventive activities of tea (Blot et al. 1996; Kohlmeier et al. 1997; Arab and Il'yasova 2003; Borrelli et al. 2004; Hoshiyama et al. 2005; Shukla 2007; Tsugane and Sasazuki 2007). Although some studies have shown an inverse correlation between tea consumption and cancer risk, others have found no association. In general, case-control studies have been more positive than prospective cohort studies. The inconsistencies in these studies may be due to differences in the type of tea consumed; smoking status, alcohol consumption, and other lifestyle factors; accuracy in assessment of tea consumption; and genetic variability in study populations. In the present review, we will discuss the association between green tea and the three most common cancers in the United States (lung, breast, and prostate) as well as gastrointestinal cancers.

16.3.1.1 Lung Cancer

The literature on epidemiological studies on tea and lung cancer prevention was recently reviewed (Arts 2008). Twenty studies were reviewed and included both prospective cohort ($n=6$) and case-control studies ($n=14$), and considered green tea ($n=10$), black tea ($n=8$), and other/any tea type ($n=2$). Overall, 4 of the 20 studies found a significant inverse correlation, whereas the majority of the remaining studies found no effect of a non-significant reduced risk. Among never-smokers, (reported in 7 studies) a protective effect was observed in 4 studies (two green tea and two black tea).

A recent population-based cohort study (Ohsaki National Insurance Cohort) examined the relationship between green tea and lung cancer risk in a cohort of 41,440 Japanese men and women (Li et al. 2008). Even after adjusting for smoking status, the authors could find no significant association between tea consumption and lung cancer risk.

Several recent studies have demonstrated an association between lung cancer risk and tea consumption in selected populations. For example, Bonner et al. (2005) have reported that green tea was protective in individuals with the 8-oxoguanine glycosylase (OGG1) Cys (326) allele. Protective effects have also been observed for green in nonsmoking women (Zhong et al. 2001; Kubik et al. 2004).

16.3.1.2 Breast Cancer

A meta-analysis of epidemiological studies (three cohort and one case-control study) on the relationship between green or black tea intake and breast cancer risk has reported a combined reduction odds ratio (OR)=0.78 (95% confidence interval (CI): 0.61–0.98) (Sun et al. 2006a). The protective effect of tea was observed most-strongly in the case-control study, whereas much weaker (or no) protective effects were observed in the cohort studies. A more recent meta-analysis of seven epidemiological studies focused on green tea consumption and breast cancer risk found similar results with regard to breast cancer incidence (Ogunleye et al. 2010). A somewhat reduced risk (pooled risk ratio (RR)=0.81, 95% CI: 0.75–0.88) was observed in case-control studies, whereas no association was observed in cohort studies. Interestingly, there was a significant reduction in risk of breast cancer recurrence (RR=0.73, 95% CI: 0.56–0.96), although this conclusion is based on only two studies and is somewhat premature.

In a hospital-based case-control study (n=1,009 cases and 1,009 controls) in southeast China, a dose-dependent reduction in risk of breast cancer was associated with increased intake of green tea (Zhang et al. 2007). Multivariant adjusted OR of developing breast cancer were 0.87 (95% CI=0.73–1.04), 0.68 (95% CI=0.54–0.86), 0.59 (95% CI=0.45–0.77), and 0.61 (95% CI=0.48–0.78) for intake of <250 g, 250–500 g, 500–750 g, and >750 g of green tea leaves per year.

A case-control study of Asian-American women in Los Angeles county suggests that genetic polymorphism in catechol-O-methyltransferase (COMT) may represent a modifying factor in the breast cancer preventive effects of tea (Wu et al. 2003). In this study of 1,152 women (589 cases and 563 age-matched controls), the authors found that whereas green tea had no protective effect (OR=0.86, 95% CI=0.46–1.62) in women with two high activity alleles of COMT, a protective effect was observed in women with one or more low activity allele of COMT (OR=0.42, 95% CI=0.22–0.80). Based on the role of COMT in the metabolism (and inactivation) of the tea catechins (Zhu et al. 2000; Lu et al. 2003; Chen et al. 2005; Bai et al. 2007), the authors hypothesized that the protective effect was due to slower metabolism and increased exposure to biologically active tea compounds in women with one or more low activity allele of COMT. The role of this polymorphism in modulating the protective effects of tea against other cancer should be investigated further.

For the most part, the results of prospective studies of tea and breast cancer have been less positive. A report on two cohorts of women in rural northern Japan (total n=35,004) showed no reduction in relative risk of breast cancer associated consumption of ≥5 cups of green tea per day compared to consumption of <1 cup per day (RR=0.84, 95% CI=0.57–1.24) (Suzuki et al. 2004). Similar results were also reported as part of the Nurses' Health Study which involved 85,987 subjects and 22 years of follow-up (Ganmaa et al. 2008).

By contrast a prospective study of 50,633 first-time outpatient visitors as part of the Hospital-based Epidemiologic Research Program at Aichi Cancer Center found a significant inverse relationship between breast cancer recurrence and intake of three or more cups of green tea regardless of initial tumor grade (hazard ratio (HR)=0.69, 95% CI=0.47–1.00) (Inoue et al. 2001). This study would suggest that selection of high risk populations (i.e. cancer survivors) for study might yield a more clearly beneficial effect of tea.

16.3.1.3 Prostate Cancer

Cohort-based epidemiological studies on tea and prostate cancer have yielded largely negative results (Lee et al. 2009). For example, a prospective cohort study of 19,561 men in Japan found no significant association between consumption five or more cups of tea per day and prostate cancer risk (HR=0.85, CI: 0.5–1.43) (Kikuchi et al. 2006). Although a small number of case-control studies have yielded positive results, overall, these studies have also been largely negative (Lee et al. 2009). Whereas two hospital-based case control studies in Japan (140 cases and 140 controls) and China (130 cases and 274 controls) found a significant inverse correlation between tea consumption and prostate cancer, three population-based case-control studies in Canada and the United States reported no association (Ellison 2000; Sharpe and Siemiatycki 2002; Jian et al. 2004; Sonoda et al. 2004). These differences may be the result of higher prevalence of green tea intake in China and Japan compared to a higher intake of black tea in the United States and Canada.

A study which may shed light on the design of future epidemiological studies of tea and prostate cancer is a prospective cohort study conducted with 49,920 Japanese men (The Japanese Public Health Center study). This study found no association between green tea consumption and incidence of localized prostate cancer, but did find a significant dose-dependent inverse correlation between incidence of advanced prostate cancer and consumption of five or more cups of green tea per day (RR=0.52, CI: 0.28–0.96) (Kurahashi et al. 2008). The authors also reported a significant trend for increased tea consumption being inversely correlated with invasive prostate cancer ($P=0.01$). These data suggest that although tea may not be useful for the prevention of primary prostate cancer in men, it may prevent the progression of localized prostate cancer to invasive prostate cancer. This hypothesis has been borne out by a recent controlled intervention study by Bettuzzi et al. discussed below.

16.3.1.4 Esophageal, Stomach, and Colon Cancer

Historically, tea consumption has been considered a risk factor for esophageal cancer. Nevertheless, of the publications reporting a positive association between tea consumption and esophageal cancer, nearly all have been attributed to the hot temperature of tea (Yang and Wang 1993).

In a population-based case-control study of 133 stomach cancer cases, 166 chronic gastritis cases, and 433 healthy controls in Yangzhong, China, an inverse association was observed between green tea drinking and risk of chronic gastritis and stomach cancer. ORs of green tea consumption were 0.52 (95% CI: 0.29–0.94) and 0.49 (95% CI: 0.31–0.77) for stomach cancer and chronic gastritis, respectively. The results provide support for earlier studies showing a protective effect of green tea against stomach cancer. This study represents the first report of the protective effect of green tea against chronic gastritis. Since this populations suffering from this condition are at increased risk of stomach cancer, this finding may be of importance in designing intervention strategies for stomach cancer and its pre-malignant lesions in high-risk populations (Setiawan et al. 2001).

A population-based case-control study was conducted in Taixing City in 2000 in which 206 patients with primary stomach cancer were recruited (Mu et al. 2005). In the same study, over 200 cases with esophageal cancer and 200 cases with liver cancer were also recruited. Green tea drinking was associated with decreased risk of stomach cancer, with an adjusted OR of 0.59 (95% CI: 0.34–1.01). Tea concentration was categorized into three levels: low (tea leaves were less than 25% of the volume of the cup), moderate (tea leaves were between 25–50% of the volume of the cup, and high volume of tea leaves was more than 50% of cup volume). Based on these categories, a strong dose-response relationship was observed between increased green tea concentration and decreased risk of stomach cancer (P value for the trend was 0.01).

A case-control study in Montevideo, Uraguay (240 cases, 960 controls) of incidence of gastric cancer and its relationship to diet found a significant decrease in OR of gastric cancer associated with tea intake (De Stefani et al. 2004). A dose-dependent decrease was observed in gastric cancer risk with OR=0.43 (95% CI: 0.27–0.69) for the middle tertile and 0.13 (95% CI: 0.05–0.32) for the highest tertile of consumption. The protective effect of tea appeared to be more pronounced in men (OR=0.23, 95% CI: 0.13–0.41) than in women (OR=0.47, 95% CI: 0.25–0.88).

A meta-analysis in 2006 of 25 studies on the association between tea consumption and colon cancer risk concluded that there was insufficient evidence to suggest a protective effect against cancer (Sun et al. 2006b). This has also generally been the conclusion of prospective cohort studies and some of the case-control studies. Several recent case-control studies, however, suggest that a more careful determination of tea consumption may allow detection of a protective association.

The associations between validated biomarkers of specific tea exposure and the risk of colorectal cancer among a cohort of 18,244 men in Shanghai, China, with 16 years of follow-up was prospectively examined (Yuan et al. 2007b). EGC, 4'-*O*-methyl-epigallocatechin (4'-*O*-MeEGC), EC, and their ring-fission metabolites were measured in baseline urine samples from 162 incident colorectal cancer cases and 806 matched controls. Compared with undetectable EGC, odds ratios for colon cancer in the lowest, intermediate and highest tertile of detectable EGC were 0.64 (95% CI: 0.33–1.24), 0.60 (95% CI: 0.30–1.20), and 0.40 (95% CI: 0.19–0.83), respectively. A similar inverse relation between 4'-*O*-MeEGC and colon cancer also was observed. The strongest protective effect was seen for regular tea drinkers who showed high levels of urinary EGC and 4'-MeEGC. No association between urinary levels of EC or its metabolite and colon cancer risk was observed.

16.3.2 Human Intervention Studies

Several human studies have been conducted examining the effects of dietary catechins on cancer. Although most have dealt with surrogate biomarkers of cancer such as effects on oxidative stress markers and carcinogen metabolism, at least two studies on cancer progression have been conducted.

For example, Hakim et al. (2003) have reported that supplementation of heavy smokers (>10 cigarettes per day) with four cups of decaffeinated green tea (73.5 mg catechins per cup) per day for 4 months reduced urinary 8-OHdG levels by 31% compared to control. In a second study, that examined the effect of genotype, this antioxidant effect was found to be dependent on GST $\mu 1$ and $\theta 1$ status: a protective effect was only in GST $\mu 1$ and $\theta 1$ positive individuals (Hakim et al. 2004). This

GST dependence suggests that the antioxidant effects *in vivo* may work indirectly *via* endogenous antioxidant systems.

Similarly, other studies have suggested a role for modulation of endogenous antioxidant systems and Phase II metabolism in the protective effects of green tea. A study in China has reported that 3 month treatment with 500 or 1,000 mg/day GTP increased urinary excretion of the mercapturic acid conjugated of AFB_1 by 10-fold and 8.4-fold, respectively compared to baseline (Tang et al. 2008). Schwartz et al. (2005) have also reported that heavy smokers treated with green tea (400–500 mg green tea powder per cup) 5 times per day for 4 weeks had 50% lower levels of B[a]P-DNA adducts and 8-OHdG compared to control subjects. In another study of healthy volunteers reported that treatment for 4 weeks with 800 mg/day PPE increased GST-π activity in blood lymphocytes (Chow et al. 2007). These results may provide a mechanistic explanation for the enhanced formation of the mercapturic acid metabolite of AFB_1.

In contrast to studies of Phase II metabolism, studies in humans have yielded less impressive results on the modulation of CYP expression and activity by tea. Treatment of human volunteers with 800 mg PPE for 4 weeks did not significantly alter the activity of CYP1A2, CYP2D6, or CYP2C9 (Chow et al. 2006). Another study has reported that treatment of healthy volunteers with 844 mg decaffeinated green tea extract (59% EGCG) for 14 days did not significantly affect CYP3A4 or CYP2D6 activity (Donovan et al. 2004). These results raise questions about the relevance of the observed CYP-modulating effect in animal studies to human disease (Sohn et al. 1994; Xu et al. 1996; Krishnan et al. 2005).

Relatively few studies have examined cancer incidence or progression directly. A double-blind study in Italy, followed 200 individuals with high-grade prostate intraepithelial neoplasia receiving either 600 mg of green tea catechins daily or placebo ($n=100$ per arm) for 12 months. Only 3% of the patients in the catechin treatment group developed prostate cancer, whereas the rate of cancer development on the placebo group was 30% (Bettuzzi et al. 2006). No adverse effect was associated with the treatment.

The efficacy of green tea extract was examined in a recent placebo-controlled Phase II trial against oral precancerous lesions (Tsao et al. 2009). Patients were evaluated and randomized into four groups ($n=9$ per arm): placebo, 500, 750, or 1,000 mg green tea extract. Patients were treated for 12 weeks and response was determined. Although there was no statistically significant effect of green tea extract on oral precancerous lesions, there was a trend for increased positive response to treatment (50% of patients had a partial or complete response) and decreased disease progression (28% of patients had disease progression from baseline) compared to placebo (18.2% response, 45% progression from baseline).

An earlier Phase I study of green tea extract in patients with advanced lung cancer found that a dose of up to 3 g/m^2/day for 16 weeks was well-tolerated by produced no objective response in tumor size (Laurie et al. 2005). Although the results of this study were negative, the study size was small ($n=17$) and the disease stage was likely inappropriate since most dietary components lack necessary efficacy to affect late-stage cancer of any type.

16.4 Protection Against Chemotherapy Side-effects

Based on the putative antioxidant effects and the ability of green tea to stimulate the expression of endogenous antioxidant systems, it has been hypothesized that green tea may alleviate some side-effects associated with cytotoxic cancer chemotherapy. Indeed several animal model studies with common chemotherapeutic agents have borne this out. Treatment of Wistar rats with 3% green tea as the sole source of drinking fluid beginning 5 days before cisplatin injection and continuing for the duration of the study reduced nephrotoxicity compared to water-treated controls (Khan et al. 2009). Blood urea nitrogen levels were reduced by 42%, and levels of catalase and superoxide dismutase were increased by 37% and 50%, respectively compared to water-treated control.

Another study found that green tea (1% as drinking fluid) prevented irinotecan-induced small intestinal toxicity in mice (Wessner et al. 2007). Mice treated with irinotecan only had decreased ratio of reduced to oxidized glutathione. Green treatment prevented this reduction. Green tea treatment did not however reduce irinotecan-induced lipid peroxidation or inflammation in the small intestinal mucosa. Further studies are necessary to clearly delineate the protective effect of green tea against irinotecan-mediated toxicity.

More recently, Sato et al. (2010) have reported that orally administered green tea reduced doxorubicin-mediated spermatogenic disorders. Treatment of male mice with low dose doxorubicin reduced sperm concentration, sperm motility, and decreased Sertoli cell index compared to untreated controls. Co-administration of 200 or 500 mg/kg GTE *via* the diet ameliorated spermatotoxicity, increased sperm motility, and prevented decreased in the Sertoli cell index. These effects correlated with an increase in telomerase activity in the testes compared to doxorubicin-treated controls.

There is significant work that remains to be done in the area of green tea as palliative therapy for cancer patients. There have been no *in vivo* studies on chemotherapy-induced alopecia or immunosuppression. Further, only a small number of clinically-used classes of chemotherapy agents have been examined.

16.5 Potential Toxicity of High-dose Oral GTP

Green tea enjoys a long history of use as a beverage and is generally regarded as safe. Moreover, numerous human intervention and bioavailability studies using low to moderate doses of green tea preparations or EGCG have reported no serious adverse effects (Lee et al. 2002; Bettuzzi et al. 2006; Chow et al. 2006). One effect of the increasing number of reports describing the potential anti-obesity, and other beneficial effects of tea and tea polyphenols has been a proliferation of green tea-based dietary supplements. Sales of green tea-based dietary supplements in the US totaled (USD) 5.6 million in 2005, an increase of 94% from 2004 (Blumenthal et al.

2006). Green tea supplements typically contain 200–400 mg EGCG and recommending dosing of 1–2 capsules up to three times daily. This results in a total recommended dose that may be up to 2,400 mg/day. Although there have been no reported adverse effects associated with green tea beverage consumption, green tea based dietary supplements represent a different dosage form, and have the potential deliver a much higher dose of catechins than green tea beverages. Indeed, studies from our laboratory have shown that oral bolus dosing results in greatly increased peak plasma concentrations of EGCG compared to dietary administration of the same total daily dose (Lambert et al. 2006). Treatment of mice with a single oral bolus dose of 500 mg/kg EGCG result in peak plasma concentrations of 898 ng/ml, whereas the same total daily dose given *via* the diet result in plasma concentrations of 231 ng/ml.

Since 1999, there have been 34 case studies linking consumption of green tea-based supplements to hepatotoxicity (Mazzanti et al. 2009). In most cases, elevations in serum transaminase levels, as well as increased serum bilirubin, were observed. Histological examination revealed inflammatory, cholestatic, or necrotic liver damage depending on the subject. No clear determinants for the type of pathology observed have been reported. In approximately 20% of these reported case studies, additional liver damage following re-challenge with the same preparation was observed. This suggests a causal relationship between hepatotoxicity and green tea. Generally, incidence of toxicity have been related to the use of concentrated extracts or "pill or capsule" dosage forms, however, there is a report of a 45-year old man who developed jaundice and elevated serum alanine aminotransferase (ALT) following six cups/day green tea infusion for 4 months (Jimenez-Saenz and Martinez-Sanchez Mdel 2006). The reasons for this difference in sensitivity are unclear, but may be related to intra-individual differences in the detoxification of GTP.

Laboratory studies of green tea-derived preparations in rodents and dogs have generally supported the potential toxicity of those preparations at high doses (Isbrucker et al. 2006; Galati et al. 2006). Oral administration of Teavigo or PPE (standardized tea polyphenol preparations) for 13 or 9 weeks, respectively, to Beagle dogs resulted in dose-dependent toxicity and death (Isbrucker et al. 2006). Vomiting and diarrhea were observed throughout both studies. In addition, 500 mg/kg, po Teavigo caused proximal tubule necrosis and elevated serum bilirubin in all dogs treated. Most male dogs (two of three) had elevated serum aspartate aminotransferase levels. Female dogs (two of three), but not male dogs, had liver necrosis. Oral administration of 2,000 mg/kg, ig Teavigo to rats resulted in lethality in 80% of animals treated (Isbrucker et al. 2006). Histological analysis revealed hemorrhagic lesions in the stomach and intestine.

Our laboratory has recently reported that high, oral dose EGCG is hepatotoxic in mice (Lambert et al. 2010). Treatment with either a single bolus (1,500 mg/kg) or repeated daily (750 mg/kg) doses of EGCG resulted in elevated ALT levels and severe, generalized liver necrosis. These effects appear to result from pro-oxidant insult as toxicity correlated with increased lipid peroxidation, elevated metallothionein, and increased histone 2A.X phosphorylation in the liver. These results build on previous observations that intraperitoneal administration of EGCG to CD-1 mice resulted in dose-dependent lethality beginning at 150 mg/kg (Galati et al. 2006).

Toxicity, especially in the liver and kidney, appears to be correlated with the bioavailability of EGCG. In the rat, where bioavailability is low (absolute bioavailability = 1.6%), toxicity is confined to the GI tract following oral administration (Chen et al. 1997). In the dog, where bioavailability is much higher, hepatotoxicity and nephrotoxicity, as well as intestinal toxicity, were observed. Toxicity was greater in fasted, than in pre-fed, dogs (Isbrucker et al. 2006). The plasma exposure (AUC_{plasma}) in the pre-fed dogs was 19.8 μg h/ml compared to 205 μg h/ml in fasted dogs following administration of 300 mg/kg, po. Recent studies in humans have also demonstrated that fasting increases the bioavailability of EGCG (Chow et al. 2005). It is unclear, however, what the safety factor of the pre-fed condition is, and whether the aforementioned human case studies involved fasting. Decreasing the biotransformation of EGCG could also result in increased exposure to unmetabolized EGCG and potentially increased hepatotoxicity.

The induction of nephrotoxicity and hepatotoxicity by green tea catechins contradicts a significant body of literature demonstrating that these compounds can protect the liver and kidney from a wide variety of toxicants. As mentioned in a previous section of this review, green tea supplementation ameliorated the nephrotoxic side effects of cisplatin in rats (Khan et al. 2009). More studies have examined the hepatoprotective effects of green tea. For example, co-treatment of ICR mice with EGCG reduced the hepatotoxic effects of carbon tetrachloride (Chen et al. 2004). These effects were correlated with an EGCG-mediated decrease in the expression of inducible nitric oxide synthase and concomitant reduction in reactive nitrogen species. Other studies have shown that green tea or green tea catechins can reduce the toxic effects of 2-nitropropane and acetaminophen among others (Sai et al. 1998; Chen et al. 2004; Oz et al. 2004). Furthermore, dietary green tea and EGCG have been shown to prevent fatty liver disease in both diet-induced and genetic animal models (Baltaziak et al. 2004; Kuzu et al. 2008; Bose et al. 2008; Bruno et al. 2008). These hepatoprotective *versus* hepatotoxic effects clearly point to the importance of dose and dosage form. Hepatotoxic effects are almost universally observed with large oral bolus or intraperitoneal doses. By contrast, hepatoprotective effects are generally observed following oral administration of lower doses (in the diet or drinking fluid) that more closely mimic typical human exposure.

These findings suggest that caution should be exercised in the use of green tea-based dietary supplements and that further studies are needed to determine the upper limit of safety for bolus dosing with tea polyphenols as well as the underlying mechanisms of toxicity.

16.6 Conclusions and Future Directions

This chapter has discussed the evidence for the cancer preventive effects of green tea that have been generated from laboratory animal models, human epidemiological studies, and human invention studies. The results of animal model studies overwhelmingly support the cancer preventive effects of GTP. Both GTP and caffeine have shown cancer preventive activity in different models. The broad range of carcinogen and non-carcinogen-driven models, the various timing of treatment, the multitude of doses used, and differences in physiology for between the test species used (mouse, rat, hamster) make it difficult to compare potential mechanisms of action and effective doses between studies. That being said, the apparent efficacy of tea preparations under such a broad range of experimental conditions suggests that the observed effects may be general to the carcinogenic process rather than specific to a single model.

Future studies in animal models should be mechanism-driven in terms of study design and model selection. Mechanistic data obtained from *in vivo* studies avoids issues of bioavailability that typically hinder interpretation of *in vitro* studies, and simultaneously allows development of biomarkers that can be used in future human clinical trials.

These studies should utilize multiple doses of green tea preparations in order to develop data on both the maximum tolerated doses of green tea preparations as well as the potential dose-response relationships. Both types of data will aid in the development of human studies.

Finally, future animal model studies should focus on systems that are most relevant to human disease. For example, the most likely point of intervention for lung cancer prevention in smokers is following smoking cessation. Such individuals are already at elevated risk of developing lung cancer and already likely have premalignant lesions or adenomas. The NNK-treated A/J mouse model which is allowed to develop adenomas (~20 weeks after NNK treatment) prior to starting green tea treatment represents a more realistic model of human disease than the A/J mouse that is pre-treated with green tea prior to injection with NNK. Such studies will ultimately result in the most translatable data to human disease.

Future epidemiological studies should be tightly linked to exposure biomarkers for green tea consumption. Such biomarkers will allow a more accurate assessment of green tea consumption and should provide clues about the bioavailability of green tea across a study population. In addition, the use of genetic analysis for expression of key antioxidant, phase II metabolic, and other cancer related genes (e.g. *GST, COMT,* and *OGG1*) may prove invaluable in determining potentially responsive populations for future intervention studies. Such considerations must be made given the high cost of human intervention trials.

Green tea has enjoyed a long history of safe use as a beverage. Recent case-reports of hepatotoxicity linked to consumption of green tea-based dietary supplements suggest, however, that pharmacological doses of green tea in certain individuals (or populations) may have deleterious effects (Mazzanti et al. 2009). Given that cancer chemoprevention involves long-term treatment of otherwise healthy individuals with the preventive agent, it is critical that these agents have extremely low risk of adverse effects. Future studies in animal models, further controlled Phase I trials, and mechanistic studies to understand risk factors for green tea polyphenol-associated hepatotoxicity are needed.

Finally, the efficacy of green tea as a cancer preventive agent will only be established by further controlled human interventions studies of cancer incidence and progression. These studies are costly but necessary to move beyond our current understanding.

To summarize, although the results of laboratory studies overwhelmingly support the efficacy of green tea preparation for cancer prevention, the epidemiological data remains mixed and there are only few human intervention studies. Green tea as a beverage appears to be part of a healthy diet, yet more work is needed to demonstrate its cancer prevention properties.

Acknowledgment This work was supported in part by a grant from the National Center for Complementary Medicine (AT004678).

References

Adhami, V. M., Siddiqui, I. A., Ahmad, N., et al. (2004). Oral consumption of green tea polyphenols inhibits insulin-like growth factor-I-induced signaling in an autochthonous mouse model of prostate cancer. *Cancer Research, 64,* 8715–8722.

Adhami, V. M., Malik, A., Zaman, N., et al. (2007). Combined inhibitory effects of green tea polyphenols and selective cyclooxygenase-2 inhibitors on the growth of human prostate cancer cells both in vitro and in vivo. *Clinical Cancer Research: An Official Journal of the American Association for Cancer Research, 13,* 1611–1619.

Arab, L., & Il'yasova D. (2003). The epidemiology of tea consumption and colorectal cancer incidence. *The Journal of Nutrition, 133,* 3310S–3108S.

Arts, I. C. (2008). A review of the epidemiological evidence on tea, flavonoids, and lung cancer. *The Journal of Nutrition, 138,* 1561S–1566S.

Azam, S., Hadi, N., Khan, N. U., et al. (2004). Prooxidant property of green tea polyphenols epicatechin and epigallocatechin-3-gallate:implications for anticancer properties. *Toxicology In Vitro: An International Journal Published in Association with BIBRA, 18,* 555–561.

Bai, H. W., Shim, J. Y., Yu, J., et al. (2007). Biochemical and molecular modeling studies of the O-methylation of various endogenous and exogenous catechol substrates catalyzed by recombinant human soluble and membrane-bound catechol-O-methyltransferases. *Chemical Research in Toxicology, 20,* 1409–1425.

Balentine, D. A., Wiseman, S. A., & Bouwens, L. C. (1997). The chemistry of tea flavonoids. *Critical Reviews in Food Science and Nutrition, 37,* 693–704.

Baltaziak, M., Skrzydlewska, E., Sulik, A., et al. (2004). Green tea as an antioxidant which protects against alcohol induced injury in rats—a histopathological examination.*Folia Morphologica, 63,* 123–126.

Bettuzzi, S., Brausi, M., Rizzi, F., et al. (2006). Chemoprevention of human prostate cancer by oral administration of green tea catechins in volunteers with high-grade prostate intraepithelial neoplasia:a preliminary report from a one-year proof-of-principle study. *Cancer Research, 66,* 1234–1240.

Blot, W. J., Chow, W. H., & McLaughlin, J. K. (1996). Tea and cancer:a review of the epidemiological evidence. *European Journal of Cancer Prevention: The Official Journal of the European Cancer Prevention Organisation (ECP), 5,* 425–438.

Blumenthal, M., Ferrier, G. K. L., & To, C. C. (2006). Total sales of herbal supplements in United States show steady growth sales in mass market channel show continued decline. *HerbalGram, 71,* 64–66.

Bold, R. J., Virudachalam, S., & McConkey, D. J. (2001). Chemosensitization of pancreatic cancer by inhibition of the 26S proteasome. *The Journal of Surgical Research, 100,* 11–17.

Bonner, M. R., Rothman, N., Mumford, J. L., et al. (2005). Green tea consumption, genetic susceptibility, PAH-rich smoky coal, and the risk of lung cancer. *Mutation Research, 582,* 53–60.

Borrelli, F., Capasso, R., Russo, A., et al. (2004). Systematic review: Green tea and gastrointestinal cancer risk. *Alimentary Pharmacology & Therapeutics, 19,* 497–510.

Bose, M., Lambert, J. D., Ju, J., et al. (2008). The major green tea polyphenol, (-)-epigallocatechin-3-gallate, inhibits obesity, metabolic syndrome, and fatty liver disease in high-fat-fed mice. *The Journal of Nutrition, 138,* 1677–1683.

Bruno, R. S., Dugan, C. E., Smyth, J. A., et al. (2008). Green tea extract protects leptin-deficient, spontaenously obese mice from hepatic steatosis and injury. *The Journal of Nutrition, 138,* 323–331.

Castro, D. J., Yu, Z., Lohr, C. V., et al. (2008). Chemoprevention of dibenzo[a,l]pyrene transplacental carcinogenesis in mice born to mothers administered green tea: Primary role of caffeine. *Carcinogenesis, 29,* 1581–1586.

Chan, M. M., Soprano, K. J., Weinstein, K., et al. (2006). Epigallocatechin-3-gallate delivers hydrogen peroxide to induce death of ovarian cancer cells and enhances their cisplatin susceptibility. *Journal of Cellular Physiology, 207,* 389–396.

Chearwae, W., Wu, C. P., Chu, H. Y., et al. (2006). Curcuminoids purified from turmeric powder modulate the function of human multidrug resistance protein 1 (ABCC1). *Cancer Chemotherapy and Pharmacology, 57,* 376–388.

Chen, L., Bondoc, F. Y., Lee, M. J., et al. (1996). Caffeine induces cytochrome P4501A2: Induction of CYP1A2 by tea in rats. *Drug Metabolism and Disposition: The Biological Fate of Chemicals, 24,* 529–533.

Chen, L., Lee, M. J., Li, H., et al. (1997). Absorption, distribution, elimination of tea polyphenols in rats. *Drug Metabolism and Disposition: The Biological Fate of Chemicals, 25,* 1045–1050.

Chen, J. H., Tipoe, G. L., Liong, E. C., et al. (2004). Green tea polyphenols prevent toxin-induced hepatotoxicity in mice by down-regulating inducible nitric oxide-derived prooxidants. *The American Journal of Clinical Nutrition, 80,* 742–751.

Chen, D., Wang, C. Y., Lambert, J. D., et al. (2005). Inhibition of human liver catechol-O-methyltransferase by tea catechins and their metabolites: Structure-activity relationship and molecular-modeling studies. *Biochemical Pharmacology, 69,* 1523–1531.

Chow, H. H., Hakim, I. A., Vining, D. R., et al. (2005). Effects of dosing condition on the oral bioavailability of green tea catechins after single-dose administration of Polyphenon E in healthy individuals. *Clinical Cancer Research: An Official Journal of the American Association for Cancer Research, 11,* 4627–4633.

Chow, H. H., Hakim, I. A., Vining, D. R., et al. (2006). Effects of repeated green tea catechin administration on human cytochrome P450 activity. *Cancer Epidemiology, Biomarkers & Prevention: A Publication of the American Association for Cancer Research, Cosponsored by the American Society of Preventive Oncology, 15,* 2473–2476.

Chow, H. H., Hakim, I. A., Vining, D. R., et al. (2007). Modulation of human glutathione s-transferases by polyphenon e intervention. *Cancer Epidemiology, Biomarkers & Prevention: A Publication of the American Association for Cancer Research, Cosponsored by the American Society of Preventive Oncology, 16,* 1662–1666.

Chung, F. L. (1999). The prevention of lung cancer induced by a tobacco-specific carcinogen in rodents by green and black Tea. *Proceedings of the Society for Experimental Biology and Medicine, 220,* 244–248.

Chung, F. L., Schwartz, J., Herzog, C. R., et al. (2003). Tea and cancer prevention: Studies in animals and humans. *The Journal of Nutrition, 133,* 3268S–3274S.

De Stefani, E., Correa, P., Boffetta, P., et al. (2004). Dietary patterns and risk of gastric cancer: A case-control study in Uruguay. *Gastric Cancer: Official Journal of the International Gastric Cancer Association and the Japanese Gastric Cancer Association, 7,* 211–220.

Donovan, J. L., Chavin, K. D., Devane, C. L., et al. (2004). Green tea (Camellia sinensis) extract does not alter cytochrome p450 3A4 or 2D6 activity in healthy volunteers. *Drug Metabolism and Disposition: The Biological Fate of Chemicals, 32,* 906–908.

Dou, Q. P. (2009). Molecular mechanisms of green tea polyphenols. *Nutrition and Cancer, 61,* 827–835.

Ellison, L. F. (2000). Tea and other beverage consumption and prostate cancer risk: A Canadian retrospective cohort study. *European Journal of Cancer Prevention: The Official Journal of the European Cancer Prevention Organisation (ECP), 9,* 125–130.

Galati, G., Lin, A., Sultan, A. M., et al. (2006). Cellular and in vivo hepatotoxicity caused by green tea phenolic acids and catechins. *Free Radical Biology & Medicine, 40,* 570–580.

Ganmaa, D., Willett, W. C., Li, T. Y., et al. (2008). Coffee, tea, caffeine and risk of breast cancer: A 22-year follow-up. *International Journal of Cancer. Journal International Du Cancer, 122,* 2071–2076.

Glei, M., & Pool-Zobel, B. L. (2006). The main catechin of green tea, (-)-epigallocatechin-3-gallate (EGCG), reduces bleomycin-induced DNA damage in human leucocytes. *Toxicology In Vitro: An International Journal Published in Association with BIBRA, 20,* 295–300.

Grove, K. A., & Lambert, J. D. (2010). Laboratory, epidemiological, and human intervention studies show that tea (Camellia sinensis) may be useful in the prevention of obesity. *The Journal of Nutrition, 140,* 446–453.

Gupta, S., Hastak, K., Ahmad, N., et al. (2001). Inhibition of prostate carcinogenesis in TRAMP mice by oral infusion of green tea polyphenols. *Proceedings of the National Academy of Sciences of the United States of America, 98,* 10350–10355.

Hakim, I. A., Harris, R. B., Brown, S., et al. (2003). Effect of increased tea consumption on oxidative DNA damage among smokers:a randomized controlled study. *The Journal of Nutrition, 133,* 3303S–3309S.

Hakim, I. A., Harris, R. B., Chow, H. H., et al. (2004). Effect of a 4-month tea intervention on oxidative DNA damage among heavy smokers: Role of glutathione S-transferase genotypes. *Cancer Epidemiology, Biomarkers & Prevention: A Publication of the American Association for Cancer Research, Cosponsored by the American Society of Preventive Oncology, 13,* 242–249.

Halder, B., Bhattacharya, U., Mukhopadhyay, S., et al. (2008). Molecular mechanism of black tea polyphenols induced apoptosis in human skin cancer cells: Involvement of Bax translocation and mitochondria mediated death cascade. *Carcinogenesis, 29,* 129–138.

Hong, J., Lambert, J. D., Lee, S. H., et al. (2003). Involvement of multidrug resistance-associated proteins in regulating cellular levels of (-)-epigallocatechin-3-gallate and its methyl metabolites. *Biochemical and Biophysical Research Communications, 310,* 222–227.

Hoshiyama, Y., Kawaguchi, T., Miura, Y., et al. (2005). Green tea and stomach cancer—a short review of prospective studies. *Journal of Epidemiology, 15*(Suppl 2), S109–S112.

Hou, Z., Sang, S., You, H., et al. (2005). Mechanism of action of (-)-epigallocatechin-3-gallate: Auto-oxidation-dependent inactivation of epidermal growth factor receptor and direct effects on growth inhibition in human esophageal cancer KYSE 150 cells. *Cancer Research, 65,* 8049–8056.

Huang, M. T., Xie, J. G., Wang, Z. Y., et al. (1997). Effects of tea, decaffeinated tea, and caffeine on UVB light-induced complete carcinogenesis in SKH-1 mice: Demonstration of caffeine as a biologically important constituent of tea. *Cancer Research, 57,* 2623–2629.

Inoue, M., Tajima, K., Mizutani, M., et al. (2001). Regular consumption of green tea and the risk of breast cancer recurrence: Follow-up study from the hospital-based epidemiologic research program at Aichi cancer center (HERPACC), Japan. *Cancer Letters, 167,* 175–182.

Isbrucker, R. A., Edwards, J. A., Wolz, E., et al. (2006). Safety studies on epigallocatechin gallate (EGCG) preparations. Part 2: Dermal, acute and short-term toxicity studies. *Food and Chemical Toxicology: An International Journal Published for the British Industrial Biological Research Association, 44,* 636–650.

Jian, L., Xie, L. P., Lee, A. H., et al. (2004). Protective effect of green tea against prostate cancer: A case-control study in southeast China. *International Journal of Cancer. Journal International Du Cancer, 108,* 130–135.

Jiang, T., Glickman, B. W., & de Boer, J. G. (2001). Protective effect of green tea against benzo[a] pyrene-induced mutations in the liver of Big Blue transgenic mice. *Mutation Research, 1,* 147–151, 480–481.

Jimenez-Saenz, M., & Martinez-Sanchez Mdel, C. (2006). Acute hepatitis associated with the use of green tea infusions. *Journal of Hepatology, 44,* 616–617.

Ju, J., Lu, G., Lambert, J. D., et al. (2007). Inhibition of carcinogenesis by tea constituents. *Seminars in Cancer Biology, 17,* 395–402.

Kaur, S., Greaves, P., Cooke, D. N., et al. (2007). Breast cancer prevention by green tea catechins and black tea theaflavins in the C3(1) SV40 T,t antigen transgenic mouse model is accompanied by increased apoptosis and a decrease in oxidative DNA adducts. *Journal of Agricultural and Food Chemistry, 55,* 3378–3385.

Khan, S. A., Priyamvada, S., Khan, W., et al. (2009). Studies on the protective effect of green tea against cisplatin induced nephrotoxicity. *Pharmacological Research: The Official Journal of the Italian Pharmacological Society, 60,* 382–391.

Kikuchi, N., Ohmori, K., Shimazu, T., et al. (2006). No association between green tea and prostate cancer risk in Japanese men: The Ohsaki cohort study. *British Journal of Cancer, 95,* 371–373.

Kim, M., Murakami, A., & Ohigashi, H. (2007). Modifying effects of dietary factors on (-)-epigallocatechin-3-gallate-induced pro-matrix metalloproteinase-7 production in HT-29 human colorectal cancer cells. *Bioscience, Biotechnology, and Biochemistry, 71,* 2442–2450.

Kohlmeier, L., Weterings, K. G., Steck, S., et al. (1997). Tea and cancer prevention: An evaluation of the epidemiologic literature. *Nutrition and Cancer, 27,* 1–13.

Kramata, P., Lu, Y. P., Lou, Y. R., et al. (2005). Effect of administration of caffeine or green tea on the mutation profile in the p53 gene in early mutant p53-positive patches of epidermal cells induced by chronic UVB-irradiation of hairless SKH-1 mice. *Carcinogenesis, 26,* 1965–1974.

Krishnan, R., Raghunathan, R., & Maru, G. B. (2005). Effect of polymeric black tea polyphenols on benzo(a)pyrene [B(a)P]-induced cytochrome P4501A1 and 1A2 in mice. *Xenobiotica, 35,* 671–682.

Kubik, A. K., Zatloukal, P., Tomasek, L., et al. (2004). Dietary habits and lung cancer risk among non-smoking women. *European Journal of Cancer Prevention: The Official Journal of the European Cancer Prevention Organisation (ECP), 13,* 471–480.

Kurahashi, N., Sasazuki, S., Iwasaki, M., et al. (2008). Green tea consumption and prostate cancer risk in Japanese men: A prospective study. *American Journal of Epidemiology, 167,* 71–77.

Kuzu, N., Bahcecioglu, I. H., Dagli, A. F., et al. (2008). Epigallocatechin gallate attenuates experimental non-alcoholic steatohepatitis induced by high fat diet. *Journal of Gastroenterology and Hepatology, 23*(8 Pt 2), e465–e470.

Lambert, J. D., & Elias, R. J. (2010). The antioxidant and pro-oxidant activities of green tea polyphenols: A role in cancer prevention. *Archives of Biochemistry and Biophysics, 501,* 65–72.

Lambert, J. D., Hong, J., Kim, D. H., et al. (2004). Piperine enhances the bioavailability of the tea polyphenol (-)-epigallocatechin-3-gallate in mice. *The Journal of Nutrition, 134,* 1948–1952.

Lambert, J. D., Hong, J., Yang, G. Y., et al. (2005). Inhibition of carcinogenesis by polyphenols: Evidence from laboratory investigations. *The American Journal of Clinical Nutrition, 81,* 284S–291S.

Lambert, J. D., Lee, M. J., Diamond, L., et al. (2006). Dose-dependent levels of epigallocatechin-3-gallate in human colon cancer cells and mouse plasma and tissues. *Drug Metabolism and Disposition: The Biological Fate of Chemicals, 34,* 8–11.

Lambert, J. D., Kennett, M. J., Sang, S., et al. (2010). Hepatotoxicity of high oral dose (-)-epigallocatechin-3-gallate in mice. *Food and Chemical Toxicology: An International Journal Published for the British Industrial Biological Research Association, 48,* 409–416.

Laurie, S. A., Miller, V. A., Grant, S. C., et al. (2005). Phase I study of green tea extract in patients with advanced lung cancer. *Cancer Chemother Pharmacol, 55,* 33–38.

Lee, M. J., Maliakal, P., Chen, L., et al. (2002). Pharmacokinetics of tea catechins after ingestion of green tea and (-)-epigallocatechin-3-gallate by humans: Formation of different metabolites and individual variability. *Cancer Epidemiology, Biomarkers & Prevention: A Publication of the American Association for Cancer Research, Cosponsored by the American Society of Preventive Oncology, 11,* 1025–1032.

Lee, A. H., Fraser, M. L., & Binns, C. W. (2009). Tea, coffee and prostate cancer. *Molecular Nutrition & Food Research, 53,* 256–265.

Leone, M., Zhai, D., Sareth, S., et al. (2003). Cancer prevention by tea polyphenols is linked to their direct inhibition of antiapoptotic Bcl-2-family proteins. *Cancer Research, 63,* 8118–8121.

Li, G. X., Chen, Y. K., Hou, Z., et al. (2010). Pro-oxidative activities and dose-response relationship of (-)-epigallocatechin-3-gallate in the inhibition of lung cancer cell growth: A comparative study in vivo and in vitro. *Carcinogenesis, 31,* 902–910.

Li, N., Chen, X., Liao, J., et al. (2002). Inhibition of 7,12-dimethylbenz[a]anthracene (DMBA)-induced oral carcinogenesis in hamsters by tea and curcumin. *Carcinogenesis, 23,* 1307–1313.

Li, Q., Kakizaki, M., Kuriyama, S., et al. (2008). Green tea consumption and lung cancer risk: The Ohsaki study. *British Journal of Cancer, 99,* 1179–1184.

Lin, D. X., Thompson, P. A., Teitel, C., et al. (2003). Direct reduction of N-acetoxy-PhIP by tea polyphenols: A possible mechanism for chemoprevention against PhIP-DNA adduct formation. *Mutation Research, 523–524,* 193–200.

Liu, T. T., Liang, N. S., Li, Y., et al. (2003). Effects of long-term tea polyphenols consumption on hepatic microsomal drug-metabolizing enzymes and liver function in Wistar rats. *World Journal of Gastroenterology: WJG, 9,* 2742–2744.

Lou, Y. R., Lu, Y. P., Xie, J. G., et al. (1999). Effects of oral administration of tea, decaffeinated tea, and caffeine on the formation and growth of tumors in high-risk SKH-1 mice previously treated with ultraviolet B light. *Nutrition and Cancer, 33,* 146–153.

Lu, G., Liao, J., Yang, G., et al. (2006). Inhibition of adenoma progression to adenocarcinoma in a 4-(methylnitrosamino)-1-(3-pyridyl)-1-butanone-induced lung tumorigenesis model in A/J mice by tea polyphenols and caffeine. *Cancer Research, 66,* 11494–11501.

Lu, G., Xiao, H., You, H., et al. (2008). Synergistic inhibition of lung tumorigenesis by a combination of green tea polyphenols and atorvastatin. *Clinical Cancer Research: An Official Journal of the American Association for Cancer Research, 14,* 4981–4988.

Lu, H., Meng, X., & Yang, C. S. (2003). Enzymology of methylation of tea catechins and inhibition of catechol-O-methyltransferase by (-)-epigallocatechin gallate. *Drug Metabolism and Disposition: The Biological Fate of Chemicals, 31,* 572–579.

Lu, Y. P., Lou, Y. R., Liao, J., et al. (2005). Administration of green tea or caffeine enhances the disappearance of UVB-induced patches of mutant p53 positive epidermal cells in SKH-1 mice. *Carcinogenesis, 26,* 1465–1472.

Maeta, K., Nomura, W., Takatsume, Y., et al. (2007). Green tea polyphenols function as prooxidants to activate oxidative-stress-responsive transcription factors in yeasts. *Applied and Environmental Microbiology, 73,* 572–580.

Maliakal, P. P., Coville, P. F., & Wanwimolruk, S. (2001). Tea consumption modulates hepatic drug metabolizing enzymes in Wistar rats. *The Journal of Pharmacy and Pharmacology, 53,* 569–577.

Mazzanti, G., Menniti-Ippolito, F., Moro, P. A., et al. (2009). Hepatotoxicity from green tea: A review of the literature and two unpublished cases. *European Journal of Clinical Pharmacology, 65,* 331–341.

Mu, L. N., Lu, Q. Y., Yu, S. Z., et al. (2005). Green tea drinking and multigenetic index on the risk of stomach cancer in a Chinese population. *International Journal of Cancer. Journal International Du Cancer, 116,* 972–983.

Ogunleye, A. A., Xue, F., & Michels, K. B. (2010). Green tea consumption and breast cancer risk or recurrence: A meta-analysis. *Breast Cancer Research and Treatment, 119,* 477–484.

Oz, H. S., McClain, C. J., Nagasawa, H. T., et al. (2004). Diverse antioxidants protect against acetaminophen hepatotoxicity. *Journal of Biochemical and Molecular Toxicology, 18,* 361–368.

Sai, K., Kai, S., Umemura, T., et al. (1998). Protective effects of green tea on hepatotoxicity, oxidative DNA damage and cell proliferation in the rat liver induced by repeated oral administration of 2-nitropropane. *Food and Chemical Toxicology: An International Journal Published for the British Industrial Biological Research Association, 36,* 1043–1051.

Saleem, M., Adhami, V. M., Siddiqui, I. A., et al. (2003). Tea beverage in chemoprevention of prostate cancer: A mini-review. *Nutrition and Cancer, 47,* 13–23.

Sang, S., Hou, Z., Lambert, J. D., et al. (2005). Redox properties of tea polyphenols and related biological activities. *Antioxidants & Redox Signaling, 7,* 1704–1714.

Sato, K., Sueoka, K., Tanigaki, R., et al. (2010). Green tea extracts attenuate doxorubicin-induced spermatogenic disorders in conjunction with higher telomerase activity in mice. *Journal of Assisted Reproduction and Genetics, 27,* 501–508.

Schwartz, J. L., Baker, V., Larios, E., et al. (2005). Molecular and cellular effects of green tea on oral cells of smokers: A pilot study. *Molecular Nutrition & Food Research, 49,* 43–51.

Setiawan, V. W., Zhang, Z. F., Yu, G. P., et al. (2001). Protective effect of green tea on the risks of chronic gastritis and stomach cancer. *International Journal of Cancer. Journal international Du Cancer, 92,* 600–604.

Sharpe, C. R., & Siemiatycki, J. (2002). Consumption of non-alcoholic beverages and prostate cancer risk. *European Journal of Cancer Prevention: The Official Journal of the European Cancer Prevention Organisation (ECP), 11,* 497–501.

Shen, G., Xu, C., Hu, R., et al. (2005). Comparison of (-)-epigallocatechin-3-gallate elicited liver and small intestine gene expression profiles between C57BL/6J mice and C57BL/6J/Nrf2 (-/-) mice. *Pharmaceutical Research, 22,* 1805–1820.

Shi, S. T., Wang, Z. Y., Smith, T. J., et al. (1994). Effects of green tea and black tea on 4-(methylnitrosamino)-1-(3-pyridyl)-1-butanone bioactivation, DNA methylation, and lung tumorigenesis in A/J mice. *Cancer Research, 54,* 4641–4647.

Shimizu, M., Shirakami, Y., Sakai, H., et al. (2008). (-)-Epigallocatechin gallate suppresses azoxymethane-induced colonic premalignant lesions in male C57BL/KsJ-db/db mice. *Cancer Prevention Research (Philadelphia, Pa), 1,* 298–304.

Shukla, Y. (2007). Tea and cancer chemoprevention: A comprehensive review. *Asian Pacific Journal of Cancer Prevention: APJCP, 8,* 155–166.

Siddiqui, I. A., Saleem, M., Adhami, V. M., et al. (2007). Tea beverage in chemoprevention and chemotherapy of prostate cancer. *Acta Pharmacologica Sinica, 28,* 1392–1408.

Sohn, O. S., Surace, A., Fiala, E. S., et al. (1994). Effects of green and black tea on hepatic xenobiotic metabolizing systems in the male F344 rat. *Xenobiotica, 24,* 119–127.

Sonoda, T., Nagata, Y., Mori, M., et al. (2004). A case-control study of diet and prostate cancer in Japan: Possible protective effect of traditional Japanese diet. *Cancer Science, 95,* 238–242.

Sun, C. L., Yuan, J. M., Koh, W. P., et al. (2006a). Green tea, black tea and breast cancer risk: A meta-analysis of epidemiological studies. *Carcinogenesis, 27,* 1310–1315.

Sun, C. L., Yuan, J. M., Koh, W. P., et al. (2006b). Green tea, black tea and colorectal cancer risk: A meta-analysis of epidemiologic studies. *Carcinogenesis, 27,* 1301–1309.

Suzuki, Y., Tsubono, Y., Nakaya, N., et al. (2004). Green tea and the risk of breast cancer: Pooled analysis of two prospective studies in Japan. *British Journal of Cancer, 90,* 1361–1363.

Tachibana, H. (2009). Molecular basis for cancer chemoprevention by green tea polyphenol EGCG. *Forum of Nutrition, 61,* 156–169.

Tang, L., Tang, M., Xu, L., et al. (2008). Modulation of aflatoxin biomarkers in human blood and urine by green tea polyphenols intervention. *Carcinogenesis, 29,* 411–417.

Tsao, A. S., Liu, D., Martin, J., et al. (2009). Phase II randomized, placebo-controlled trial of green tea extract in patients with high-risk oral premalignant lesions. *Cancer Prevention Research (Philadelphia, Pa), 2,* 931–941.

Tsugane, S., & Sasazuki, S. (2007). Diet and the risk of gastric cancer: Review of epidemiological evidence. *Gastric Cancer: Official Journal of the International Gastric Cancer Association and the Japanese Gastric Cancer Association, 10,* 75–83.

Tulayakul, P., Dong, K. S., Li, J. Y., et al. (2007). The effect of feeding piglets with the diet containing green tea extracts or coumarin on in vitro metabolism of aflatoxin B_1 by their tissues. *Toxicon, 50,* 339–348.

Vaidyanathan, J. B., & Walle, T. (2003). Cellular uptake and efflux of the tea flavonoid (-)epicatechin-3-gallate in the human intestinal cell line Caco-2. *The Journal of Pharmacology and Experimental Therapeutics, 307,* 745–752.

Weisburg, J. H., Weissman, D. B., Sedaghat, T., et al. (2004). In vitro cytotoxicity of epigallocatechin gallate and tea extracts to cancerous and normal cells from the human oral cavity. *Basic & Clinical Pharmacology & Toxicology, 95,* 191–200.

Wessner, B., Strasser, E. M., Koitz, N., et al. (2007). Green tea polyphenol administration partly ameliorates chemotherapy-induced side effects in the small intestine of mice. *The Journal of Nutrition, 137,* 634–640.

Wolf, A., Bray, G. A., & Popkin, B. M. (2008). A short history of beverages and how our body treats them. *Obesity Reviews: An Official Journal of the International Association for the Study of Obesity, 9,* 151–164.

Wolfram, S. (2007). Effects of green tea and EGCG on cardiovascular and metabolic health. *Journal of the American College of Nutrition, 26,* 373S–388S.

Wortelboer, H. M., Usta, M., van Zanden, J. J., et al. (2005). Inhibition of multidrug resistance proteins MRP1 and MRP2 by a series of alpha, beta-unsaturated carbonyl compounds. *Biochemical Pharmacology, 69,* 1879–1890.

Wu, A. H., Tseng, C. C., Van Den Berg, D., et al. (2003). Tea intake, COMT genotype, and breast cancer in Asian-American women. *Cancer Research, 63,* 7526–7529.

Xiao, H., Hao, X., Simi, B., et al. (2008). Green tea polyphenols inhibit colorectal aberrant crypt foci (ACF) formation and prevent oncogenic changes in dysplastic ACF in azoxymethane-treated F344 rats. *Carcinogenesis, 29,* 113–119.

Xu, G., Ren, G., Xu, X., et al. (2010). Combination of curcumin and green tea catechins prevents dimethylhydrazine-induced colon carcinogenesis. *Food and Chemical Toxicology: An International Journal Published for the British Industrial Biological Research Association, 48,* 390–395.

Xu, M., Bailey, A. C., Hernaez, J. F., et al. (1996). Protection by green tea, black tea, and indole-3-carbinol against 2- amino-3-methylimidazo[4,5-f]quinoline-induced DNA adducts and colonic aberrant crypts in the F344 rat. *Carcinogenesis, 17,* 1429–1434.

Yang, C. S., & Wang, Z. Y. (1993). Tea and cancer. *Journal of the National Cancer Institute, 85,* 1038–1049.

Yang, C. S., Maliakal, P., & Meng, X. (2002). Inhibition of carcinogenesis by tea. *Annual Review of Pharmacology and Toxicology, 42,* 25–54.

Yang, C. S., Lambert, J. D., Ju, J., et al. (2007). Tea and cancer prevention: Molecular mechanisms and human relevance. *Toxicology and Applied Pharmacology, 224,* 265–273.

Yang, C. S., Wang, X., Lu, G., et al. (2009). Cancer prevention by tea: Animal studies, molecular mechanisms and human relevance. *Nature Reviews Cancer, 9,* 429–439.

Yang, G. Y., Liao, J., Kim, K., et al. (1998). Inhibition of growth and induction of apoptosis in human cancer cell lines by tea polyphenols. *Carcinogenesis, 19,* 611–616.

Yang, G. Y., Liao, J., Li, C., et al. (2000). Effect of black and green tea polyphenols on c-jun phosphorylation and H(2)O(2) production in transformed and non-transformed human bronchial cell lines: Possible mechanisms of cell growth inhibition and apoptosis induction. *Carcinogenesis, 21,* 2035–2039.

Yen, G. C., Ju, J. W., & Wu, C. H. (2004). Modulation of tea and tea polyphenols on benzo(a) pyrene-induced DNA damage in Chang liver cells. *Free Radical Research, 38,* 193–200.

Yuan, J. H., Li, Y. Q., & Yang, X. Y. (2007a). Inhibition of epigallocatechin gallate on orthotopic colon cancer by upregulating the Nrf2-UGT1A signal pathway in nude mice. *Pharmacology, 80,* 269–278.

Yuan, J. M., Gao, Y. T., Yang, C. S., et al. (2007b). Urinary biomarkers of tea polyphenols and risk of colorectal cancer in the Shanghai cohort study. *International Journal of Cancer. Journal International Du Cancer, 120,* 1344–1350.

Zhang, M., Holman, C. D., Huang, J. P., et al. (2007). Green tea and the prevention of breast cancer: A case-control study in Southeast China. *Carcinogenesis, 28,* 1074–1078.

Zhong, L., Goldberg, M. S., Gao, Y. T., et al. (2001). A population-based case-control study of lung cancer and green tea consumption among women living in Shanghai, China. *Epidemiology, 12,* 695–700.

Zhu, B. T., Patel, U. K., Cai, M. X., et al. (2000). O-Methylation of tea polyphenols catalyzed by human placental cytosolic catechol-O-methyltransferase. *Drug Metabolism and Disposition: The Biological Fate of Chemicals, 28,* 1024–1030.

Chapter 17
Mylabris Phalerata (Chinese Blister Beetle) on Hematological Malignancies

Masahiro Kizaki and Morihiko Sagawa

Abstract The therapeutic approach to hematological malignancies is based on chemotherapy using anticancer drugs, but such treatment has serious side effects, and some complications (e.g. serious infection and bleeding) associated with use of these drugs can be fatal. Therefore, investigators have sought out new less toxic agents targeted against the molecules responsible for the pathogenesis of the hematological malignancies. Natural compounds appear to be safer than some recently released anticancer drugs, and such compounds are promising for the development of novel compounds with improved clinical activities. Cantharidin (CTD) is the active ingredient of *Mylabris phalerata* (Chinese blister beetle), and is one of many natural products used in traditional Chinese medicine for cancer treatment. CTD is a selective inhibitor of PP1 (protein phosphatase 1) and PP2A, and as such is necessary for growth inhibition of tumor cells. In addition, the cytotoxic activity of CTD is likely to be associated with PP1 and PP2A activity. Therefore, a number of investigations of the effects of CTD on cancer cells have been carried out to date. Although CTD and its derivatives have been synthesized and examined in terms of their participation in antitumor processes in various cancer cell lines, there has been little progress in terms of clinical applications. Thus, it will be necessary to address the molecular modes of action of CTD in tumor cells, including hematological malignant cells. To address the molecular modes of action of CTD in tumor cells, several genes functionally related to cell proliferation or apoptosis were recently identified by cDNA microarray analysis in CTD-treated cells. It is possible that CTD will eventually contribute to the development of molecular-targeted therapies and individualized treatment strategies for the patients with hematological malignancies.

M. Kizaki (✉)
Department of Hematology, Saitama Medical Center, Saitama Medical University, 1981 Kamoda, Kawagoe, Saitama 350-8550, Japan
e-mail: makizaki@saitama-med.ac.jp

W. C. S. Cho (ed.), *Evidence-based Anticancer Materia Medica,* Evidence-based Anticancer Complementary and Alternative Medicine,
DOI 10.1007/978-94-007-0526-5_17,

17.1 Introduction

Natural product extracts have been proven to be a rich source of small molecules that display antitumor activity (Mann 2002; Gordaliza 2007). The search for improved cytotoxic agents continues to be important for the discovery of novel anticancer drugs. The structural diversity of natural compounds and their potential bioactivity have led to the isolation of several products isolated from plants, marine flora and microorganisms expected to provide "lead" compounds for the development of their therapeutic potential *via* molecular modification. The molecular modification of the functional groups of such lead compounds can generate structural analogs with anticancer effects. Modification processes have thus been able to advance the compounds towards their introduction into the clinical setting. For example, the main constituent of the plant *Catharanthus reseus* was the frontrunner among anticancer agents known as the vinca alkaloids (vincristine and vinblastine). Both of these drugs, which were introduced into clinical use in the late 1960s, and both have contributed to long-term remissions and cures of testicular teratoma, malignant lymphoma, lymphoblastic leukemia, and many other cancers (Lee 1999; Mukherjee et al. 2001). An example of a marine-derived anticancer agent developed for clinical use would be cytarabine (Adrian 2007), which is currently used to treat patients with lymphoma and leukemia, in particular acute myeloid leukemia (AML). Another agent, gemcitabine, was derived from a marine organism; gemcitabine has been shown to process significant therapeutic activity in patients with various types of solid tumors (Coseri 2009). In general, agents derived from natural compounds have contributed significantly to the successful treatment of several different types of cancers. Natural products or their structural relatives now account for approximately 50% of all drugs used as cancer chemotherapies. In addition, natural products from traditional Chinese medicine (TCM) can recently contribute to the development of molecular-targeted therapy for various malignancies (Chen and Liang 2010; Youns et al. 2010). Therefore, a very important area of research within the field of hematology/oncology is the routine examination of terrestrial plants and microorganisms aimed at the discovery of novel anticancer agents with more potent, more selective and less toxic compounds (Kizaki 2006).

17.2 Cantharidin (CTD)

CTD is a derivative of *Mylabris phalerata* (Chinese blister beetle) and a potent inhibitor of the protein phosphatase (PP) 2A (Li and Casida 1992; Honkanen 1993; Eldridge and Casida 1995; McCluskey et al. 2001; To 2004; Shan et al. 2006; Rauh et al. 2007). The basic structure of CTD consists of a 7-oxobicycloheptane, a dicarboxylic acid, and an anhydride (Fig. 17.1) (Honkanen 1993; Shan et al. 2006; Rau et al. 2007). The use of dried Chinese blister beetles as an anticancer agent extends back through more than 2,000 years of TCM (Wang 1989; Rauh et al. 2007). To

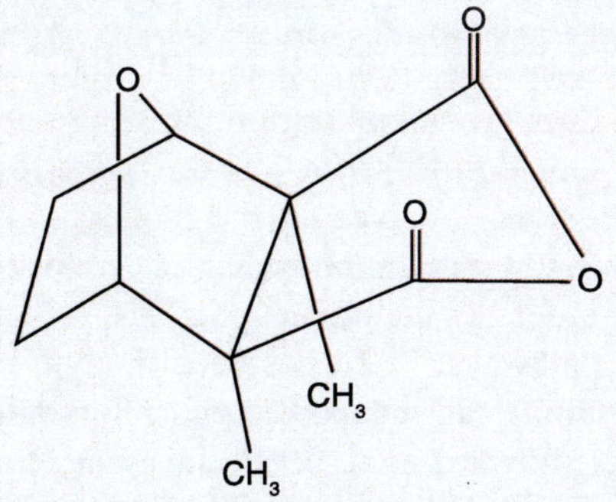

Fig. 17.1 Chemical structure of cantharidin

date, CTD has been reported in the literature to exert an effect on the control of cell-cycle regulation and on the cellular growth of tumor cells with transformed SV40 antigen (Clarke et al. 1993; Sontag et al. 1993; Janssens and Goris 2001). Recent studies have shown that CTD induces apoptosis in the following types of cancer; human hepatoma, pancreatic cancer, colon cancer, oral buccal carcinoma, and human leukemia cells (Peng et al. 2002; Chen et al. 2002, 2008; Williams et al. 2003; Huh et al. 2004; Huan et al. 2006; Kok et al. 2006a, b; Li et al. 2010). In clinical trials, CTD was found to be toxic to mucous membrane tissues, including those in the gastrointestinal tract, urethra, and kidney with a dose-limiting factor of renal toxicity (Wang 1989). In addition, it has reported that there are cases of CTD poisoning following ingestion of the beetles (Tagwireyi et al. 2000). The symptoms of CTD poisoning are directed towards the gastrointestinal tract and the genitourinary tract. In this case report, the patient presented with lower abdominal pain, hematuria, proteinuria and oligouria (Tagwireyi et al. 2000). Norcantharidin was synthesized to reduce renal toxicity, and is currently being clinically evaluated for use in colon cancer patients (Wang 1989). Furthermore, CTD and its analogues have exhibited therapeutic effects against primary hepatoma and esophageal carcinoma without being associated with the suppression of bone marrow (Wang 1989; Li et al. 2010). Recent large systemic review and meta-analysis in China provide the important information that TCM including CTD can be possible therapeutic option for the patients with hepatoma (Wu et al. 2009). However, this should be necessary to confirm by well-conducted prospective randomized clinical trial.

17.3 Effects of CTD on Tumor Cells

17.3.1 Protein Phosphatase Activity

The molecular mechanisms of CTD responsible for its underlying anticancer effects remain unknown. However, CTD is known to be a PP1 (protein phosphatase 1) inhibitor as well as a PP2A (protein phosphatase 2A) inhibitor (Li and Casida 1992;

Honkanen 1993; Eldridge and Casida 1995; McCluskey et al. 2001; To et al. 2004; Shan et al. 2006; Rauh et al. 2007). Such activity appears necessary for CTD to induce growth inhibition. PP2A is a ubiquitous enzyme involved in the dephosphorylation of the serine and threonine residues of cellular phosphoproteins, and thereby is also involved in the moderation of cellular proliferation (Janssens and Goris 2001). Protein phosphatases are also involved in the regulation of multiple cellular processes including apoptosis, cell cycle regulation, and various signal transduction pathways. PP2A has been thought to be a tumor suppressor, because inhibition of PP2A can induce the phosphorylation and activation of various substrate kinases (Millward et al. 1999; Janssens et al. 2005). Members of the MAPK family (i.e. ERK, JNK, and p38 MAP kinase) are direct substrates of PP2A (Millward et al. 1999). In general, the activation of MAPKs promotes the growth of tumor cells; however, continuous activation of these kinases can also inhibit proliferation and induce apoptosis in many tumor cells (Zhang and Liu 2002). PP2A inhibits the signal transduction pathway of ERK, thus affecting the calcium/calmodulin- and ceramide-dependent pathways of cellular growth in tumor cells (Li et al. 2010). It has also been reported that MAPK-family kinases such as JNK and ERK become active after CTD stimulation, which results in an increase in the caspase-3-mediated apoptosis of tumor cells (Huh et al. 2004; Schweyer et al. 2007). PP4 is a family of serine/threonine phosphatases that regulate a variety of cellular functions not regulated by PP2A. It has been reported that NF-κB signaling as well as the mammalian target of rapamycin (mTOR) pathways are regulated by PP4, and CTD possesses similar inhibitory activity to that of PP4 (Cohen et al. 2005). In addition, it has been reported that CTD inhibits JNK, but neither ERK nor p38 in pancreatic cancer cells, suggesting that CTD exerts its anticancer effects *via* a JNK-dependent pathway (Li et al. 2010). The findings to date suggest that CTD activates the MAP-kinase pathways, which results in a significant increase in caspase-mediated apoptosis in tumor cells. In addition, it is well known that okadaic acid, another PP2A inhibitor, induces the apoptosis of various types of tumor cell including leukemia, myeloma, hepatoma, and intestinal epithelial cells (Ishida et al. 1992; Kang et al. 1996; Lamboley et al. 2000; Ray et al. 2005).

17.3.2 Microarray Analysis

To identify candidate genes that affect the sensitivity of tumor cells to CTD, several microarray analyses were performed. Efferth (2005) and Rauh et al. (2007) identified 21 of 9,706 genes, the mRNA expression of which in 60 tumor cell lines correlated with the sensitivity of tumor cells to CTD. The majority of these genes are involved in DNA damage response, DNA repair, and apoptosis (Efferth 2005). From these CTD-related 21 genes identified by Efferth, important genes related to DNA repair and induction of cellular apoptosis is listed (Table 17.1). Among the panel of CTD-regulated genes, *PPP1R13B* is an interesting gene, because CTD is an inhibitor of PP1 and PP2A. PPP1R13B is the regulatory subunit 13B of PP1

Table 17.1 Important genes induced by cantharidin. (from Efferth 2005 with modification)

Symbols	Gene name	Function
SUSP1	SUMO-1-specific protease	Maturation and activation of Sumo-1. Related to DNA repair and induction of apoptosis
RPA2	Replication protein A2	DNA damage recognition, DNA replication, recombination and repair
CASP4	Caspase 4, apoptosis-related cysteine peptidase	Apoptosis execution. Stress-inducing agents activate endoplasmic reticulum-localized caspase 4
PPP1R13B	Protein phosphatase 1, regulatory (inhibitor) subunit 13B	Interaction with protein phosphatase 1. DNA repair and regulation of apoptosis
PDLIM1	PDZ and LIM domain 1	Cytoskeletal adaptor for proteins *via* its LIM and PDZ domains. Regulation of apoptosis *via* interaction with Fas-, p53-, NF-κB-, and Myc-signaling
UNR	N-RAS-related gene	RNA-binding protein; coordinated up-regulation with *N-RAS*. Regulation of apoptosis
HSBP1	Heat shock binding protein 1	Negative regulator of stress responses by binding to HSP70. DNA repair and regulation of apoptosis

Out of 21 genes identified by Efferth, important genes related to DNA repair and induction of cellular apoptosis are listed. These genes were identified by COMPARE and false discovery rate analyses whose mRNA expression in the panel of 60 human tumor cell lines of the Developmental Therapeutics program of the National Cancer Institute (USA) correlated with IC_{50} values for cantharidin (Efferth 2005)

and plays a central role in the regulation of induction of cellular apoptosis *via* interaction with p53, suggesting that PPP1R13B has related to CTD-induced DNA repair and apoptosis. These results suggest that oxidative stress response genes reduce the activity of CTD by inducing DNA strand breaks, which may induce the apoptosis of tumor cells in a p53- and Bcl-2-dependent manner. It has also been reported that CTD induced the expression of Bax protein, but down-regulated the expression of Bcl-2 and survivin in A549 cells resulting in the induction of apoptosis (Liu and Chen 2009). Another group used HL-60 myeloid leukemia cells treated with CTD to demonstrate the up- and down-regulation of mRNA expression levels of 2,087 genes by cDNA microarrays (Zhang et al. 2004). That group reported that the CTD-treated cells did not exhibit any decreases in the expression of genes coding for proteins involved in DNA repair, DNA replication, or proteins with oncogenic activity. Furthermore, these cells overexpressed genes that encode growth inhibitory proteins such as *BTG2* and *MCP-3*, and proapoptotic genes. They also observed a down-regulation in the expression of multidrug resistance-associated protein genes in CTD-treated HL-60 cells, suggesting that CTD may be related to the increased expression of genes that modulate drug sensitivity in tumor cells.

17.3.3 DNA Damage and Repair Induced by CTD

The microarray analyses identified several apoptosis- and cell cycle regulated-genes involved in CTD-induced growth inhibition of tumor cells. In addition, DNA damage and repair correlated with the IC_{50} values for CTD in many tumor cell lines (Pang et al. 2007). Recently, Efferth et al. reported that CTD induces apoptosis in various leukemic cells *via* a p53-dependent mechanism (Efferth et al. 2005). The phosphorylation of p53 stabilizes the protein; therefore, PP1 and PP2A inhibitors such as CTD may synergistically enhance p53 activity. CTD inhibits the expression of anti-apoptotic Bcl-2 protein in leukemic cells, suggesting that a DNA damage-triggered mitochondrial pathway is also involved. Moreover, CTD causes both DNA single- and double-strand breaks in a time-dependent manner (Rauh et al. 2007). The nuclear DNA PolB (polymerase β) is a key enzyme in base excision repair (Efferth et al. 2005). This enzyme was found to be correlated with increased cell survival in CTD-treated cells, suggesting that the increased DNA strand breakage and DNA repair were related to a decrease in cellular sensitivity to CTD. Interestingly, it has reported that the synthetic integration of CTD-derived demethylcantharidin with classical platinum-based cytotoxic drug shows superior anticancer effect (Pang et al. 2007). This novel compound caused additional DNA damage in cancer cells suggesting that drug-herb interaction can provide novel efficient anticancer agent.

It has been also reported that the CTD-induced growth inhibition of tumor cells was dependent on the induction of oxidative stress, resulting in the induction of apoptosis and cell cycle arrest (Rauh et al. 2007). This oxidative stress initiated by CTD may cause direct DNA damage and p53-dependent apoptosis in tumor cells. However, Li et al. (2010) reported that CTD increased reactive oxygen species (ROS) levels in pancreatic cancer cells, and growth inhibition remained unaffected by the treatment with CTD. These results suggest that CTD may induce apoptosis and cell cycle arrest in pancreatic cancer cells in an oxidative stress-independent manner.

17.4 Effects of CTD on Malignant Hematological Cells

17.4.1 Therapeutic Approaches to the Treatment of Leukemia

Significant advances in molecular biology and therapies for hematological malignancies have been made over the last decade. Therapeutic approaches to the treatment of hematological malignancies such as leukemia, malignant lymphoma, and multiple myeloma are basically chemotherapies to eradicate malignant cells. In the past two decades, clinical research aimed at improving the cure rate for patients with hematological malignancies has focused primarily on increasing cytotoxic drug delivery with the aim of maximizing the number of tumor cells eradicated based on a concept of referred to as "total cell kill" (Skipper 1974). However, severe side effects and complications due to anticancer drugs remain major problems in clinical practice. In addition, relapses are usually refractory to chemotherapy and are associated with a poor prognosis. Therefore, chemotherapy for hematological malignancies is limited by the development of drug resistance in tumor cells, adverse side effects, and myelosuppression. These serious clinical issues point to some of the limitations of current therapeutic strategies for the treatment of hematological malignancies. Therefore, novel more effective therapeutic approaches with less toxicity are still needed. Recent advances in the clarification of the molecular pathogenesis of leukemia have led to development of novel therapeutic approaches for the treatment of both AML and chronic myeloid leukemia (CML).

AML is a heterogeneous group of malignant disorders of hematopoietic progenitor cells marked by an accumulation of granulocyte and monocyte precursors in the bone marrow and peripheral blood. Despite scientific advances in our understanding of the epidemiologic, genetic, and biological features of AML, the disease remains fatal in a majority of patients, especially older individuals. In the 1980s, the use of all-*trans* retinoic acid for differentiation-inducing therapy in patients with acute promyelocytic leukemia was proposed by Professor Wang in Shanghai (China); subsequent therapeutic strategies for the treatment of leukemia yielded dramatic improvements in clinical outcome (Huang et al. 1988). Therapeutic strategies for inducing cellular differentiation and apoptosis in acute promyelocytic leukemia cells using all-*trans* retinoic acid and arsenic trioxide first described in China are one recent successful examples of the clinical application of natural compounds (Tallman et al. 2002). Recently, more specifically targeted agents have been developed for the treatment of AML, including anti-CD33 antibodies and immunoconjugate drugs, inhibitors of multidrug resistance proteins, farnesyl transferase inhibitors, tyrosine kinase inhibitors, histone deacetylase, and proteasome inhibitors (Grant 2009). However, these targeted-therapy candidates have yet to be translated into clinical application. In addition to conducting therapeutic trials, it will be important to identify other novel and highly specific therapeutic agents in parallel with our evolving understanding of the biology of AML.

CML is a clonal disease characterized by the presence of the Philadelphia chromosome and its oncogenic product, BCR-ABL, a constitutively active tyrosine kinase. Over the past three decades, several effective strategies for the treatment of CML have been developed (Valent 2010). IFN-α (interferon-alpha) is a cytoreductive agent that has been widely used to control CML. However, most patients treated with IFN-α do not exhibit a long-lasting response (Goldman 2009); the standard curative approach to CML is hematopoietic stem cell transplantation (HSCT) (Venepalli et al. 2010). However, HSCT can only be offered to a small subset of patients. Notably, HSCT is associated with enhanced disease-free survival, but also with transplant-related mortality and occasionally high morbidity (Venepalli et al. 2010). The approach to treating CML was revolutionized by the introduction of imatinib mesylate, a BCR-ABL tyrosine kinase inhibitor as a molecular-targeted therapy (Goldman 2001). The clinical use of specific BCR-ABL inhibitors has resulted in a significantly improved prognosis, response rate, overall survival, and patient outcome in CML patients compared to that achieved with previous therapeutic regimens. However, the complete eradication of CML in patients receiving imatinib mesylate was limited by the emergence of resistance, mostly due to mutations in the ABL kinase domain, and to a lesser extent by molecular residual disease after treatment. The second-generation BCR-ABL tyrosine kinase inhibitors, nilotinib and dasatinib, have shown significant activity in clinical trials in patients intolerant or resistant to imatinib therapy, except in those patients with the T315I *BCR-ABL* mutation (Agrawal et al. 2010).

17.4.2 Effects of CTD on Leukemia

Epidemiological investigation and laboratory studies have indicated that bioactive natural compounds play an important role in the treatment of many cancers. Therefore, investigators have actively sought out new agents that will potentially yield positive clinical outcomes while inducing less toxicity than previous agents; these candidate agents are derived from a variety of chemical compounds, and from various natural products. Along these lines, we have carried out and reported studies of various bioactive agents derived from natural compounds that induce apoptosis, a basic molecular mechanism that takes place in human leukemia and myeloma cells (Ito et al. 2005; Nakazato et al. 2005a, b; Shimizu et al. 2006; Xian et al. 2007; Sagawa et al. 2008; Nakaya et al. 2010).

In the clinical setting, CTD has been used for the treatment of hepatocellular carcinoma and leukemia. However, renal toxicity and suppressive effects on bone marrow limit its use. To reduce toxicity, several modified CTD analogues with antitumor effects have been chemically synthesized (Lin et al. 2000). Recently, Kok et al. (2006b) synthesized a number of CTD analogues and screened them for possible cytotoxic effects using a panel of cancer cell lines. The focus of that study was on the electron distribution of molecules and the core structure of CTD with diimide and the 6-position trifluoromethroxy group. A new CTD analogue designated as CAN 032 was synthesized, and then CAN 037 was subsequently generated by modifying the methyl group at the 6-position (Kok et al. 2006b). This new compound strongly induced apoptosis in the myeloid leukemia cell line KG-1 *via* the induction of caspase-3, -8, and -9 activity. Furthermore, CTD induced apoptosis in the human lymphoid leukemia cell line CCRF-CEM in a p53-dependent manner. That study by Kok and colleagues demonstrated that CTD causes oxidative stress provoking DNA damage and p53-dependent cell death (Kok et al. 2006b).

Recent stem cell biology studies have classified leukemic stem cells (LSCs) as associated with a number of different types of leukemia; LSCs are a major focus of current research interest (Lane and Gilliland 2010). LSCs are an important target for the treatment of leukemia, and failure to eradicate these very primitive cancer cells is a common cause of relapse in leukemia patients. Therefore, improved understanding of the biology of LSCs, and of the differences between normal hematopoietic and leukemic stem cells is likely to lead to the development of novel therapies as well as to increase in patient survival. Recently, it has been shown that CTD targets primary AML stem and progenitor cells in contrast to conventional chemo-therapeutic agents such as cytarabine and daunorubicin, which primarily target differentiated and cycling leukemic cells (Dorn et al. 2009). Dorn et al. reported that CTD reduced levels of the LSC target gene *HLF* at the protein level, and CTD also reduced the incidence of other leukemia-associated gene mutations such as activating mutations of *FLT-3* (Dorn et al. 2009). Constitutive active FLT-3 activates STAT5, causing an expansion of the hematopoietic stem cell pool and inhibiting myeloid differentiation *via* C/EBPα down-regulation (Schuringa et al. 2004; Chung et al. 2005; Wierenga et al. 2006). It was found that CTD inhibited IL-3-induced STAT5 phosphorylation in human AML stem and progenitor cells, suggesting that STAT5 is the direct target of CTD. From these experimental results, it appears likely that CTD could have a beneficial therapeutic effect on LSC-associated pathways, leading to potent eradication of LSCs and thereby a reduction in tumor burden.

17.4.3 Therapeutic Approaches to Multiple Myeloma

Multiple myeloma is characterized by the latent accumulation of secretory plasma cells with a low proliferative index and an extended life span in the bone marrow. Conventional therapy for multiple myeloma involves combinations of vincristine,melphalan, cyclophosphamide,doxorubicin (Adriamycin), and prednisone or dexamethasone (Palumbo and Rajkumar 2010). Patients younger than 65 years are usually given high-dose melphalan with autologous stem cell support, and older patients or those who cannot tolerate such intensive treatment are given standard-dose oral melphalan and dexamethasone. However, these treatments are associated with low remission rates, short survival times, and the development of drug resistance (Kumar et al. 2008). Chemo-resistance remains a major therapeutic challenge in the treatment of multiple myeloma.The precise mechanism underlying chemo-resistance in multiple myeloma is not clear, but one of the main contributors to both chemo-resistance and pathogenesis is thought to be the activation of NF-κB and STAT3 and the dysregulation of apoptosis (Hideshima and Anderson 2002). Recently, novel agents such as bortezomib, thalidomide, and lenalidomide, which target myeloma cells and their microenvironments, have shown remarkable activity against refractory and chemo-resistant cases in early clinical trials, and prolonged progression-free and overall survival of multiple myeloma patients (Weber 2003; Yasui et al. 2006). Progression and chemo-resistance are thought to involve interleukin (IL)-6, the expression of which is induced by NF-κB, *via* its regulation of the growth and survival of myeloma cells (Hideshima and Anderson 2002). IL-6 leads to the constitutive activation of STAT3, which in turn results in the expression of high levels of anti-apoptotic Bcl-xL and Mcl-1 proteins (Catlett-Falcone et al. 1999; Zhang et al. 2002). Thus, both the constitutive activation of NF-κB and STAT3 play an important role in chemo-resistance, and it is expected that the inhibition of NF-κB and STAT3 may overcome such chemo-resistance.

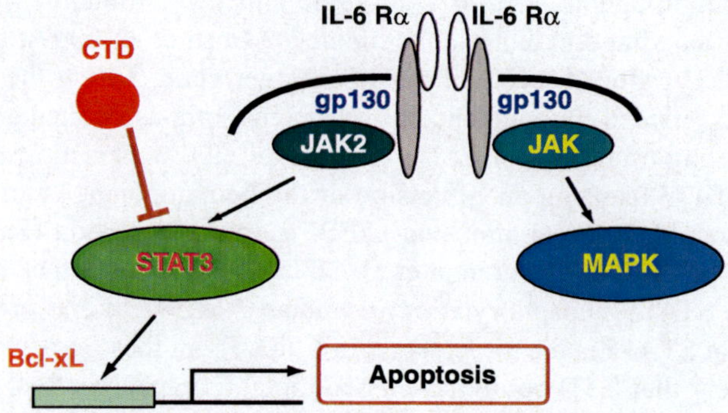

Fig. 17.2 Mechanisms of cantharidin (CTD)-induced apoptosis in human myeloma cells. CTD inhibited the phosphorylation of STAT3 (signal transduction and activator of transcription 3) and down-regulated the expression of anti-apoptotic Bcl-xL protein. STAT3 directly bound and activated the transcription of the *Bcl-xL* gene promoter. Thus, CTD induced apoptosis of human myeloma cells *via* the down-regulation of Bcl-xL and modulation of the STAT3 signaling pathway

17.4.4 Effects of CTD on Myeloma Cells

The use of natural agents may increasingly allow us to overcome treatment resistance without incurring some of the debilitating side effects of conventional chemotherapy. We have reported that CTD inhibited the cellular growth of human myeloma cell lines as well as that of myeloma cells freshly isolated from patients (Sagawa et al. 2008). Cultivation with CTD induced the apoptosis of myeloma cells in a cell-cycle-independent manner. Treatment with CTD induced caspase-3, -8, and -9 activity, which was in turn completely blocked by the respective caspase inhibitors. CTD inhibited the phosphorylation of STAT3 at the tyrosine 705 residue and down-regulated the expression of the anti-apoptotic Bcl-xL protein. STAT3 directly bound and activated the transcription of the *Bcl-xL* gene promoter, resulting in the induction of the expression of Bcl-xL in myeloma cells (Fig. 17.2) (Sagawa et al. 2008). Thus, it is clear that CTD-induced apoptosis in human myeloma cells may be mediated by the induction of anti-apoptotic proteins, and that CTD may has the potential to become a new therapeutic agent in the field of signal transduction therapies.

17.5 Conclusion

Patients with hematological malignancies such as leukemia and multiple myeloma often face a fatal clinical outcome. High-dose chemotherapy followed by hematopoietic stem cell transplantation has produced increasingly high remission rates, but this approach often causes serious clinical side effects and incurs the risk of early mortality, especially in elderly patients. Most patients ultimately relapse; therefore, new therapeutic approaches based on novel insights into the pathogenesis of hematological malignancies and that target molecules involved in cellular growth are needed. Natural products may play an important role in the development of novel drugs, in particular those for the treatment of cancer. The advantage of natural products for clinical application is the potential lack of toxicity. Therefore, it is hoped that compounds that induce apoptosis in cancer cells might be developed as new potent anticancer agents for the management of hematological malignancies, particularly in older patients. In order to yield successful clinical applications, target molecules from natural products will need to be identified for the development of rational treatment strategies.

Recently, many studied of the potential anticancer effects of CTD and its derivatives have been reported. CTD is recognized as a potent and selective inhibitor of PP1 and PP2A in tumor cells. The effects of CTD and its derivatives on cancer cell signal transduction pathways have investigated and the molecular mechanisms of CTD have gradually been revealed; however, the precise mechanisms of the above mentioned anticancer activities of CTD remain unclear. The molecular mechanisms of CTD and its derivatives still need to be investigated *in vitro* and *in vivo*; future studies will help elucidate the details of CTD-induced anticancer effects. Additional chemical modifications and the development of new CTD analogues are also needed, since there is great potential for molecules to be developed with reduced toxicity that retain antitumor- and target-related efficacy.

References

Adrian, T. E. (2007). Novel marine-derived anti-cancer agents. *Current Pharmaceutical Design, 13,* 3417–3426.

Agrawal, M., Garg, R. J., Kantarjian, H., et al. (2010). Chronic myeloid leukemia in the tyrosine kinase inhibitor era: what is the "best" therapy? *Current Oncology Reports, 12,* 302–323.

Catlett-Falcone, R., Landowski, T. H., Oshiro, M. M., et al. (1999). Constitutive activation of Stat3 signaling confers resistance to apoptosis in human U266 myeloma cells. *Immunity, 10,* 105–115.

Chen, Y. J., Kuo, C. D., Tsai, Y. M., et al. (2008). Norcantharidin induces anoikis through Jun-N-terminal kinase activation in CT26 colorectal cancer cells. *Anti-Cancer Drugs, 19,* 55–64.

Chen, Y. N., Chen, J. C., Yin, S. C., et al. (2002). Effector mechanisms of norcantharidin-induced mitotic arrest and apoptosis in human hepatoma cells. *International Journal of Cancer. Journal International Du Cancer, 100,* 158–165.

Chen, Z. F., & Liang, H. (2010). Progress in TCM metal-based antitumor agents. *Anti-Cancer Agents in Medicinal Chemistry, 10,* 412–423.

Chung, K. Y., Morrone, G., Schuringa, J. J., et al. (2005). Enforced expression of an Flt3 internal tandem duplication in human CD34+ cells confers properties of self-renewal and enhanced erythropoiesis. *Blood, 105,* 77–84.

Clarke, P. R., Hoffmann, I., Draetta, G., et al. (1993). Dephosphorylation of cdc25-C by a type-2A protein phosphatase: Specific regulation during the cell cycle in Xenopus egg extracts. *Molecular Biology of the Cell, 4,* 397–411.

Cohen, P. T., Philp, A., & Vázquez-Martin, C. (2005). Protein phosphatase 4—from obscurity to vital function. *FEBS Letters, 579,* 3278–3286.

Coseri, S. (2009). Natural products and their analogues as efficient anticancer drugs. *Mini Reviews in Medicinal Chemistry, 9,* 560–571.

Dorn, D. C., Kou, C. A., Png, K. J., et al. (2009). The effect of cantharidins on leukemic stem cells. *International Journal of Cancer. Journal International Du Cancer, 124,* 2186–199.

Efferth, T. (2005). Microarray-based prediction of cytotoxicity of tumor cells to cantharidin. *Oncology Reports, 13,* 459–463.

Efferth, T., Rauh, R., Kahl, S., et al. (2005). Molecular modes of action of cantharidin in tumor cells. *Biochemical Pharmacology, 69,* 811–818.

Eldridge, R., & Casida, J. E. (1995). Cantharidin effects on protein phosphatases and the phosphorylation state of phosphoproteins in mice. *Toxicology and Applied Pharmacology, 130,* 95–100.

Goldman, J. M. (2009). Treatment strategies for CML. *Best Practice & Research. Clinical Haematology, 22,* 303–313.

Goldman, J. M., & Melo, J. V. (2001). Targeting the BCR-ABL tyrosine kinase in chronic myeloid leukemia. *The New England Journal of Medicine, 344,* 1084–1086.

Gordaliza, M. (2007). Natural products as leads to anticancer drugs. *Clinical & Translational Oncology: Official Publication of the Federation of Spanish Oncology Societies and of the National Cancer Institute of Mexico, 9,* 767–776.

Grant, S. (2009). New agents for AML and MDS. *Best Practice & Research. Clinical Haematology, 22,* 501–507.

Hideshima, T., & Anderson, K. C. (2002). Molecular mechanisms of novel therapeutic approaches for multiple myeloma. *Nature Reviews. Cancer, 2,* 927–937.

Honkanen, R. E. (1993). Cantharidin, another natural toxin that inhibits the activity of serine/threonine protein phosphatases types 1 and 2A. *FEBS Letters, 330,* 283–286.

Huan, S. K., Lee, H. H., Liu, D. Z., et al. (2006). Cantharidin-induced cytotoxicity and cyclooxygenase 2 expression in human bladder carcinoma cell line. *Toxicology, 223,* 136–143.

Huang, M. E., Ye, Y. C., Chen, S. R., et al. (1988). Use of all-trans retinoic acid in the treatment of acute promyelocytic leukemia. *Blood, 72,* 567–572.

Huh, J. E., Kang, K. S., Chae, C., et al. (2004). Roles of p38 and JNK mitogen-activated protein kinase pathways during cantharidin-induced apoptosis in U937 cells. *Biochem Pharmacol, 67,* 1811–1818.

Ishida, Y., Furukawa, Y., Decaprio, J. A., et al. (1992). Treatment of myeloid leukemic cells with the phosphatase inhibitor okadaic acid induces cell cycle arrest at either G1/S or G2/M depending on dose. *Journal of Cellular Physiology, 150,* 484–492.

Ito, K., Nakazato, T., Xian, M. J., et al. (2005). 1'-acetoxychavicol acetate is a novel NF-κB inhibitor with significant activity against multiple myeloma in vitro and in vivo. *Cancer Research, 65,* 4417–4424.

Janssens, V., & Goris, J. (2001). Protein phosphatase 2A: A highly regulated family of serine/threonine phosphatases implicated in cell growth and signaling. *The Biochemical Journal, 353,* 417–439.

Janssens, V., Goris, J., & Van Hoof, C. (2005). PP2A: The expected tumor suppressor. *Current Opinion in Genetics & Development, 15,* 34–41.

Kang, H. S., Lee, B. S., Yang, Y., et al. (1996). Roles of protein phosphatase 1 and 2A in an IL-6-mediated autocrine growth loop of human myeloma cells. *Cellular Immunology, 168,* 174–183.

Kizaki, M. (2006). New therapeutic approach for myeloid leukemia: implications of apoptosis via modulation of reactive oxygen species production by natural compounds. *International Journal of Hematology, 83,* 283–288.

Kok, S. H., Chui, C. H., Lam, W. S., et al. (2006a). Apoptotic activity of a novel synthetic cantharidin analogue on human hepatoma cell lines. *International Journal of Molecular Medicine, 17,* 945–949.

Kok, S. H., Chui, C. H., Lam, W. S., et al. (2006b). Apoptogenic activity of a synthetic cantharimide in leukemia: implication on its structural activity relationship. *International Journal of Molecular Medicine, 18,* 1217–1221.

Kumar, S. K., Rajkumar, S. V., Dispenzieri, A., et al. (2008). Improved survival in multiple myeloma and the impact of novel therapies. *Blood, 111,* 2516–2526.

Lamboley, C., Bringuier, A. F., Feldmann, G., et al. (2000). Induction of apoptosis in normal cultured rat hepatocytes and in Hep3B, a human hepatoma cell line. *Cell Biology and Toxicology, 16,* 185–200.

Lane, S. W., & Gilliland, D. G. (2010). Leukemia stem cells. *Seminars in Cancer Biology, 20,* 71–76.

Lee, K. H. (1999). Novel antitumor agents from higher plants. *Medicinal Research Reviews, 19,* 569–596.

Li, W., Xie, L., Chen, Z., et al. (2010). Cantharidin, a potent and selective pp2A inhibitor, induces an oxidative stress-independent growth inhibition of pancreatic cancer cells through G2/M cell-cycle arrest and apoptosis. *Cancer Science, 101,* 1226–1233.

Li, Y. M., & Casida, J. E. (1992). Cantharidin-binding protein: identification as protein phosphatase 2A. *Proceedings of the National Academy of Sciences of the United States of America, 89,* 11867–11870.

Lin, P. Y., Shi, S. J., Shu, H. L., et al. (2000). A simple procedure for preparation of N-Thiazoyl and N-Thiadizoylcantharidinimides and evaluation of their cytotoxicities against human hepatocellular carcinoma cells. *Bioorganic Chemistry, 28,* 266–272.

Liu, D., & Chen, Z. (2009). The effects of cantharidin and cantharidin derivatives on tumour cells. *Anti-Cancer Agents in Medicinal Chemistry, 9,* 392–396.

Mann, J. (2002). Natural products in cancer chemotherapy: Past, present and future. *Nature Reviews. Cancer, 2,* 143–148.

McCluskey, A., Ackland, S. P., Gardiner, E., et al. (2001). The inhibition of protein phosphatases 1 and 2A: A new target for rational anti-cancer drug design? *Anti-Cancer Drug Design, 16,* 291–303.

Millward, T. A., Zolnierowicz, S., Hemmings, B. A., et al. (1999). Regulation of protein kinase cascades by protein phosphatase 2A. *Trends in Biochemical Sciences, 24,* 186–191.

Mukherjee, A. K., Basu, S., Sarkat, N., et al. (2001). Advances in cancer therapy with plant based natural products. *Current Medicinal Chemistry, 8,* 1467–1486.

Nakaya, A., Sagawa, M., Muto, A., et al. (2010). The gold compound auranofin induces apoptosis of human multiple myeloma cells through both down-regulation of STAT3 and inhibition of NF-κB activity. *Leukemia Research, 35,* 243–249.

Nakazato, T., Ito, K., Ikeda, Y., et al. (2005a). Green tea component, catechin, induces apoptosis of human malignant B cells via production of reactive oxygen species. *Clinical Cancer Research: An Official Journal of the American Association for Cancer Research, 11,* 6040–6049.

Nakazato, T., Ito, K., Miyakawa, Y., et al. (2005b). Catechin, a green tea component, rapidly induces apoptosis of myeloid leukemic cells via modulation of reactive oxygen species production *in vitro* and inhibits tumor growth *in vivo*. *Haematologica, 90,* 317–325.

Palumbo, A., & Rajkumar, S. V. (2010). Multiple myeloma: Chemotherapy or transplantation in the era of new drugs. *European Journal of Haematology, 84,* 379–390.

Pang, S. K., Yu, C. W., Au-Yeung, S. C., et al. (2007). DNA damage induced by novel demethylcantharidin-integrated platinum anticancer complexes. *Biochemical and Biophysical Research Communications, 363,* 235–240.

Peng, F., Wei, Y. Q., Tian, L., et al. (2002). Induction of apoptosis by norcantharidin in human colorectal carcinoma cell lines: involvement of the CD95 receptor/ligand. *Journal of Cancer Research and Clinical Oncology, 128,* 223–230.

Rauh, R., Kahl, S., Boechzelt, H., et al. (2007). Molecular biology of cantharidin in cancer cells. *Chinese Medicine, 2,* 8.

Ray, R. M., Bhattacharya, S., & Johnson, L. R. (2005). Protein phosphatase 2A regulates apoptosis in intestinal epithelial cells. *The Journal of Biological Chemistry, 280,* 31091–31100.

Sagawa, M., Nakazato, T., Uchida, H., et al. (2008). Cantharidin induces apoptosis of human multiple myeloma cells via inhibition of the JAK/STAT pathway. *Cancer Science, 99,* 1820–1826.

Schuringa, J. J., Chung, K. Y., Morrone, G., et al. (2004). Constitutive activation of STAT5A promotes human hematopoietic stem cell self-renewal and erythroid differentiation. *The Journal of Experimental Medicine, 200,* 623–635.

Schweyer, S., Bachem, A., Bremmer, F., et al. (2007). Expression and function of protein phosphatase PP2A in malignant testicular germ cell tumours. *The Journal of Pathology, 213,* 72–81.

Shimizu, T., Nakazato, T., Xian, M. J., et al. (2006). Resveratrol induces apoptosis of human malignant B cells by activation of caspase-3 and p38 MAP kinase pathway. *Biochem Pharmacol, 71,* 742–750.

Shan, H. B., Cai, Y. C., Liu, Y., et al. (2006). Cytotoxicity of cantharidin analogues targeting protein phosphatase 2A. *Anti-Cancer Drugs, 17,* 905–911.

Skipper, H. E. (1974). Thoughts on cancer chemotherapy and combination modality therapy. *JAMA: The Journal of the American Medical Association, 230,* 1033–1035.

Sontag, E., Fedorov, S., Kamibayashi, C., et al. (1993). The interaction of SV40 small tumor antigen with protein phosphatase 2A stimulates the map kinase pathway and induces cell proliferation. *Cell, 75,* 887–897.

Tagwireyi, D., Ball, D. E., Loga, P. J., et al. (2000). Cantharidin poisoning due to "Blister beetle" ingestion. *Toxicon, 38,* 1865–1869.

Tallman, M. S., Nabhan, C., Feusner, J. H., et al. (2002). Acute promyelocytic leukemia: evolving therapeutic strategies. *Blood, 99,* 759–767.

To, K. K., Wang, X., Yu, C. W., et al. (2004). Protein phosphatase 2A inhibition and circumvention of cisplatin cross-resistance by novel TCM-platinum anticancer agents containing demethylcantharidin. *Bioorganic & Medicinal Chemistry, 12,* 4565–4573.

Valent, P. (2010). Standard treatment of Ph+ CML in 2010: How, when and where not to use what BCR/ABL1 kinase inhibitor? *European Journal of Clinical Investigation, 40,* 918–931.

Venepalli, N., Rezvani, K., Mielke, S., et al. (2010). Role of allo-SCT for CML in 2010. *Bone Marrow Transplant, 45,* 1579–1586.

Wang, G. S. (1989). Medical uses of mylabris in ancient China and recent studies. *Journal of Ethnopharmacology, 26,* 147–162.

Weber, D. (2003). Thalidomide and its derivatives: New promise for multiple myeloma. *Cancer Control: Journal of the Moffitt Cancer Center, 10,* 375–383.

Wierenga, A. T., Schepers, H., Moore, M. A., et al. (2006). STAT5-induced self-renewal and impaired myelopoiesis of human hematopoietic stem/progenitor cells involves down-regulation of C/EBPalpha. *Blood, 107,* 4326–4333.

Williams, L. A., Moller, W., Merisor, E., et al. (2003). *In vitro* antiproliferation/cytotoxic activity of cantharidin (Spanish Fly) and related derivatives. *The West Indian Medical Journal, 52,* 10–13.

Wu, P., Dugoua, J. J., Eyawa, O., et al. (2009). Traditional Chinese medicines in the treatment of hepatocellular cancers: A systemic review and mata-analysis. *Journal of Experimental & Clinical Cancer Research: CR, 28,* 112.

Xian, M., Ito, K., Nakazato, M., et al. (2007). Zerumbone, a bioactive sesquiterpene, induces G2/M cell cycle arrest and apoptosis in leukemia cells via a Fas- and mitochondria-mediated pathway. *Cancer Science, 98,* 118–126.

Yasui, H., Hideshima, T., Richardson, P. G., et al. (2006). Recent advances in the treatment of multiple myeloma. *Current Pharmaceutical Biotechnology, 7,* 381–393.

Youns, M., Hoheisel, J. D., & Efferth, T. (2010). Traditional Chinese medicines (TCMs) for molecular targeted therapies of tumors. *Current Drug Discovery Technologies, 7,* 37–45.

Zhang, B., Gojo, I., Fenton, R. G., et al. (2002). Myeloid cell factor-1 is a critical survival factor for multiple myeloma. *Blood, 99,* 1885–1893.

Zhang, W., & Liu, H. T. (2002). MAPK signal pathways in the regulation of cell proliferation in mammalian cells. *Cell Research, 12,* 9–18.

Zhang, J. P., Ying, K., Xiao, Z. Y., et al. (2004). Analysis of gene expression profiles in human HL-60 cell exposed to cantharidin using cDNA microarray. *International Journal of Cancer. Journal International Du Cancer, 108,* 212–218.

Chapter 18
An Evidence-based Perspective of *Bufo Gargarizans* (Asiatic Toad) for Cancer Patients

Paulus S. Wang, Jiun-Yih Yeh, Ching-Han Yu and Shyi-Wu Wang

Abstract Bufalin and cinobufagin are cardiotonic steroids extracted from Chansu, a galenical preparation of the dried white venom of Chinese *Bufo gargarizans* (Asiatic toad). Chansu is also one of the major components of Kyushin, a traditional Chinese medicine. It has been shown that bufalin and cinobufagin increase vascular resistance, vasoconstriction, and blood pressure *via* an inhibition of Na^+,K^+-ATPase. This evidence indicates that bufalin and cinobufagin are endogenous digitalis-like factors. Since bufalin, cinobufagin, and digoxin share the similarity in the chemical structure, it is not surprising that bufalin and cinobufagin have digoxin-like effects. It is well-known that bufalin, cinobufagin, and digoxin as well as ouabain are specific blockers of the sodium pump. The digoxin-like immunoactivity has been observed in Chansu. It has been shown that bufalin and cinobufagin possess antitumor effects on human lung adenocarcinoma cells, leukemia cells, pancreatic cancer cells, gastric cancer cells, prostate cancer cells, endometrial cancer cells, ovarian cancer cells, and hepatocellular carcinoma *via* induction of growth inhibition, cell cycle arrest and/or apoptosis. Reactive oxygen species-dependent Bax translocation, PI3K/Akt and apoptotic modulators including Bax, cytochrome c, and caspases are involved in bufalin-induced apoptosis or anticancer pathways. In the present chapter, we shall review antitumor effects of bufalin on human cancer cells. Although in most studies, human carcinoma cells or cell lines were employed, we expect that more *in vivo* studies and clinical trials will be performed in the near future.

P. S. Wang (✉)
Department of Physiology, School of Medicine, National Yang-Ming University,
Shih-Pai, Taipei, 11221, Taiwan
e-mail: pswang@ym.edu.tw

W. C. S. Cho (ed.), *Evidence-based Anticancer Materia Medica,* Evidence-based Anticancer Complementary and Alternative Medicine,
DOI 10.1007/978-94-007-0526-5_18,

18.1 Introduction

Bufalin is a cardiotonic steroid isolated from the Chinese toad venom Chansu, a galenical preparation of the dried white venom of Chinese *Bufo gargarizans* (Asiatic toad) (Hong et al. 1992; Panesar 1994). Chansu is one of the major components of Liu-Shen-Wan and Kyushin (Hong et al. 1992; Ojiri et al. 1992; Chan et al. 1995). Liu-Shen-Wan has been used for the treatment of tonsillitis, sore throat, and furuncle because of its local anesthetic and antibiotic actions (Chan et al. 1995). Kyushin is used for the treatment of palpitation and anhelation, and it has been shown to have a cardiotonic effect (Morishita et al. 1986, 1992). The cardiotonic effect of Kyushin is due to the action of bufadienolides such as bufalin, cinobufagin, and resibufogenin (Hong et al. 1992; Morishita et al. 1992).

It has been demonstrated that bufalin inhibits vasodilation and increases vasoconstriction, vascular resistance, and blood pressure *via* inhibition of Na^+,K^+-adenosine triphosphatase (Na^+,K^+-ATPase) (Eliades et al. 1989; Pamnani et al. 1991, 1994; Bagrov et al. 1993) in spite of the increase of sodium excretion (Yates and McDougall 1993).

Bufalin and digoxin as well as ouabain are specific inhibitors of the sodium pump (Tao et al. 1995). Digoxin is a cardiac glycoside purified from the plant *Digitalis lanata*, which has been used clinically in the treatment of congestive heart conditions for more than 200 years (Doherty et al. 1978; Antman and Smith 1985; Rietbrock and Woodcock 1985; Heller 1990).

In addition to cardiotonic effect and excitation of respiration, the inhibition of steroidogenesis and anticancer effects are involved in the actions of digoxin and bufalin. Digoxin and/or digitoxin may directly inhibit the secretion of testosterone (Lin et al. 1998; Wang et al. 1999), progesterone (Chen et al. 2001, 2002), corticosterone (Wang et al. 2004; Pu et al. 2006), cortisol (Kau et al. 2009), and aldosterone (Kau et al. 2009) *via* action mechanisms independent of Na^+,K^+-ATPase. Like digoxin, bufalin can also inhibit the production of testosterone *via* a decrease of testicular cAMP accumulation and luteinizing hormone (LH) response to gonadotropin releasing hormone (GnRH) (Wang et al. 1997).

The anticancer effects caused by digoxin and/or bufalin have also been reported. It has been shown that digoxin directly inhibits the proliferation of prostate cancer cell lines including LNCaP, DU145, and PC3 *via* a mechanism involving sustained elevation of the concentration of intracellular Ca^{2+} and of apoptosis (Yeh et al. 2001). Bufalin and cinobufagin also suppress cell proliferation and cause apoptosis in human prostate cancer cells, *via* an action of sustained election of the $[Ca^{2+}]_i$ and apoptotic modulators including Bax, cytochrome c, and caspases (Yeh et al. 2003; Yu et al. 2008). Not only human prostate cancer cells but also leukemia (Zhang et al. 1991, 1992; Watabe et al. 1996, 1997; Hashimoto et al. 1997; Kawazoe et al. 1999a, b; Chen et al. 2009), ovarian cancer cells (Takai et al. 2008), endometrial cancer cells (Takai et al. 2008), gastric cancer (Li et al. 2009), pancreatic cancer (Meng et al. 2009), lung cancer (Meng et al. 2009), and hepatocellular carcinoma (Han et al. 2007; Qi et al. 2010) are inhibited in proliferation by bufalin and/or cinobufagin.

In this chapter, the anticancer effects of bufalin are reviewed. The herb-drug interaction, toxicity, and side effects of bufalin are also discussed.

18.2 Natural Sources

It is well known that Chansu (also called toad venom or toad poison) is a popular Chinese traditional medicine, which is made of the skin secretions of giant toads including Asiatic toad and *Bufo melanostictus* (Fig. 18.1) (Zhang et al. 2007). Bufalin, cinobufagin, and resibufogenin are the three major components of Chansu, and their contents in the crude drug could be as high as 5–10% of the dry weight (Fig. 18.2). Asiatic toad ranges in snout-vent length from 56 to 102 mm. It has

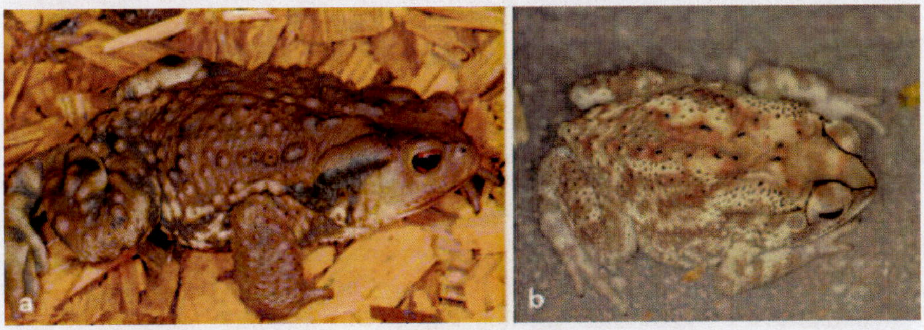

Fig. 18.1 **a** *Bufo gargarizans* (Asiatic toad). **b** *Bufo melanostictus*. (Photos used with permission from Dr. Peter Janzen, World Association of Zoos and Aquariums, Germany)

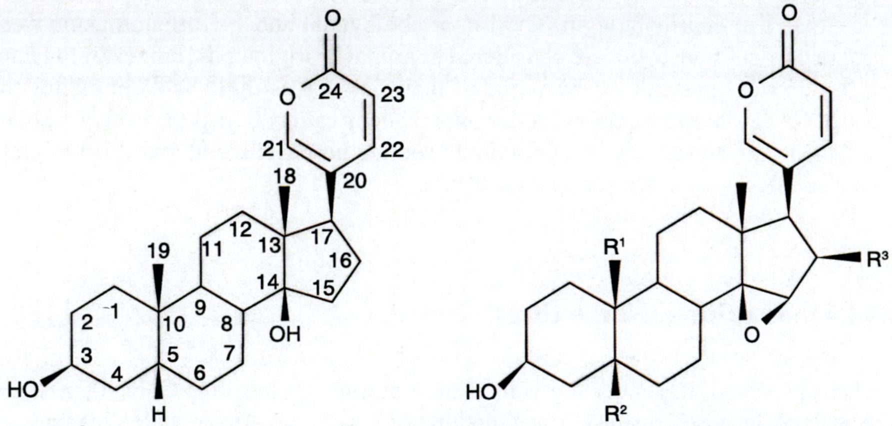

Bufalin

Cinobufagin R1 = Me: R2 = H; R3 = OAc
Resinobufagin R1 = Me; R2 = R3 = H

Fig. 18.2 Three major bufadienolides compounds: bufalin, cinobufagin, and resinobufagin spines on the dorsal skin tubercles and a black band extended from the outer surface of the parotoid onto the body flank. Dorsal skin tubercles are large, and coloration is dark-gray, olive-gray or olive-brownish, with three wide longitudinal bands. Wide dark band extends from the inner surface of the parotid onto the body flank. This band is interrupted posteriorly into large spots. Females are larger than males. The hindlegs seem to be relatively longer in males, whereas females have a wider head. The countries that Asiatic toad distribution include China, Japan, Korea, Democratic People's Republic of Korea, and Republic of Russian Federation (http://amphibiaweb.org/index.html).

18.3 Chemical Structures

Bufadienolides, the major active constituents of Chansu, are C-24 steroids, the characteristic structural feature of which is a doubly unsaturated six membered lactone ring (α-pyrone) on position 17β. Furthermore, these compounds are characterized by the *trans*-junction of rings B and C and usually the *cis*-junction of rings C and D (Krenn and Kopp 1998; Steyn and Heerden 1998; Mijatovic et al. 2007; Zhang et al. 2007). Bufadienolides and the more polar conjugates, the bufotoxins, are present in the bodies of toads of the genus *Bufo*. The toad bufadienolides occur not only in the unconjugated form, but several C-3 conjugates are also known: sulfates, dicarboxylic esters, and amino acid—dicarboxylic acid esters. The arginine-suberoyl esters, e.g. bufalitoxin, are known as the bufotoxins. There are several variations of these compounds where the suberyl ester has been replaced by succinyl, glutaryl, adipyl, or pimelyl residues and the amino acid arginine by glutamine, histidine, 1-methylhistidine, or 3-methylhistidine. The major toad bufadienolides are derivatives of the steroids. A 14-deoxy derivative found in the bile of *B. marinus* is 1,160 times less toxic than marinobufagin. The 14á-artebufogenin and 14â-artebufogenin were isolated from Chansu, but are considered as artefacts originated from resibufogenin. It has been suggested that, by virtue of their potency as digitalislike inhibitors of Na^+,K^+-ATPase and therefore active monovalent cation transport, bufadienolides and their derivatives may be important in sodium homeostasis in toads that migrate between fresh and salt water environments.

18.4 Traditional Functions

Chansu is one of the major ingredients in traditional medicine in China (Liu-Shen-Wan) and Japan (Kyushin). The functions of Liu-Shen-Wan include clearance of toxic heat, sedation of internal heat, elimination of phlegm-heat, sedation of agitation and disturbed spirit. It is also used for a variety of severe heat problems including acute tonsillitis, acute mumps, or sore throat; also high fever with delirium.

The term "heat" presented here is a translated Chinese language, and it means the inflammatory symptoms in Western medicine system.

Bufadienolides represent a group of compounds that share the capacity to bind to the extra-cellular surface of the main ion transport protein, the membrane-inserted sodium pump (Na^+,K^+-ATPase), in the cell. It has been shown that bufalin blocks vasodilation and increases vasoconstriction, vascular resistance, and blood pressure *via* an inhibition of Na^+,K^+-ATPase (Eliades et al. 1989; Pamnani 1991; Bagrov et al. 1993). Digoxin has been valued as a potent and highly selective Na^+,K^+-ATPase inhibitor, and it is capable of elevating the intracellular calcium to sustained levels in order to boost myocardial contractility or inotropism (Schmidt et al. 1993). Because of the similarity in the chemical structure between bufalin and digoxin, it is not surprising that bufalin has digoxin-like function (Panesar 1994). Besides the cardiotonic effect, the comprehensive pharmacological evaluation indicated that the bufadienolides possess broad bioactivity including cardiotoxic, anti-neoplastic, neurotoxic, immunomodulatory, anti-microbial, sleep potentiation activity, and local anesthetization (Das et al. 2000; Cunha Filho et al. 2005; Garg et al. 2007; Xu et al. 2007; Meng et al. 2009; Gao et al. 2010).

18.5 Effects on Steroidogenesis

18.5.1 Testosterone

Testosterone production in the testicular interstitial cells or Leydig cells is stimulated by gonadotropin released from the anterior pituitary gland *via* the hypothalamus-pituitary-testis axis. After the binding of gonadotropin with receptors located in the plasma membrane of Leydig cells, the activation of cAMP pathway and intracellular calcium channels increase the activities of P450scc (the key enzyme for the rate-limiting step of steroidogenesis) and the steroidogenic acute regulatory protein as well as other steroidogenic enzymes to enhance the production of testosterone (Hadley and Levine 2007).

In rats, both the basal and human chorionic gonadotropin (hCG) evoked secretion of testosterone can be inhibited by the administration of bufalin, digoxin, or digitoxin (Lin et al. 1998; Wang et al. 1999). In the *in vitro* study, the administration of bufalin, digoxin, or digitoxin inhibits the production of testosterone and cAMP in response to hCG (Wang et al. 1997, 1999; Lin et al. 1998).

Not only testosterone but also LH response to GnRH is decreased by the administration of bufalin either *in vivo* or *in vitro* (Wang et al. 1997). Digoxin and digitoxin inhibit cytochrome P450 side chain cleavage enzyme (cytochrome P450scc) activity (conversion of cholesterol to pregnenolone) in rat testicular interstitial or Leydig cells but not the activities of other steroidogenic enzymes (Lin et al. 1998; Wang et al. 1999). These results suggest that the inhibitory effects of cardiotonic steroids including bufalin, digoxin, and digitoxin on the production of testosterone are due to, at least in part, (1) an inefficiency of post-cAMP events, (2) a decrease of P450scc activity, (3) a decrease of testicular cAMP accumulation, as well as (4) a reduction of LH response to GnRH.

18.5.2 Progesterone, Glucocorticoid, and Mineralocorticoid

In rat granulosa and luteal cells, administration of digoxin or digitoxin, but not ouabain, decreased (1) the basal and hCG-evoked release of progesterone, (2) the stimulatory effects of forskolin (an activator of adrenylyl cyclase) and 8-bromo-cyclic 3′:5′-adenosine monophosphate (8-Br-cAMP, a permeable analogue of cAMP) on progesterone release, and (3) P450scc activity (Chen et al. 2001, 2002). These results suggest that digoxin and digitoxin inhibit the progesterone release by rat ovarian cells *via* a Na^+,K^+-ATPase-independent mechanism involving the inhibition of post-cyclic AMP pathway; and P450scc functions (Chen et al. 2001, 2002).

A single injection of digoxin inhibits the corticosterone secretion in response to adrenocorticotropin (ACTH) *in vivo* (Wang et al. 2004). Administration of digoxin or digitoxin, but not ouabain, inhibits corticosterone production in response to ACTH, forskolin, 8-Br-cAMP, cyclopiazonic acid (CPA, a specific inhibitor of Ca^{2+}-ATPase in the sarcoplasmic reticulum), P450scc activity, and the activity of 11β-hydroxylase (Wang et al. 2004; Pu et al. 2006) in rat adrenocortical zona fasciculata-reticularis cells. These results suggest that digoxin and digitoxin decrease the production and secretion of corticosterone by acting directly on adrenocortical zona fasciculata-reticularis cells *via* a Na^+,K^+-ATPase-independent mechanism involving the inhibition of the activities of adrenylyl cyclase, cytochrome P450scc, and 11β-hydroxylase, as well as the function of cAMP and intracellular calcium (Wang et al. 2004; Pu et al. 2006).

Acute effects of digoxin and ouabain on plasma levels of adrenal corticosteroid hormones in the primates *in vivo* have also been examined (Kau et al. 2009). A single injection of digoxin decreases the basal secretion of aldosterone and cortisol, and their responses to ACTH or KCl in monkeys, whereas administration of ouabain does not affect the levels of plasma aldosterone or cortisol. These results suggest that the Na^+,K^+-ATPase pathway might not be involved in the inhibitory action of digoxin on aldosterone or cortisol secretion in monkeys (Kau et al. 2009).

18.6 Anticancer Effects and Cancer Therapy

18.6.1 Leukemia

Bufalin has been used on the study of leukemia for decades. The literature about the potential effects of bufalin as a therapeutic agent on leukemia draws the most attention among all of the cancer types. In the treatment of leukemia, bufalin has been used as the (1) differentiation inducing agent, (2) inhibitor of Na^+,K^+-ATPase, (3) inhibitor of topoisomerase II, and (4) inducer of apoptosis. These aspects are introduced in detail as following.

18.6.1.1 Differentiation Inducing Agent

Induction of differentiation is the main therapeutic approach for leukemia. Bufalin alone is only the modest inducer of cell differentiation for the primary culture cells of acute myeloid leukemia patients. However, bufalin significantly enhances the effects of differentiation induced by all-*trans* retinoic acid (ATRA) in acute promyelocytic leukemia cells (Yamada et al. 1998). In addition, treatment of 10 nM bufalin in combination with etoposide (VP16), ATRA, vitamin D_3, rTNF-α, or γ-interferon may be useful in the induction of differentiation of human leukemia (Zhang et al. 1991, 1992). The enhanced effect of bufalin on vitamin D-induced differentiation in leukemia cells is due to through the activation of ERK and increase of nuclear vitamin D receptor expression (Amano et al. 2009). Inhibition of cPKC decreases the bufalin-induced differentiation of human monocytic cells (Kurosawa et al. 2001). Therefore, the expressions of vitamin D receptor and cPKC might be involved in the cell differentiation induced by bufalin.

18.6.1.2 Na^+,K^+-ATPase Inhibitor

As an inhibitor of Na^+,K^+-ATPase, bufalin may induce both cell differentiation (Numazawa et al. 1994, 1995) and cell apoptosis of leukemia cells (Kawazoe et al. 1999b). The increase of intracellular Na^+ concentration resulted from the Na^+,K^+-ATPase inhibition possibly alters the expression of proto-oncogene modulated by bufalin (Numazawa et al. 1996). Bufalin causes cell apoptosis in human colon adenocarcinoma COLO320DM cells, in human lymphoblastic leukemia MOLT-3 cells, and in human monocytic leukemia THP-1 cells, but not in normal human leukocytes and not in murine leukemia cells. Plasma membrane potential is decreased after bufalin treatment at 1 μM. Overexpression of Bcl-2 attenuates the bufalin-induced apoptosis. The inhibition of Na^+,K^+-ATPase may act upstream of the Bcl-2 protein (Kawazoe et al. 1999a, b).

18.6.1.3 Topoisomerase II Inhibitor

As an inhibitor of topoisomerase II, bufalin remarkably inhibits the activity of topoisomerase II at the concentration of 10 nM in human leukemia ML1 cells (Jing et al. 1994). The mRNA and protein expression of topoisomerase IIα (topo IIα) are down-regulated by bufalin, and then induced the DNA fragmentation. At the concentration of 0.1 μM, bufalin potentiates the inhibitory effects of cisplatin and

ATRA on HL-60 cells (Hashimoto et al. 1997). The mechanism of inhibiting the topo IIα may be due to the translocation of CK2 to transmit the bufalin signal to the nucleus, and then formed a complex of CK2 and topo IIα. The activity of topo IIα is mediated and leads to the induction of apoptosis (Watabe et al. 1997).

18.6.1.4 Apoptosis Inducing Agent

Bufalin induces apoptosis by altering the expression of apoptosis-related genes *c-myc* and *bcl-2* (Masuda et al. 1995). The activation of mitogen-activated protein kinase (MAPK) may be involved in bufalin-induced apoptosis in U937 cells (Watabe et al. 1996). Over-expression of Bcl-2 inhibits the bufalin-induced MAPK activation and the subsequent AP-1 activation and cell apoptosis in U937 cells (Watabe et al. 1997). Using the inhibitor of the biosynthesis of AP-1 and the dominant negative c-Jun reduces the activation of AP-1 and the induction of apoptosis after bufalin treatment. Bufalin induces apoptosis *via* the activation of AP-1 through MAPK cascade including JNK in human leukemia U937 cells (Watabe et al. 1996, 1997). In addition, bufalin induces HL-60 cell apoptosis *via* the activation of the transcription factor NF-κB and AP-1 (Chen et al. 2009).

Other signaling transductions are involved in the bufalin-induced apoptosis pathway. For example, the expression of Tiam1 is elevated after the treatment of bufalin at 0.1 μM. Furthermore, the activity of Rac1 and PAK (p21-activated kinase) is also increased after bufalin treatment. Transfection of antisense RNA for Tiam1 reduces the bufalin-induced apoptosis in leukemic U937 cells. Tiam1 may play an important role in bufalin-induced apoptosis *via* the activation of Rac 1, PAK, and JNK pathway (Kawazoe et al. 1999b). Inhibitors of cPKCβ and nPKCδ attenuate the DNA ladder formation induced by bufalin. Bufalin-induced cell differentiation and apoptosis are interlinked *via* the regulation by PKC (protein kinase C) (Kurosawa et al. 2001). On the other hand, bufalin shows both anti-proliferative effects if applied alone and enhances the activity in combination with daunorubicin in multidrug-resistant human CCRF-CEM leukemia cells (Efferth et al. 2002).

Bufalin reveals the profound effects on the leukemia therapy *in vitro*. The mechanisms of bufalin-induced apoptosis have been deeply understood. It is well prepared to process in the *in vivo* studies or even in clinical trials. Hopefully, bufalin may contribute clinically to patients with leukemia.

18.6.2 Prostate Cancer

Other cardiac glycosides have been reported to induce apoptosis in prostate cancer, such as digoxin, digitoxin, and ouabain (Yeh et al. 2001). Bufalin and cinobufagin at the concentration of 1–10 μM inhibit the proliferation and induce the cell apoptosis of androgen-dependent (LNCaP) and -independent (DU145 and PC3) prostate cancer cell lines. The sustained elevation of the intracellular calcium concentration may be related with bufalin- and cinobufagin-induced DNA fragmentation and caspases activation (Yeh et al. 2003). Furthermore, a sequence of apoptotic modulators, including Bax, cytochrome c, and caspases, are involved in bufalin- and cinobufagin-caused apoptosis in prostate cancer cells. The antisense of p53 and the siRNA of Fas are used to determine the role of p53 and Fas in the bufalin-induced apoptosis in prostate cancer cell lines. The upstream mediators might be p53 and Fas in androgen-dependent LNCaP cells and Fas in androgen-independent DU145 and PC3 cells (Yu et al. 2008). In the *in vivo* study, bufalin significantly inhibited the tumor growth on DU145- and PC3-xenographed SCID mice. Therefore, these cardiac glycosides show the potential on treatment for patients with prostate cancer.

18.6.3 Gynaecological Cancer

To detect the cytotoxic effect of bufalin on normal human cells, the normal human endometrial epithelial cells are obtained from the subjects and are treated with bufalin. In comparison of the anti-proliferative effects on normal endometrial cancer cells and gynaecological cancer cell lines, cancer cells are 100 times more sensitive to bufalin than the normal cells (Takai et al. 2008). In the aspect of cell cycle distribution, exposure to bufalin decreases the proportion of cells in the S-phase and increases the cell numbers in the G0/G1 phase of the cell cycle in both endometrial and ovarian cancer cell lines. The expression of cell cycle-related proteins, cyclin A, and cyclin D3, are down-regulated after 48 h treatment. However, the protein level of p21, the cyclin dependent kinase inhibitor, is up-regulated (Takai et al. 2008). In the aspect of cell apoptosis, treatment with 1 ng/ml bufalin results in early apoptotic (annexin V+) cells and necrotic (annexin V+/Propidium iodide+) cells in both endometrial and ovarian cancer cell lines. After the treatment, approximately half of the cells lose the mitochondrial transmembrane potential which is also the indicator of apoptosis. The expressions of anti-apoptotic protein, Bcl-2, and Bcl-xL are decreased. On the other hand, the pro-apoptotic caspase-9 is activated after being treated with 1 ng/ml bufalin for 24 and 48 h (Takai et al. 2008). Although there is only one published article about the treatment of bufalin on gynaecological cancer *in vitro*, the significant effects of bufalin on cancer cell line than on normal endometrial cells deserves further investigation.

18.6.4 Lung Cancer

Bufalin reduces the cell viability of human lung adenocarcinoma (ASTC-a-1) cells and induces the caspase-3 activation in ASTC-a-1 (Pan et al. 2009). The production of reactive oxygen species (ROS) and cell apoptosis are also induced by 0.1 μM bufalin treatment. The antioxidant N-acetyl-cysteine (a ROS scavenger) inhibits the generation of ROS, the translocation of Bax, and the activation of caspase-3, which leads to the protection of bufalin-induced apoptosis. Therefore, the induction of ROS is involved in the bufalin-caused apoptosis in ASTC-a-1 cells (Sun et al. 2009).

In the clinical study, there are two patients with advanced non-small cell lung cancer (NSCLC) who join the pilot study of huachansu (10 and 40 ml/m^2). No grade III or IV adverse effects including leukopenia, thrombocytopenia, loss of appetite, constipation, diarrhea, rash, myalgia, dizziness, oral ulcer, dyspnea, premature ventricular contraction, and hypertension are found in both patients after the infusion of huachansu. However, the patients still have progressive disease after receiving two cycles of infusion (one cycle is composed of 14 daily treatment followed by 7 days off). It needs a larger sample size to draw a conclusion of the therapeutic effects of huachansu on the patients with NSCLC (Meng et al. 2009).

18.6.5 Hepatocellular Carcinoma

The effects of huachansu on hepatoma have been investigated *in vitro*, *in vivo*, and in clinical trials. The components of huachansu used in these studies have been determined by liquid chromatography coupled to tandem mass spectrometry. Once the components are consistent from lot to lot, the mixture of Chinese medicine is still reliable in the Western medicine (Meng et al. 2009).

18.6.5.1 *In Vitro* Study

Cinobufacini, the aqueous extracts from the venom glands of Asiatic toad, significantly inhibits cell proliferation of hepatoma cell lines, HepG2 and Bel-7402. The indication of early apoptotic signal, Annexin V, is increased after the 0.1 mg/ml cinobufacini treatment for 48 h. Cinobufacini also disrupts the mitochondrial membrane potential in both cell lines. The expression of mitochondrial proteins Bax increases whereas that of Bcl-2 decreases, resulting in an increase in the Bax/Bcl-2 ratio. Subsequently, the cytochrome c release, caspase-9 and -3 activation, and poly(ADP-ribose) polymerase (PARP) cleavage occur in the cinobufacini-treated groups. These results suggest that cinobufacini induces apoptosis of hepatocellular carcinoma cells *via* a mitochondria-mediated apoptosis pathway (Qi et al. 2008, 2010).

18.6.5.2 *In Vivo* Study

The tumor volume of orthotopic hepatoma is significantly reduced after the intraperitoneal injection of bufalin at the dose of 1.5 (BF1), 1 (BF2), and 0.5 (BF3) mg/kg for 15–24 days. Furthermore, the tumor volumes of BF1 and BF2 are significantly reduced than the Adriamycin (ADM) group which is treated with Adriamycin at the dose of 8 mg/kg. The mean survival time is also prolonged in BF groups than normal saline group and ADM group. The DNA fragmentation detected by TUNEL assay was observed in the tumor tissue of BF groups. The mRNA and protein expressions of Bcl-2 are decreased, whereas the expressions of Bax are elevated in BF groups but not in normal saline or ADM groups. On the other hand, there is no evidence of toxicity to the heart, lungs, liver, kidneys, and brain in the bufalin-treated nude mice (Han et al. 2007).

18.6.5.3 Clinical Study

Huachansu is the water extract of dried toad skin, which has been applied in the clinical trials of China. However, these clinical results haven't been proved in the Western medicine-based approach, though this pilot study was conducted by Fudan University Cancer Hospital and The University of Texas MD Anderson Cancer Center. The dose usually used in China is 15 ml/m^2 or 20–25 ml. In order to determine the toxicities of huachansu, the doses used in the pilot study are 10, 20, 40, 60, and 90 ml/m^2 (which is 8 times higher than the generally dose in China). One treatment cycle is composed of 14 daily treatments followed by 7 days off. The components of bufalin, cinobufagin, cinobufotalin, and resibufogenin in the huachansu are 14.30 ± 0.03, 3.35 ± 0.10, 21.5 ± 0.2, and 24.5 ± 2.2 ng/ml, respectively. Most of the patients participate in this study are with hepatocellular carcinoma at the disease stage IV. There is no grade III/IV toxicity, e.g. leucopenia, thrombocytopenia, loss of appetite, constipation, diarrhea, rash, myalgia, dizziness, oral ulcer, and dyspnea found in the patients after the infusion. In addition, no cardiovascular toxicities, premature ventricular contraction, and hypertension, are observed. As a result, no dose-related toxicity is found in this study (Meng et al. 2009). The levels of bufalin in the plasma reach the maximum concentration 2 h after the infusion, and then drop to the undetectable level after 24 h. Six patients (40%) have stable disease and 9 patients (60%) have progressive disease after infusion. One patient who received 10 ml/m^2 huachansu shows 20% reduction of tumor mass and stable disease for more than 11 months. The alpha-fetoprotein also drops from 138 to 70 µg/ml after 4 cycles of treatment and then climbs again after 8 cycles of treatment (107 µg/ml). One patient with metastatic hepatocellular carcinoma has stable disease that lasted for 8 months in response to huachansu treatment alone. In conclusion, no dose-limiting toxicities are found in patients who accept the huachansu treatment 8X higher than the typical used in China. Six out of 15 patients have prolonged stable disease or 20% tumor shrinkage (Meng et al. 2009). Therefore, huachansu shows the therapeutic potential for patients with hepatoma. The proper infusion components and delivery cycle need to be further investigated.

18.6.6 Gallbladder Carcinoma

The most effective chemo-therapeutic drug for gallbladder carcinoma (GBC) is gemcitabine. The regimen of gemcitabine (1,000 mg/m^2)-oxaliplatin (120 mg/m^2) (GEMOX) to treat patients with advanced biliary tract adenocarcinoma shows that the response is 36% and the median progression free survival is 5.7 months (Andre et al.

2004). Therefore, the combination of GEMOX and huachansu (20 ml/m^2) infusion every 3–4 weeks is used to treat the patients with advanced GBC (grade III to V of total 25 patients). All patients received gemcitabine (1,000 mg/m^2) on day 1 and 8, oxaliplatin (120 mg/m^2) on day 1, and huachansu (20 ml/m^2) on day 3–11. These treatments were repeated every 3–4 weeks. Most of the patients experience the grade I and II toxicities, including leukopenia, neutropenia, anemia, thrombocytopenia, diarrhea, nausea/emesis, peripheral, neuropathy, and pain. Only some of them show the grade III or IV toxicities, including leucopenia, neutropenia, and thrombocytopenia. No treatment-related death is found. Evaluated by the Chinese version of the European Organization for Research and Treatment of Cancer Quality of Life Questionnaire-Core 30, the scores significantly increase in physical, emotional, and cognitive functions but decreased in pain and vomiting one month after the completion of treatment (6 cycles) ($P<0.05$). The responses of patients to this regimen show that 8 patients are partial response (34.8%), 7 have stable disease (30.4%), and 8 have progressive disease (34.8%). The disease control rate is 65.2%, and the median progression free survival is 5.8 months. In conclusion, the combination treatment of GEMOX and huachansu is tolerable and effective in improving the quality of life (QoL) and controlling the disease of patients with advanced GBC (Qin et al. 2008).

18.6.7 Other Cancer Types

18.6.7.1 Gastric Cancer

Bufalin inhibits the proliferation of MGC803 gastric cancer cells. The cell cycle distribution is arrested at G2/M phase after the treatment. In addition, bufalin at the concentration of 80 nM elevates the percentage of hypodiploid cells. The ratio of Bax/Bcl-2 and the active form of caspase-3 increases in bufalin-treated cells. By up-regulating the Casitas B-lineage lymphoma family of ubiquitin lipases, bufalin induces the degradation of PI3K/Akt pathway, and then results in the apoptosis of gastric cancer MGC803 cells (Li et al. 2009).

18.6.7.2 Osteosarcoma

The overexpression of dihydrofolate reductase (DHFR) enzyme enhances the proliferation of osteosarcoma. Methotrexate, one of the most common therapeutic drugs for osteosarcoma, is the inhibitor of DHFR. Human osteosarcoma U-2OS and methotrexate-resistant U-2OS cell lines are employed to determine the relation of bufalin-induced apoptosis and the expression of DHFR. The result shows that bufalin induces the cell cycle arrest at the G2/M phase and causes the cell apoptosis *via* the increase of p53 and the ratio of Bax/Bcl-2. However, the expression of DHFR in methotrexate-resistant U-2OS cells is not altered by bufalin treatment. Therefore, the apoptosis-inducing effects of bufalin are not affected by the high expression of DHFR (Yin et al. 2007).

18.6.7.3 Pancreatic Cancer

There were two patients with advanced pancreatic cancer joining the pilot study of huachansu (20 and 90 ml/m^2). No grade III or IV adverse effects including leukopenia, thrombocytopenia, loss of appetite, constipation, diarrhea, rash, myalgia, dizziness, oral ulcer, dyspnea, premature ventricular contraction, and hypertension were found in both patients. Although the Karnofsky performance status decreases from 80 to 70 after the first cycle of huachansu infusion, the patients still have progressive disease. It may need a larger database to draw the conclusion of the therapeutic effects of huachansu on the patients with pancreatic cancer (Meng et al. 2009).

Huachansu, especially the component of bufalin, induces the cell differentiation in leukemia cells, inhibits the proliferation, causes the cell cycle arrest and cell apoptosis in leukemia, prostate cancer, gynecological cancer, lung cancer, hepatoma, gallbladder cancer, gastric cancer, and osteosarcoma. The relationship of the antitumor signal transduction pathway induced by huachansu is illustrated in Fig. 18.3. In the clinical trials, these cardiac glycosides, huachansu, don't induce the intolerable adverse effects even at the dose 8 times higher than the generally used. Some patients show the stable disease after the huachansu infusion. However, the infusion components and cycles need to be further designed to enlarge the antitumor effects of huachansu. Some clinical trials have been performed in Chinese research institutes, such as Guangxi Medical University, The Second Military Medical University, and Guizhou People's Hospital. However, these clinical data haven't been demonstrated or confirmed by the Western medicine-based approach. Based on the positive results from these pilot studies, hopefully, the clinical trials of Phase II and III will be performed implemented in the near future.

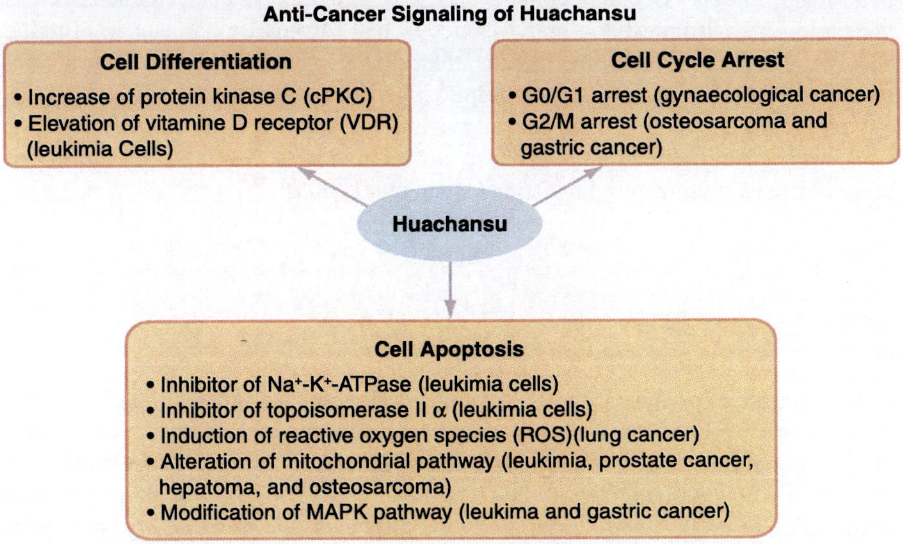

Fig. 18.3 Three major aspects involved in the signaling pathways of huachansu in cancer therapy

18.7 Herb-drug Interaction, Toxicity, and Adverse Effects

Chansu has been widely used to treat cancer in China (Su and Nu 2001; Meng et al. 2009). Bufalin, cinobufagin, and resibufogenin are the 3 major cardiac glycosides contribute the antitumor activity of huachansu (Su et al. 2003). It has been suggested that bufalin is the major bufadienolide with the most powerful anticancer effect (Masuda et al. 1995; Haas et al. 2000; Kamano et al. 2002). In a study of efficacy, safety, and QoL of gemcitabine-oxaliplatin (GEMOX) combined with huachansu in patients with metastatic gallbladder carcinoma, the toxicities including leucopenia, anemia, neutropenia, thrombocytopenia, diarrhea, nausea/vomiting, and peripheral neuropathy have been observed (Qin et al. 2008). Moderate myelosuppression is the main side effect after the combined therapy of huachansu and GEMOX. The symptoms induced by grade non-hematologic toxicity including nausea/emesis, diarrhea, pain, and peripheral neuropathy were from mild to moderate. No treatment-related death occurred and patient QoL was improved after chemotherapy. Overall, GEMOX combined with huachansu injection is well tolerated and effective with moderate myelosuppression as the main toxicity (Qin et al. 2008).

Meng et al. (2009) have performed a study of huachansu in patients with hepatocellular carcinoma, NSCLC, or pancreatic cancer, and found that mild adverse events can be observed at each dose level. Under the treatment of huachansu, some patients had stable disease, and one patient with hepatocellular carcinoma had a 20% reduction in tumor mass which extended for more than 11 months. All toxicity such as leucopenia, thrombocytopenia, loss of appetite, constipation, diarrhea, rash, myalgia, dizziness, oral ulcer, dyspnea, premature ventricular contraction, and hypertension were grade I or II, no grade III or IV toxicities were observed. The grade I and II side effects associated with huachansu treatment have been characterized as hematologic, gastrointestinal, mucocutaneous, and cardiovascular aspects. Totally, 73% patients had no toxicities greater than grade I. Furthermore, no significant cardiac toxicity has been observed despite high doses (90 ml/m^2, standard dose is approximately 15 ml/m^2) of huachansu is administrated (Meng et al. 2009).

It has been suggested that huachansu is well tolerated, even at doses 8 times of those normally administered in China (Meng et al. 2009).

18.8 Perspectives

Despite the effect of myocardial contraction, cardiotonic steroids (e.g. bufalin, cinobufagin, and resibufogenin) have been demonstrated to possess anticancer effects *via* the inhibition of proliferation and enhancement of apoptosis of cancer cells (Han et al. 2007; Mijatovic et al. 2007; Schoner et al. 2007; Yin et al. 2007; Takai et al. 2008; Li et al. 2009; Sun et al. 2009). However, the most *in vitro* outcomes of anticancer studies might not completely reflect the *in vivo* expressions of whole body systems in response to the anticancer therapy of cardiotonic steroids.

It has been known that cancer is due to uncontrolled cell proliferation and deregulation of apoptosis (Kaufmann and Earnshaw 2000; Hu et al. 2009). Apoptosis, also called programmed cell death, is a physiological process which plays a significant role in regulating many normal functions and tissue homeostasis (Kerr et al. 1994). Two main signaling pathways, the extrinsic pathway and the intrinsic pathway, are involved in apoptosis, and the former is characterized by the activation of ligand-bound death receptors of the TNFR (tumor-necrosis factor receptor) superfamily. The latter is a signal transduction pathway associated with the mitochondria and the Bcl-2 family. Finally, both pathways activate a cascade of proteolytic enzymes named caspases to induce apoptosis (Finkel 1999; Park et al. 2009). Despite the plant and animal sources of cardiotonic steroids, the endogenous steroids with ouabain has been isolated from human plasma, bovine adrenal gland, and bovine hypothalamus, and the endogenous digitalis activities have been isolated from human urine, and bovine adrenal gland (Schoner and Scheiner-Bobis 2007). Therefore, it is possible that both the intrinsic pathway and the extrinsic pathway might be activated after the treatment with cardiotonic steroids.

It has been demonstrated that the steroidogenesis of the testosterone in the testis (Wang et al. 1997, 1999; Lin et al. 1998), the progesterone in the ovary (Chen et al. 2001, 2002), the corticosterone (Wang et al. 2004; Pu et al. 2006), and the aldosterone (Kau et al. 2009) in adrenal cortex can be reduced by bufalin and digoxin *via* Na^+,K^+-ATPase-independent mechanisms involving the inhibition of the activities of adrenylyl cyclase, P450scc, cAMP function, intracellular level calcium, as well as the release of tropic hormones from anterior pituitary glands. The linkage between the decreased steroidogenesis and anticancer effects caused by bufalin and digoxin is still unclear. Although the direct inhibitory effects of cardiotonic steroids on cancer cells have been ascertained, the possible indirect effects of cardiotonic steroids on the proliferation and migration of cancer cells caused by the reduction of steroidogenesis will be an interesting study in the future.

Acknowledgments Many thanks to Mr. Wu-Chen Li, Miss Yi-Ting Chen, and Miss Wan-Yun Chen for their technical assistance in preparing this chapter.

References

Amano, Y., Cho, Y., Matsunawa, M., et al. (2009). Increased nuclear expression and transactivation of vitamin D receptor by the cardiotonic steroid bufalin in human myeloid leukemia cells. *The Journal of Steroid Biochemistry and Molecular Biology, 114,* 144–151.

Andre, T., Tournigand, C., Rosmorduc, O., et al. (2004). Gemcitabine combined with oxaliplatin (GEMOX) in advanced biliary tract adenocarcinoma: A GERCOR study. *Annals of Oncology, 15,* 1339–1343.

Antman, E. M., & Smith, T. W. (1985). Digitalis toxicity. *Annual Review of Medicine, 36,* 357–367.

Bagrov, A. V., Roukoyatkina, N. I., Fedorova, O. V., et al. (1993). Digitalis-like and vasoconstrictor effects of endogenous digoxin-like factor(s) from the venom of Bufo marinus toad. *European Journal of Pharmacology, 234,* 165–172.

Chan, W. Y., Ng, T. B., & Yeung, H. W. (1995). Examination for toxicity of a Chinese drug, the toad glandular secretory product Chansu, in pregnant mice and embryos. *Biology of the Neonate, 67,* 376–380.

Chen, J. J., Wang, P. S., Chien, E. J., et al. (2001). Direct inhibition of digitalis on progesterone release from rat granulosa cells. *British Journal of Pharmacology, 132,* 1761–1768.

Chen, J. J., Chien, E. J., & Wang, P. S. (2002). Progesterone attenuates the inhibitory effects of cardiotonic digitalis on pregnenolone production in rat luteal cells. *Journal Cell of Biochemistry, 86,* 107–117.

Chen, A., Yu, J., Zhang, L., et al. (2009). Microarray and biochemical analysis of bufalin-induced apoptosis of HL-60 Cells. *Biotechnology Letters, 31,* 487–494.

Cunha Filho, G. A., Schwartz, C. A., Resck, I. S., et al. (2005). Antimicrobial activity of the bufadienolides marinobufagin and telocinobufagin isolated as major components from skin secretion of the toad Bufo rubescens. *Toxicon, 45,* 777–821.

Das, M., Mallick, B. N., Dasgupta, S. C., et al. (2000). A sleep inducing factor from common Indian toad (Bufo melanostictus, Schneider) skin extract. *Toxicon, 38,* 1267–1281.

Doherty, J. E., DeSoyza, N., Kane. J. J., et al. (1978). Clinical pharmacokinetics of digitalis glycosides. *Progress in Cardiovascular Diseases, 21,* 141–158.

Efferth, T., Davey, M., Olbrich, A., et al. (2002). Activity of drugs from traditional Chinese medicine toward sensitive and MDR1- or MRP1-overexpressing multidrug-resistant human CCRF-CEM leukemia cells. *Blood Cells, Molecules & Diseases, 28,* 160–168.

Eliades, D., Swindall, B., Johnston, J., et al. (1989). Hemodynamic effects of bufalin in the anesthetized dog. *Hypertension, 13,* 690–695.

Finkel, E. (1999). Dose cancer therapy trigger cell suicide? *Science, 286,* 2256–2258.

Gao, H., Zehl, M., Kaehlig, H., et al. (2010). Rapid structural identification of cytotoxic bufadienolide sulfates in toad venom from Bufo melanostictus by LC-DAD-MS(n) and LC-SPE-NMR. *Journal of Natural Products, 73,* 603–608.

Garg, A. D., Kanitkar, D. V., Hippargi, R. V., et al. (2007). Antimicrobial activity of skin secretions isolated from Indian toad, Bufo melanostictus Schneider 1799. *Nature Precedings.* doi:10.1038/npre.2007.1204.1.

Haas, M., Askari, A., & Xie, Z. (2000). Involvement of Src and epidermal growth factor receptor in the signal-transduction function of Na+/K+-ATPase. *The Journal of Biological Chemistry, 275,* 27832–27837.

Hadley, M. E., & Levine, J. E. (2007). Hormones and male reproductive physiology. *Endocrinology* (6th ed., pp. 387–407). Upper Saddle River: Pearson Education Inc.

Han, K., Huang, G., Gu, W., et al. (2007). Anti-tumor activities and apoptosis-regulated mechanisms of bufalin on the orthotopic transplantation tumor model of human hepatocellular carcinoma in nude mice. *World Journal of Gastroenterology, 13,* 3374–3379.

Hashimoto, S., Jing, Y., & Kawazoe, N.,(1997). Bufalin reduces the level of topoisomerase II in human leukemia cells and affects the cytotoxicity of anticancer drugs. *Leukemia Research, 21,* 875–883.

Heller, M. (1990). Cardiac glycosides: New/old ideas about drugs. *Biochemistry and Pharmacology, 40,* 919–925.

Hong, Z., Chan, K., & Yeung, H. W. (1992). Simultaneous determination of bufadienolides in the traditional Chinese medicine preparation, Liu-Shen-Wan, by liquid chromatography. *The Journal of Pharmacy and Pharmacology, 44,* 1023–1026.

Hu, Y. W., Liu, C. Y., Du, C. M., et al. (2009). Induction of apoptosis in human hepatocarcinoma SMMC-7721 cells *in vitro* by flavonoids from Astragalus complanatus. *Journal of Ethnopharmacology, 123,* 293–301.

Jing, Y., Watabe, M., Hashimoto, S., et al. (1994). Cell cycle arrest and protein kinase modulating effect of bufalin on human leukemia ML1 cells. *Anticancer Research, 14,* 1193–1198.

Kamano, Y., Nogawa, T., Yamachita, A., Hayashi, M., & Inoue, M. (2002). Isolation and structure of a 20, 21-epoxybufenolide series from "Ch'an Su". *Journal of Natural Products, 65,* 1001–1015.

Kau, M. M., Kau, S. F., Wang, J. R., et al. (2009). Acute effects of digoxin on plasma aldosterone and cortisol in monkeys. *Metabolism, 58,* 55–61.

Kaufmann, S. H., & Earnshaw, W. C. (2000). Induction of apoptosis by cancer chemotherapy. *Experimental Cell Research, 256,* 42–49.

Kawazoe, N., Aiuchi, T., Masuda, Y., et al. (1999a). Induction of apoptosis by bufalin in human tumor cells is associated with a change of intracellular concentration of Na+ ions. *Journal of Biochemistry, 126,* 278–286.

Kawazoe, N., Watabe, M., Masuda, Y., et al. (1999b). Tiam1 is involved in the regulation of bufalin-induced apoptosis in human leukemia cells. *Oncogene, 18,* 2413–2421.

Kerr, J. F., Winterford, C. M., & Harnon, B. V. (1994). Apoptosis. Its significance in cancer and cancer therapy. *Cancer, 73,* 2013–2026.

Krenn, L., & Kopp, B. (1998). Bufadienolides from animal and plant sources. *Phytochemistry, 48,* 1–29.

Kurosawa, M., Tani, Y., Nishimura, S., et al. (2001). Distinct PKC isozymes regulate bufalin-induced differentiation and apoptosis in human monocytic cells. *American Journal of Physiology Cell Physiology, 280,* C459–C464.

Li, D., Qu, X., Hou, K., et al. (2009). PI3K/Akt is involved in bufalin-induced apoptosis in gastric cancer cells. *Anti-cancer Drugs, 20,* 59–64.

Lin, H., Wang, S. W., Tasi, S. C., et al. (1998). Inhibitory effect of digoxin on testosterone secretion through mechanisms involving decreases of cyclic AMP production and cytochrome P450scc activity in rat testicular interstitial cells. *British Journal of Pharmacology, 125,* 1635–1640.

Masuda, Y., Kawazoe, N., Nakajo, S., et al. (1995). Bufalin induces apoptosis and influences the expression of apoptosis-related genes in human leukemia cells. *Leukemia Research, 19,* 549–556.

Meng, Z., Yang, P., Shen, Y., et al. (2009). Pilot study of huachansu in patients with hepatocellular carcinoma, nonsmall-cell lung cancer, or pancreatic cancer. *Cancer, 115,* 5309–5318.

Mijatovic, T., Van Quaquebeke, E., Delest, B., et al. (2007). Cardiotonic steroids on the road to anti-cancer therapy. *Biochimica et Biophysica Acta, 1776,* 32–57.

Morishita, S., Saito, T., Mishima, Y., et al. (1986). Pharmacological studies of Senso (Ch'an Su) containing drugs. *Folia Pharmacologica Japonica, 87,* 361–378.

Morishita, S., Shoji, M., Oguni, Y., et al. (1992). Pharmacological actions of "Kyushin", a drug containing toad venom: Cardiotonic and arrhythmo-genic effects, and excitatory effect on respiration. *The American Journal of Chinese Medicine, 20,* 245–256.

Numazawa, S., Shinoki, M. A., Ito, H., et al. (1994). Involvement of Na+, K(+)-ATPase inhibition in K562 cell differentiation induced by bufalin. *Journal of Cellular Physiology, 160,* 113–120.

Numazawa, S., Honma, Y., Yamamoto, T., et al. (1995). A cardiotonic steroid bufalin-like factor in human plasma induces leukemia cell differentiation. *Leukemia Research, 19,* 945–953.

Numazawa, S., Inoue, N., Nakura, H., et al. (1996). A cardiotonic steroid bufalin-induced differentiation of THP-1 cells. Involvement of Na+, K(+)-ATPase inhibition in the early changes in proto-oncogene expression. *Biochemical Pharmacology, 52,* 321–329.

Ojiri, Y., Chibana, T., Noguchi, K., et al. (1992). Cardiovascular effect of a senso (toad venom)-containing drug in anesthetized dog. II. Influence of propranolol. *The American Journal of Chinese Medicine, 20,* 147–155.

Pamnani, M. B., Chen, S., Bryant, H. J., et al. (1991). Effects of three sodium-potassium adenosine triphosphatase inhibitors. *Hypertension, 18,* 316–324.

Pamnani, M. B., Chen, S., Yuan, C. M., et al. (1994). Chronic blood pressure effects of bufalin, a sodium-potassium ATPase inhibitor, in rats. *Hypertension, 23*(Suppl), I106–I109.

Pan, W., Qu, J., Chen, T., et al. (2009). FLIM and emission spectral analysis of caspase-3 activation inside single living cell during anticancer drug-induced cell death. *European Biophysics Journal, 38,* 447–456.

Panesar, N. S. (1994). Bufalin radioimmunoassays: In search of the endogenous digitalis- like substance. *Journal of Immunoassay, 15,* 371–391.
Park, S. E., Park, C., Kim, S. H., et al. (2009). Korean red ginseng extract induces apoptosis and decreases telomerase activity in human leukemia cells. *Journal of Ethnopharmacology, 121,* 304–312.
Pu, H. F., Wang, S. W., Tseng, C. I., et al. (2006). Mechanisms of digoxin and digitoxin on the production of corticosterone in zona fasciculata-reticularis cells of ovariectomized rats. *Journal of Cellular Biochemistry, 97,* 303–313.
Qi, F. H., Li, A. Y., Gao, B., et al. (2008). Apoptosis-inducing effect of cinobufacini, Bufo bufo gargarizans Cantor skin extract, on human hepatoma cell line BEL-7402. *Drug Discoveries & Therapeutics, 2,* 339–343.
Qi, F., Li, A., Zhao, L., et al. (2010). Cinobufacini, an aqueous extract from Bufo bufo gargarizans Cantor, induces apoptosis through a mitochondria-mediated pathway in human hepatocellular carcinoma cells. *Journal of Ethnopharmacology, 128,* 654–661.
Qin, T. J., Zhao, X. H., Yun, J., et al. (2008). Efficacy and safety of gemcitabine-oxaliplatin combined with huachansu in patients with advanced gallbladder carcinoma. *World Journal of Gastroenterology, 14,* 5210–5216.
Rietbrock, N., & Woodcock, B. (1985). Two hundred years of foxglove therapy. *Trends in Pharmacological Sciences, 6,* 267–269.
Schmidt, T. A., Allen, P. D., Colucci, W. S., et al. (1993). No adaptation to digitalisation as evaluated by digitalis receptor (Na, K-ATPase) quantification in explanted hearts from donor without heart disease and from digitalised recipients with endstage heart failure. *The American Journal of Cardiology, 70,* 110–114.
Schoner, W., & Scheiner-Bobis, G. (2007). Endogenous and exogenous cardiac glycosides: Their roles in hypertension, salt metabolism, and cell growth. *American Journal of Physiology Cell Physiology, 293,* C509–C536.
Steyn, P. S., & Heerden, R. R. (1998). Bufadienolides of plant and animal origin. *Natural Product Reports, 15,* 397–413.
Su, Y. H., & Nu, X. (2001). Evaluation of pharmacodynamic effect of pharmaceutical agents of Chansu. *Journal of Beijing University TCM, 24,* 51–54.
Su, Y. H., Huang, X. Q., Zhang, D. Z., et al. (2003). HPLC separation and determination of bufadienolide in cinobufacini injection. *Chinease Traditional Patent Medicines, 25,* 24–27.
Sun, L., Chen, T., Wang, X., et al. (2009). Bufalin induced reactive oxygen species dependent Bax translocation and apoptosis in ATSC-a-1 cells. *eCAM.* doi:10.1093/ecam/nepo82.
Takai, N., Ueda, T., Nishida, M., et al. (2008). Bufalin induces growth inhibition, cell cycle arrest and apoptosis in human endometrial and ovarian cancer cells. *International Journal of Molecular Medicine, 21,* 637–643.
Tao, Q. F., Soszynski, P. A., Hollenberg, N. K., et al. (1995). A sensitive [Na, K] ATPase assay specific for inhibitors acting through the digitalisbinding site. *Journal of Cardiovascular Pharmacology, 25,* 859–863.
Wang, S. W., Chiao, Y. C., Tsai, S. C., et al. (1997). Inhibition of bufalin on pituitary and testicular function in rats. *The Journal of Pharmacology and Experimental Therapeutics, 283,* 528–532.
Wang, S. W., Lin, H., Hwang, J. J., et al. (1999). Inhibition of testosterone secretion by digitoxin in rat testicular interstitial cells. *Journal of Cellular Biochemistry, 74,* 74–80.
Wang, S. W., Pu, H. F., Kan, S. F., et al. (2004). Inhibitory effects of digoxin and digitoxin on corticosterone production in rat zona fasciculata-reticularis cells. *British Journal of Pharmacology, 142,* 1123–1130.
Watabe, M., Masuda, Y., Nakajo, S., et al. (1996). The cooperative interaction of two different signaling pathways in response to bufalin induces apoptosis in human leukemia U937 cells. *The Journal of Biological Chemistry, 271,* 14067–14072.
Watabe, M., Kawazoe, N., Masuda, Y., et al. (1997). Bcl-2 protein inhibits bufalin-induced apoptosis through inhibition of mitogen-activated protein kinase activation in human leukemia U937 cells. *Cancer Research, 57,* 3097–3100.

Xu, W., Luo, H., Zhang, Y., et al. (2007). Simultaneous determination of five main active bufadienolides of Chansu in rat plasma by liquid chromatography tandem mass spectrometry. *Journal of Chromatography, 859,* 157–163.

Yamada, K., Hino, K., Tomoyasu, S., et al. (1998). Enhancement by bufalin of retinoic acid-induced differentiation of acute promyelocytic leukemia cells in primary culture. *Leukemia Research, 22,* 589–595.

Yates, N. A., & McDougall, J. G. (1993). Effects of direct renal arterial infusion of bufalin and ouabain in conscious sheep. *British Journal of Pharmacology, 108,* 627–630.

Yeh, J. Y., Huang, W. J., Kan, S. F., et al. (2001). Inhibitory effects of digitalis on the proliferation of androgen-dependent and independent prostate cancer cells. *The Journal of Urology, 166,* 1937–1942.

Yeh, J. Y., Huang, W. J., Kan, S. F., et al. (2003). Effects of bufalin and cinobufagin on the proliferation of androgen dependent and independent prostate cancer cells. *Prostate, 54,* 112–124.

Yin, J., Shen, J., Su, W., et al. (2007). Bufalin induces apoptosis in human osteosarcoma U-2OS and U-2OS methotrexate300-resistant cell lines. *Acta Pharmacologica Sinica, 28,* 712–720.

Yu, C. H., Kan, S. F., Pu, H. F., et al. (2008). Apoptotic signaling in bufalin- and cinobufagin-treated androgen-dependent and -independent human prostate cancer cells. *Cancer Science, 99,* 2467–2476.

Zhang, L. S., Nakaya, K., Yoshida, T., et al. (1991). Bufalin as a potent inducer of differentiation of human myeloid leukemia cells. *Biochemical and Biophysical Research Communications, 178,* 686–693.

Zhang, L., Nakaya, K., Yoshida, T., et al. (1992). Induction by bufalin of differentiation of human leukemia cells HL60, U937, and ML1 toward macrophage/monocyte-like cells and its potent synergistic effect on the differentiation of human leukemia cells in combination with other inducers. *Cancer Research, 52,* 4634–4641.

Zhang, J., Sun, Y., Liu, J. H., et al. (2007). Microbial transformation of three bufadienolides by Nocardia sp. and some insight for the cytotoxic structure-activity relationship (SAR). *Bioorganic & Medicinal Chemistry Letters, 17,* 6062–6065.

Chapter 19
An Evidence-based Perspective of Herbal Remedies for Cancer Patients

Harendra S. Parekh and Jing Jing

Abstract In spite of great advances in cancer therapy, conventional treatments, most notably chemotherapy, are inherently associated with a plethora of side-effects induced by damage to otherwise healthy cells/tissues. Herbal remedies are, however, yielding important breakthroughs in cancer prevention and treatment, and are presently used first line in numerous cultures across the world. Novel effective anticancer herbal constituents have been extensively studied, and shown to be effective in preventing tumor growth while improving quality of life. Furthermore, clinical studies indicate that herbal remedies when used alone or as an adjunct to conventional anticancer treatment can effectively treat cancer as well as reduce side-effects associated with chemotherapy/radiotherapy. Hence, there is an increasingly convincing rationale for employing herbal remedies against neoplastic disease. However, before acceptance of herbal remedies into wide scale clinical use can be achieved, two critical issues need to be addressed; these include a clearer understanding of their precise mechanism of action, along with supporting empirical clinical data on efficacy. In this chapter, firstly, we describe the four primary modes of action that include inducing apoptosis, boosting of the immune system, overcoming multi-drug resistance and inhibiting angiogenesis by herbal remedies. Then, evidence of experimental studies on the families of effective herbal remedies reported to-date, such as alkaloids, flavonoids, polysaccharides, and glycosides are provided. Furthermore, the application, be it *via* oral or parenteral administration in clinical cases and observations from treating various cancer types along with relevant meta-analyses/systematic reviews are summarized. In closing, future prospects of the road ahead and clinical acceptance of herbal remedies are discussed.

H. S. Parekh (✉)
School of Pharmacy, Pharmacy Australia Centre of Excellence, The University of Queensland, Brisbane, QLD, Australia
e-mail: h.parekh@uq.edu.au

W. C. S. Cho (ed.), *Evidence-based Anticancer Materia Medica,* Evidence-based Anticancer Complementary and Alternative Medicine,
DOI 10.1007/978-94-007-0526-5_19,

19.1 Introduction

19.1.1 Defining a "Herbal Medicine" and a "Herbal Remedy"

There is an important but often overlooked distinction that exists between the terms "herbal medicine" and "herbal remedy". Herbal medicine is a practice based upon the use of medicinal plants and their extracts, with the scope sometimes extending to include fungal and bee-based products, as well as minerals, shells, and various animal components (Acharya and Shrivastava 2008). This practice of medicine is used extensively, and its popularity, especially in the West, is growing at a significant pace as patients turn to more holistic forms of therapy in the management of diseases and ailments, which conventional therapies often fail to address. Traditional Chinese and Indian (Ayurvedic) medicine are by far the most popular practices, having evolved over many thousands of years, and they are considered without question to be among the world's oldest medical systems still in wide use today.

An herbal remedy, on the other hand, is one that encompasses the pharmacologically active component(s) of plant-based remedies used in the practice of herbal medicine (Treasure 2005). The pharmacology of such remedies as well as the pharmaceutical products available is frequently updated, based on extensive ongoing research undertaken to elucidate their mechanisms of action. A classic example of this is the ginseng plant, where more than 28 ginsenosides have been extracted to-date, and these are associated with a wide range of therapeutic actions in the central nervous, cardiovascular and endocrine systems (Cho 2009). China, having an unrivaled predominance in the traditional herbal medicine and remedy market, boasts a library of no less than 12,000 medical plants. In this chapter we bring to the fore a 'clinical perspective' on the use of the most salient Chinese herbs in cancer treatment and management.

19.1.2 Role of Herbal Remedies in Cancer Prevention and Treatment

Cancer presents a serious and ever-spiraling burden on public healthcare systems. According to the World Cancer Report 2008, it was projected to become the leading cause of death worldwide, in just two years, i.e. by 2010 (Boyle et al. 2008). Herbal medicine/remedies are increasing their presence in the therapeutic armament against cancer contributing not only in disease prevention and treatment but also working to reduce resistance to therapy, facilitate recovery from chemo/radiotherapy, and prevent tumor regeneration or the development of metastases, as well as in effective pain management. By virtue of the fewer side effects reported with their use herbal medicines/remedies are now extensively used, and accepted as a "complementary" (adjunct to conventional medicine) or "alternative medicine" (used alone) by many clinicians and cancer patients alike (Huntley and Rees 2010). They exist in a variety of dosage forms (e.g. tablets, capsules, powders, extracts, and fresh or dried plants) and it is estimated that >80% of cancer patients report using at least some forms of complementary/alternative medicine alongside their treatment regimen (Lerner and Kennedy 1992).

Given the popularity and success of traditional therapies in this setting, we will attempt to showcase the most noteworthy, describing them from the viewpoint of their proposed mode of action, their use in a given class/type of cancer, and where applicable in the stage of disease (early/late-stage/palliative) when they would be enlisted. Only recently, and following rigorous and credible *in vitro*/*in vivo* studies has real evidence supporting the use of traditional medicines become available. The aim of such studies being to elucidate their precise modes of action in relation to various cancer types, and emerging from these are clinical studies where convincing scientific evidence now further supports a place for unconventional therapies in the global fight against cancer.

19.2 Herbal Medicines and Their Primary Modes of Action

It is apparent that extracts or isolated compounds from a broad selection of Chinese herbs act along multiple biochemical pathways, affecting tumor survival, growth, and metastasis. Understanding the processes by which these agents act is a key to the discovery and development of new anticancer agents and to identifying optimal treatment strategies for cancer sufferers. The primary mechanisms by which herbal medicines have been shown to target different stages of cancer development include an interplay between the induction of apoptosis, inhibition of angiogenesis, overcoming multidrug resistance, and boosting the immune system (Parekh et al. 2009).

19.2.1 Inducing Apoptosis

Apoptosis or "programmed cell death" is a highly regulated process that involves activation or suppression of a series of genes, enzymes and key receptors leading to cell death (Douglas and Laster 1991; MacFarlane 2003). Agents with the ability to induce apoptosis in a selective manner have the potential to preferentially eliminate cancer cells, and are therefore appropriate for selection in antitumor therapy.

Caspases are a class of cysteine proteases, which operate as signal transducers and catabolic enzymes (Los et al. 1999). The caspase-cascade signalling system is a vital underlying process that directly influences apoptosis (Launay et al. 2005). Two major apoptotic cascades have been identified and are triggered by specific initiator caspases; they are the mitochondrial pathway (*via* caspase-9) and the death receptor pathway (*via* caspase-8) (Ferri and Kroemer 2001; Kwon et al. 2007). Initiator caspases are activated by binding to adaptor molecules which subsequently trigger release of 'effector' caspases. Different apoptotic pathways share the same effector caspases (i.e. caspase-3, -6, and -7), which cleave a wide range of substrates, ultimately leading to apoptosis. For example, Takrisodokyeum, a remedy consisting of twelve herbs, induces apoptosis through activation of caspase-3 in HL-60 (leukemic) cells (Kwon et al. 2003). Recently, the petroleum-ether extract drawn from the root of *Onosma paniculatum,* which is traditionally used as a herbal remedy to treat cancer (Chen et al. 2002), also exhibited caspase-3 dependent induction of apoptosis in a range of tumor cells (Rinner et al. 2010).

The mitochondrial pathway leading to cell death is mainly regulated by the Bcl-2 protein family, one which is divided into pro-survival and pro-apoptotic subclasses. Diminution in the ratio of the pro-survival/-apoptotic family proteins will drive cells down the path of apoptosis (Lanave et al. 2004). To this end the flavonoid, licochalcone-A, isolated from the licorice root induces apoptosis in MCF-7 breast cancer and HL-60 cell lines, this shown to result from reduced expression of *Bcl-2* pro-survival genes (Rafi et al. 2000, 2002). *In vitro* studies using matrine, a primary active component from the dry roots of *Sophora flavescence* indicated that MKN45 gastric cancer cells were indeed driven down the apoptotic pathway, and this appears to proceed *via* activation of caspase-3 and -7, leading to a rise in the presence of pro-apoptotic molecules of the Bcl-2 family (Luo et al. 2007).

Camptothecin and its analogues, are widely used and reported herbal agents, showing cytotoxic effects against a variety of cancer cells by inducing activation of apoptotic caspases (Wenzel et al. 2004). Glioblastoma cells, when exposed to camptothecin exhibited increased activation of caspase-3, -8, and -9 (Xia et al. 2005). Topotecan, a camptothecin analogue induced cell death in non-small cell lung cancer cells in a caspase-8-dependent manner, although without activation of caspase-9 (Ferreira et al. 2000). The active component emodin, isolated from *Polygonum cuspidatum*, effectively induces apoptosis in Bu 25TK cervical cancer cells. This occurs through activation of the caspase-9 cascade, but independent of caspase-8 activation (Srinivas et al. 2003). These examples suggest that apoptosis can indeed be induced *via* a combination of the mitochondrial and death receptor pathways, or purely *via* one of the two pathways.

Death receptors are classified as type-I transmembrane proteins with their activation also leading to initiation of the caspase cascade (Ashkenazi and Dixit 1998; Kumar et al. 2005). The receptors' intra-cytoplasmic death domain component is essential for signal transduction and ultimately, apoptosis (Itoh and Nagata 1993). Hence ligands that bind and activate this pathway have become a source of attention in recent times as potential therapeutic agents in cancer.

19.2.2 Inhibiting Angiogenesis

Angiogenesis refers to the formation of new blood capillaries from pre-existing vasculature (Hanahan and Weinberg 2000). The entire process is tightly regulated by a fine balance between angiogenic activators and inhibitors in healthy patients (Risau 1997). However, excessive angiogenesis results when cancer cells produce abnormally large amounts of angiogenesis factors. Expanding neovasculature supports tumors in their growth and helps them to form hematogenous metastases in distant organs (Gimbrone et al. 1972). Many pro-angiogenic factors have been identified, and among them, vascular endothelial growth factor (VEGF) has been identified to be the most potent tumor-related angiogenesis inducer (Ribatti et al. 2001). VEGF acts by binding to several high-affinity trans-membranes endothelial cell receptors, most notably VEGF receptors types 1 and 2 (Flt-1 and Flk-1 in mice, or KDR in humans). This binding of VEGF, to its receptors leads to intracellular receptor phosphorylation, which initiates various intracellular downstream receptor pathways causing the proliferation, migration and blood vessel formation of endothelial cells (Eskens and Verweij 2006; Fayette et al. 2005). Therefore, the most impressive data to-date has been generated by those agents that target the production of VEGF and/or its receptors.

A range of compounds extracted from herbal medicines, exert their effects, at least in-part, by lowering VEGF and its receptors in tumor and/or endothelial cells, thereby inhibiting new blood vessel formation. For example, fungal polysaccharopeptide of *Coriolus versicolor* suppresses *VEGF* gene expression, resulting in a deprivation of angiogenic stimulation to the growth of S180 solid tumors (Ho et al. 2004). Taspine, isolated from *Radix et Rhizoma Leonticis* down-regulates VEGF secretion in human non-small cell lung cancer cells (A549) and human umbilical vein endothelial cells with a concomitant reduction in *VEGF* and *Flk-1/KDR* mRNA, also observed in human umbilical vein endothelial cells (Zhang et al. 2008). Curcumin possesses similar properties with it shown to inhibit VEGF synthesis and secretion in Ehrlich ascites tumor cells; while down-regulating expression of *VEGF* and *VEGF* receptor mRNA, in immortalized mouse fibroblast NIH 3T3 cells (Gururaj et al. 2002). The evidence presented here certainly suggests that these compounds could be further developed and integrated into therapeutic regimes for treatment of a range of human cancers.

19.2.3 Overcoming Multi-drug Resistance (MDR)

Cancer cells exposed to one or more cytotoxic agents can, over the course of repeat treatments, develop resistance to their effects. This occurs through a number of mechanisms and by activating several of these pathways can ultimately lead to MDR. In such a scenario, clinicians have fewer and fewer efficacious therapeutic options, leading to a poor prognosis (Dean et al. 2001). MDR is of particular concern in patients receiving chemotherapy, with the primary cause attributed to the over-expression of P-gp (P-glycoprotein), a 170 kD transmembrane glycoprotein (Higgins 2007). P-gp actively pumps out a wide range of structurally and functionally diverse amphipathic anticancer drugs from within tumor cells thereby decreasing their intracellular concentration and so their potency (Ling 1997). This has led to the concept of combining existing anticancer drugs with non-toxic, potent P-gp inhibitors in an attempt to overcome or resist MDR in cancer cells.

Herbal remedies can be largely seen as an untapped resource offering up potential inhibitors to the major drug efflux pathways contributing to MDR. The terpernoids,

(R)-(+)-citronellal, abietic acid and glycyrrhetic acid can significantly increase the cellular influx of P-gp substrates (as determined *via* [^{3}H]-digoxin) and decrease the efflux transport across Caco-2 cell monolayers (Yoshida et al. 2006a). Additionally, (R)-(+)-citronellal also increases the bioavailability of oral digoxin in rats, based on the blockade of P-gp-mediated efflux from intestinal epithelia to the lumen in the absorption process (Yoshida et al. 2006b). Therefore, these findings suggest that the intestinal absorption of P-gp substrates may be favorably affected by these herb-based P-gp inhibitors. In the context of cancer therapy, this is a particularly useful approach to overcoming MDR *via* P-gp modulation in an effort to more effectively shrink tumors, and so improve patient prognoses. An example in case is tetrandrine, a potent inhibitor of P-gp mediated MDR. *In vitro* and *in vivo* studies have identified that MDR acquired by tumor cells can be partially reversed following co-administration of tetrandrine alongside the conventional anticancer drug doxorubicin, and no evidence of an increase in doxorubicin-associated toxicity was observed (Fu et al. 2002). These findings therefore have potential future clinical significance.

19.2.4 Boosting the Immune System

One of the important roles of the immune system is to identify and eliminate pathogens, and this includes when normal cells turn cancerous. Clearly, some pathogenic cells/organisms have discovered ways to continually evade detection by the immune system, going on to establish themselves as solid tumors (Yamamoto et al. 1995). Cytokines, such as TGF-β and IL-6 are secreted by tumor cells, and both have been implicated in the stealth-like properties of cancer cells (Seliger 2005). These cytokines function by inhibiting proliferation of cytotoxic T-lymphocytes as well as the production of both IFN-γ and TNF-α, which play key roles in the prevention of tumorigenesis, detection and elimination of established tumors (Werneck et al. 2008). The mechanism underlying the antitumor effects of *Citrus unshiu* extract is strongly suggested to be *via* boosting cytokines IFN-γ and TNF-α, thereby enhancing immune-mediated antitumor responses (Lee et al. 2010).

19.3 Experimental Perspectives of Effective Herbal Ingredients in Cancer Therapy

19.3.1 Alkaloids

19.3.1.1 Berberine

Berberine, an isoquinoline alkaloid is a very widely distributed alkaloid commonly used in China as a botanical drug, while being the major active compound in

Coptidis rhizoma (Chinese term Huanglian, common uses include treatment of inflammatory diseases and various cancers) (Tang et al. 2009). Berberine is a potent apoptosis inducing agent with proven activity in breast cancer (MCF-7), colorectal cancer (HCT-116 and SW480) and lung cancer (A549) cell lines. Studies have identified that it exerts its effects preferentially through the "mitochondrial-dependent" pathway, as was explained in Sect. 19.2.1 (Piyanuch et al. 2007; Wong et al. 2009; Patil et al. 2010).

Formation of metastases as well as the metastatic potential of cancers casts a major shadow over the longer term success of present treatments. It is not uncommon for a cancer, thought to have been totally eradicated, to re-appear months, or even years later, possibly in a distant organ to that originally infected. Berberine presents a remarkable advancement in this regard having been successfully tested in anoikis-resistant breast cancer cells; these are cells possessing a high metastatic tendency, leading to recurrent diagnoses and a poor prognosis. Anoikis, or detachment-induced apoptosis, is likely to prevent cancer progression and metastasis by blocking the signals critical to the survival of localized cancer cells (Shen and Kramer 2004), with resistance to anoikis regarded as a prerequisite to metastasis occurring. To demonstrate this unique property, berberine has been trialed in combination with doxorubicin, which is commonly used in breast cancer treatment. Data supports the presence of a synergistic relationship, with berberine markedly inhibiting the growth of cells, and to a greater extent than doxorubicin treatment alone, in anoikis-resistant MCF-7 and MDA-MB-231 breast cancer cells (Kim et al. 2010a).

In addition to its role in dual chemo-therapeutic approaches, berberine also has reported benefits when employed alongside radiotherapy, where it has been shown to augment the effects of radiotherapy, both *in vitro* and *in vivo*. Studies using lung cancer as a model demonstrate that this underlying synergistic effect is, at least in part, due to the induction of apoptosis (Peng et al. 2008). These findings collectively suggest that berberine ought to be trialed as an adjuvant therapy in select cases of chemotherapy and radiotherapy, and in particular cases/cancers where resistance to conventional therapy is increasingly prevalent.

19.3.1.2 Camptothecin and Its Analogues

Camptothecin was first isolated and identified in the 1960's, from the wood of *Camptotheca acuminata*, a tree native to the rocky slopes of Northern China (Wall et al. 1966). At the same time its cytotoxic activity was demonstrated against murine leukemia in L1210 cells (Wall et al. 1966). In much later studies, camptothecin and its analogues were found to be effective against a broader spectrum of cancers and have been used in the treatment of ovarian, cervical, colorectal, and small cell lung carcinomas (Arbuck and Takimoto 1998; Ackermann et al. 2007).

The molecular target of camptothecin and its analogues has been firmly established as being human DNA-topoisomerase I (Zunino and Pratesi 2004). They act by stabilizing the DNA-topoisomerase I cleavable complex, causing accumulation of single-stranded breaks in the DNA (Shao et al. 1999). Collision of this cleaved strand of DNA with a DNA replication fork causes an irreversible double-strand DNA break, ultimately leading to cell death (Tsao et al. 1993).

Initial human clinical trials of camptothecin and its analogues were abandoned due to severe toxicity, in particular hemorrhagic cystitis. Renewed interest in this class of compounds has developed as a result of the synthesis of more water soluble and less toxic analogues, particularly 7-ethyl-10-[4-(1piperidino)-1-piperidino] carbonyl-oxycamptothecin (CPT-11) and 9-aminocamptothecin (9-AC). Both CPT-11 and 9-AC demonstrated a marked survival advantage in models of advanced renal cancer and have been identified as agents warranting further clinical assessment (El-Galley et al. 2003).

19.3.1.3 Matrine

Sophora flavescens, also referred to as "Kushen", is an important source of therapeutic alkaloids, being grown in China, Japan and across many parts of Europe (Liu et al. 2010). Its dry root has long been regarded as a potent anticancer herb, especially in China (Liu et al. 2007). Matrine is just one of the primary active components extracted from the dry roots of *Sophora flavescence* and *in vitro* studies reveal that it has the potential to be developed into an antitumor drug. It functions by triggering apoptosis *via* the mitochondrial pathway, as has been demonstrated in a range of gastric (MKN45 see Sect. 19.2.1, SGC-7901) and leukemic (K562) cells (Jiang et al. 2007; Dai et al. 2009).

19.3.2 Flavonoids

Flavonoids are considered nature's antioxidants, commonly found in fruits, vegetables, seed, and bark (Choi et al. 2002), with isoflavones representing a subclass of the flavonoid family. Genistein is one of several well-known isoflavones, and widely reported to possess pro-oestrogenic properties (Harborne and Williams 2000). The majority of prostate cancers cases are found to be hormone-dependent, relying on the continuous supply of androgen for their survival, hence androgen (receptor) blockade is a common therapeutic intervention strategy (Swanson 2006). Flavonoids, and in particular isoflavones are being pursued as possible candidates in prostate cancer prevention/treatment with *in vivo* studies revealing several of these agents, including soy isoflavones, biochanin A and genistein are able to diminish prostate-specific antigen production in breast and prostate cancer cell lines (Rosenberg et al. 2002). Given these findings and the poor prognosis of many mid-late stage cases of prostate cancer, isoflavones present us with attractive options and require further clinical investigation.

19.3.3 Glycosides

Ginsenosides are a class of steroid glycoside found exclusively in the plant genus *Panax* (commonly referred to as 'ginseng'). Over 28 ginsenosides have been isolated from American and Asian ginseng and identified to possess multiple beneficial pharmacological properties (Li and Fitzloff 2002). Among them, ginsenoside-Rg3 is the most widely reported having been shown to halt the process of angiogenesis in establishing tumors (Zhang et al. 2006). It has also been found to inhibit proliferation of colon cancer cells (Tyagi et al. 2002), and when administered in combination with docetaxol it enhances the susceptibility of prostate cancer cells to treatment, this compared to cells treated with docetaxol alone (Kim et al. 2010b). In 2000, ginsenoside-Rg3 appeared on the Chinese market for the first time as the anticancer drug "Shen-Yi", available in easy-to-swallow capsule form. It has been marketed in China as a medicine that can prevent the invasion and metastasis of tumor cells to distant tissues/organs, and we discuss its effects further below (Yun et al. 2001).

19.3.4 Polysaccharides

Polysaccharides from natural sources are a class of macromolecule that can profoundly trigger our immune system (Tzianabos 2000), therefore having application as immune modulators in cancer prevention and treatment. The root of *Astragalus membranaceus* is among the most popular health-promoting herbs, which has immune potentiating effects associated with polysaccharide-containing fractions (astragalus polysaccharides) (Shao et al. 2004). *In vitro* and *in vivo* studies show that *Astragalus membranaceus* exhibits antitumor effects, and these are purported to be achieved through activation of host antitumor immune responses (Cho and Leung 2007). Recently, a particular polysaccharide fraction, isolated from astragalus polysaccharides also showed immunostimulatory activity in rats predisposed to stomach cancer (Li et al. 2009).

19.3.5 Miscellaneous

Arsenic trioxide (As_2O_3), known as Pi Shuang in traditional Chinese medicine, is a US FDA approved agent for treating patients with acute promyelocytic leukemia (APL) (Jeyapalan 2001). As_2O_3 induces apoptosis in a wide variety of cancer cell lines including, but not limited to APL, neuroblastoma, gallbladder, lung, renal, and glioma to name just a handful (Akao et al. 1999; Hyun et al. 2003; Kanzawa et al. 2003; Ai et al. 2006; Jin et al. 2006; Emadi and Gore 2010). As_2O_3 also appears to be a promising therapeutic agent for the management of autoimmune diseases. Laboratory studies in MRL/lpr mice show that As_2O_3 protects younger animals against developing lympho-proliferative and autoimmune syndromes, while it induces near total remission in older, formerly diseased mice (Bobe et al. 2006).

19.4 Clinical Perspectives of Herbal Remedies in Combination Cancer Therapy

Many reports exist where Chinese herbs have been used in randomized human trials as an adjunct to existing conventional chemotherapy and radiotherapy, and in various types of cancers. These published studies are evidence that Chinese herbal remedies when combined with conventional therapy can increase tumor response, improve performance status and survival, or reduce chemotherapy-induced toxicity. Numerous meta-analyses/systematic reviews conducted to-date also elaborate on the efficacy and safety of herbal remedy use in cancer patients and report this, among others, in terms of overall survival, progression-free survival, and adverse effect reduction. Noteworthy clinical studies alongside meta-analyses/systematic reviews in the most prevalent cancer types are explored in detail below.

19.4.1 Lung Cancer

Lung cancer remains a major global health problem due to its prevalence and poor prognosis once diagnosed (Chen et al. 2010; Stephens et al. 1994). It is classified into two broad histological types that account for the majority of diagnosed cases; non-small cell lung cancer (NSCLC) (80%) and small cell lung cancer (20%) (Hirsch et al. 1988). Chemotherapy is first-line treatment for those sensitive to its effects, although combination chemotherapy is commonly employed where patients are in the advanced/late stages of the disease (Buccheri and Ferrigno 1994). However, as expected this is almost always associated with a high incidence of treatment-related toxicity, which limits the frequency of repeat dosing regimens (Mell et al. 2007).

With Chinese remedies having been reported to be beneficial adjuncts in cancer treatment they have been incorporated into a number of clinical trials. Seven trials, conducted in hospitals across China, involving a total of 717 patients, are summarized in Table 19.1. The primary comparators for those studies outlined in Table 19.1 were the treatment group (herbal remedy+chemotherapy) and control group (chemotherapy alone). As outlined in Table 19.1, four of the studies showed statistically significant improvements ($P<0.05$) in the outcomes of short term therapy when validated by using the World Health Organization criteria for response rate (RR) (Julka et al. 2008). The remaining three studies, although highlighting no statistically significant advantage in RR, did show improvements

Table 19.1 Response rates (RR) for combined therapy *versus* chemotherapy alone in seven human lung cancer clinical trials

Herbal remedy	Treatment group/Control group	
	No. of patients	RR (%)
'Fufang Kushen' injection (Ding et al. 2006)	32/30	53.1/33.3*
'Fufang Kushen' injection (Bai and Yan 2008)	60/60	73.3/53.3*
CHM[1] (Meng et al. 2006)	28/26	46.4/38.4*
CHM[2] (Zhu 2002)	120/85	68.3/58.8*
'AiYiShu' injection (Chang et al. 2008)	36/25	41.7/36.0**
CHM[3] (Zhang et al. 2006)	16/16	43.7/37.5**
CHM[4] (Qin et al. 2007)	33/30	54.6/43.3**

$*P<0.05$; $**P>0.05$

CHM[1,2,3,4]: Chinese herbal mixture with four different compositions

in quality of life (QoL) markers, as validated using the Karnofsky performance status.

QoL is usually measured by a range of indicators and Karnofsky performance status incorporates the most internationally recognized and validated range of tools. It aims to assess the overall well-being of patients over the course of their therapy. A systematic review of fifteen Chinese clinical trials involving a total of 862 participants, demonstrated that select herbal remedies show marked improvements in QoL indicators in patients receiving co-therapy for NSCLC (Chen et al. 2010). Meta-analysis of randomized trials has shown that, when taken together, astragalus-based remedies (see Sect. 19.3.4) may increase the effectiveness of platinum-based chemotherapy in advanced NSCLC (McCulloch et al. 2006).

19.4.2 Breast Cancer

Despite great efforts to improve breast cancer awareness, leading to better screening and detection rates, as well as treatment, it remains the world's second most common cause of cancer-related death in women (Greenlee et al. 2001). Conventional therapies for patients diagnosed with the disease can include a combination of surgery, radiotherapy, chemotherapy, and hormonal therapy; the choice of therapy depends on a number of factors namely, the stage in which a diagnosis is made, menopausal status, age, and overall health (Dog and Micozzi 2005). Although conventional therapy remains the cornerstone of breast cancer treatment, multiple surveys completed by patients have shown that up to 80% of women with the cancer report using some form of complementary and/or alternative medicines, the most popular of which were herbal remedies (Field et al. 2009).

One such survey, with over 360 respondents, revealed the two commonest reasons for herbal remedy use were to "improve physical wellbeing" and "boost the immune system" (Kremser et al. 2008). One school of thought is that herbal remedies play a valuable role in easing conventional therapy-induced side-effects, a significant impediment to both the administration of higher doses as well as more frequent dosing. Clinical studies go on to corroborate such claims by traditional medicine practitioners and their patients. For example, one such human trial used a traditional prescription of Chinese medicine commonly referred to as the "eight dainties granula"—this being employed as an adjuvant therapy in a cohort of breast cancer patients (treatment group), while the control group received chemotherapy alone (Sun 2009). Leukocyte levels were monitored immediately prior to, and following treatment as this marker is considered imperative to a functional immune system and their numbers as well as function is often significantly lowered and impaired, leading to increased incidences of infection in patients undergoing chemotherapy. The results of the study were conclusive in demonstrating an appreciable benefit to employing this particular herbal remedy alongside chemotherapy as the leukocyte count remained near constant ($P > 0.05$) in the treatment group (before 6.4×10^9/L; after 6.3×10^9/L); while a significant ($P < 0.05$) decrease was observed in the control group (before 6.3×10^9/L; after 4.3×10^9/L). This suggests, as alluded to by the authors that the "eight dainties granula" resists decreases in peripheral blood leukocytes induced by chemotherapy.

19.4.3 Colorectal Cancer

Colorectal cancer has a worldwide annual new incidence rate of some one million cases, with more than half this number of patients succumbing to the disease on an annual basis (Winawer 2007). Surgical resection is by far the most predominant form of intervention, performed with curative intentions in mind (Faria et al. 2005). However, sooner or later at least half of these patients will go on and relapse (Carlsson et al. 1987; Stipa et al. 1991; Kievit 2002). To address this high relapse rate chemotherapy is often chosen as an adjuvant treatment in colorectal cancer patients to eliminate microscopic traces of the disease (Chau and Cunningham 2002).

The place of herbal remedies in this setting has been as an adjunct to chemotherapy, especially in patients relapsing following surgical intervention. Chemotherapy induced side-effects have been shown to be markedly alleviated in co-therapy. In a recent randomized clinical study including 210 patients (Xu 2010), the orally administered herbal prescription "NiJiangLing", was combined with FOLFOX4 chemotherapy. RR, QoL and side effect comparisons showed improvements in RR and QoL markers, as well as a lower incidence of gastrointestinal side-effects such as nausea, vomiting, and diarrhoea in patients receiving the herbal remedy in combination with chemotherapy, over chemotherapy alone.

This synergistic effect corroborates findings of a previous randomized clinical trial comprising 35 patients (Zeng 2009), where "Sen-Yi" capsules (primary component being ginsenoside-Rg3) were also combined with FOLFOX4, and compared to chemotherapy alone (control group) (RRs outlined in Table 19.2). Patients selected

Table 19.2 Response rates (RR) for combined therapy *versus* chemotherapy alone, in two human colorectal cancer clinical trials

Herbal remedy	Treatment group/Control group	
	No. of patients	RR (%)
'JiangNiLing' (Xu and Wang 2010)	61/60	54.1/40.0*
'Sen-Yi' capsules (Zeng et al. 2009)	35/32	45.7/40.6*

*$P < 0.05$

for this latter study were diagnosed following pathological assessment with recurrent, metastatic colorectal cancer. Compared to controls, there were statistically significant improvements in RR and decreased levels of the pro-angiogenic factor-hVEGF (defined in Sect. 19.2.2) in the treatment group. This certainly points toward ginsenoside-Rg3 acting in a synergistic fashion and by blocking angiogenesis.

19.4.4 Gastric Cancer

Gastric cancer is one of the most common causes of death from cancer in China and amongst other Asian neighbors, including Japan and Korea (Yang 2006; Leung et al. 2008). Although surgery is the standard treatment option, early detection and treatment is the only way to effectively reduce mortality. Gastric cancer is most often than not diagnosed only in the mid-late stages, and even if treated patients suffer from exceptionally high rates of relapse and metastasis following surgery. There are also problems of poor sensitivity to radiotherapy/chemotherapy, therefore herbal remedies have a crucial role to play in preventing relapse and metastasis in gastric cancer.

A randomized clinical trial in 2008 recruiting 123 (stage IV) patients given the herbal remedy "Yang Wei Kang Liu" (YWKLF) was shown to improve survival rates (Li et al. 2008). Of the 123 patients, 77 were assigned to the treatment group, receiving MFP (methotrexate, 5-fluorouracil, and low-dose cisplatin) chemotherapy in combination with orally administered YWKLF granules, while 46 patients were assigned to receive chemotherapy alone (control group). The RRs and median survival times in the control group *versus* treatment groups were 13:34 and 14:22 months, respectively. The investigators also went on to explore potential modes of action of YWKLF, by focusing on its predicted ability to activate apoptotic pathways in MGC-803 human gastric cancer cells, which was indeed found to be the case and so defines its precise mode of action.

19.4.5 Esophageal Cancer

Esophageal cancer is highly treatable by surgical intervention if diagnosed early, but is usually incurable once in the advanced stages (Clark et al. 1994; Steup et al.

1996). Chemotherapy/radiotherapy is used primarily to relieve dysphagia, with the primary goal being to eliminate residual cancer cells following surgery. Chemotherapy/radiotherapy also serves to limit proliferation of the cancer in patients with advanced stage disease, and in some cases is used as a preoperative measure (Cooper et al. 1999).

A systematic review of 406 esophageal cancer patients from five separate randomized controlled trials provides evidence that Chinese remedies combined with radiotherapy or chemotherapy are indeed superior to radiotherapy/chemotherapy alone, and in at least five aspects; these being prolonged survival times, improved QoL, maintaining an adequate level of immune function, preventing metastasis to distant organs, and absence/reduction of adverse events (Wu et al. 2009a).

19.4.6 Prostate Cancer

Prostate cancer is the most common cancer amongst men, worldwide. Patients with localized disease may be treated with surgery or radiation, while approximately 25–40% of these patients go on to relapse within 5 years (Clark 2008) and the treatment for patients with confirmed metastatic disease is purely palliative (Martel et al. 2003). The high incidence of prostate cancer coupled with a long latency period (slow progress) affords a particularly attractive intervention strategy with target herbal remedies. The eight-herb formulation PC-SPES, has its name derived from "PC", i.e. prostate cancer, and *spes*- the Latin term for "hope" (Marks et al. 2002). The compound has gained popularity as a therapeutic in prostatic cancer, since first being introduced in 1996. However, the US FDA, in 2002, advised against the use of this Chinese phyto-medicine owing to the relatively high concentrations of diethylstilbestrol and warfarin in PC-SPES, which resulted in estrogenic and hemorrhagic complications. As a result of this decision the remedy was no longer marketed in the US (Park et al. 2008).

19.4.7 Liver Cancer

Primary liver cancer is a major health problem worldwide and survival rates remain very low. This trend is not helped by a cohort effect resulting from infections with hepatitis B and C (Bosch et al. 2004). Therapies developed along the principles of Western medicine such as liver transplantation (Ringe et al. 1996) and local ablation therapy (Maeda et al. 2003) are considered first-line followed by a number of other interventions dependent upon patient response; these include chemo-embolization, per-cutaneous ethanol injection, radiofrequency ablation, systemic chemotherapy, and inclusion in clinical trials (Jelic and Grp 2009). The aforementioned treatment options are often limited in efficacy and carry the risk of significant adverse effects.

Besides, these therapies are also too costly, and so inaccessible to patients in the developing world, e.g. Africa, where the highest prevalence of liver cancer exists (Ogunbiyi 2001; Stickel and Schuppan 2007). Therefore, plant-derived compounds may be worth exploring.

Several traditional Chinese remedies have been used and evaluated in clinical trials for transcatheter arterial chemo-embolization in hepatocellular cancer, the most frequent form of liver cancer among populations of many Asian countries. Two independent meta-analyses including 3,236 and 2,428 patients respectively, both displayed compelling evidence of the use of traditional Chinese remedies to enhance the efficacy of transcatheter arterial chemo-embolization in hepatocellular cancer patients (Cho and Chen 2009a; Wu et al. 2009b).

19.4.8 Leukemia

Leukemia represents a group of hematological malignancies characterized by clonal expansion of hematopoietic cells with uncontrolled proliferation, decreased rates of apoptosis, and blocked differentiation. APL is one such type of leukemia characterized by an accumulation of abnormal promyelocytes in bone marrow, with a severe bleeding tendency and the presence of chromosomal translocation (Bennett et al. 1976; Rowley et al. 1977). Arsenic trioxide, described earlier, has been used to treat patients relapsing with APL and this led to the revival of an ancient drugs use in modern medicine (Zhou et al. 2007). In September 2000, As_2O_3 was approved by the US FDA for treatment of patients with APL (see in Sect. 19.3.5). Recent trials in the US demonstrate that the addition of As_2O_3 to standard treatment regimens improves survival outcomes in patients with APL, and may also allow a reduction in cytotoxic chemo-therapeutic doses (Emadi and Gore 2010).

19.4.9 Nasopharyngeal Carcinoma (NPC)

NPC is a malignancy with remarkable racial and geographic distribution. While it is rare in most parts of the world, this disease is the major cause of cancer death in southern China and South Eastern Asia (Trimeche et al. 2008). Many published studies have reported the use of Chinese remedies in combination with conventional cancer therapy for nasopharyngeal carcinoma. A systematic review covering both English- and Chinese-language studies and collating data published over more than four decades (1966–2007) assessed the efficacy of Chinese remedies as a concomitant therapy for NPC patients. The result of meta-analysis suggested that the remedies are indeed efficacious as a concomitant therapy for NPC (Cho and Chen 2009b).

Clinical trials discussed in this section form just the tip of the iceberg, and although promising, further randomized clinical trials are necessary to convince us of the role herbal remedies have to play in the future management and treatment of cancer. Systematic reviews and meta-analyses further demonstrate the potential of herbal remedies and justify broader trials of select remedies combined with conventional cancer therapy in improving clinical efficacy. However, although significant, some of the methods used by the chief investigators in conducting these trials and in analyzing data suggest a high level of bias. Hence, it is essential for those working in the field to design, develop and conduct more rigorously controlled clinical trials, before any firm conclusions can be drawn on the true benefit of combination therapies.

19.5 Herbal Remedies in Terminal/Palliative Cancer Care

Palliative/terminal cancer care focuses on the palliation of physical symptoms caused by cancer and any treatment, which is generally reserved for patients in end-stage disease. Pain is a common and debilitating symptom of cancer, being associated with poor quality of life indicators, such as depression and disability (Carr and Pujol 2010). Effective pain control however remains a primary goal and challenge in palliative cancer care given the significant number of side-effects associated with (non-)opioid-based analgesics (Pham and Primack 2003). There are numerous examples of effective pain management with herbal remedies in a range of cancers, such as that of the liver, stomach, and pancreas (Xu 1996). A non-opioid analgesic containing "capsaicin", derived from the chilli pepper plant has in recent times been reported for topical treatment of post-mastectomy pain syndrome with remarkable results.

Post-mastectomy pain syndrome is a chronic neuropathic pain syndrome occurring in breast cancer patients who have undergone a mastectomy or lumpectomy with axillary dissection (Stevens et al. 1995) and associate with a much poorer QoL (Carpenter et al. 1998). Capsaicin was showed to produce effective analgesia by binding to the vanilloid receptor-TRPV1, which acts as a molecular integrator of chemical and physical painful stimuli (Hayman and Kam 2008). In a randomized parallel trial of 25 patients the application of topical 0.075% capsaicin cream was reported to be more effective than the vehicle cream alone in relieving post-mastectomy pain syndrome (Watson and Evans 1992).

19.6 Future Prospects

19.6.1 The Road Ahead for Herbal Remedies—New Strategies in the Fight Against Cancer

When the multitude of conventional options to anticancer therapy fail to provide effective treatment/relief, or are exhausted, patients often turn to alternative medicine, and herbal remedies are one such option whose popularity and acceptance is growing at a significant pace in the West. The favorable characteristics of medicinal herbs, especially their ability to act on multiple biochemical targets/pathways render them highly potent and effective anticancer agents in their own right. Given their scope they remain without doubt a rich and largely untapped source for new anticancer drug targets that both researchers and oncologist will need to exploit if we are to have any realistic chance of winning the continuing struggle for an effective "cure" against cancer.

With these challenges in mind, the road ahead can be mapped along three primary levels, each being key to the future role of herbal medicine usage in mainstream oncology. The first requires use of modern scientific techniques, such as DNA microarray and proteomics, to assist in deciphering the precise mechanisms by which given herbal remedies exert their anticancer effects. And with expanding knowledge of their primary modes of action, new possibilities arise, particularly as potential targets in the drug discovery process, and as alluded to above these candidates can serve as the foundation for further basic and clinical research. The second level requires teams in the field to conduct high-quality pre-clinical research reporting on efficacy while accurately documenting the incidences, nature and severity of all herb-drug and herb-herb related adverse events. With patient safety being of paramount importance, knowledge acquired should aim to guide clinicians toward safer practices and also assist policy-makers in planning stringent regulations for herbal remedy preparation and usage. Finally, herbal remedies should be effectively integrated into the practice of palliative cancer care, most notably, in effective pain management to ensure patients receive optimal therapy in late-stage disease.

Claims by traditional practitioners as to the immense benefits herbal remedies offer have been received with much cynicism by the West, although there is a noticeable wind of change and acceptance of the true therapeutic value of this ancient practice of medicine is now most certainly emerging. With the door of opportunity now ajar there is no doubt that through international collaboration and rigorous studies scientists can indeed cement a place for herbal medicine in the global formulary of cancer therapies.

19.6.2 Clinical Acceptance of Herbal Remedies—"Evidence" Is the Key to Success

Ample evidence exists in the literature, purporting on benefits herbal remedies bring to the treatment and management of various cancers. Nonetheless, uncertainty remains over their wide-spread use and acceptance by health professionals and cancer patients alike. This situation, as mentioned above, stems to a large degree from the lack of reliable evidence-based research in this area (Ernst 2000).

Cancer patients, depending on the stage/type of disease suffer a multitude of symptoms (physical and psychological), thus accurate reporting before, during and following treatment with herbal remedies is absolutely vital to creating a true picture of benefit *versus* risk. Accuracy of evidence-based reporting is crucial to this outcome, and through rigorously monitored randomized controlled trials and systematic reviews this can be achieved. Such evidence would be invaluable, contributing to the promotion of better outcomes, leading to greater patient acceptance of such treatment options. There is an urgent call for further large scale studies investigating herbal remedy use in effective pain management and advanced stage/palliative care, as these patient groups are often more poorly managed, with greater emphasis placed on aggressive treatments regimens to rid the cancer (Costantini et al. 1999). Key to this outcome is funding support as well as strategic collaborative research with knowledge exchange ensuring sound trial design and execution. All this should be driven toward the common goal of creating evidence that regulatory authorities, clinicians, and policy-makers alike can safely rely upon.

Important challenges certainly lie ahead for herbal remedy research and medical oncology; with this comes the responsibility of scientists and clinicians working towards a marriage of these two disciplines, where international recognition and acceptance is at the heart. This can be achieved through international dialogue and cooperation ensuring any skepticism and doubt of the past is cast well into the shadows.

References

Acharya, D., & Shrivastava, A. (2008). *Indigenous herbal medicines: Tribal formulations and traditional herbal practices.* Jaipur: Aavishkar.

Ackermann, S., Beckmann, M. W., Thiel, F., et al. (2007). Topotecan in cervical cancer. *International Journal of Gynecological Cancer, 17,* 1215–1223.

Ai, Z. L., Lu, W. Q., Qin, X. Y., et al. (2006). Arsenic trioxide induces gallbladder carcinoma cell apoptosis via downregulation of Bcl-2. *Biochemical and Biophysical Research Communications, 348,* 1075–1081.

Akao, Y., Nakagawa, Y., & Akiyama, K. (1999). Arsenic trioxide induces apoptosis in neuroblastoma cell lines through the activation of caspase 3 in vitro. *FEBS Letters, 455,* 59–62.

Arbuck, S. G., & Takimoto, C. H. (1998). An overview of topoisomerase I—targeting agents. *Seminars in Hematology, 35,* 3–12.

Ashkenazi, A., & Dixit, V. M. (1998). Death receptors: Signaling and modulation. *Science, 281,* 1305–1308.

Bai, X., & Yan, M. (2008). Fufang Kushen injection in the treatment of 60 cases lung cancer. *Shaanxi Journal of TCM, 29,* 1317–1318.

Bennett, J. M., Catovsky, D., Daniel, M. T., et al. (1976). Proposals for classification of acute leukemias. *British Journal of Haematology, 33,* 451–458.

Bobe, P., Bonardelle, D., Benihoud, K., et al. (2006). Arsenic trioxide: A promising novel therapeutic agent for lymphoproliferative and autoimmune syndromes in MRL/lpr mice. *Blood, 108,* 3967–3975.

Bosch, F. X., Ribes, J., Diaz, M., et al. (2004). Primary liver cancer: Worldwide incidence and trends. *Gastroenterology, 127,* S5–S16.

Boyle, P., & Levin, B., International Agency for Research on Cancer, et al. (2008). *World cancer report 2008.* Lyon, LYON: International Agency for Research on Cancer; Geneva, GVA: Distributed by WHO Press.

Buccheri, G., & Ferrigno, D. (1994). A randomised trial of MACC chemotherapy with or without lonidamine in advanced non-small cell lung cancer. *European Journal of Cancer, 30,* 1424–1431.

Carlsson, U., Lasson, A., & Ekelund, G. (1987). Recurrence rates after curative surgery for rectal carcinoma, with special reference to their accuracy. *Diseases of the Colon and Rectum, 30,* 431–434.

Carpenter, J. S., Andrykowski, M. A., Sloan, P., et al. (1998). Postmastectomy/postlumpectomy pain in breast cancer survivors. *Journal of Clinical Epidemiology, 51,* 1285–1292.

Carr, B. I., Pujol, L. (2010). Pain at presentation and survival in hepatocellular carcinoma. *The Journal of Pain, 11,* 988–993.

Chang, Y., Ge, M. D., Tang, A. M., et al. (2008). "AiYiShu" injection in the treatment of lung cancer: A clinical assessment. *Chinese Journal of Cancer Prevention Treatment, 15,* 1279–1280.

Chau, I., & Cunningham, D. (2002). Adjuvant therapy in colon cancer: Current status and future directions cancer. *Cancer Treatment Reviews, 28,* 223–236.

Chen, X., Yang, L., Oppenheim, J. J., et al. (2002). Cellular pharmacology studies of shikonin derivatives. *Phytotherapy Research, 16,* 199–209.

Chen, S., Flower, A., Ritchie, A., et al. (2010). Oral Chinese herbal medicine (CHM) as an adjuvant treatment during chemotherapy for non-small cell lung cancer: A systematic review. *Lung Cancer, 68,* 137–145.

Cho, W. C. (2009). Ginseng, Panax ginseng CA Meyer. In J. N. Govil & V. K. Singh (Eds.), *Recent progress in medicinal plant* (Vol. 28, pp. 19–30). Texas: Studium Press.

Cho, W. C., & Chen, H. Y. (2009a). Investigation of the efficacy on transcatheter arterial chemoembolization combined with or without Chinese herbal therapy for hepatocellular carcinoma: Meta-analysis. *European Journal of Cancer, 7*(Suppl.), 393–394.

Cho, W. C., & Chen, H. Y. (2009b). Clinical efficacy of traditional chinese medicine as a concomitant therapy for nasopharyngeal carcinoma: A systematic review and meta-analysis. *Cancer Investigation, 27,* 334–344.

Cho, W. C., & Leung, K. N. (2007). In vitro and in vivo anti-tumor effects of Astragalus membranaceus. *Cancer Letters, 252,* 43–54.

Choi, J., Conrad, C. C., Malakowsky, C. A., et al. (2002). Flavones from scutellaria baicalensis georgi attenuate apoptosis and protein oxidation in neuronal cell lines. *Biochimica et Biophysica Acta, 1571,* 201–210.

Clark, P. E. (2008). Diet and dietary supplement intervention trials for the prevention of prostate cancer recurrence: A review of the randomized controlled trial evidence. *Journal of Urology, 180,* 2314–2322.

Clark, G. W. B., Peters, J. H., Ireland, A. P., et al. (1994). Nodal metastasis and sites of recurrence after enbloc esophagectomy for adenocarcinoma. *The Annals of Thoracic Surgery, 58,* 646–654.

Cooper, J. S., Guo, M. D., Herskovic, A., et al. (1999). Chemoradiotherapy of locally advanced esophageal cancer: Long-term follow-up of a prospective randomized trial. *The Journal of the American Medical Association, 281,* 1623–1627.

Costantini, M., Toscani, F., Gallucci, M., et al. (1999). Terminal cancer patients and timing of referral to palliative care: A multicenter prospective cohort study. *Journal of Pain Symptom Management, 18,* 243–252.

Dai, Z. J., Gao, J., Ji, Z. Z., et al. (2009). Matrine induces apoptosis in gastric carcinoma cells via alteration of Fas/FasL and activation of caspase-3. *Journal of Ethnopharmacology, 123,* 91–96.

Dean, M., Hamon, Y., & Chimini, G. (2001). The human ATP-binding cassette (ABC) transporter superfamily. *Journal of Lipid Research, 42,* 1007–1017.

Ding, L., Cui, C. M., & Xing, B. (2006). Fufang Kushen injection combined with chemotherapy in the treatment of small cell lung cancer: A clinical assessment. *Chinese Journal of Misdiagnosis, 6,* 4575–4576.

Dog, T. L., & Micozzi, M. S. (2005). Breast Cancer. In T. L. Dog (Ed.), *Women's health in complementary and integrative medicine* (pp. 247–270). Saint Louis: Churchill Livingstone.

Douglas, R., & Laster, K. (1991). A Matter of life and death—the victorian executive and the decision to execute 1842–1967. *The Australian & New Zealand Journal of Criminology, 24,* 144–160.

El-Galley, R., Keane, T. E., & Sun, C. (2003). Camptothecin analogues and vinblastine in the treatment of renal cell carcinoma: An in vivo study using a human orthotopic renal cancer xenograft. *Urologic Oncology, 21,* 49–57.

Emadi, A., & Gore, S. D. (2010). Arsenic trioxide—an old drug rediscovered. *Blood Reviews, 24,* 191–199.
Ernst, E. (2000). The role of complementary and alternative medicine in cancer. *The Lancet Oncology, 1,* 176–180.
Eskens, F. A., & Verweij, J. (2006). The clinical toxicity profile of vascular endothelial growth factor (VEGF) and vascular endothelial growth factor receptor (VEGFR) targeting angiogenesis inhibitors: A review. *European Journal of Cancer, 42,* 3127–3139.
Faria, S. C., Tamm, E. P., Varavithya, V., et al. (2005). Systematic approach to the analysis of cross-sectional imaging for surveillance of recurrent colorectal cancer. *European Journal of Radiology, 53,* 387–396.
Fayette, M., Soria, J. C., & Armand, J. P. (2005). Use of angiogenesis inhibitors in tumour treatment. *European Journal of Cancer, 41,* 1109–1116.
Ferreira, C. G., Span, S. W., Peters, G. J., et al. (2000). Chemotherapy triggers apoptosis in a caspase-8-dependent and mitochondria-controlled manner in the non-small cell lung cancer cell line NCI-H460. *Cancer Research, 60,* 7133–7141.
Ferri, K. F., & Kroemer, G. (2001). Organelle-specific initiation of cell death pathways. *Nature Cell Biology, 3,* E255–E263.
Field, K. M., Jenkins, M. A., Friedlander, M. L., et al. (2009). Predictors of the use of complementary and alternative medicine (CAM) by women at high risk for breast cancer. *European Journal of Cancer, 45,* 551–560.
Fu, L. W., Zhang, Y. M., Liang, Y. J., et al. (2002). The multidrug resistance of tumour cells was reversed by tetrandrine in vitro and in xenografts derived from human breast adenocarcinorna MCF-7/adr cells. *European Journal of Cancer, 38,* 418–426.
Gimbrone, M. A., Cotran, R. S., Folkman, J., et al. (1972). Tumor dormancy in-vivo by prevention of neovascularization. *The Journal of Experimental Medicine, 136,* 261–276.
Greenlee, R. T., Hill-Harmon, M. B., Murray, T., et al. (2001). Cancer statistics, 2001. *Cancer Journal for Clinicians, 51,* 15–36.
Gururaj, A. E., Belakavadi, M., Venkatesh, D. A., et al. (2002). Molecular mechanisms of anti-angiogenic effect of curcumin. *Biochemical and Biophysical Research Communications, 297,* 934–942.
Hanahan, D., & Weinberg, R. A. (2000). The hallmarks of cancer. *Cell, 100,* 57–70.
Harborne, J. B., & Williams, C. A. (2000). Advances in flavonoid research since 1992. *Phytochemistry, 55,* 481–504.
Hayman, M., & Kam, P. C. (2008). Capsaicin: A review of its pharmacology and clinical applications. *Current Anaesthesia & Critical Care, 19,* 338–343.
Higgins, C. F. (2007). Multiple molecular mechanisms for multidrug resistance transporters. *Nature, 446,* 749–757.
Hirsch, F. R., Matthews, M. J., Aisner, S., et al. (1988). Histopathologic classification of small cell lung cancer—changing concepts and terminology. *Cancer, 62,* 973–977.
Ho, J. C., Konerding, M. A., Gaumann, A., et al. (2004). Fungal polysaccharopeptide inhibits tumor angiogenesis and tumor growth in mice. *Life Sciences, 75,* 1343–1356.
Huntley, A., & Rees, M. (2010). Complementary and alternative medicine for the menopause. *Maturitas, 66,* 109–110.
Hyun, P. W., Hee, C. Y., et al. (2003). Arsenic trioxide inhibits the growth of A498 renal cell carcinoma cells via cell cycle arrest or apoptosis. *Biochemical and Biophysical Research Communications, 300,* 230–235.
Itoh, N., & Nagata, S. (1993). A Novel protein domain required for apoptosis—mutational analysis of human Fas antigen. *The Journal of Biological Chemistry, 268,* 10932–10937.
Jelic, S., & Grp, E. G. (2009). Hepatocellular carcinoma: ESMO Clinical Recommendations for diagnosis, treatment and follow-up. *Annals of Oncology, 20,* 41–45.
Jeyapalan, J. (2001). Ascorbic acid and arsenic trioxide for myeloma. *The Lancet Oncology, 2,* 527.
Jiang, H., Hou, C., Zhang, S., et al. (2007). Matrine upregulates the cell cycle protein E2F-1 and triggers apoptosis via the mitochondrial pathway in K562 cells. *European Journal of Pharmacology, 559,* 98–108.

Jin, H. O.,Yoon, S. I., Seo, S. K., et al. (2006). Synergistic induction of apoptosis by sulindac and arsenic trioxide in human lung cancer A549 cells via reactive oxygen species-dependent down-regulation of survivin. *Biochemical Pharmacology, 72,* 1228–1236.

Julka, P. K., Doval, D. C., Gupta, S., et al. (2008). Response assessment in solid tumours: A comparison of WHO, SWOG and RECIST guidelines. *The British Journal of Radiology, 81,* 444–449.

Kanzawa, T., Kondo, Y., Ito, H., et al. (2003). Induction of autophagic cell death in malignant glioma cells by arsenic trioxide. *Cancer Research, 63,* 2103–2108.

Kievit, J. (2002). Follow-up of patients with colorectal cancer: Numbers needed to test and treat. *European Journal of Cancer, 38,* 986–999.

Kim, J. B., Yu, J. H., Ko, E., et al. (2010a). The alkaloid berberine inhibits the growth of Anoikis-resistant MCF-7 and MDA-MB-231 breast cancer cell lines by inducing cell cycle arrest. *Phytomedicine: International Journal of Phytotherapy and Phytopharmacology, 17,* 436–440.

Kim, S. M., Lee, S. Y., Cho, J. S., et al. (2010b). Combination of ginsenoside Rg3 with docetaxel enhances the susceptibility of prostate cancer cells via inhibition of NF-kappa B. *European Journal of Pharmacology, 631,* 1–9.

Kremser, T., Evans, A., Moore, A., et al. (2008). Use of complementary therapies by Australian women with breast cancer. *Breast, 17,* 387–394.

Kumar, R., Herbert, P. E., & Warrens, A. N. (2005). An introduction to death receptors in apoptosis. *International Journal of Surgery, 3,* 268–277.

Kwon, K. B., Kim, E. K., Shin, B. C., et al. (2003). Induction of apoptosis by takrisodokyeum through generation of hydrogen peroxide and activation of caspase-3 in HL-60 cells. *Life Sciences, 73,* 1895–1906.

Kwon, K. B., Park, B. H., & Ryu, D. G. (2007). Chemotherapy through mitochondrial apoptosis using nutritional supplements and herbs: A brief overview. *Journal of Bioenergetics and Biomembranes, 39,* 31–34.

Lanave, C., Santamaria, M., & Saccone, C. (2004). Comparative genomics: The evolutionary history of the Bcl-2 family. *Gene, 333,* 71–79.

Launay, S., Hermine, O., Fontenay, M., et al. (2005). Vital functions for lethal caspases. *Oncogene, 24,* 5137–5148.

Lee, S., Ra, J., Song, J. Y., et al. (2010). Extracts from citrus unshiu promote immune-mediated inhibition of tumor growth in a murine renal cell carcinoma model. *Journal of Ethnopharmacology, 133,* 973–979.

Lerner, I. J., & Kennedy, B. J. (1992). The prevalence of questionable methods of cancer treatment in the United States. *Cancer Journal for Clinicians, 42,* 181–191.

Leung, W. K., Wu, M. S., Kakugawa, Y., et al. (2008). Screening for gastric cancer in Asia: Current evidence and practice. *The Lancet Oncology, 9,* 279–287.

Li, W. K., & Fitzloff, J. F. (2002). HPLC determination of ginsenosides content in ginseng dietary supplements using ultraviolet detection. *Journal of Liquid Chromatography & Related Technologies, 25,* 2485–2500.

Li, J., Sun, G. Z., Lin, H. S., et al. (2008). The herb medicine formula "Yang Wei Kang Liu" improves the survival of late stage gastric cancer patients and induces the apoptosis of human gastric cancer cell line through Fas/Fas ligand and Bax/Bcl-2 pathways. *International Immunopharmacology, 8,* 1196–1206.

Li, R., Chen, W. C., Wang, W. P., et al. (2009). Extraction, characterization of Astragalus polysaccharides and its immune modulating activities in rats with gastric cancer. *Carbohydrate Polymers, 78,* 738–742.

Ling, V. (1997). Multidrug resistance: Molecular mechanisms and clinical relevance. *Cancer Chemotherapy and Pharmacology, 40,* S3–S8.

Liu, J. Y., Hu, J. H., Zhu, Q. G., et al. (2007). Effect of matrine on the expression of substance P receptor and inflammatory cytokines production in human skin keratinocytes and fibroblasts. *International Immunopharmacology, 7,* 816–823.

Liu, X. J., Cao, M. A., Li, W. H., et al. (2010). Alkaloids from Sophora flavescens Aition. *Fitoterapia, 81,* 524–527.

Los, M., Wesselborg, S., & Schulze-Osthoff, K. (1999). The role of caspases in development, immunity, and apoptotic signal transduction: Lessons from knockout mice. *Immunity, 10,* 629–639.

Luo, C., Zhu, Y. L., Jiang, T. J., et al. (2007). Matrine induced gastric cancer MKN45 cells apoptosis via increasing pro-apoptotic molecules of Bcl-2 family. *Toxicology, 229,* 245–252.

MacFarlane, M. (2003). TRAIL-induced signalling and apoptosis. *Toxicology Letters, 139,* 89–97.

Maeda, T., Beppu, T., Hayashi, N., et al. (2003). Molecular detection of occult cancer cells in patients with liver tumors treated by local ablation therapy. *International Congress Series, 1255,* 119–123.

Marks, L. S., DiPaola, R. S., Nelson, P., et al. (2002). PC-SPES: Herbal formulation for prostate cancer. *Urol, 60,* 369–375.

Martel, C. L., Gumerlock, P. H., Meyers, F. J., et al. (2003). Current strategies in the management of hormone refractory prostate cancer. *Cancer Treatment Reviews, 29,* 171–187.

McCulloch, M., See, C., Shu, X. J., et al. (2006). Astragalus-based Chinese herbs and platinum-based chemotherapy for advanced non-small-cell lung cancer: Meta-analysis of randomized trials. *Journal of Clinical Oncology, 24,* 419–430.

Mell, L. K., Malik, R., Komaki, R., et al. (2007). Effect of amifostine on response rates in locally advanced non-small-cell lung cancer patients treated on randomized controlled trials: A meta-analysis. *International Journal of Radiation Oncology, Biology, Physics, 68,* 111–118.

Meng, C., Shi, S. Z., Dou, J. Q., et al. (2006). Herbal combination chemotherapy in advanced non-small cell lung cancer. *Liaoning Journal of Traditional Chinese Medicine, 33,* 1436–1437.

Ogunbiyi, J. O. (2001). Hepatocellular carcinoma in the developing world. *Seminars in Oncology, 28,* 179–187.

Parekh, H. S., Liu, G., & Wei, M. Q. (2009). A new dawn for the use of traditional Chinese medicine in cancer therapy. *Molecular Cancer, 8,* 21.

Park, J., Shin, D. W., Ahn, T. Y., et al. (2008). Complementary and alternative medicine in men's health. *Journal of Men's Health, 5,* 305–313.

Patil, J. B., Kim, J., & Jayaprakasha, G. K. (2010). Berberine induces apoptosis in breast cancer cells (MCF-7) through mitochondrial-dependent pathway. *European Journal of Pharmacology, 645,* 70–78.

Peng, P. L., Kuo, W. H., Tseng, H. C., et al. (2008). Synergistic tumor-killing effect of radiation and berberine combined treatment in lung cancer: The contribution of autophagic cell death. *International Journal of Radiation Oncology, Biology, Physics, 70,* 529–542.

Pham, P. T. K., & Primack, A. (2003). Cancer. In W. S. John (Ed.), *Complementary and alternative medicine* (2nd ed., pp. 93–152). Saint Louis: Mosby.

Piyanuch, R., Sukhthankar, M., Wandee, G., et al. (2007). Berberine, a natural isoquinoline alkaloid, induces NAG-1 and ATF3 expression in human colorectal cancer cells. *Cancer Letters, 258,* 230–240.

Qin, Z., Wei, P. K., Xu, L., et al. (2007). Xiaotan Sanjie and chemotherapy under medical treatment research of non-cellule lung cancer of later period. *Journal of Traditional Chinese Medicine, 48,* 822–823.

Rafi, M. M., Rosen, R. T., Vassil, A., et al. (2000). Modulation of bcl-2 and cytotoxicity by licochalcone-A, a novel estrogenic flavonoid. *Anticancer Research, 20,* 2653–2658.

Rafi, M. M., Vastano, B. C., Zhu, N. Q., et al. (2002). Novel polyphenol molecule isolated from licorice root (Glycrrhiza glabra) induces apoptosis, G2/M cell cycle arrest, and Bcl-2 phosphorylation in tumor cell lines. *Journal of Agricultural and Food Chemistry, 50,* 677–684.

Ribatti, D., Nico, B., Morbidelli, L., et al. (2001). Cell-mediated delivery of fibroblast growth factor-2 and vascular endothelial growth factor onto the chick chorioallantoic membrane: Endothelial fenestration and angiogenesis. *Journal of Vascular Research, 38,* 389–397.

Ringe, B., Canelo, R., & Lorf, T. (1996). Liver transplantation for primary liver cancer. *Transplant, 28,* 1174–1175.

Rinner, B., Kretschmer, N., Knausz, H., et al. (2010). A petrol ether extract of the roots of Onosma paniculatum induces cell death in a caspase dependent manner. *Journal of Ethnopharmacology, 129,* 182–188.

Risau, W. (1997). Mechanisms of angiogenesis. *Nature, 386,* 671–674.
Rosenberg, Z. R., Jenkins, D. J., Brown, T. J., et al. (2002). Flavonoids can block PSA production by breast and prostate cancer cell lines. *Clinica Chimica Acta, 317,* 17–26.
Rowley, J. D., Golomb, H. M., & Dougherty, C. (1977). 15–17 Translocation, a Consistent Chromosomal Change in Acute Promyelocytic Leukemia. *Lancet, 1,* 549–550.
Seliger, B. (2005). Strategies of tumor immune evasion. *Biodrugs, 19,* 347–354.
Shao, R. G., Cao, C. X., Zhang, H. L., et al. (1999). Replication-mediated DNA damage by camptothecin induces phosphorylation of RPA by DNA-dependent protein kinase and dissociates RPA: DNA-PK complexes. *The EMBO Journal, 18,* 1397–1406.
Shao, B. M., Xu, W., Dai, H., et al. (2004). A study on the immune receptors for polysaccharides from the roots of Astragalus membranaceus, a Chinese medicinal herb. *Biochemical and Biophysical Research Communications, 320,* 1103–1111.
Shen, X. D., & Kramer, R. H. (2004). Adhesion-mediated squamous cell carcinoma survival through ligand-independent activation of epidermal growth factor receptor. *The American Journal of Pathology, 165,* 1315–1329.
Srinivas, G., Anto, R. J., Srinivas, P., et al. (2003). Emodin induces apoptosis of human cervical cancer cells through poly(ADP-ribose) polymerase cleavage and activation of caspase-9. *European Journal of Pharmacology, 473,* 117–125.
Stephens, R. J., Girling, D. J., Machin, D., et al. (1994). Treatment-related deaths in small cell lung cancer trials—can patients at risk be identified. *Lung Cancer, 11,* 259–274.
Steup, W. H., DeLeyn, P., Deneffe, G., et al. (1996). Tumors of the esophagogastric junction—Long-term survival in relation to the pattern of lymph node metastasis and critical analysis of the accuracy or inaccuracy of pTNM classification. *Journal of Thoracic and Cardiovascular Surgery, 111,* 85–94.
Stevens, P. E., Dibble, S. L., & Miaskowski, C. (1995). Prevalence, characteristics, and impact of postmastectomy pain syndrome: An investigation of women's experiences. *Pain, 61,* 61–68.
Stickel, F., & Schuppan, D. (2007). Herbal medicine in the treatment of liver diseases. *Digestive and Liver Disease, 39,* 293–304.
Stipa, S., Nicolanti, V., Botti, C., et al. (1991). Local recurrence after curative resection for colorectal cancer—frequency, risk-factors and treatment. *Journal of Surgical Oncology, 48,* 155–160.
Sun, Z. (2009). Eight dainties granula combination with chemotherapy in the treatment of advanced breast cancer. *Chinese Modern Medicines, 16,* 121–122.
Swanson, G. P. (2006). Management of locally advanced prostate cancer: Past, present, future. *The Journal of Urology, 176,* S34–S41.
Tang, J., Feng, Y., Tsao, S., et al. (2009). Berberine and coptidis rhizoma as novel antineoplastic agents: A review of traditional use and biomedical investigations. *Journal of Ethnopharmacology, 126,* 5–17.
Treasure, J. (2005). Herbal medicine and cancer: An introductory overview. *Seminars in Oncology Nursing, 21,* 177–183.
Trimeche, M., Braham, H., Ziadi, S., et al. (2008). Investigation of allelic imbalances on chromosome 3p in nasopharyngeal carcinoma in Tunisia: High frequency of microsatellite instability in patients with early-onset of the disease. *Oral Oncology, 44,* 775–783.
Tsao, Y. P., Russo, A., Nyamuswa, G., et al. (1993). Interaction between replication forks and topoisomerase-I-DNA cleavable complexes—studies in a cell free Sv40 DNA-replication system. *Cancer Research, 53,* 5908–5914.
Tyagi, A. K., Singh, R. P., et al. (2002). Silibinin strongly synergizes human prostate carcinoma DU145 cells to doxorubicininduced growth inhibition, G2-Marrest, and apoptosis. *Clinical Cancer Research, 8,* 3512–3519.
Tzianabos, A. O. (2000). Polysaccharide immunomodulators as therapeutic agents: Structural aspects and biologic function. *Clinical Microbiology Reviews, 13,* 523–533.
Wall, M. E., Wani, M. C., Cook, C. E., et al. (1966). Plant antitumor agents I: Isolation and structure of camptothecin a novel alkaloidal leukemia and tumor inhibitor from camptotheca acuminata. *Journal of the American Chemical Society, 88,* 3888–3890.

Watson, C. P. N., & Evans, R. J. (1992). The postmastectomy pain syndrome and topical capsaicin: A randomized trial. *Pain, 51,* 375–379.

Wenzel, U., Nickel, A., Kuntz, S., et al. (2004). Ascorbic acid suppresses drug-induced apoptosis in human colon cancer cells by scavenging mitochondrial superoxide anions. *Carcinogenesis, 25,* 703–712.

Werneck, M. B. F., Lugo-Villarino, G., Hwang, E. S., et al. (2008). T-bet plays a key role in NK-Mediated control of melanoma metastatic disease. *Journal of Immunology, 180,* 8004–8010.

Winawer, S. J. (2007). Colorectal cancer screening. *Best Practice & Research Clinical Gastroenterology, 21,* 1031–1048.

Wong, B. S., Hsiao, Y. C., Lin, T. W., et al. (2009). The in vitro and in vivo apoptotic effects of Mahonia oiwakensis on human lung cancer cells. *Chemico Biological Interactions, 180,* 165–174.

Wu, P., Dugoua, J. J., et al. (2009a). Traditional Chinese medicines in the treatment of hepatocellular cancers: A systematic review and meta-analysis. *Journal of Experimental & Clinical Cancer Research, 28,* 112.

Wu, T. X., Yang, X. Z., Zeng, X. X., et al. (2009b). Traditional Chinese Medicinal herbs in the treatment of patients with esophageal cancer: A systematic review. *Gastroenterology Clinics of North America, 38,* 153–167.

Xia, S. L., Rosen, E. M., & Laterra, J. (2005). Sensitization of glioma cells to fas-dependent apoptosis by chemotherapy-induced oxidative stress. *Cancer Research, 65,* 5248–5255.

Xu, W. X., & Fan, Z. Z. (1996). Clinical progress of Chinese medicine in the treatment of cancer pain. *Shaanxi journal of traditional Chinese medicine, 17,* 332–334.

Xu, Y. X., & Wang, S. L. (2010). JiangNiLing combination with chemotherapy in the treatment of advanced colorectal cancer patients. *China medical herald, 7,* 84–85.

Yamamoto, N., Zou, J. P., Li, X. F., et al. (1995). Regulatory mechanisms for production of Ifn-gamma and Tnf by antitumor T-cells or macrophages in the tumor-bearing state. *Journal of Immunology, 154,* 2281–2290.

Yang, L. (2006). Incidence and mortality of gastric cancer in China. *World Journal of Gastroenterology, 12,* 17–20.

Yoshida, N., Koizumi, M., Adachi, I., et al. (2006a). Inhibition of P-glycoprotein-mediated transport by terpenoids contained in herbal medicines and natural products. *Food and Chemical Toxicology, 44,* 2033–2039.

Yoshida, N.,Takagi, A., Kitazawa, H., et al. (2006b). Effects of citronellal, a monoterpenoid in Zanthoxyli fructus, on the intestinal absorption of digoxin in vitro and in vivo. *Journal of Pharmaceutical Sciences US, 95,* 552–560.

Yun, T. K., Lee, Y. S., Lee, Y. H., et al. (2001). Anticarcinogenic effect of Panax ginseng CA Meyer and identification of active compounds. *Journal of Korean Medical Science, 16,* S6–S18.

Zeng, D., Ling, Y., Yang, Q. L., et al. (2009). Clinical study on an angiogenesis inhibitor: Ginsenoside Rg3 combination with FOL FOX4 chemotherapy in the treatment of advanced colorectal cancer. *Journal of Clinical Medicine in Practice, 13,* 57–58.

Zhang, Q., Kang, X., & Zhao, W. (2006). Antiangiogenic effect of low-dose cyclophosphamide combined with ginsenoside Rg3 on Lewis lung carcinoma. *Biochemical and Biophysical Research Communications, 342,* 824–828.

Zhang, Y. M., He, L. C., Meng, L., et al. (2008). Suppression of tumor-induced angiogenesis by taspine isolated from Radix et Rhizoma Leonticis and its mechanism of action in vitro. *Cancer Letters, 262,* 103–113.

Zhou, G. B., Zhang, J., Wang, Z. Y., et al. (2007). Treatment of acute promyelocytic leukaemia with all-trans retinoic acid and arsenic trioxide: a paradigm of synergistic molecular targeting therapy. *Philosophical transactions of the Royal Society of London. Series B, Biological sciences, 362,* 959–971.

Zhu, S. J. (2002). Integrative medicine treated 120 cases of advanced lung cancer. *Chinese journal of the practical Chinese with modern medicine, 4,* 374.

Zunino, F., & Pratesi, G. (2004). Camptothecins in clinical development. *Expert opinion on investigational drugs, 13,* 269–284.

Index

W. C. S. Cho (ed.), *Evidence-based Anticancer Materia Medica,* Evidence-based Anticancer Complementary and Alternative Medicine,
DOI 10.1007/978-94-007-0526-5,